Analytical Techniques in Pharmaceutical and Biomedical Analysis

Analytical Techniques in Pharmaceutical and Biomedical Analysis

Editors

Franciszek Główka
Marta Karaźniewicz-Łada

MDPI • Basel • Beijing • Wuhan • Barcelona • Belgrade • Manchester • Tokyo • Cluj • Tianjin

Editors
Franciszek Główka
Poznan University of Medical
Sciences
Poland

Marta Karaźniewicz-Łada
Poznan University of Medical
Sciences
Poland

Editorial Office
MDPI
St. Alban-Anlage 66
4052 Basel, Switzerland

This is a reprint of articles from the Special Issue published online in the open access journal *Molecules* (ISSN 1420-3049) (available at: https://www.mdpi.com/journal/molecules/special_issues/analyt_tech).

For citation purposes, cite each article independently as indicated on the article page online and as indicated below:

LastName, A.A.; LastName, B.B.; LastName, C.C. Article Title. *Journal Name* **Year**, *Volume Number*, Page Range.

ISBN 978-3-0365-7024-2 (Hbk)
ISBN 978-3-0365-7025-9 (PDF)

© 2023 by the authors. Articles in this book are Open Access and distributed under the Creative Commons Attribution (CC BY) license, which allows users to download, copy and build upon published articles, as long as the author and publisher are properly credited, which ensures maximum dissemination and a wider impact of our publications.

The book as a whole is distributed by MDPI under the terms and conditions of the Creative Commons license CC BY-NC-ND.

Contents

Anna Przybylska, Marcin Gackowski and Marcin Koba
Application of Capillary Electrophoresis to the Analysis of Bioactive Compounds in Herbal Raw Materials
Reprinted from: *Molecules* 2021, 26, 2135, doi:10.3390/molecules26082135 1

Marcin Gackowski, Anna Przybylska, Stefan Kruszewski, Marcin Koba, Katarzyna Mądra-Gackowska and Artur Bogacz
Recent Applications of Capillary Electrophoresis in the Determination of Active Compounds in Medicinal Plants and Pharmaceutical Formulations
Reprinted from: *Molecules* 2021, 26, 4141, doi:10.3390/molecules26144141 25

Jangmi Choi, Min-Ho Park, Seok-Ho Shin, Jin-Ju Byeon, Byeong ill Lee, Yuri Park and Young G. Shin
Quantification and Metabolite Identification of Sulfasalazine in Mouse Brain and Plasma Using Quadrupole-Time-of-Flight Mass Spectrometry
Reprinted from: *Molecules* 2021, 26, 1179, doi:10.3390/molecules26041179 59

Aurélien Millet, Nihel Khoudour, Dorothée Lebert, Christelle Machon, Benjamin Terrier, Benoit Blanchet and Jérôme Guitton
Development, Validation, and Comparison of Two Mass Spectrometry Methods (LC-MS/HRMS and LC-MS/MS) for the Quantification of Rituximab in Human Plasma
Reprinted from: *Molecules* 2021, 26, 1383, doi:10.3390/molecules26051383 73

Xiaoting Gu, Dongwu Wang, Xin Wang, Youping Liu and Xin Di
Fast Screening of Biomembrane-Permeable Compounds in Herbal Medicines Using Bubble-Generating Magnetic Liposomes Coupled with LC–MS
Reprinted from: *Molecules* 2021, 26, 1742, doi:10.3390/molecules26061742 91

Katarzyna Lipska, Anna Gumieniczek, Rafał Pietraś and Agata A. Filip
HPLC-UV and GC-MS Methods for Determination of Chlorambucil and Valproic Acid in Plasma for Further Exploring a New Combined Therapy of Chronic Lymphocytic Leukemia
Reprinted from: *Molecules* 2021, 26, 2903, doi:10.3390/molecules26102903 103

Paulina Markowska, Zbigniew Procajło, Joanna Wolska, Jerzy Jan Jaroszewski and Hubert Ziółkowski
Development, Validation, and Application of the LC-MS/MS Method for Determination of 4-Acetamidobenzoic Acid in Pharmacokinetic Pilot Studies in Pigs
Reprinted from: *Molecules* 2021, 26, 4437, doi:10.3390/molecules26154437 121

Stanislawa Koronkiewicz
Photometric Determination of Iron in Pharmaceutical Formulations Using Double-Beam Direct Injection Flow Detector
Reprinted from: *Molecules* 2021, 26, 4498, doi:10.3390/molecules26154498 135

Ioanna Chrisikou, Malvina Orkoula and Christos Kontoyannis
Analysis of IV Drugs in the Hospital Workflow by Raman Spectroscopy: The Case of Piperacillin and Tazobactam
Reprinted from: *Molecules* 2021, 26, 5879, doi:10.3390/molecules26195879 147

Mahesh Attimarad, Katharigatta N. Venugopala, Bandar E. Al-Dhubiab,
Rafea Elamin Elgack Elgorashe and Sheeba Shafi
Development of Ecofriendly Derivative Spectrophotometric Methods for the Simultaneous
Quantitative Analysis of Remogliflozin and Vildagliptin from Formulation
Reprinted from: *Molecules* 2021, 26, 6160, doi:10.3390/molecules26206160 175

Elżbieta Gniazdowska, Wojciech Goch, Joanna Giebułtowicz and Piotr J. Rudzki
Replicates Number for Drug Stability Testing during Bioanalytical Method Validation—An
Experimental and Retrospective Approach
Reprinted from: *Molecules* 2022, 27, 457, doi:10.3390/molecules27020457 191

Mahmood Karimi Abdolmaleki, Deepak Ganta, Ali Shafiee, Carlo Alberto Velazquez and
Devang P. Khambhati
Efficient Heparin Recovery from Porcine Intestinal Mucosa Using Zeolite Imidazolate
Framework-8
Reprinted from: *Molecules* 2022, 27, 1670, doi:10.3390/molecules27051670 211

Chang-Seob Seo and Hyeun-Kyoo Shin
Simultaneous Analysis of 19 Marker Components for Quality Control of Oncheong-Eum Using
HPLC–DAD
Reprinted from: *Molecules* 2022, 27, 2992, doi:10.3390/molecules27092992 223

Tingting Huang, Ting Huang, Yongyi Zou, Kang Xie, Yinqin Shen, Wen Zhang, et al.
Wooden-Tip Electrospray Mass Spectrometry Characterization of Human Hemoglobin in
Whole Blood Sample for Thalassemia Screening: A Pilot Study
Reprinted from: *Molecules* 2022, 27, 3952, doi:10.3390/molecules27123952 235

Kyoungmin Lee, Wokchul Yoo and Jin Hyun Jeong
Analytical Method Development for 19 Alkyl Halides as Potential Genotoxic Impurities by
Analytical Quality by Design
Reprinted from: *Molecules* 2022, 27, 4437, doi:10.3390/molecules27144437 245

Gellért Balázs Karvaly, István Vincze, Alexandra Balogh, Zoltán Köllő, Csaba Bödör and
Barna Vásárhelyi
A High-Throughput Clinical Laboratory Methodology for the Therapeutic Monitoring of
Ibrutinib and Dihydrodiol Ibrutinib
Reprinted from: *Molecules* 2022, 27, 4766, doi:10.3390/molecules27154766 267

Sreenath Nair, Abigail Davis, Olivia Campagne, John D. Schuetz and Clinton F. Stewart
Development and Validation of a Sensitive and Specific LC-MS/MS Method for IWR-1-Endo, a
Wnt Signaling Inhibitor: Application to a Cerebral Microdialysis Study
Reprinted from: *Molecules* 2022, 27, 5448, doi:10.3390/molecules27175448 283

Norberto Masciocchi, Vincenzo Mirco Abbinante, Marco Zambra, Giuseppe Barreca
and Massimo Zampieri
Thermal and Structural Characterization of Two Crystalline Polymorphs of Tafamidis Free Acid
Reprinted from: *Molecules* 2022, 27, 7411, doi:10.3390/molecules27217411 301

Nataša Gros, Tadej Klobučar and Klara Gaber
Accuracy of Citrate Anticoagulant Amount, Volume, and Concentration in Evacuated Blood
Collection Tubes Evaluated with UV Molecular Absorption Spectrometry on a Purified Water
Model
Reprinted from: *Molecules* 2023, 28, 486, doi:10.3390/molecules28020486 315

Jin Seok Lee, Yu Ran Nam, Hyun Jong Kim and Woo Kyung Kim
Quantification and Validation of an HPLC Method for Low Concentrations of Apigenin-7-O-Glucuronide in *Agrimonia pilosa* Aqueous Ethanol Extract Topical Cream by Liquid–Liquid Extraction
Reprinted from: *Molecules* **2023**, *28*, 713, doi:10.3390/molecules28020713 . . . . **331**

Review

Application of Capillary Electrophoresis to the Analysis of Bioactive Compounds in Herbal Raw Materials

Anna Przybylska *, Marcin Gackowski and Marcin Koba

Department of Toxicology and Bromatology, Faculty of Pharmacy, L. Rydygier Collegium Medicum in Bydgoszcz, Nicolaus Copernicus University in Torun, A. Jurasza 2 Street, PL-85089 Bydgoszcz, Poland; marcin.gackowski@cm.umk.pl (M.G.); kobamar@cm.umk.pl (M.K.)
* Correspondence: aniacm@cm.umk.pl

Abstract: The article is a summary of scientific reports from the last 16 years (2005–2021) on the use of capillary electrophoresis to analyze polyphenolic compounds, coumarins, amino acids, and alkaloids in teas or different parts of plants used to prepare aqueous infusions, commonly known as "tea" or decoctions. This literature review is based on PRISMA guidelines and articles selected in base of criteria carried out using PICOS (Population, Intervention, Comparison, Outcome, Study type). The analysis showed that over 60% of articles included in this manuscript comes from China. The literature review shows that for the selective electrophoretic separation of polyphenolic and flavonoid compounds, the most frequently used capillary electromigration technique is capillary electrophoresis with ultraviolet detection. Nevertheless, the use of capillary electrophoresis-mass spectrometry allows for the sensitive determination of analytes with a lower limit of detection and gives hope for routine use in the analysis of functional foods. Moreover, using the modifications in electrochemical techniques allows methods sensitivity reduction along with the reduction of analysis time.

Keywords: capillary electrophoresis; herbal; raw material; tea; polyphenols; flavonoids; amino acids; coumarins; alkaloids

Citation: Przybylska, A.; Gackowski, M.; Koba, M. Application of Capillary Electrophoresis to the Analysis of Bioactive Compounds in Herbal Raw Materials. *Molecules* **2021**, *26*, 2135. https://doi.org/10.3390/molecules 26082135

Academic Editors: Franciszek Główka and Marta Karaźniewicz-Łada

Received: 1 March 2021
Accepted: 6 April 2021
Published: 8 April 2021

Publisher's Note: MDPI stays neutral with regard to jurisdictional claims in published maps and institutional affiliations.

Copyright: © 2021 by the authors. Licensee MDPI, Basel, Switzerland. This article is an open access article distributed under the terms and conditions of the Creative Commons Attribution (CC BY) license (https:// creativecommons.org/licenses/by/ 4.0/).

1. Introduction

Is common use, the word "tea" means an aqueous infusion prepared from dried herbal materials. It is one of the most popular beverages in the world, which is made from a variety of plants, other than *Camellia sinensis*. Real tea is prepared from leaves and buds of *Camellia sinensis*. It is estimated that the highest consumption of tea is in China [1].

Tea, especially green tea, is a rich composition of compounds that prevent obesity and exhibit anti-cancer, anti-diabetic properties, and reduces the risk of cardiovascular problems [2–4]. *Evodiae Fructus* has been used in traditional Chinese medicine for over 2000 years, and the limonin as a major metabolite responsible for anti-HIV effects [5]. Tea is also a part of the Mediterranean diet used by Italians and Greeks, which may contribute to the prevention of chronic diseases such as metabolic syndrome and obesity [3]. Moreover, studies have shown that long-term consumption of polyphenols can reduce chronic inflammation and oxidative stress and at the same time inhibit the growth, reproduction, and diffusion of cancer cells. Meta-analysis found that green tea consumption potentially reduced cancer risk by 15% [6]. The rich properties of teas come from profile of biologically active compounds. Additionally, teas and aqueous infusions are rich in catechins, epicatechins, and epicatechin gallate and epigallocatechin gallate amino acids and organic acids [3,7,8]. The production process of herbal raw materials is not standardized, which means that the profile of compounds present in leaves or inflorescences is different and depends on the fermentation processes [9]. Studies with Chinese black, green, and blue (oolong tea) teas indicate that black tea is the richest in gallic acid (582.43 µg g^{-1}), vanillic acid (4.31 µg g^{-1}), trans-p-coumaric acid (10.15 µg g^{-1}), and caffeic acid (1.01 µg g^{-1}) [9]. On the other hand, the highest concentration of (−)—epicatechin was found in green tea infusions (83.95 µg g^{-1}), almost twice as much as in oolong tea [9]. In turn, fresh leaves of *Camelia sinensis* are richer source of gallic acid (2.37 mg g^{-1}) and caffeic acid (92.0 µg g^{-1}) compared to black, green, and oolong tea and fermented leaves of *Camelia sinensis* (831.3 and 37.5 µg g^{-1}) [4]. Research shows that average intake of total polyphenols with the total diet in Poland is 989.3 mg day^{-1} [10]. The research published in 2018 shows that the daily consumption of polyphenolic compounds along with water infusions prepared from hawthorn fruits (*Crataegus fructus*) and inflorescences (*Crataegus inflorescences*) is approximately 4.7% (46.58 mg day^{-1}) and 15.12% (149.56 mg day^{-1}), respectively [7]. Knowledge of the content of biologically active compounds in infusions prepared from fresh or fermented plant materials is extremely important from the pharmacological and dietary point of view.

Nowadays, various analytical techniques including high-performance liquid chromatography (HPLC) [4,11–14], gas chromatography (GC) [15], inductively coupled plasma mass spectrometry (ICP-MS) [16,17], tin layer chromatography (TLC) [13,18], inductively coupled plasma optical emission spectrometry (ICP-OES) [12], atomic absorption spectrometry among (AAS) [19], and inductively coupled plasma-atomic emission spectrometry (ICP-AES) [17] are used for qualitative and quantitative analysis of various bioactive components in "teas". However, high performance liquid chromatography (HPLC) and gas chromatography (GC) are most commonly used. See Table 1. However, HPLC has several limitations compared to GC or capillary electrophoresis (CE). First of all, the time of analysis and optimization of the method, which often requires a large consumption of solvents and long separation time [5]. An important limitation of GC is that it allows the determination of only those compounds that are volatile and not highly polar, which is of great importance in the analysis of polyphenols, coumarins, organic acids, and vitamins.

Table 1. Number of papers published in the Web of Science database using various analytical techniques.

Keyword	Web of Science							
	HPLC	GC	CE	ICP-MS	TLC	ICP-OES	AAS	ICP-AES
"tea"	2912	1665	356	260	149	145	142	134
	(51%)	(29%)	(6%)	(5%)	(3%)	(2%)	(2%)	(2%)

HPLC—High-Performance Liquid Chromatography, GC—Gas Chromatography, CE—capillary electrophoresis, ICP-MS—Inductively Coupled Plasma Mass Spectrometry, TLC—Tin Layer Chromatography, ICP-OES—Inductively Coupled Plasma Optical Emission Spectrometry, AAS—Atomic Absorption Spectrometry, ICP-AES—Inductively Coupled Plasma-Atomic Emission Spectrometry.

High-performance CE is one of the methods used to separate, identify, and quantify various chemical compounds in food products. In recent years, it has become an increasingly popular technique for the separation of analytes. CE is estimated to be the most developing separation technique in the last 20 years. Now CE is increasingly used for powerful separation of different compounds in food and more and more often many researchers use it in the analysis of herbal raw materials. Moreover, according to the World Health Organization, it is recommended to use CE to check the authenticity of herbal raw materials [20]. The availability of high-performance CE enables its wide application in environmental samples [21], pharmaceutical analysis [22], as well as food analysis [23,24].

Due to the higher resolution, CE is more preferred than HPLC in the determination of several components of plant raw materials [25]. Compared to HPLC, GC, or HPTLC, CE is a relatively new analytical technique used for the chemical analysis of multicomponent samples. For the first time in the world, CE was used by A. W. Tiselius in 1937 to separate proteins [26]. After that time, using the PubMed and Web of Science databases, the greatest expansion of works devoted to this technique was noticeable after 1990.

Whereas in CE, the essence of separation is the electric field. The movement of the mixture components is determined by the electroosmotic flow (EOF) of the buffer and the action of the electric field [27]. The liquid flow in the capillary is due to the potential difference. The speed of an electrically charged particle depends on its charge, size, shape, and resistance. The analyzed compounds, which have an electric charge, are characterized by electric mobility, which causes them to separate. CE resolution is influenced by the total charge in the capillary walls and matrix. Therefore, the pH of the buffer is most important as it dictates the number of theoretical plates. The phenomenon of electrophoresis consists in the migration of charged particles in an electric field towards the electrode with the opposite sign what was shown in Scheme 1. The most important element of the apparatus is the capillary. Their dimensions and type of material (fused silica, or quartz) may affect the analysis conditions. The use of quartz capillaries has the advantage that they can also function as measuring cells of UV/Vis or fluorescent detectors. The flow of electric current through the narrow capillary causes the generation of Joule heat, which is a disadvantageous effect due to the blurring of the component zones of the mixture. Low conductivity, which depends on the material from which the capillary is made, allows the use of a very high electric field (100–500 V cm^{-1}), which in turn generates a small amount of Joule heat. In turn, the undeniable advantages of CE include high efficiency, low consumption of reagents, small sample volumes, short analysis times, and the possibility of measuring in a wide range of pH [27,28]. However, an extremely important factor that determines the choice of using CE is that it enables the simultaneous determination of cations and anions [28–30]. Due to the use of capillaries with very small diameters, the volume of the introduced sample may reach several nL. For this reason, it is very important in capillary electrophoresis to select an appropriate detection method. The most common are UV/Vis diode array detectors, fluorescence, and amperometric detectors [31]. Unfortunately, the use of small sample volumes can reduce the sensitivity and precision of CE results, what can be avoided by using solid phase extraction (SPE) or optimization of the pre-conditioning and washing step prior to injection [31]. The choice of detector in chemical analysis plays an important role in expressing selectivity. It is obvious that detectors greatly influence the limits of detection (LOD) [32].

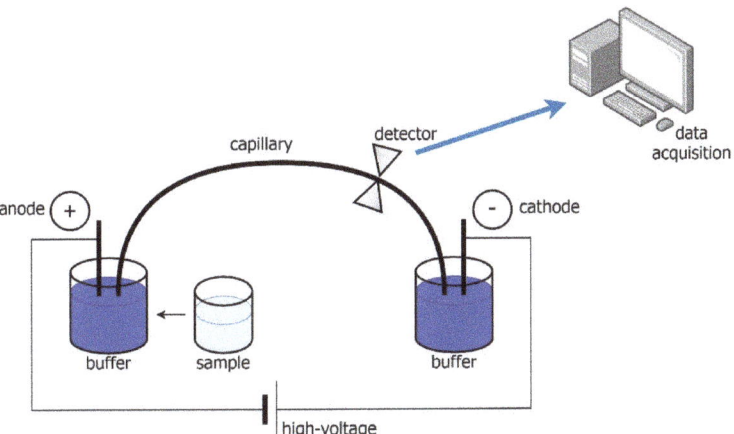

Scheme 1. Capillary electrophoresis system.

An undeniable advantage of CE is the ability to adapt various types of CE depending on the chemical structure of the analyzed compounds. There are several techniques of CE: capillary zone electrophoresis (CZE), non-aqueous CE (NACE), micellar electrokinetic chromatography (MEKC), capillary electrochromatography (CEC), capillary isotachophoresis (CITP), capillary isoelectric focusing (CIEF), chiral CE (CCE), capillary gel electrophoresis (CGE), and microemulsion electrokinetic capillary chromatography (MEEKC) [33].

This article is a valuable summary on the analysis of bioactive compounds in plant materials and their infusions, commonly used "teas". This issue is important because there is still no data on the amount of consumed substances with water infusions or decoctions. According to data from the last 16 years, CE enables the determination of a wide range of chiral compounds and enantiomers, such as polyphenols, coumarins, alkaloids, vitamins, and ions. This paper discusses in detail the physical (length and diameter of capillary, voltage) and chemical (type of buffer, pH) parameters of CE needed for the selective separation of biologically active compounds. This article summarizes the current application of CE in the analysis of bioactive compounds in plant raw materials, but also indicates the prospects for further applications of this technique.

2. Results

2.1. Literature Analysis

In the first search in the PubMed and the Web of Science database, 721 records potentially meeting the included criteria were found, 154 and 567, respectively. Then, after reviewing the bibliography, remove duplicates (n = 194), and selected articles were subjected to subsequent verification by co-authors. After that, the articles were selected based on the title and abstract. Papers describing the application of different methods than CE or the using CE technique to the analysis of compounds (polyphenols, organic acids, and amino acids) present in spices, vegetables and fruits, blood serum and urine or drugs were eliminated. There were 467 abstracts and papers, but they were not qualified for this review. Fifty-nine articles found in the PubMed and the Web of Science databases were used to review the analysis of various bioactive compounds using CE in *Camelia sinensis* and infusions prepared from dried plant materials commonly known as teas.

2.2. Analysis of Plant Material Used in Included Articles

The literature review showed that as much as 62% articles included in this paper comes from China, 20% articles from Europe, 15% from Asia, and only 3% papers from South America. The material covered by the research were parts of plants used as infusions and decoctions belonging to the following families: Onagraceae, Fabaceae, Chloranthaceae, Asteraceae, Hypericaceae, Rubiaceae, Lamiaceae, Compositae, Rosaceae, Pentaphylacaceae, Umbelliferae, Schisandraceae, Rutaceae, Oleaceae, Malvaceae, Nymphaeaceae, Ranunculaceae, Berberidaceae, Ranunculacese, Leguminosae, Gesneriacese, Gentianaceae, Seriphidium, Caprifoliaceae, Phellodendron, Papaveraceae, Nymphaeaceae, Lenguminosae, Apiaceae, and Sapindaceae. Parts of the plants covered by this review are widely used in traditional medicine. CE has been mostly used to analyze bioactive substances in various types of herbs medicine (e.g., *Salvia* species, *Hibiscus sabdariffa*, *Melissae herba*, or *Scutellaria baicalensis*). In cited works, green tea and tea of unknown species were used six times. Oolong tea, black tea, and jasmine tea, twice; *Camelia sinensis*, once; and rooibos, honey bush, red tea, and mate tea were repeated only three times.

2.3. Capillary Zone Electrophoresis

2.3.1. Capillary Zone Electrophoresis with UV Detection

The most widely used method for the selective separation of bioactive compounds in plant raw materials was CE with UV detection (CE-UV). See Table 2. Thanks to UV detectors, it is possible to determine analytes directly or indirectly. The latter method is used when the analyzed substances poorly absorb ultraviolet light [34]. For this, chromophore ions such as phthalic acid, 3,5-dinitrobenzoic acid, or 2,3-pyrazine dicarboxylic acid were added to the buffer for the separation of organic compounds [34]. CE-UV is also used as reference method [35].

CE with UV detection was used for polyphenols compounds analysis in chamomile flowers [36] *Cynanchum chinense* [27], *Salvia* species [37–39], *Radix Scutellariae* [20], *Gentiana lutea* [40], mate herb [41], black tea [42], rooibos and honeybush tea [43], flavonoids in *Crataegus monogyna* [44], *Coreopsis tinctoria* [45], and flavone derivatives in *Seriphidium santolinum* [46], fewflower Lysionotus herb [47]. Moreover, this techniques was used to determinate organic acids in sage [48], *Pterocephalus hookeri* [49] and chamomile, linden, and mint [50]. Moreover, CE-UV was used to identify alkaloids in tea [51], *Sophora tonki*-

nensis [52], *Chelidonium majus* [53], *Cortex Phellodendri Chinensis* [54], *Nelumbo nucifera* [55], anthraquinones in slimming tea [56], coumarins in *Cacalia Tangutica* [57], and nitrate, nitrite, and bromate in tea infusions [30].

To obtain a good separation of the analytes, appropriate conditions of analysis should be selected: such as capillary length, voltage, pH, composition, and buffer concentration. The most appropriate buffer to obtain a good resolution and separation of polyphenolic compounds was borate buffer.

To begin with, 40 mM L^{-1} borate buffer was used to separate of (−)-epicatechin, catechin, vanillic acid, rosmarinic acid, caffeic acid, and gallic acid from *Salvia* species [38]. A 50 and 100 mM L^{-1} sodium tetraborate solution was used in isolation of apigenin-7-glucoside and apigenin from chamomile [36]. Rosmarinic acid (RA) and carnosic acid (CA) are antioxidant compounds in *Salvia* species and were determined with use of a background electrolyte (BGE), which constituted of 20 mM L^{-1} borate at 9.6 pH. The average amount of real samples ranged within 1.08–14.40 mg g^{-1} DW for RA and 11.0 mg g^{-1} for CA [37]. This method was validated and LOD for RA was 1.72 µg mL^{-1} and 1.86 µg mL^{-1} for CA. CE-UV (254 nm) was used for determination of two flavonoids 7-*O*-α-L-rhamnopyranosyl-kaempferol-3-*O*-β-D-glucopyranoside (GL) and 7-*O*-α-L-rhamnopyranosyl-kaempferol-3-*O*-α-L-rhamnopyranoside (RH) [27]. Authors determined GL and RH in *Cynanchum chinense* using 30 mM L^{-1} borate at 9.50 pH. Under this condition authors obtained LOD 2.1 and 1.6 µg mL^{-1} for GL and RH but concentration of this two flavonoids were in the range 0.151–0.284 mg g^{-1} for GL and 0.502–20.412 for RH [27].

In chiral separation, the addition of β-cyclodextrin (β-CD) to the buffer is often used to prove the distribution of the analyzed ions. This kind of modification was used in determination of seven phenolic acids in *Salvia* species [39] in within 17 min, oleanolic and ursolic acid in *Pterocephalus hookeri* [49] in less than 11 min, gentisin, isogentisin, and amarogentin in *Gentiana lutea* [40] and alkaloids in *Sophora tonkinensis* [52]. Cao et al. determined the highest concentration of RA in *Salvia castanea* from Yunnan (21.42 ± 0.73 mg g^{-1}) but the lowest in *Salvia miltiorrhiza* from Anhui (1.30 ± 0.33 mg g^{-1}) [39]. In turn, *Salvia miltiorrhiza* (Hebei) was the richest (44.61 ± 0.72 mg g^{-1}) in salvianolic acid B but *Salvia castanea* from Yunnan was only 1.66 ± 0.07 mg g^{-1} this acid [39]. Application of CE-UV allowed on determination gentisin in *Gentiana lutea* in the range of 4.27–9.72 µg mL^{-1} but amarogentin was not detected in two samples. In turn, the highest concentration of isogentisin (12.30 µg mL^{-1}) was found in the sample without amarogentin [40].

The efficient separation of phenolic compounds such as rutin (R), caffeic acid (CAA) and 3,4-dihydroxybenzoic acid (3,4-DHBA) in mate herb with the 100 mM L^{-1} boric acid (pH 9.0) allowed the analysis time to be reduced to 4 min [41]. Moreover, authors obtained low limit of detection of 0.14, 0.05, and 0.05 mg L^{-1} for R, CAA and 3,4-DHBA, respectively. The content of R and CAA before hydrolysis was in the range of 343.55–516.74 and 3.77–7.58 mg 100 g^{-1}, respectively, but 3,4-DHBA was not identified. Additionally, no R was found after the hydrolysis process. In turn, concentration of CAA and 3,4-DHBA in mate herb was in the range 443.14–737.49 and 24.85–35,59 mg 100 g^{-1}, respectively [41].

Three buffers with different range pH were used to optimize the separation of epicatechin, epigallocatechin gallate, morin, chrysin, and hesperidin: sodium tetraborate (pH 8.0–10.0), sodium hydrogen phosphate (pH 7.5–9.0), and sodium acetate (pH 6.0–7.0) [42]. Therefore, sodium tetraborate buffer with 1-butyl-3-methyl imidazolium hexafluorophosphate (BMIM-PF_6) as an additive was chosen for analysis [42]. Under this separation conditions, five different compounds could be completely separated in 8 min. Authors showed that average concentration of epigallocatechin gallate (EGCG) and epicatechin (EC) in green tea samples was higher compared to black tea. Average content of EGCG and EC was 28.36 and 15.60 mg g^{-1} for green tea and 14.02 and 6.65 mg g^{-1} for black tea [42]. In the other studies, selected polyphenols compounds were isolated in tea rooibos (*Aspalathus linearis*), and honeybush (*Cyclopia substernata* and *Cyclopia maculata*) [43]. In this research, optimal separation for honeybush and rooibos phenolics was achieved in 21 and 32 min. In pursuit of the most optimal separation conditions, Urbonavičiute et al.

originally conducted studies using buffer 50 mM L^{-1} Na$_2$B$_4$O$_7$ at pH 9. However, at pH 8.2, the separation of rutin, vitexin-2″-O-rhamnoside and hyperoside was the best [44]. In quantitative analysis of polyphenols in *Coreopsis tinctoria* authors used running buffer at different pH values, between 8.0 and 10.0. The best selectivity obtained with pH 9.0 borate buffer [45]. The contents of flavonoids and phenolic acids were decreased in the following: flower > bud > seed > leaf and >stem. Authors observed that okanin 4′-O-glucoside was the most abundant in dried flower—33.8–60.9 mg g^{-1} [45]. In addition, the optimal BGE composition for the selective separation of arteanoflavone, eupatilin, hispidulin, and 5, 7,4′-trihydroxy-6,3′,5′- trimethoxyflavone in Chinese herbs was 25 mM L^{-1} borate and 6 mM L^{-1} β-cyclodextrin (β-CD) [46]. In turn, using of 30 mM L^{-1} borax solution (pH 10.2) allowed an identification of nevadensin in Fewflower Lysionotus Herb on the level of 2.82 mg g^{-1} (RSD 3.22%) [47]. Application of 40 mM L^{-1} borate buffer and shorter fused silica capillary (53 cm) in comparison to the previously cited Adimicilar et al. studies, higher detection limits for carnosic acid (CA) and rosmarinic acid (RA) in sage were 2.79 and 3.18 µg mL^{-1}, respectively [48]. Studies have shown a higher concentration of CA (5.26 mg g^{-1}) compared to RA (3.14 mg g^{-1}).

For the purpose of quantifying of succinic (SA), malic (MA), tartaric (TA), citric (CITA), and lactic acid (LA) in chamomile, linden, and mint, a 0.5 M H$_3$PO$_4$ and 0.5 mM L^{-1} cetyltrimethylammonium bromide (pH 6.25) was used [50]. Application of indirect UV detection caused that average content of SA in decoction from chamomile was in the level of 9.98 mg L^{-1} but in infusion was not detected. In turn, MA was identified in all analyzed samples in wide range of 18.20–111.53 mg L^{-1} and TA was only in infusion and decoction mint (19.02 and 24.76 mg L^{-1}) [50]. The separation of alkaloids in tea was obtained using 9.2 pH 15 mM L^{-1} borax (BGE) [51]. Under this condition, the LOD was 3.0 for caffeine, 2.1 for theobromine, and 1.6 µg mL^{-1} for theophylline and the concentration of these analytes in the real samples were 34.30, 2.87, and 2.64 mg mL^{-1}, respectively. Zhou et al. developed for the quantification of eight isoquinoline alkaloids in *Chelidonium majus* [53]. Low pH value of Tris-H$_3$PO$_4$ buffer (pH 2.5) containing 50% methanol and 2 mM HP-β-CD allowed identification of sanguinarine, coptisine, chelerythrine, berberine, chelidonine, protopine, allocryptopine, and stylopine with low limits of detection in 9 min. See Table 2.

Table 2. Application of capillary electrophoresis with ultraviolet detection (CZE-UV).

Ref	Instrumental Variables					Chemical Variables		Compounds	λ (nm)	LOD (µg mL^{-1})
	V (kV)	L$_{cap}$ (cm)	ID (µm)	OD (µm)	T (°C)	BGE	pH			
[30]	−20	61.5 FSC	50	nd	nd	50 mM L^{-1} phosphate buffer	3.5	nitrate, nitrite, bromate	200	0.60 0.99 2.14
[35]	25	51.5 FSC	nd	nd	nd	50 mM L^{-1} sodium tetraborate	9.3	vitexin derivative, quercetin derivatives, chlorogenic acid, caffeic acid, isorhamnetin derivative, rosmarinic acid, protocatechuin acid, salvianolic acid, derivative, luteolin derivatives, narigeninin derivatives, dicaffeoylqunic acid derivatives, tannins derivatives	200–400	*
[36]	20	75 QC	50	nd	RT	50 and 100 mM L^{-1} sodium tetraborate	nd	apigenin-7-glucoside, apigenin	254	nd
[27]	20.0	50 FSC	75	nd	RT	30 mM L^{-1} borate buffer	9.50	GL, RH	254	2.1 1.6
[37]	28	67 FSC	50	nd	25	20 mM L^{-1} borate buffer	9.6	carnosic acid, rosmarinic acid	210	1.7 1.9

Table 2. Cont.

Ref	Instrumental Variables					Chemical Variables		Compounds	λ (nm)	LOD (µg mL^{-1})
	V (kV)	L$_{cap}$ (cm)	ID (µm)	OD (µm)	T (°C)	BGE	pH			
[38]	20.0	43 FSC	75	nd	25	40 mM L^{-1} borate buffer	9.2	(−)-epicatechin, catechin, vanillic acid, rosmarinic acid, caffeic acid, gallic acid	280	0.2 0.2 0.5 0.8 0.7 1.7
[39]	20	70 FSC	75	375	nd	20 mM L^{-1} sodium tetraborate with 12 mM L^{-1} β-CD	9.0	protocatechuic acid, salvianolic acid, rosmarinic acid, salvianolic acid A, danshemsu, salvianolic acid B, protocatechuic acid	280	0.18 0.14 0.24 0.28 0.26 0.36 0.18
[20]	12.0	75 FSC	75	nd	RT	50 mM L^{-1} sodium borate, 5% acetonitrile, 1 M NaOH	9.3	baicalin	280	nd
[40]	25	60 FSC	50	nd	30	100 mM L^{-1} sodium tetraborate, 10 mM L^{-1} β-CD	9.3	gentisin, isogentisin, amarogentin	260 260 242	0.69 1.22 1.24
[41]	30	40 nd	50	nd	25	100 mM L^{-1} boric acid	9.0	rutin, caffeic acid, 3,4-dihydroxybenzoic acid	217	0.14 0.05 0.05
[42]	17	57 FSC	75	375	25	25 mM L^{-1} borate buffer	9.0	hesperidin, chrysin, epicatechin, epigallocatechin gallate, morin,	214	0.44 0.50 0.48 0.54 0.47
[43]	30	88.5 FSC	nd	nd	20	200 mM L^{-1} borate buffer	8.8 and 9.25	phenylpyruvic acid-2-O-glucoside, luteolin-7-O-glucoside, isovitexin, isoorientin, vitexin, orientin, chrysoeriol, luteolin, scolymoside, hesperidin, eriocitrin, quercetin, isoquercitrin, hyperoside, rutin, nothofagin, aspalathin, phloretin-3′,5′-di-C-glucoside, hydroxycinnamic acid, iriflophnone-3-C-glucoside-4-O-glucoside, iriflphnone-3-C-glucoside, mangiferin, isomangiferin	283 330 384	0.73–14.9
[44]	25	71 FSC	50	nd	25	50 mM L^{-1} Na$_2$B$_4$O$_7$	8.2	vitexin, rutin, vitexin-2″-O-rhamnoside, hyperoside	280	0.43 0.88 0.53 0.43
[45]	25	50 FSC	75	365	25	50 mM L^{-1} borax, 15% (v/v) acetonitrile	9.0	tiaxifolin-7-O-glucoside, flavanomarein, quecetagetin-7-O-glucoside, okanin 4′-O-glucoside, chlorogenic acid, okanin	280	3.23 0.95 1.46 2.21 0.58 0.63
[46]	20.0	50 FSC	75	nd	20	25 mM L^{-1} borate, 6 mM L^{-1} β-CD	10.2	arteanoflavone, eupatilin, hispidulin, TTMF	254	0.945 0.762 1.002 1.036
[47]	16	52 nd	75	nd	27	30 mM L^{-1} borax solution	nd	nevadensin	335	nd

Table 2. Cont.

Ref	Instrumental Variables					Chemical Variables		Compounds	λ (nm)	LOD (μg mL^{-1})
	V (kV)	L$_{cap}$ ^ (cm)	ID (μm)	OD (μm)	T (°C)	BGE	pH			
[48]	28	53 FSC	50	nd	nd	40 mM L^{-1} borate buffer	9.6	carnosic acid, rosmarinic acid	210	2.79 3.18
[49]	20	62.5 FSC	75	nd	25	50 mM L^{-1} borax and 8 mM L^{-1} β-CD	9.53	ursolic acid, oleanolic acid	214	3.8 3.4
[50]	−10	72 FSC	50	nd	25	50 mM H$_3$PO$_4$ and 0.5 mM L^{-1} cetyltrimethylammonium bromide	6.25	succinic acid, malic acid, tartaric acid, citric acid, lactic acid	200	0.97 3.10 2.18 2.53 89.73
[51]	20	72 FSC	75	375	nd	15 mM L^{-1} borax	9.2	caffeine, theobromine, theophylline	274	3.0 2.1 1.6
[52]	25	64.5 FSC	50	nd	25	50 mM L^{-1} phosphate buffer, 1% hydroxypropyl-β-CD, 3.3% isopropanol	2.5	cytosine, sophocarpine, matrine, lehmannine, sophoranol, oxymatrine, oxysophocarpine	200	nd
[53]	20	35 FSC	50	375	20	500 mM L^{-1} Tris-H$_3$PO$_4$ buffer, 50% (v/v) methanol, 2 mM L^{-1} HP-β-CD	2.5	sanguinarine, coptisine, chelerythrine, berberine, chelidonine, protopine, allocryptopine, stylopine	205	4.27 4.28 1.87 3.78 0.69 0.75 0.65 0.90
[54]	nd	nd FSC	nd	nd	nd	20 mM L^{-1} Na$_2$HPO$_4$, 10% methanol	7.0	berberine, jatrorrhizine, palmatine	nd	0.3 0.3
[55]	25	52 FSC	50	nd	30	100 mM L^{-1} ammonium acetate in methanol, acetonitrile, water (70/25/5; $v/v/v$) with 0.6% acetic acid	nd	(−)-caaverine, (+)-isoliensinine, (+)-norarmepavine, (−)-armepavine, (−)-nuciferine, (−)-nornuciferine, (+)-pronuciferine	225	1.53 1.28 0.82 0.55 0.45 1.16 1.28
[56]	30	48.5 FSC	75	nd	20	10 mM L^{-1} Na$_2$HPO$_4$, 6 mM L^{-1} Na$_3$PO$_4$, 15% methanol (v/v)	11.8	physcion, aloe-emodin, chrysophanol, emodin, aurantio-obtusin, rhein	254	0.94 0.42 0.33 0.43 1.40 1.03
[57]	15	35 FSC	50	365	25	20 mM L^{-1} borax buffer	10.5	HC, HMC, 7-GC	230	3.75 0.94 3.75
[58]	15	65 nd	50	375	RT	50 mM L^{-1} copper sulfate and 0.5% acetic acid (v/v)	4.5	L-phenylalanine, L-histidine, L-leucine, L-glutamic, L-proline	254	5.0 M 3.0 M 1.0 M 2.0 M 3.0 M

nd—no data, V—voltage (KV), L$_{cap}$—length of capillary (cm), BGE—background electrolyte, ^—material of capillary: FSC—fused silica capillary, QC—quartz capillary, ID—inner diameter of capillary (μm), OD—outer diameter of capillary (μm), T—temperature (°C), RT—room temperature, λ—wavelength (nm), LOD—limit of detection (μg mL^{-1}), *—reference method, β-CD—β-cyclodextrin, GL—7-O-α-L-rhamnopyranosyl-kaempferol-3-O-β-D-glucopyranoside, RH—7-O-α-L-rhamnopyranosyl-kaempferol-3-O-α-L-rhamnopyranoside, TTMF—5,7,4′-trihydroxy-6,3′,5′-trimethoxyflavone, HC—7-hydroxy-coumarin, HMC—7-hydroxy-8-methoxy-coumarin, 7-GC—7-O-β-D-glucosyl-coumarin.

2.3.2. Capillary Zone Electrophoresis with Electrochemical Detection

In recent years, the use of CE with electrochemical detection for the separation and subsequent determination of polyphenols, flavonoids oolong tea, and Chinese herbal tea [25,59] and alkaloids in *Plumula Nelumbinis* [60]. See Table 3.

Researchers have high hopes for application of amperometric detection (AD) with CE, because of high sensitivity and selectivity and miniaturization of the detection system [60]. In the case of CE-AD, the miniaturization process consists in the use of reagent volumes from several dozen to several hundred µL [60]. For the determination of alkaloids, Wan et al. used only 100 µL of sample and 1 mL of the running buffer in the study [60]. In the cited article, the alkaloids were separated from *Plumula Nelumbinis* within 12 min in a 40 cm long fused silica capillary with a 50 mM L^{-1} borate buffer (pH 9.2). In this study, the authors used a three-electrode detection cell with a carbon detection electrode with a diameter of 300 µm, a platinum auxiliary electrode and a saturated calomel electrode as a reference electrode [60]. Authors determined neferine, liensinine, isoliensinine, rutin, and hyperoside in the range of 2.24–3.67, 5.14–8.99, 1.11–2.84, 1.35–3.49, and 0.14–0.29 mg g^{-1}. Amperometric detection with Cu disc electrode with diameter of 300 µm as working electrode, saturated calomel reference electrode and a Pt electrode was used for detection of L-theanine (L-THE), L-glutamine (L-GLU), sucrose (SUC), glucose (GLU), fructose (FRU), ascorbic acid (ASC), and (−)-epigallocatechin gallate (EGCG) [25]. The authors argue that due to complexation, amino acids, such as L-THE, L-GLU may respond well to the Cu disc electrode. Thanks to the application this kind of detector concentration of L-THE, L-GLU, SUC, GLU, FRU, and EGCG in oolong tea originated from China were 0.03, 0.03, 0.72, 0.14, 0.33, and 0.10 mM L^{-1}. In turn, ASC was not detected in analyzed samples [25]. Due to the similar structure of the antioxidant compounds, the separation of kaempferol (K), apigenin (A), rutin (R), ferulic acid (FA), quercetin (Q) and luteolin (LUT) is difficult. This problem was solved by adding 0.20 mM L^{-1} β-CD to the buffer [59]. Under the optimum CZE-AD conditions, it was found of K, A, R, FA, Q, and LUT in the average level of 0.74, 0.36, 0.61, 0.53, 0.35, and 0.55 µg 100 mL^{-1}, respectively [59].

Three electrodes, 500 µm diameter carbon disc working electrode, a Pt auxiliary electrode and a Ag/AgCl reference electrode in combination with an amperometric detector was used for determination of catechin (CAT), rutin (R), hyperoside (H), quercetin (Q) and quercitrin (QU) in *Agirimonia pilosa* [61]. Under these conditions, it was found that the *Agrimonia pilosa* stems were a richer source of R and H and poorer of CAT, Q and QU. The highest concentration of H was determined in stems (576.0 µg g^{-1}) [61]. The use of an ultra-small sample volume, the consumption of low solvent volumes and a simple pre-treatment of the sample were proposed in studies with the herb *Acanthopanax senticosus* [62]. The best resolution and the higher peak currents for isofraxadin (ISOF) and rutin (R) could be possible with use of mixture 7.5 mM NaH_2PO_4 and 7.5 mM borax (pH 6.0). The lowest concentration of ISOF was found in leaf (1.2 µ g^{-1}) and root (1.2 µ g^{-1}) of *A. senticosus* and R was isolated only in leaf (13.0 µ g^{-1}). Moreover, in the Zhou et al. research 33 µm carbon fiber microdisk electrode (CFE) has been applied to identification of aristolochic acid I (AA-I) and aristolochic acid II (AA-II) [63]. In this paper, the optimum condition to separation of this two compounds was 20 mM L^{-1} phosphate buffer solution with pH 10.0. Modification of CE-ED technique allowed for obtained low LOD, which is equal for AA-I and AA-II 0.04 and 0.01 µM, respectively. The concentration of AA-I in the root of *Aristolochia debilis* was more than twice higher than in AA-II [63].

Table 3. Application of capillary electrophoresis with electrochemical detection.

Ref	Instrumental Variables					Chemical Variables		Compounds	Method	LOD ($\mu g\ mL^{-1}$)
	V (kV)	Lcap^ (cm)	ID (μm)	OD (μm)	T (°C)	BGE	pH			
[60]	30	40 FSC	25	360	nd	50 mM L^{-1} borate buffer	9.2	neferine, liensinine, isoliensinine, rutin, hyperoside	CE-AD	0.67 μM 0.50 μM 0.62 μM 0.57 μM 0.84 μM
[25]	18	75 FSC	25	360	nd	30 mM L^{-1} borate, 40 mM L^{-1} phosphate	8.5	L-theanine, L-glutamine, sucrose, glucose, fructose, ascorbic acid, (−)-epigallocatechin gallate	CE-AD	0.53 μM 0.47 μM 0.24 μM 0.07 μM 0.53 μM 2.39 μM 0.86 μM
[59]	15.0	60 FSC	25	360	RT	50 mM L^{-1} KH_2PO_4 250 mM L^{-1} $Na_2B_4O_7$ and 0.20 mM L^{-1} β-CD	7.6	kaempferol, apigenin, rutin, ferulic acid, quercetin, luteolin	CE-AD	0.92×10^{-4} 0.79×10^{-4} 2.59×10^{-4} 1.90×10^{-4} 1.37×10^{-4} 1.13×10^{-4}
[61]	19.5	60 FSC	25	370	nd	60 mM L^{-1} $Na_2B_4O_7$, 120 mM L^{-1} NaH_2PO_4	8.8	catechin, rutin, hyperoside, quercetin, quercitrin	CE-ED	0.02 0.05 0.02 0.03 0.02
[62]	30.0	40 FSC	25	360	RT	7.5 mM L^{-1} NaH_2PO_4, 7.5 mM L^{-1} borax	6.0	isofraxidin, rutin, chlorogenic acid	CE-ED	0.10 μM 0.20 μM 0.15 μM
[63]	12.5	40 FSC	25	360	RT	20 mM L^{-1} phosphate buffer solution	10.0	aristolochic acid I, aristolochic acid II	CE-ED	0.04 μM 0.01 μM
[64]	16	75 FSC	25	nd	nd	50 mM L^{-1} borate buffer	9	genistin, genistein, rutin, kaempferol, quercetin	CE-ED	0.12 0.11 0.20 0.17 0.28

nd—no data, V—voltage (KV), Lcap—length of capillary (cm), BGE—background electrolyte, ^—material of capillary: FSC—fused silica capillary, ID—inner diameter of capillary (μm), OD—outer diameter of capillary (μm), T—temperature (°C), RT—room temperature, λ—wavelength (nm), LOD—limit of detection ($\mu g\ mL^{-1}$), CE-AD—capillary electrophoresis with amperometric detector, CE-ED—capillary electrophoresis with electrochemical detection.

2.3.3. Microfluidic Analysis with Contactless Conductivity Detection

In recent years, more and more attention has been paid to new, more improved solutions with the use of lab-on-a-chip microfluidic devices [65]. They enable fast and highly sensitive analysis and improvement of the repeatability of analyzes. The miniaturization of analytical processes has brought great hope in recent years. This is due not only to the small volumes of the reagents, but also to the reduced analysis time and low energy consumption [66].

Microfluidic analytical system (MFAS) was used for determination of epigallocatechin gallate (EGCG), epicatechin (EC), epicatechin gallate (ECG), and epigallocatechin (EGC) in green tea [65]. In this paper, authors manufacturing a chip based on polydimethylsiloxane and glass. Under optimum conditions (20 mM borate buffer electrolyte, pH 9.2) LOD for EC, EGC, ECG, and EGCG was 3.5, 3.5, 3.2, and 2.3 $\mu g\ mL^{-1}$. The concentration of EGCG, GCG, EC, ECG, and EGC was 167 ± 15, 44 ± 8, 162 ± 15, 67± 9, and 4 ± 2, respectively [65]. Application of conductively detector is very sensitive and suitable. Tang et al. stabilized the ionic strength by adding lactic acid and β-alanine as background electrolyte components in the CE method with capacitively coupled contactless conductivity detection (CE-C^4D) [67]. C^4D is suggested when the molecules are without or weak chromophores and when their detection with optical systems is impossible. This procedure enabled the determination of eight metal ions (Mg^{2+}, Mn^{2+}, Cd^{2+}, Co^{2+}, Pb^{2+}, Ni^{2+}, Zn^{2+}, and Cu^{2+}) in dried *Forsythiae Fructus* (Oleaceae) in 10 min, which possesses anti-inflammatory, antioxidant as well as hepatoprotective, neuroprotective and cardiovascular protective effects [67]. The highest concentration of analyzed compounds was found for magnesium (2.38–4.14 mg g^{-1}) what is important from the nutritional point of view [68].

In recent years, unique solutions have also been sought, such as the use of home microchip electrophoresis with an integrated Pt detector [66]. This type of solution was used to analyze guanosine (G), methionine (M), glycine (GLY), 3,4-dihydroxybenzaldehyde (DHB), and homogentisic acid (HA) in *Pinellia ternata* used in traditional Chinese medicine. Using an innovative solution, Shih et al. determined five ingredients were determined within 5 min using the special platform [66]. Using this techniques authors determined G, M, GL, DHB, and HA on the average level of 141.5, 20.5, 25.1, 44.3, and 62.6 µg g^{-1}. Strychalski et al. developed simple and robust analytical technique, well-suited to microfluidic, called GEMBE (gradient elution moving boundary electrophoresis) [69]. This technique uses ultrasmall capillary or microchannel (few mm–several cm in length). In this study 5.5 cm long fused silica capillary (OD 363.5 and ID 13.5 µm) was used [69]. Detection point was only approximately 15 mm from the capillary inlet. The essence of this method is also the use of much lower volumes of solutions (200 µL) [69]. In a complex matrix such as tea, CE chips with electrochemical detection are selective micro-flow microfluidic platforms [70]. Two analytical solutions were tested in these studies: class-selective electrochemical index determination (CSEID) and individual antioxidant determination (IAD). For CSEID optimal results were obtained for the separation of flavonoids and phenolic acids in less than 100 s using MES (2-(N-morpholino)ethanesulfonic acid) at pH 5 and for IAD separation of nine phenolic compounds was provided in a borate buffer at pH 9 in only 260 s. In this paper, authors determined (+)-catechin, rutin, ferulic acid, chlorogenic acid, vanillic acid, quercetin, caffeic acid, gallic acid, and protocatechuic acid in the total level of 290 ± 2 for CSEID and 321 ± 11 µg mL^{-1} in green tea [70].

2.3.4. Capillary Zone Electrophoresis with Fluorescence Detection

The literature review shows that in CE-UV detection is also replaced by fluorescence detection (Table 4). Combination of CE with laser induced fluorescence (LIF) has provided improvement in detection limit, compared with UV detector [71]. This type of detection was used to determine the amino acids in tea infusions, oolong tea and jasmine tea [72–74] and riboflavin in *Camelia sinensis* [71] and green tea name Zhuyeqing [75]. CE derivatization methods were used to determine γ-aminobutyric acid (GABA) and alanine in aqueous extract of Chinese tea after derivatization with o-phtaldialdehyde/2-mercaptoethanol (OPA/2-ME) to produce fluorescently labeled analytes [72]. The labeled derivatization with 20 mM L^{-1} OPA and 26.67 mM L^{-1} 2-ME at pH 10.0 gave the most sensitive detection and optimum buffer was composed with 30 mM L^{-1} sodium tetraborate (pH 10.0) for determination of these two amino acids in tea. Under this conditions authors obtained LOD 0.004 and 0.02 µM for GABA and alanine. In turn, GABA was determined in jasmine green tea, oolong tea and GABA-rich tea on the level of 2.5, 6.0, and 157.2 mg 100 g^{-1}, respectively, and alanine was identified at 22.7, 13.3, and 51.8 mg 100 g^{-1}, respectively [73]. Moreover, amino acids was analyzed with use of combination of CE and light-emitting diode-induced fluorescence detection (LED-IF) [74]. Authors claimed that by using 0.5% PEO solution (prepared in 10 mM L^{-1} Na$_2$B$_4$O$_7$ at 9.3 pH) and 60 cm capillary length, GABA, GL, and aspartic acid (ASP) were marketed within 16 min. Moreover, fluorescence detector was used in the analysis of riboflavin (RF) concentration in *Camelia sinensis* [71,75]. RF was determined in green tea by CE with in-column optical fiber laser-induced fluorescence detection (CE-LIF). The concentration of RF in samples were between 0.05 and 20 µM with LOD 3.0 nM [75].

Table 4. Application of capillary electrophoresis with fluorescence detection.

Ref	Instrumental Variables					Chemical Variables		Compounds	λ (nm)	LOD (µM)
	V (kV)	Lcap ^ (cm)	ID (µm)	OD (µm)	T (°C)	BGE	pH			
[72]	21	80 FSC	50	360	23	30 mM L^{-1} sodium tetraborate, 20 mM L^{-1} OPA/2-ME	10.0	γ-aminobutyric acid, alanine	495	0.004 0.02
[73]	21	80 FSC	50	365	23	50 mM L^{-1} sodium tetraborate	10.0	γ-aminobutyric acid, alanine	495	700 800
[74]	15	60 FSC	75	365	nd	10 mM L^{-1} Na$_2$B$_4$O$_7$	9.3	γ-aminobutyric acid, glycine, aspartic acid	410	360–28,300
[71]	25	50 FSC	50	nd	25	30 mM L^{-1} phosphate buffer	9.9	riboflavin	488	1.08 ng mL^{-1}
[75]	18	55 FSC	100	nd	RT	water/acetonitrile (9:1, v/v), 10 mM L^{-1} borate buffer	9.6	riboflavin	474	300,0
[76]	10	30 FSC	75	nd	25	5% acetonitrile, 0.25% acetic acid, 35 mM L^{-1} ammonium acetate in methanol	nd	berberine, palmatine, jatrorrhizine	488	6.0 ng mL^{-1} 7.5 ng mL^{-1} 380 ng mL^{-1}
[77]	25	50.2 QC	75	nd	30	5 mM borate buffer, 10^{-4} mM fluorescin sodium salt, 20% methanol	9.4	naringin, sophoricoside, esculin, genistein, isofraxidin, esculetin	488 520	20.2 5.0 7.5 7.5 6.8 9.7

nd—no data, V—voltage (KV), Lcap—length of capillary (cm), BGE—background electrolyte, ^—material of capillary: FSC—fused silica capillary, QC—quartz capillary, ID—inner diameter of capillary (µm), OD—outer diameter of capillary (µm), T—temperature (°C), RT—room temperature, λ—wavelength (nm), LOD—limit of detection (µg mL^{-1}), OPA—o-phthaldialdehyde; 2-ME—2-mercaptoethanol.

In the other studies, application of non-aqueous capillary electrophoresis (NACE) was coupled with laser-induced native fluorescence detection for analysis of three alkaloids in methanol extract prepared from dried *Rhizoma coptidis* and *Caulis mahoniae* [76]. The authors consistently changed the analysis conditions to obtain the shortest possible analysis time and to generate the lowest Joule heat possible not to cause overlapping of the mixture component zones. In the course of the conducted experiments, it was noted that with the increasing percentage of ammonium acetate in BGE, the migration time increased. The strongest detector signal was obtained with 35 mM L^{-1} ammonium acetate. That is way, the authors obtained the optimal conditions using 5% acetonitrile, 0.25% acetic acid and 35 mM L^{-1} ammonium acetate in methanol [76]. The use of fluorescence detector allowed to obtain a lower LOD compared to UV detection. For palmitine (PAL), authors received LOD on the level of 7.5, for berberine (BER) 6.0 and jatrorhizine (JATR) 380.0 ng mL^{-1}. Under this conditions, BER and PAL was identification on the average level of 4.95 ± 0.23 and 2.92 ± 0.17% for *R. coptidis* and 1.02 ± 0.08 and 0.59 ± 0.03% for *C. mahoniae* [76]. Previously, the aqueous electrolyte was not suitable for the MS system, therefore, non-aqueous CE with UV and MS detection was used to determine the alkaloids in *Nelumbo nucifaera* [55]. In turn, using the fluorescence detection has its limitations as not all compounds are capable of fluorescence, which is necessary for the application of this type of detector in quantification. Only Wang et al. determined naringin, esculin, genistein, isofraxidin, and esculetin in Fructus Sophorae japonicae and Herb sarcandrae belonging to the Chloranthaceae family [77].

2.4. Micellar Electrokinetic Chromatography

Another kind of CE is micellar electrokinetic chromatography (MEKC) used mainly for the separation of mixture components whose analytes are electrically inert and charged. An important difference that distinguishes this type of CE is the use of a surfactant in the buffer to form micelles. The result is a pseudostationary phase (micellar phase) and a mobile liquid phase.

This technique was used for the analysis of polyphenols in *Scutellaria baicalensis*, tea samples, oolong tea, and green tea, *Salvia officinallis* [78], tea samples [79], and *Arnica montana* [80]. In turn, catechins and methylxanthines in green tea samples and coumarins in *Aesculus hippocastanum* and *Heracleum sphondyliu* [81]. Moreover, amino acids in black, jasmine, green tee [82], and indoleamines (melatonine) in *Camelia sinensis* and *Tilia cordata* [83]. The full characteristics of the method used are presented in the Table 5. The most common surfactant used in the MEKC technique is SDS (sodium dodecyl sulfate). Buffer system containing 15 mM L^{-1} borate, 40 mM L^{-1} phosphate and 15 mM L^{-1} SDS with 15% acetonitrile and 7.5% 2-propanol was used to separate baicalin (B), baicalein (BC) and wogonin (W) in *Scutellaria baicalensis* originated from China, where baseline separation was obtained within 15 min [78]. In this research, in analyzed samples concentration of B, BC and W were determined in the range of 24.74–143.56 mg g^{-1}, 1.53–15.12 mg g^{-1}, and 0.37–4.80 mg g^{-1}. Similarly, SDS has also been used to separate catechins from green tea [79,84]. But, for the separation of six major green tea catechins and enantiomers of theanine was used Heptakis (2,6-di-O-methyl)-β-cyclodextrin [84] and hydroxypropyl-β-cyclodextrin (HP-β-CD) as chiral selector [79]. Moreover, Gomez et al. used a 10 mM L^{-1} sodium tetraborate (pH 9.2) as a BGE and mixture of 20 mM L^{-1} SDS anionic micelles and 20 mM L^{-1} β-CD with 10% acetonitrile in identification of antioxidative melatonin [85] in *Camelia sinensis* and *Tilia cordata*. This allowed the detection to be reduced to low ppb levels [83]. The average content of melatonin in green tea was 386 ng g^{-1} [83]. To isolate the coumarins (isopimpineline, bergapten, phellopterin, esculin, and esculetin) from *A. hippocastanum* and *H. sphondylium* originated from Poland, Dresler et al. replaced SDS with 65 mM L^{-1} SC (sodium salt) [81]. Using the 50 mM L^{-1} sodium tetraborate and 60 mM L^{-1} SC and 20% methanol (v/v) authors ensured good resolution of the analyzed compounds. Similarly, during the analysis of amino acids in aqueous infusions of tea, 20 mM L^{-1} sodium borate (pH 8.5) and 20 mM L^{-1} Brij 35 with 10% acetonitrile was used [82]. Due to the long analysis time and the poor separation efficiency, popular SDS was changed to Brij 35. Under this separation conditions, 15 different amino acids could be completely separated in 11 min.

2.5. Capillary Isotachophoresis

Capillary isotachophoresis (CITP) is an anion-cation separation technique in which, unlike CZE, two buffer systems are used. One of them is called the leading electrolyte (LE) with high ion mobility compared to the analyte and terminating electrolyte (TE) with reduced ion mobility. This kind of technique of CE was used in two articles to analyze selected antioxidants in *Melissae herba* and in *Herba Epilobi* which are traditionally used for the symptomatic treatment of gastrointestinal disturbances and for urological problem in men, respectively [87–89], which was shown in Table 6. Using 10 mM L^{-1} HCl and 0.2% hydroxyethylcellulose as LE buffer and 50 mM L^{-1} H_3BO_3 as TE buffer, CA, RA, *p*-coumaric acid (*p*A), chlorogenic acid (CLA), FA and QU were determined in *Melissa herb* at the level 1.65 ± 0.80, 43.54 ± 1.73, 1.00 ± 1.04, 0.30 ± 4.65, 3.70 ± 2.17, and 1.25 ± 2.97 mg g^{-1}, respectively [88]. Using of CITP method authors obtained limit of detection at level of 0.018 for CA, 0.027 for RA, 0.030 μg mL^{-1} for the *p*A, 0.032 for CLA, 0.020 for FA and 0.035 μg mL^{-1} for QU. In turn, the combination of capillary isotachophoresis (ITP) and capillary zone electrophoresis (CZE) was used to determinate phenolic acids in Herba Epilobi, which could allow to obtain LOD on the level 0.05 for cinnamic acid, 0.010 for *p*A, 0.021 for FA, 0.026 for syringic acid, 0.034 for CA, 0.041 for protocatechuic acid, 0.044 for vanillic acid, and 0.061 μg mL^{-1} for CLA [87].

Table 5. Application of micellar electrokinetic chromatography.

Ref	Instrumental Variables					Chemical Variables		Compounds	λ (nm)	LOD (µg mL^{-1})
	V (kV)	Lcap^ (cm)	ID (µm)	OD (µm)	T (°C)	BGE	pH			
[78]	20	50 FSC	75	nd	25	15 mM L^{-1} borate, 40 mM L^{-1} phosphate, 15 mM L^{-1} SDS, 15% (v/v) acetonitrile, 7.5% (v/v) 2-propanol	nd	baicelin, baicalein, wogonin	280	1.79 1.19 0.78
[79]	15.0	8.5 nd	nd	nd	25	25mM L^{-1} borate–phosphate buffer, 25 mM L^{-1} HP-β-CD, 90 mM L^{-1} SDS	2.5	(−)-EGC; (−)-GCG; (−)-C; (−)-ECG; (−)-EGCG; (+)-C; (−)-EC; CAF; TB; (−)-GC; (+)-GC;	200	0.05 0.15 0.1 0.1 0.2 0.4 0.7
[80]	25	62 FSC	50	nd	40	50 mM L^{-1} borax, 25 mM SDS, 30% acetonitrile	6.75	kaempferol-3-O-glucoside, 6-methoxy-kaempferol 3-O-glucoside, hispidulin, quercetin 3-O-glucoside, patuletin-3-O-glucoside, quercetin 3-O-glucuronic acid, chlorogenic acid, 3,5-dicaffeoylquinic acid, 4,5-dicaffeoylquinic acid	254	≤4.5
[81]	30	64.5 nd	50	nd	27	50 mM L^{-1} sodium tetraborate, 60 mM L^{-1} SC, 20% methanol (v/v)	9.0	coumarin, scoparone, xanthotoxin, byakangelicin, isoscopoletin, esculin, isopimpinellin, bergapten, esculetin, phellopterin, xanthotoxol, umbelliferone	214	0.575 0.927 1.042 1.290 1.558 1.273 1.512 1.358 0.860 1.987 1.599 1.422
[82]	20	50.2 FSC	75	375	25	20 mM L^{-1} sodium borate, 20 mM Brij 35, acetonitrile 10%	8.5	lysine, phenylalanine, leucine, methionine, valine, theanine, histidine, γ-aminobutyric acid, threonine, alanine, serine, glycine, cysteine, glutamic acid, aspartic acid	nd	0.5 ng mL^{-1} 0.2 ng mL^{-1} 0.2 ng mL^{-1} 0.5 ng mL^{-1} 0.2 ng mL^{-1} 0.5 ng mL^{-1} 10 ng mL^{-1} 0.1 ng mL^{-1} 0.5 ng mL^{-1} 0.3 ng mL^{-1} 0.5 ng mL^{-1} 0.2 ng mL^{-1} 100 ng mL^{-1} 3 ng mL^{-1} 5 ng mL^{-1}
[83]	20	50 FSC	75	375	25	10 mM L^{-1} sodium tetraborate, 20 mM β-CD, 20 mM SDS	9.2	serotonin, melatonin, tryptophan, indole-3-acetic acid	220	4.16 ng g^{-1} 0.79 ng g^{-1} 0.72 ng g^{-1} 0.55 ng g^{-1}
[84]	30	48.5 FSC	8.5	nd	30	25 mM L^{-1} borate phosphate buffer, 65 mM L^{-1} SDS, 28 mM L^{-1} Heptakis (2,6-di-O-methyl)-β-CD	2.5	(−)-EGC; (−)-C; (−)-ECG; L-THE; D-THE; (−)-EGCG; (−)-EC; (+)-C; (−)-GCG;	220	0.1 0.2 0.2 0.2 0.2

Table 5. Cont.

Ref	Instrumental Variables				Chemical Variables		Compounds	λ (nm)	LOD (µg mL^{-1})
	V (kV)	Lcap^ (cm)	ID (µm)	OD (µm)	T (°C)	BGE	pH			
[86]	30	60 FSC	50	nd	25	0.03 M L^{-1} borate buffer, 10% acetonitrile, 0.01 mM L^{-1} SDS	10.2	apigenin	390	0.48 µmol L^{-1}

nd—no data; V—voltage (KV), L$_{cap}$—length of capillary (cm), BGE—background electrolyte, ^—material of capillary: FSC—fused silica capillary, ID—inner diameter (µm), OD—outer diameter (µm), T—temperature (°C), RT—room temperature, λ—wavelength (nm), LOD—limit of detection (µg mL^{-1}), SDS—sodium dodecyl sulfate; HP-β-CD—hydroxypropyl-β-cyclodextrin; (−)-EC—(−)-epicatechin; (−)-ECG—(−)-epicatechin gallate; (−)-EGC—(−)-epigallocatechin; (−)-EGCG—(−)-epigallocatechin gallate; (+)-C—(+)-catechin; (−)-C—(−)-catechin; (−)-GCG—(−)-gallocatechin gallate; (+)-GC—(+)-gallocatechin; (−)-GC—(−)-gallocatechin; CAF—caffeine; TB—theobromine; THE—theanine; nd—no data, OPA—o-phthaldialdehyde; 2-ME—2-mercaptoethanol.

Table 6. Process of separation of analytes with application capillary isotachophoresis.

Ref	Instrumental Variables				Chemical Variables		Compounds	λ (nm)	LOD (µg mL^{-1})
	V (kV)	Lcap^ (cm)	ID (µm)	OD (µm)	T (°C)	BGE	pH			
[87]	nd	60 FEP	300	nd	nd	LE: 10 mM L^{-1} HCl, 0.02 M IMI, 0.2% HEC TE: 0.01 M HEPES	7.2 8.2	protocatechuic acid, caffeic acid, cinnamic acid, vanillic acid, coumaric acid, syringic acid, ferulic acid, chlorogenic acid	270	0.041 0.034 0.005 0.044 0.010 0.026 0.021 0.061
[88]	3	16 FEP	300	nd	nd	25 mM L^{-1} MOPSO, 50 mM L^{-1} Tris buffer, 40 mM L^{-1} H$_3$BO$_3$, 0.2% HEC	8.1	caffeic acid, rosmarinic acid, p-coumaric acid, chlorogenic acid, ferulic acid, quercitrin	320	0.018 0.027 0.030 0.032 0.020 0.035

nd—no data, V—voltage (KV), Lcap—length of capillary (cm), BGE—background electrolyte, ^—material of capillary: FEP—fluorinated ethylene-propylene (FEP) copolymer, ID—inner diameter (µm), OD—outer diameter (µm), T—temperature (°C), RT—room temperature, λ—wavelength (nm), LOD—limit of detection (µg mL^{-1}), LE—leading electrolyte; TE—terminating electrolyte, IMI—imidazole, HEC—hydroxyethylcellulose, HEPES—4-(2-Hydroxyethyl)-1-piperazine ethanesulfonic acid, MOPSO—3-(N-morpholino)-2-hydroxypropanesulfonic acid.

2.6. Capillary Electrochromatography

Capillary electrochromatography (CEC) is an electrokinetic separation technique [90]. Using the CEC technique it is possible to separate uncharged and charged substances. CEC combines elements of two techniques: capillary electrophoresis (CZE) and high performance liquid chromatography (HPLC). In CEC it is possible to use packed, monolithic, and open-tubular columns (OTC). However, in recent years, monolithic and open-tubular columns have been used more frequently [90]. OTC allows the use of innovative microporous materials, nanoparticles, and biomaterials as stationary phase elements, which gives wide analytical possibilities. In turn, monolithic columns have higher efficiency and resolution compared to OTC [90]. CEC combines the advantages of both techniques, HPLC and CE. On the one hand, the retention of the analytes depends on their interaction with the surface of the stationary phase particles. On the other hand, in the case of electrically charged elements, it also depends on their electrophoretic mobility. This, in turn, is strongly influenced by the strength of the electric field, the composition of the mobile phase, the ionic strength and the pH of the buffer [91].

The methanolic extracts of *Adinandra nitida* was analyzed with use of monolithic columns of CEC [92]. Using 10 mM L^{-1} ammonium formate (3.0) as BGE, separation of flavonoids, e.g., EC and A, could be accomplished in 25 min on a monolithic rod of macroporous poly(butyl methacrylate-co-ethylene dimethacrylate). In turn, during the analysis of 11 coumarins, flavones, and flavanone (Table 7) in *Chamomilla recutita*, the Hypersil SCX/C18 column with phosphate buffer (pH 2.8) at 50 mM L^{-1} with 50% acetonitrile was used [93]. These conditioned parameters could separate all compounds in less than 7.5 min under isocratic conditions, and moreover, the LOD for A with UV detection at 337 nm was 35.0 µg mL^{-1}. In other studies, good resolution of (+)—catechins, (−)—epicatechins, (−)—epigallocatechins, theophylline, and caffeine in black and green teas were used with a capillary column (ID 100 µm) filled with C18 bidentate particles at 24.5 cm. The mobile phase was a mixture of 5 mM L^{-1} ammonium acetate buffer (pH 4.0) with H$_2$O/acetonitrile (80:20, v/v) and LOD for all analyzed compounds with UV detection at 200 nm was 1.0 µg mL^{-1} [94]. Moreover, capillary electrochromatography was used to determinate six coumarins in *Fructus cnidii* ethanolic extracts [95]. The separation efficiency of the methods was performed in an in-house packed column with a monolithic outlet frit with 10 mM L^{-1} ammonium acetate buffer (pH 4.0) and 50% acetonitrile in 15 min. Limit of detection for bergapten, imperatorin, osthole, 2′-acetylangelicin, oroselone, and O-acetylcoumbianetin was 2.5, 5.0, 1.0, 2.5, 2.5, and 2.5 µg mL^{-1}, respectively. Another type of detection was used in Liu et al. research, where authors determined evadiamine, rutaecarpine, and limonin in *Evidiae fructus* fruit [5]. As stationary phases was used home-developed monolithic columns with methyl-vinylimidazole functionalized organic polymer monolilth. In this study CEC-MS and CEC-UV were compared. Authors obtained LODs of three analyzed compounds in the range of 2.0–12.5 µg mL^{-1} by UV detector and 0.12–3.1 µg mL^{-1} by MS detector. Studies have confirmed that the use of CE with MS detection increases the sensitivity of the method several times, which allows for the determination of alkaloids and limonoids in plant materials [5].

Table 7. Process of separation of analytes with application capillary electrochromatography.

Ref	Instrumental Variables					Chemical Variables		Compounds	λ (nm)	LOD (µg mL^{-1})
	V (kV)	Lcap ^ (cm)	ID (µm)	OD (µm)	T (°C)	BGE	pH			
[93]	25	50 FSC	75	nd	25	50 mM L^{-1} phosphate buffer (pH 2.8) and 50% acetonitrile		herniarin, umbelliferone, caffeic acid, chlorogenic acid, apigenin, naringenin, apigenin-7-O-glucoside, luteolin, luteolin-7-O-glucoside, quercetin, rutin	337	35.0 35.0 35.0 35.0 35.0 35.0
[94]	10	33 FSC	75/100	375	20	acetate buffer	4.0	epigallocatechin, theophylline, catechin, epicatechin, caffeine	200	1.0 1.0 1.0
[92]	10	nd FSC	75	375	nd	10 mM L^{-1} ammonium formate	3.0	epicatechin, apigenin	260	50.0 ng 50.0 ng
[95]	−6	20 FSC	100	375	nd	10 mM L^{-1} ammonium acetate buffer and acetonitrile (50/50, v/v)	4.0	bergapten, imperatorin, osthole, 2′-acetylangelicin, oroselone, O-acetylcolumbianetin	320	2.5 5.0 1.0 2.5 2.5 2.5
[5]	−25	50 FSC	100	365	nd	30% acetonitrile and 1% ammonia aqueous solution	8.2	limonin, evodiamine, rutaecarpine	CEC-MS	3.10 0.63 0.15

nd—no data, V—voltage (KV), Lcap—length of capillary (cm), BGE—background electrolyte, ^—material of capillary: FSC—fused silica capillary, ID—inner diameter (µm), OD—outer diameter (µm), T—temperature (°C), RT—room temperature, λ—wavelength (nm), LOD—limit of detection (µg mL^{-1}), LE—leading electrolyte; TE—terminating electrolyte.

2.7. Capillary Electrophoresis-Mass Spectrometry

CE can be coupled with mass spectrometry detector (CE-MS) what was shown in Table 8 [96,97]. The essence of CE-MS is the electrokinetic separation of analyte groups as a result of the mobility of ions in the electric field [97]. Thanks to the use of MS, it is possible to explain the molecular structure of metabolites that cannot be obtained by other detection methods, e.g., UV detector [5]. MS detection distinguishes the target analyte signal from the sample of the composite matrix and eliminates high background noise. The use of an MS detector may result in a higher sensitivity and selectivity of the analysis compared to the UV detector [5].

Table 8. Process of separation of analytes with application capillary electrophoresis with MS detection.

Ref	Instrumental Variables					Chemical Variables		Compounds	LOD (fg)
	V (kV)	Lcap ^ (cm)	ID (μm)	OD (μm)	T (°C)	BGE	pH		
[96]	30	70 FSC	30	150	nd	1% acetic acid and methanol/water (1/1)	nd	coptisine, berberine, palmatine, jatrorrhizine	18.0 22.0 24.0 16.0
[97]	25	80 FSC	50	nd	25	200 mM L^{-1} boric acid and ammonia	9.0	chlorogenic acid, delphinidin-3-O-glucoside, cyaniding-3-O-rutinoside, cyaniding-3-O-sambubioside, delphinidin-3-sambubioside, Cy-3,5-O-diglucoside	nd

nd—no data, V—voltage (KV), Lcap—length of capillary (cm), BGE—background electrolyte, ^—material of capillary: FSC—fused silica capillary, ID—inner diameter (μm), OD—outer diameter (μm), T—temperature (°C), RT—room temperature, λ—wavelength (nm), LOD—limit of detection (μg mL^{-1}), LE—leading electrolyte; TE—terminating electrolyte.

Studies with use of four alkaloids standards (coptisine, berberine, palmatine, and jatrorhizine) found in *Rhizoma coptidis* proved that 1000 times lower LOD with the use of CE-MS compared to UHPLC-MS [96]. SPE-CE-ESI-MS was used to develop a method of separation and identification of anthocyanins in *Hibiscus sabdariffa* showing antihypertensive and cardioprotective effects [97]. Using 200 mM L^{-1} boric acid and ammonia (9.0) as BGE, separation of chlorogenic acid, delphinidin-3-O-glucoside, cyaniding-3-O-rutinoside, cyaniding-3-O-sambubioside, and delphinidin-3-sambubioside in *Hibiscus sabdariffa* could be accomplished in below 20 min [97].

3. Materials and Methods

This review is based on PRISMA guidelines. The articles selection criteria for the review were carried out using PICOS (Population, Intervention, Comparison, Outcome, Study type) process. For the purpose of this review, articles from 2005 to 2021 were used. Searching of literature for this publication was performed between November 2020 and January 2021 using the PubMed and Web of Science database. The search strategy was with use of the following keywords:

1. "capillary electrophoresis" and "raw material",
2. "capillary electrophoresis" and "tea", and
3. "capillary electrophoresis" and "herb".

In the PubMed base a combination of terms "All fields" and in Web of Science base terms "Topic" was used, which searches titles, abstracts, author keywords, keywords Plus. Only articles in English, available full texts and articles containing publications focused on the analysis of bioactive compounds in plant raw materials by CE are included in this review. See Table 9. Moreover, the search was limited to the matrix, which was a plant materials commonly used as aqueous infusions (tea) or decoctions in traditional medicine. The exclusion criteria were opinion letters, conferences abstracts, papers not written in English (for examples Chinese). Publications in which ornamental horticulture flowers, vegetables, and spices were used as plant material were rejected. Additionally, articles with urine and blood serum, tablets and capsules as the matrix have been eliminated. Studies in which mycotoxins were analyzed using CE were also not taken into account. Duplicates were removed and next, found articles were sorted by title, abstract and then main text. The articles were excluded if they does not meet the inclusion criteria. Selection of appropriate works taking into account inclusion and exclusion criteria were controlled by three authors of this paper (A. P., M. G., and M. K.). Selection of the publications by them was based on the qualitative and quantitative evaluation of articles from the PubMed and Web of Science database, especially title of paper, first name of author and year publication.

Table 9. PICOS (Population, Intervention, Comparison, Outcome, Study type).

	Inclusion Criteria	Exclusion Criteria
Population	raw material plant using in traditional medicine	garden and ornamental plants, spices, vegetables, and fruits
Intervention	use of CE method	other methods
Comparison	capillary electrophoresis vs. other methods	not applicable
Outcome	analysis of bioactive compounds (polyphenols, coumarins, vitamins, alkaloids)	different outcomes
Study type	original research articles, full articles, English language	review articles, reports, abstracts, articles with no quantitative information or details

4. Conclusions

In this review, the authors summarized the last sixteen years of scientific research using capillary electrophoresis to identify and quantify bioactive compounds in raw materials commonly used as "tea" in China, Europe, Asia, and South America. By far China is an area in the world where the number of scientific reports about analysis of polyphenols, coumarins, alkaloids, or amino acids in dried herbal raw materials is greater than in the rest of the world.

CE's versatility is primarily due to its many techniques. Among all the capillary electrophoresis methods, the most popular CE is capillary zone electrophoresis with UV detection. With use of this technique, it is possible to analyze a numerous bioactive compounds in dried raw materials in less than 20 min and low limit of detection. Nevertheless, the use of CE-MS allows for the more sensitive determination of elements with a low limit of detection and gives hope for routine use in the analysis of functional foods. Unfortunately, a major limitation in using the MS detector in conjunction with CE may be its incompatibility with some types of CE. Chiral separations using CE-MS are also limited due to the incompatibility of the chiral selectors with the MS detector. Research aimed at developing chiral selectors compatible with MS seems to be the direction of future research by scientists. Furthermore, one of the limitation in the use of capillary electrophoresis is the choice of a chiral selector during optimizing enantiomeric separation. The type and concentration of cyclodextrins, which are used most often, is one of the most important parameters for proper separation. Moreover, the use of some modifications in electrochemical techniques allows to reduce the sensitivity of the methods along with the reduction of the analysis time.

Author Contributions: The article was prepared by all authors. M.K. and A.P.—reviewed article idea; A.P.—designed the article, wrote draft manuscript which was edited by M.K. and M.G.; A.P., M.K., and M.G.—collected and selected documents using in review. All authors have read and agreed to the published version of the manuscript.

Funding: This research received no external funding.

Institutional Review Board Statement: Not applicable.

Informed Consent Statement: Not applicable.

Conflicts of Interest: The authors declare no conflict of interest.

Abbreviations

A	apigenin
pA	p-coumaric acid
AA-I	aristolochic acid I
AA-II	aristolochic acid II
AAS	Atomic Absorption Spectrometry
AD	amperometric detection
ASC	ascorbic acid
ASP	aspartic acid
B	baicalin
BC	baicalein
BER	berberine
BGE	background electrolyte
BMIM-PF6	1-butyl-3-methyl imidazolium hexafluorophosphate
CA	carnosic acid
CAA	caffeic acid
CAF	caffeine
CAT	catechin
CCE	Chiral capillary electrophoresis
β-CD	β-cyclodextrin
CE	capillary electrophoresis
CE-AD	capillary electrophoresis with amperometric detector
CEC	capillary electrochromatography
CE-C^{4}D	capillary electrophoresis with capacitively coupled contactless conductivity detection
CE-ED	capillary electrophoresis with electrochemical detection
CE-LIF	capillary electrophoresis with laser-induced fluorescence detection
CGE	capillary gel electrophoresis
CIEF	capillary isoelectric focusing
CITA	citric acid
CITP	capillary isotachophoresis
CLA	chlorogenic acid
CSEID	class-selective electrochemical index determination
CZE	capillary zone electrophoresis
DHB	3,4-dihydroxybenzaldehyde
3,4-DHBA	3,4-dihydroxybenzoic acid
EC	epicatechin
ECG	epicatechin gallate
EGC	epigallocatechin
EGCG	(−)-epigallocatechin gallate
EOF	electroosmotic flow
ESI	electrospray ionization
FA	ferulic acid
FRU	fructose
FSC	fused silica capillary
G	guanosine
GABA	γ-aminobutyric acid
GC	Gas Chromatography
7-GC	7-O-β-D-glucosyl-coumarin
GEMBE	gradient elution moving boundary electrophoresis
GL	7-O-α-L-rhamnopyranosyl-kaempferol-3-O-β-D-glucopyranoside
GLU	glucose
L—GLU	L—glutamine
GLY	glycine
H	hyperoside
HA	homogentisic acid
HC	7-hydroxy-coumarin
HEC	hydroxyethylcellulose

HEPES	4-(2-Hydroxyethyl)-1-piperazine ethanesulfonic acid
HMC	7-hydroxy-8-methoxy-coumarin
HP-β-CD	hydroxypropyl-β-cyclodextrin
HPLC	High-Performance Liquid Chromatography
ISOF	isofraxadin
IAD	individual antioxidant determination
ICP-AES	Inductively Coupled Plasma-Atomic Emission Spectrometry
ICP-OES	Inductively Coupled Plasma Optical Emission Spectrometry
ICP-MS	Inductively Coupled Plasma Mass Spectrometry
ID	inner diameter of capillary (μm)
IMI	imidazole
JATR	jatrorrhizine
K	kaempferol
Lcap	length of capillary (cm)
LA	lactic acid
LE	leading electrolyte
LED-IF	light-emitting diode induced fluorescence detection
LIF	laser induced fluorescence
LOD	limit of detection
LUT	luteolin
M	methionine
MA	malic acid
2-ME	2-mercaptoethanol
MEEKC	microemulsion electrokinetic capillary chromatography
MEKC	micellar electrokinetic chromatography
MES	2-(N-morpholino)ethanesulfonic acid
MFAS	microfluidic analytical system
MOPSO	3-(N-morpholino)-2-hydroxypropanesulfonic acid
MS	mass spectrometry
NACE	non-aqueous capillary electrophoresis
OD	outer diameter of capillary (μm)
OPA	o-phthaldialdehyde
OTC	open-tubular columns
PAL	palmitine
PEO	ply(ethylene oxide)
PICOS	Population, Intervention, Comparison, Outcome, Study type
SDS	sodium dodecyl sulfate
SC	sodium salt
TE	terminating electrolyte
Q	quercetin
QC	quartz capillary
QU	quercitrin
R	rutin
RA	rosmarinic acid
RH	7-O-α-L-rhamnopyranosyl-kaempferol-3-O-α-L-rhamnopyranoside
RT	room temperature
SA	succinic acid
SPE	solid phase extraction
D-Th	D-Theanine
T	temperature (°C)
TA	tartaric acid
TB	theobromine
L-THE	L-theanine
TLC	Tin Layer Chromatography
TTMF	5,7,4′-trihydroxy-6,3′,5′-trimethoxyflavone
V	voltage (kV)
W	wogonin
λ	wavelength (nm)

References

1. Tian, T.; Lv, J.; Jin, G.; Yu, C.; Guo, Y.; Bian, Z.; Yang, L.; Chen, Y.; Shen, H.; Chen, Z.; et al. Tea consumption and risk of stroke in Chinese adults: A prospective cohort study of 0.5 million men and women. *Am. J. Clin. Nutr.* **2020**, *111*, 197–206. [CrossRef] [PubMed]
2. Pereira, C.G.; Barreira, L.; Bijttebier, S.; Pieters, L.; Neves, V.; Rodrigues, M.J.; Rivas, R.; Varela, J.; Custódio, L. Chemical profiling of infusions and decoctions of Helichrysum italicum subsp. picardii by UHPLC-PDA-MS and in vitro biological activities comparatively with green tea (Camellia sinensis) and rooibos tisane (Aspalathus linearis). *J. Pharm. Biomed. Anal.* **2017**, *145*, 593–603. [CrossRef]
3. Nani, A.; Murtaza, B.; Sayed Khan, A.; Khan, N.A.; Hichami, A. Antioxidant and anti-inflammatory potential of polyphenols contained in mediterranean diet in obesity: Molecular mechanisms. *Molecules* **2021**, *26*, 985. [CrossRef] [PubMed]
4. Chupeerach, C.; Aursalung, A.; Watcharachaisoponsiri, T.; Whanmek, K.; Thiyajai, P.; Yosphan, K.; Sritalahareuthai, V.; Sahasakul, Y.; Santivarangkna, C.; Suttisansanee, U. The Effect of Steaming and Fermentation on Nutritive Values, Antioxidant Activities, and Inhibitory Properties of Tea Leaves. *Foods* **2021**, *10*, 117. [CrossRef]
5. Liu, Y.; Zhou, W.; Mao, Z.; Chen, Z. Analysis of Evodiae Fructus by capillary electrochromatography-mass spectrometry with methyl-vinylimidazole functionalized organic polymer monolilth as stationary phases. *J. Chromatogr. A* **2019**, *1602*, 474–480. [CrossRef] [PubMed]
6. Gianfredi, V.; Nucci, D.; Abalsamo, A.; Acito, M.; Villarini, M.; Moretti, M.; Realdon, S. Green tea consumption and risk of breast cancer and recurrence—A systematic review and meta-analysis of observational studies. *Nutrients* **2018**, *10*, 1886. [CrossRef]
7. Przybylska, A.; Bazylak, G. Bioactive compounds in aqueous infusions of dietary supplements and herbal blends containing dried hawthorn fruits or hawthorn inflorescences (Crataegus spp.). *J. Agric. Environ. Sci.* **2018**, *7*, 131–142. [CrossRef]
8. Tafrihi, M.; Imran, M.; Tufail, T.; Gondal, T.A.; Caruso, G.; Sharma, S.; Sharma, R.; Atanassova, M.; Atanassov, L.; Valere, P.; et al. The wonderful activities of the genus mentha: Not only antioxidant properties. *Molecules* **2021**, *26*, 1118. [CrossRef]
9. Pinto, G.; Illiano, A.; Carpentieri, A.; Spinelli, M.; Melchiorre, C.; Fontanarosa, C.; Di Serio, M.; Amoresano, A. Quantification of polyphenols and metals in Chinese tea infusions by mass spectrometry. *Foods* **2020**, *9*, 835. [CrossRef]
10. Witkowska, A.M.; Zujko, M.E.; Waśkiewicz, A.; Terlikowska, K.M.; Piotrowski, W. Comparison of various databases for estimation of dietary polyphenol intake in the population of polish adults. *Nutrients* **2015**, *7*, 9299–9308. [CrossRef]
11. Tang, X.; Huang, Z.; Chen, Y.; Liu, Y.; Liu, Y.; Zhao, J.; Yi, J. Simultaneous determination of six bioactive compounds in evodiae fructus by high-performance liquid chromatography with diode array detection. *J. Chromatogr. Sci.* **2014**. [CrossRef] [PubMed]
12. Lalas, S.; Athanasiadis, V.; Karageorgou, I.; Batra, G.; Nanos, G.D.; Makris, D.P. Nutritional characterization of leaves and herbal tea of moringa oleifera cultivated in Greece. *J. Herbs Spices Med. Plants* **2017**. [CrossRef]
13. Kaltbach, P.; Ballert, S.; Kabrodt, K.; Schellenberg, I. New HPTLC methods for analysis of major bioactive compounds in mate (Ilex paraguariensis) tea. *J. Food Compos. Anal.* **2020**. [CrossRef]
14. Cabooter, D.; Broeckhoven, K.; Kalili, K.M.; de Villiers, A.; Desmet, G. Fast method development of rooibos tea phenolics using a variable column length strategy. *J. Chromatogr. A* **2011**. [CrossRef] [PubMed]
15. Ma, W.; Zhu, Y.; Shi, J.; Wang, J.; Wang, M.; Shao, C.; Yan, H.; Lin, Z.; Lv, H. Insight into the volatile profiles of four types of dark teas obtained from the same dark raw tea material. *Food Chem.* **2021**. [CrossRef]
16. Chen, S.; Liu, L.; Tang, D. Determination of total and inorganic selenium in selenium-enriched rice, tea, and garlic by high-performance liquid chromatography–inductively coupled plasma mass spectrometry (HPLC-ICP-MS). *Anal. Lett.* **2020**. [CrossRef]
17. Zhao, L.C.; Jiang, Y.J.; Guo, X.P.; Li, X.; Wang, Y.D.; Guo, X.B.; Lu, F.; Liu, H.J. Optimization of ICP-AES and ICP-MS techniques for the determination of major, minor and micro elements in lichens. *Guang Pu Xue Yu Guang Pu Fen Xi/Spectrosc. Spectr. Anal.* **2016**. [CrossRef]
18. Moreira Szokalo, R.A.; Redko, F.; Ulloa, J.; Flor, S.; Tulino, M.S.; Muschietti, L.; Carballo, M.A. Toxicogenetic evaluation of Smallanthus sonchifolius (yacon) as a herbal medicine. *J. Ethnopharmacol.* **2020**. [CrossRef] [PubMed]
19. Długaszek, M.; Kaszczuk, M. Assessment of the nutritional value of various teas infusions in terms of the macro- and trace elements content. *J. Trace Elem. Med. Biol.* **2020**. [CrossRef]
20. Sun, G.; Shi, C. The overall quality control of Radix Scutellariae by capillary electrophoresis fingerprint. *J. Chromatogr. Sci.* **2008**, *46*, 454–460. [CrossRef]
21. Alothman, Z.A.; Badjah, A.Y.; Locatelli, M. Multi-Walled Carbon Nanotubes Solid-Phase Extraction and Capillary Electrophoresis Methods for the Analysis of 4-Cyanophenol and 3-Nitrophenol in Water. *Molecules* **2020**, *25*, 3896. [CrossRef]
22. Müller, L.S.; Muratt, D.T.; Molin, T.R.D.; Urquhart, C.G.; Viana, C.; de Carvalho, L.M. Analysis of pharmacologic adulteration in dietary supplements by capillary zone electrophoresis using simultaneous contactless conductivity and UV detection. *Chromatographia* **2018**, *81*, 689–698. [CrossRef]
23. D'Orazio, G.; Asensio-Ramos, M.; Fanali, C.; Hernández-Borges, J.; Fanali, S. Capillary electrochromatography in food analysis. *TrAC-Trends Anal. Chem.* **2016**, *82*, 250–267. [CrossRef]
24. Colombo, R.; Papetti, A. Pre-concentration and analysis of mycotoxins in food samples by capillary electrophoresis. *Molecules* **2020**, *25*, 3441. [CrossRef] [PubMed]
25. Yang, Z.; Li, Z.; Zhu, J.; Wang, Q.; He, P.; Fang, Y. Use of different buffers for detection and separation in determination of physio-active components in oolong tea infusion by CZE with amperometric detection. *J. Sep. Sci.* **2010**. [CrossRef] [PubMed]

26. Tiselius, A. Electrophoresis of serum globulin: Electrophoretic analysis of normal and immune sera. *Biochem. J.* **1937**, *31*, 1464–1477. [CrossRef] [PubMed]
27. Liu, H.; Gao, Y.; Wang, K.; Hu, Z. Determination of active components in Cynanchum chinense R. Br. by capillary electrophoresis. *Biomed. Chromatogr.* **2006**. [CrossRef]
28. Pang, H.; Wu, L.; Tang, Y.; Zhou, G.; Qu, C.; Duan, J.A. Chemical analysis of the herbal medicine salviae miltiorrhizae radix et rhizoma (Danshen). *Molecules* **2016**, *21*, 51. [CrossRef]
29. Yang, X.; Zhao, Y.; Lv, Y. Chemical composition and antioxidant activity of an acidic polysaccharide extracted from Cucurbita moschata duchesne ex poiret. *J. Agric. Food Chem.* **2007**, *55*, 4684–4690. [CrossRef]
30. Tembo, Z.N.; Şeker Aygun, F.; Erdoğan, B.Y. Simultaneous determination of nitrate, nitrite and bromate by capillary zone electrophoresis in tea infusions grown in the Black Sea region of Turkey. *Sep. Sci. PLUS* **2021**. [CrossRef]
31. Ganzera, M. Quality control of herbal medicines by capillary electrophoresis: Potential, requirements and applications. *Electrophoresis* **2008**, *29*, 3489–3503. [CrossRef]
32. Xu, X.; Li, L.; Weber, S.G. Electrochemical and optical detectors for capillary and chip separations. *TrAC-Trends Anal. Chem.* **2007**. [CrossRef]
33. Hurtado-Fernández, E.; Gómez-Romero, M.; Carrasco-Pancorbo, A.; Fernández-Gutiérrez Alberto, A. Application and potential of capillary electroseparation methods to determine antioxidant phenolic compounds from plant food material. *J. Pharm. Biomed. Anal.* **2010**, *53*, 1130–1160. [CrossRef]
34. Liu, Q.; Wang, L.; Hu, J.; Miao, Y.; Wu, Z.; Li, J. Main organic acids in rice wine and beer determined by capillary electrophoresis with indirect UV detection using 2, 4-Dihydroxybenzoic acid as chromophore. *Food Anal. Methods* **2017**. [CrossRef]
35. Mazina, J.; Vaher, M.; Kuhtinskaja, M.; Poryvkina, L.; Kaljurand, M. Fluorescence, electrophoretic and chromatographic fingerprints of herbal medicines and their comparative chemometric analysis. *Talanta* **2015**, *139*, 233–246. [CrossRef] [PubMed]
36. Gavrilin, M.V.; Senchenko, S.P. Use of capillary electrophoresis for estimating the quality of chamomile flowers. *Pharm. Chem. J.* **2009**, *43*, 582–584. [CrossRef]
37. Adımcılar, V.; Kalaycıoğlu, Z.; Aydoğdu, N.; Dirmenci, T.; Kahraman, A.; Erim, F.B. Rosmarinic and carnosic acid contents and correlated antioxidant and antidiabetic activities of 14 Salvia species from Anatolia. *J. Pharm. Biomed. Anal.* **2019**, *175*. [CrossRef] [PubMed]
38. Ben Hameda, A.; Gajdošová, D.; Havel, J. Analysis of Salvia officinalis plant extracts by capillary electrophoresis. *J. Sep. Sci.* **2006**. [CrossRef]
39. Cao, J.; We, J.; Tian, K.; Su, H.; Wan, J.; Li, P. Simultaneous determination of seven phenolic acids in three Salvia species by capillary zone electrophoresis with β-cyclodextrin as modifier. *J. Sep. Sci.* **2014**. [CrossRef]
40. Citová, I.; Ganzera, M.; Stuppner, H.; Solich, P. Determination of gentisin, isogentisin, and amarogentin in Gentiana lutea L. by capillary electrophoresis. *J. Sep. Sci.* **2008**. [CrossRef] [PubMed]
41. Bizzotto, C.S.; Meinhart, A.D.; Rybka, A.C.P.; Sobrinho, M.R.; Junior, S.B.; Ballus, C.A.; Godoy, H.T. Quantification of phenolic compounds by capillary zone electrophoresis in extracts of four commercial types of mate herb before and after acid hydrolysis. *Food Res. Int.* **2012**, *48*, 763–768. [CrossRef]
42. Memon, A.F.; Solangi, A.R.; Memon, S.Q.; Mallah, A.; Memon, N. Quantitative separation of hesperidin, chrysin, epicatechin, epigallocatechin gallate, and morin using ionic liquid as a buffer additive in capillary electrophoresis. *Electrophoresis* **2018**. [CrossRef] [PubMed]
43. Arries, W.J.; Tredoux, A.G.J.; de Beer, D.; Joubert, E.; de Villiers, A. Evaluation of capillary electrophoresis for the analysis of rooibos and honeybush tea phenolics. *Electrophoresis* **2017**. [CrossRef] [PubMed]
44. Urbonavičiute, A.; Jakštas, V.; Kornyšova, O.; Janulis, V.; Maruška, A. Capillary electrophoretic analysis of flavonoids in single-styled hawthorn (Crataegus monogyna Jacq.) ethanolic extracts. *J. Chromatogr. A* **2006**. [CrossRef] [PubMed]
45. Deng, Y.; Lam, S.C.; Zhao, J.; Li, S.P. Quantitative analysis of flavonoids and phenolic acid in Coreopsis tinctoria Nutt. by capillary zone electrophoresis. *Electrophoresis* **2017**. [CrossRef] [PubMed]
46. Qi, S.; Li, Y.; Deng, Y.; Cheng, Y.; Chen, X.; Hu, Z. Simultaneous determination of bioactive flavone derivatives in Chinese herb extraction by capillary electrophoresis used different electrolyte systems-Borate and ionic liquids. *J. Chromatogr. A* **2006**, *1109*, 300–306. [CrossRef] [PubMed]
47. Ren, X.; Liu, H.; Wang, J. Determination of nevadensin in fewflower lysionotus herb by capillary electrophoresis. *IOP Conf. Ser. Earth Environ. Sci.* **2020**, *446*. [CrossRef]
48. Başkan, S.; Öztekin, N.; Erim, F.B. Determination of carnosic acid and rosmarinic acid in sage by capillary electrophoresis. *Food Chem.* **2007**, *101*, 1748–1752. [CrossRef]
49. Yang, P.; Li, Y.; Liu, X.; Jiang, S. Determination of free isomeric oleanolic acid and ursolic acid in Pterocephalus hookeri by capillary zone electrophoresis. *J. Pharm. Biomed. Anal.* **2007**, *43*, 1331–1334. [CrossRef]
50. Truică, G.; Teodor, E.D.; Radu, G.L. Organic acids assesments in medicinal plants by capillary electrophoresis. *Rev. Roum. Chim.* **2013**, *58*, 809–814.
51. Li, M.; Zhou, J.; Gu, X.; Wang, Y.; Huang, X.J.; Yan, C. Quantitative capillary electrophoresis and its application in analysis of alkaloids in tea, coffee, coca cola, and theophylline tablets. *J. Sep. Sci.* **2009**. [CrossRef]
52. Ding, P.L.; Yu, Y.Q.; Chen, D.F. Determination of quinolizidine alkaloids in Sophora tonkinensis by HPCE. *Phytochem. Anal.* **2005**. [CrossRef] [PubMed]

53. Zhou, Q.; Liu, Y.; Wang, X.; Di, X. Microwave-assisted extraction in combination with capillary electrophoresis for rapid determination of isoquinoline alkaloids in *Chelidonium majus* L. *Talanta* **2012**, *99*, 932–938. [CrossRef] [PubMed]
54. Zhang, W.; Li, Y.; Chen, Z. Selective and sensitive determination of protoberberines by capillary electrophoresis coupled with molecularly imprinted microextraction. *J. Sep. Sci.* **2015**. [CrossRef]
55. Do, T.C.M.V.; Nguyen, T.D.; Tran, H.; Stuppner, H.; Ganzera, M. Analysis of alkaloids in Lotus (Nelumbo nucifera Gaertn.) leaves by non-aqueous capillary electrophoresis using ultraviolet and mass spectrometric detection. *J. Chromatogr. A* **2013**, *1302*, 174–180. [CrossRef] [PubMed]
56. Wang, N.; Su, M.; Liang, S.; Sun, H. Investigation of six bioactive anthraquinones in slimming tea by accelerated solvent extraction and high performance capillary electrophoresis with diode-array detection. *Food Chem.* **2016**, *199*, 1–7. [CrossRef]
57. Yue, M.E.; Jiang, T.F.; Liu, X.; Shi, Y.P. Separation and determination of coumarins from Cacalia tangutica by capillary zone electrophoresis. *Biomed. Chromatogr.* **2005**. [CrossRef]
58. Jiang, X.; Xia, Z.; Wei, W.; Gou, Q. Direct UV detection of underivatized amino acids using capillary electrophoresis with online sweeping enrichment. *J. Sep. Sci.* **2009**. [CrossRef]
59. Chi, L.; Li, Z.; Dong, S.; He, P.; Wang, Q.; Fang, Y. Simultaneous determination of flavonoids and phenolic acids in Chinese herbal tea by beta-cyclodextrin based capillary zone electrophoresis. *Microchim. Acta* **2009**, *167*, 179–185. [CrossRef]
60. Wan, D.; Han, Y.; Li, F.; Mao, H.; Chen, G. Far infrared-assisted removal of extraction solvent for capillary electrophoretic determination of the bioactive constituents in Plumula Nelumbinis. *Electrophoresis* **2019**. [CrossRef]
61. Xu, X.; Qi, X.; Wang, W.; Chen, G. Separation and determination of flavonoids in Agrimonia pilosa Ledeb. by capillary electrophoresis with electrochemical detection. *J. Sep. Sci.* **2005**. [CrossRef]
62. Zhou, X.; Zheng, C.; Huang, J.; You, T. Identification of herb Acanthopanax senticosus (Rupr. et Maxim.) harms by capillary electrophoresis with electrochemical detection. *Anal. Sci.* **2007**, *23*, 705–711. [CrossRef]
63. Zhou, X.; Zheng, C.; Sun, J.; You, T. Analysis of nephroloxic and carcinogenic aristolochic acids in Aristolochia plants by capillary electrophoresis with electrochemical detection at a carbon fiber microdisk electrode. *J. Chromatogr. A* **2006**, *1109*, 152–159. [CrossRef]
64. Chu, Q.; Fu, L.; Wu, T.; Ye, J. Simultaneous determination of phytoestrogens in different medicinal parts of Sophora japonica L. by capillary electrophoresis with electrochemical detection. *Biomed. Chromatogr.* **2005**. [CrossRef]
65. Nikovaev, A.V.; Kartsova, L.A.; Filimonov, V.V. A microfluidic chip for the determination of polyphenolic antioxidants. *J. Anal. Chem.* **2015**. [CrossRef]
66. Shih, T.T.; Lee, H.L.; Chen, S.C.; Kang, C.Y.; Shen, R.S.; Su, Y.A. Rapid analysis of traditional Chinese medicine Pinellia ternata by microchip electrophoresis with electrochemical detection. *J. Sep. Sci.* **2018**. [CrossRef]
67. Tang, M.; Xu, J.; Xu, Z. Simultaneous determination of metal ions by capillary electrophoresis with contactless conductivity detection and insights into the effects of BGE component. *Microchem. J.* **2019**, *147*, 857–862. [CrossRef]
68. Tarleton, E.K.; Littenberg, B.; MacLean, C.D.; Kennedy, A.G.; Daley, C. Role of magnesium supplementation in the treatment of depression: A randomized clinical trial. *PLoS ONE* **2017**. [CrossRef]
69. Strychalski, E.A.; Henry, A.C.; Ross, D. Microfluidic analysis of complex samples with minimal sample preparation using gradient elution moving boundary electrophoresis. *Anal. Chem.* **2009**. [CrossRef]
70. Kovaehev, N.; Canals, A.; Escarpa, A. Fast and selective microfluidic chips for electrochemical antioxidant sensing in complex samples. *Anal. Chem.* **2010**. [CrossRef]
71. Tezcan, F.; Erim, F. Determination of vitamin B2 content in black, green, sage, and rosemary tea infusions by capillary electrophoresis with laser-induced fluorescence detection. *Beverages* **2018**, *4*, 86. [CrossRef]
72. Lin, Y.P.; Su, Y.S.; Jen, J.F. Capillary electrophoretic analysis of γ-aminobutyric acid and alanine in tea with in-capillary derivatization and fluorescence detection. *J. Agric. Food Chem.* **2007**, *55*, 2103–2108. [CrossRef]
73. Su, Y.I.S.; Lin, Y.P.; Cheng, F.U.C.; Jen, J.F. In-capillary derivatization and stacking electrophoretic analysis of γ-aminobutyric acid and alanine in tea samples to redeem the detection after dilution to decrease matrix interference. *J. Agric. Food Chem.* **2010**, *58*, 120–126. [CrossRef]
74. Hsieh, M.M.; Chen, S.M. Determination of amino acids in tea leaves and beverages using capillary electrophoresis with light-emitting diode-induced fluorescence detection. *Talanta* **2007**, *73*, 326–331. [CrossRef]
75. Hu, L.; Yang, X.; Wang, C.; Yuan, H.; Xiao, D. Determination of riboflavin in urine and beverages by capillary electrophoresis with in-column optical fiber laser-induced fluorescence detection. *J. Chromatogr. B Anal. Technol. Biomed. Life Sci.* **2007**, *856*, 245–251. [CrossRef]
76. Liu, Q.; Liu, Y.; Li, Y.; Yao, S. Nonaqueous capillary electrophoresis coupled with laser-induced native fluorescence detection for the analysis of berberine, palmatine, and jatrorrhizine in Chinese herbal medicines. *J. Sep. Sci.* **2006**. [CrossRef]
77. Wang, W.; Tang, J.; Wang, S.; Zhou, L.; Hu, Z. Method development for the determination of coumarin compounds by capillary electrophoresis with indirect laser-induced fluorescence detection. *J. Chromatogr. A* **2007**, *1148*, 108–114. [CrossRef]
78. Yu, K.; Gong, Y.; Lin, Z.; Cheng, Y. Quantitative analysis and chromatographic fingerprinting for the quality evaluation of Scutellaria baicalensis Georgi using capillary electrophoresis. *J. Pharm. Biomed. Anal.* **2007**, *43*, 540–548. [CrossRef]
79. Gotti, R.; Furlanetto, S.; Lanteri, S.; Olmo, S.; Ragaini, A.; Cavrini, V. Differentiation of green tea samples by chiral CD-MEKC analysis of catechins content. *Electrophoresis* **2009**. [CrossRef]

80. Ganzera, M.; Egger, C.; Zidorn, C.; Stuppner, H. Quantitative analysis of flavonoids and phenolic acids in Arnica montana L. by micellar electrokinetic capillary chromatography. *Anal. Chim. Acta* **2008**, *614*, 196–200. [CrossRef]
81. Dresler, S.; Bogucka-Kocka, A.; Kováčik, J.; Kubrak, T.; Strzemski, M.; Wójciak-Kosior, M.; Rysiak, A.; Sowa, I. Separation and determination of coumarins including furanocoumarins using micellar electrokinetic capillary chromatography. *Talanta* **2018**, *187*, 120–124. [CrossRef]
82. Yan, J.; Cai, Y.; Wang, Y.; Lin, X.; Li, H. Simultaneous determination of amino acids in tea leaves by micellar electrokinetic chromatography with laser-induced fluorescence detection. *Food Chem.* **2014**, *143*, 82–89. [CrossRef]
83. Gomez, F.J.V.; Hernández, I.G.; Cerutti, S.; Silva, M.F. Solid phase extraction/cyclodextrin-modified micellar electrokinetic chromatography for the analysis of melatonin and related indole compounds in plants. *Microchem. J.* **2015**, *123*, 22–27. [CrossRef]
84. Fiori, J.; Pasquini, B.; Caprini, C.; Orlandini, S.; Furlanetto, S.; Gotti, R. Chiral analysis of theanine and catechin in characterization of green tea by cyclodextrin-modified micellar electrokinetic chromatography and high performance liquid chromatography. *J. Chromatogr. A* **2018**, *1562*, 115–122. [CrossRef] [PubMed]
85. Chrustek, A.; Olszewska-Słonina, D. Melatonin as a powerful antioxidant. *Acta Pharm.* **2021**, *71*, 335–354. [CrossRef]
86. Głowacki, R.; Furmaniak, P.; Kubalczyk, P.; Borowczyk, K. Determination of total apigenin in herbs by micellar electrokinetic chromatography with UV detection. *J. Anal. Methods Chem.* **2016**, *2016*. [CrossRef]
87. Šafra, J.; Pospíšilová, M.; Spilková, J. Determination of phenolic acids in Herba Epilobi by ITP-CE in the column-coupling configuration. *Chromatographia* **2006**, *64*, 37–43. [CrossRef]
88. Šafra, J.; Pospíšilová, M.; Honegr, J.; Spilková, J. Determination of selected antioxidants in Melissae herba by isotachophoresis and capillary zone electrophoresis in the column-coupling configuration. *J. Chromatogr. A* **2007**, *1171*, 124–132. [CrossRef]
89. Sõukand, R.; Mattalia, G.; Kolosova, V.; Stryamets, N.; Prakofjewa, J.; Belichenko, O.; Kuznetsova, N.; Minuzzi, S.; Keedus, L.; Prūse, B.; et al. Inventing a herbal tradition: The complex roots of the current popularity of Epilobium angustifolium in Eastern Europe. *J. Ethnopharmacol.* **2020**, *247*, 112254. [CrossRef]
90. Mao, Z.; Chen, Z. Advances in capillary electro-chromatography. *J. Pharm. Anal.* **2019**, *9*, 227–237. [CrossRef]
91. Yang, F.Q.; Zhao, J.; Li, S.P. CEC of phytochemical bioactive compounds. *Electrophoresis* **2010**, *31*, 260–277. [CrossRef] [PubMed]
92. Zhang, L.; Zhang, J.; Wang, H.; Zhang, L.; Zhang, W.; Zhang, Y. Analysis of flavonoids in leaves of Adinandra nitida by capillary electrochromatography on monolithic columns with stepwise gradient elution. *J. Sep. Sci.* **2005**. [CrossRef] [PubMed]
93. Fonseca, F.N.; Tavares, M.F.M.; Horváth, C. Capillary electrochromatography of selected phenolic compounds of Chamomilla recutita. *J. Chromatogr. A* **2007**, *1154*, 390–399. [CrossRef]
94. Uysal, U.D.; Aturki, Z.; Raggi, M.A.; Fanali, S. Separation of catechins and methylxanthines in tea samples by capillary electrochromatography. *J. Sep. Sci.* **2009**. [CrossRef]
95. Chen, D.; Wang, J.; Jiang, Y.; Zhou, T.; Fan, G.; Wu, Y. Separation and determination of coumarins in Fructus cnidii extracts by pressurized capillary electrochromatography using a packed column with a monolithic outlet frit. *J. Pharm. Biomed. Anal.* **2009**. [CrossRef] [PubMed]
96. Liu, J.X.; Zhang, Y.W.; Yuan, F.; Chen, H.X.; Zhang, X.X. Differential detection of Rhizoma coptidis by capillary electrophoresis electrospray ionization mass spectrometry with a nanospray interface. *Electrophoresis* **2014**. [CrossRef]
97. Segura-Carretero, A.; Puertas-Mejía, M.A.; Cortacero-Ramírez, S.; Beltrán, R.; Alonso-Villaverde, C.; Joven, J.; Dinelli, G.; Fernández-Gutiérrez, A. Selective extraction, separation, and identification of anthocyanins from Hibiscus sabdariffa L. using solid phase extraction-capillary electrophoresis-mass spectrometry (time-of-flight/ion trap). *Electrophoresis* **2008**. [CrossRef]

Review

Recent Applications of Capillary Electrophoresis in the Determination of Active Compounds in Medicinal Plants and Pharmaceutical Formulations

Marcin Gackowski [1,*], Anna Przybylska [1], Stefan Kruszewski [2], Marcin Koba [1], Katarzyna Mądra-Gackowska [3] and Artur Bogacz [4]

1. Department of Toxicology and Bromatology, Faculty of Pharmacy, L. Rydygier Collegium Medicum in Bydgoszcz, Nicolaus Copernicus University in Torun, A. Jurasza 2 Street, PL–85089 Bydgoszcz, Poland; aniacm@cm.umk.pl (A.P.); kobamar@cm.umk.pl (M.K.)
2. Biophysics Department, Faculty of Pharmacy, L. Rydygier Collegium Medicum in Bydgoszcz, Nicolaus Copernicus University in Torun, Jagiellońska 13 Street, PL–85067 Bydgoszcz, Poland; skrusz@cm.umk.pl
3. Department of Geriatrics, Faculty of Health Sciences, L. Rydygier Collegium Medicum in Bydgoszcz, Nicolaus Copernicus University in Torun, Skłodowskiej Curie 9 Street, PL–85094 Bydgoszcz, Poland; katarzyna.madra@cm.umk.pl
4. Department of Otolaryngology and Oncology, Faculty of Medicine, L. Rydygier Collegium Medicum in Bydgoszcz, Nicolaus Copernicus University in Torun, Skłodowskiej Curie 9 Street, PL–85094 Bydgoszcz, Poland; arturbogacz@cm.umk.pl
* Correspondence: marcin.gackowski@cm.umk.pl

Abstract: The present review summarizes scientific reports from between 2010 and 2019 on the use of capillary electrophoresis to quantify active constituents (i.e., phenolic compounds, coumarins, protoberberines, curcuminoids, iridoid glycosides, alkaloids, triterpene acids) in medicinal plants and herbal formulations. The present literature review is founded on PRISMA guidelines and selection criteria were formulated on the basis of PICOS (Population, Intervention, Comparison, Outcome, Study type). The scrutiny reveals capillary electrophoresis with ultraviolet detection as the most frequently used capillary electromigration technique for the selective separation and quantification of bioactive compounds. For the purpose of improvement of resolution and sensitivity, other detection methods are used (including mass spectrometry), modifiers to the background electrolyte are introduced and different extraction as well as pre-concentration techniques are employed. In conclusion, capillary electrophoresis is a powerful tool and for given applications it is comparable to high performance liquid chromatography. Short time of execution, high efficiency, versatility in separation modes and low consumption of solvents and sample make capillary electrophoresis an attractive and eco-friendly alternative to more expensive methods for the quality control of drugs or raw plant material without any relevant decrease in sensitivity.

Keywords: capillary electrophoresis; herbal drugs; medicinal plants; quality control; quantitative analysis; pharmaceutical analysis

---

1. Introduction

From early times, people recognized plants for their therapeutic properties. Herbal medicine has been used in the management of many ailments for thousands of years and is the oldest method of healthcare in history [1]. Former herbal "drugs" were discovered by trial and error on human subjects; owing to this, the rich tradition of herbalism contains invaluable biomedical information that is continuously being uncovered by contemporary scientists. In contrast, the dynamic progress of drug production in the 20th century was grounded on the isolation of an active substance with a well-defined mechanism of action. Traditional phytochemistry and pharmaceutical chemistry lines led to the development of many safe and effective drugs used in the clinic today for the treatment of, e.g., diabetes, autoimmune diseases, degenerative disorders and cancer. A well-known example of herbalism is traditional Chinese medicine (TCM), which has its followers and is still practiced around the world. Moreover, its potency in numerous illnesses is proved by clinical researchers [2,3]. The use of herbal drugs has been gaining public interest and acceptance. On the one hand, due to poverty and limited access to healthcare in developing countries, people use herbal drugs as a first line of treatment. On the other hand, the most important reason for using herbal therapies in the West is that people believe that herbs will help us live healthier lives and are generally safe for consumption. Individuals often use over-the-counter herbal medicines without medical consultation as home remedies and spend billions of dollars on them. The growing market entails not only abuse but also the adulteration of medicinal products, which can lead to serious health consequences [4]. All drugs, whether of a plant or synthetic origin, should meet safety requirements and be effective [5,6]. The content of bioactive constituents is one of the key parameters in assessing the quality of a herbal medicine. In terms of adulteration, which is a very common problem that is mainly linked to the lack of officially established regulations by governmental organizations regarding the control of herbal preparations marketed as dietary supplements, more effective ways are needed to improve control at the production and marketing stages [7–9]. In this light, rapid, simple, accurate qualitative and quantitative methods are essential to evaluate whether herbal formulations fulfill pharmacopoeial requirements.

Quality control of herbal medicines is a real challenge for analysts because of the complex matrix and several characteristic compounds which must be identified and quantified. Apart from this, an elaboration of the analytic method must be completed and a validation protocol fulfilled. Thus, sophisticated, laborious techniques should be employed, such as high-performance liquid chromatography (HPLC), gas chromatography (GC), high performance thin layer chromatography (HPTLC) and capillary electrophoresis (CE). Among the abovementioned techniques, HPLC is the most prevalent one. Liquid chromatography with a diode-array detector (DAD) in conjunction with a mass spectrometer (MS) is an efficient measure to analyze both known and unknown compounds in a complex matrix [10]. HPLC is also the most frequently used technique for the determination of active constituents in TCMs [11].

Capillary electrophoresis has been commonly applied in the analysis of food, environmental monitoring, clinical diagnostics and pharmaceutical analysis. The latter application has become increasingly popular in recent years due to its high separation efficiency, minimal consumption of required solvents and its small sample volume, low running cost, reproducibility, and versatility in separation modes, making it an attractive, eco-friendly and powerful tool suitable for drug control purposes. Thus, CE has found its place in official pharmacopoeias and pharmaceutical control regulations [12,13]. What is more, researchers have found many solutions to handle the unwanted phenomenon, that is the limitation of poor concentration sensitivity [14]. Since the early 1980s, after Jorgenson and Lukacs demonstrated that the effectiveness of the electrophoresis may be increased if it took place in open-tubular glass capillaries with a diameter of ~75 µm, CE has developed into a flexible and versatile technique, which make it a very attractive alternative to other chromatographic techniques [15,16].

In CE, analytes are separated in a capillary column with electroosmotic flow (EOF) as the driving force for bulk fluid movement and the action of the electric field. CE requires only simple instrumentation, consisting of a high voltage power supply, two buffer reservoirs, a sample introduction system, a capillary tube, a detector and an output device. See Scheme 1. The capillary is flooded with a solution of background electrolyte (BGE) at a specific pH, which is usually a buffer able to selectively influence the effective mobility. Different capillaries (fused silica or quartz) with internal diameters ranging from 25 to 100 µm and a length of 20 to 100 cm also affect the analysis conditions. Capillaries are placed together with electrodes in reservoirs flooded with the same buffer. In electrophoresis, a mixture of different substances in solution is introduced, usually as a relatively narrow zone, into the separating system, and is induced to move under the influence of an applied potential. The basis for the separation of analytes is in the differences in the electrophoretic mobility of ions as a consequence of the variance in the size and shape of charged particles. Under the influence of an applied electric field, the diverse substances migrate at different rates; thus, after some time, the mixture separates into spatially discrete zones of individual substances [15,17,18]. The majority of capillaries for CE are fabricated from fused silica with characteristic silalol groups on the surface. Those groups dissociate, forming a negative charge in the inner surface of the capillary, attracting a positive charge from the buffer, and finally leads to the formation of an electric double layer. The dispersed cation layer (and its hydration sphere) adjacent to the silica surface tends to migrate towards the cathode, resulting in concomitant fluid migration through the capillary. Anions and cations are separated by electrophoretic migration and eloctroosmotic flow, while neutral species only coelute with the EOF [11,15]. In the terms of expanding sensitivity, introducing different additives such as methanol or acetonitrile is a common phenomenon. Those compounds work by altering viscosity and the polarity of the running buffer, which, in consequence, affects EOF and the electrophoretic mobility of the analyte [15]. As for improvements to the resolution of different compounds, cyclodextrines, for instance, are added to BGE. The use of an appropriate system for the detection of tested substances makes it possible to record the results of the analyses in the form of an electrophoregram [11,18,19].

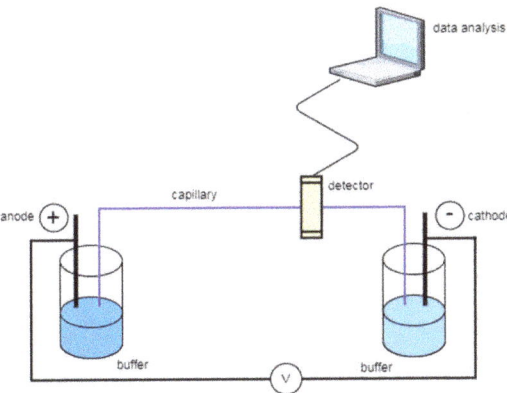

Scheme 1. Capillary electrophoresis system.

Over the last few decades, capillary electrophoresis has attracted attention, because the combination of both chromatographic and electrophoretic mechanisms of migration permits the adoption of different separation formats suited to the chemical structure of the analyzed compounds. The following techniques of CE are distinguished: capillary zone electrophoresis (CZE), non-aqueous CE (NACE), micellar electrokinetic chromatography (MEKC), capillary electrochromatography (CEC), capillary isotachophoresis (CITP), capillary isoelectric focusing (CIEF), chiral CE (CCE), capillary gel electrophoresis (CGE), and microemulsion electrokinetic capillary chromatography (MEEKC) [17]. To one the bases of charge density, size, hydrophobicity and chirality, analysts can employ CE to different categories of chemicals [15].

The following review summarizes the utilization of CE for the quantification of active constituents in medicinal plants and commercial herbal products, covering the most important applications between 2010 and 2019 (publications in English only). This scrutiny discusses in detail selected physical and chemical (type of buffer, pH) parameters of CE essential for the selective separation of bioactive constituents. Moreover, there is a greater focus on the influence of different pre-concentration and extraction techniques and additives to the background electrolyte for the improvement of resolution and sensitivity. Special emphasis is placed if the reported methods were applied to real samples (medicinal plants, commercial products) and if they were validated. Apart from reporting the current applications of CE, this paper indicates prospects for the further application of this technique.

2. Results

2.1. Literature Analysis

In the first search in the PubMed and Web of Science database, 682 records potentially meeting the inclusion criteria were found, 363 and 319, respectively. Then, after reviewing the bibliography, duplicates were removed ($n = 103$) and the selected articles were subjected to a subsequent verification by the co-authors. After this, the articles were selected based on the title and abstract. Subsequently, papers describing the application of different methods to CE, CE used only for qualitative information or using CE technique to the analysis of bioactive constituents present in garden and ornamental plants, vegetables and fruits, edible products, beverages, human plasma, blood serum and urine, were eliminated. There were 466 abstracts and papers that were not qualified for this review. One hundred and thirteen articles found in the PubMed and Web of Science databases were used to review the analysis of various bioactive compounds using CE in medicinal plants and herbal drugs.

2.2. Capillary Zone Electrophoresis

The largest number of reported methods for the determination of secondary metabolites in plant material and active substances in herbal medicines recorded in the current systematic review is based on the technique of capillary zone electrophoresis with UV detection. Although the fact that most analytes were determined by molecular absorption, other detection methods, such as fluorometric or electrochemical methods (conductometry, amperometry and potentiometry), were also applied (See Table 1).

In order to obtain a satisfactory separation and quantification of analytes, it is essential to optimize several parameters such as type of capillary, pH, voltage, injection mode, buffer composition and concentration, additives (type and concentration), etc. This scrutiny reveals that the most suitable BGE to achieve good separation and quantification of different analytes in CZE is borate buffer.

2.2.1. Separation in CZE

In some cases, adequate separation or quantification with borate buffer as a background electrolyte may be difficult, especially in plant extracts rich in different secondary metabolites or herbal preparations containing many herbs. In this case, the supplementation of the running buffer with some modifiers is a simple and effective way to improve separation efficiency in CE. The positive influence of organic solvents as BGE modifiers on the quality of the separation is expressed as a number of completely resolved peaks. Honegr and Pospíšilova evaluated the influence of methanol, acetonitrile, 2–propanol, and a mixture of 2–propanol and acetonitrile [20]. Liang et al. proved that the addition of β–cyclodextrin (β–CD) and methanol significantly improved the resolution of eight lignans in *Forsythia suspensa*. Excellent separation was accomplished within 15 min with borate buffer as BGE with the addition of 2 mM β–CD and 5% methanol (v/v) at the voltage of 20 kV, temperature of 35 °C and detection wavelength of 234 nm [21]. An increasingly common approach to increase the resolution between racemic natural products is the addition of cyclodextrins to a running buffer, such as chiral selectors. In addition, microchips are also becoming a popular strategy [22]. In the determination of arecoline by Xiang et al., the additive IL, 1–butyl–3–methylimidazolium tetrafluoroborate (BMImBF$_4$) was responsible not only for improvement of separation selectivity but also in the detection sensitivity of the analyte. This additive made the resistance of the separation buffer much lower than that of the sample solution, which resulted in an enhanced field-amplified electrokinetic injection CE [23].

2.2.2. Detection Sensitivity

The sensitivity of CE methods is limited by the use of conventional on-line UV detection, which as can be seen in Table 1, is the most common. The path length is rather short due to the capillary diameter, which has a negative influence on the detection sensitivity. This negative phenomenon is usually compensated by the high efficiency and by using low UV wavelengths, but there are also some other ways to overcome this problem. For instance, Song et al. elaborated a CZE method for the determination of aconite alkaloids, where they dissolved the extracts in acetonitrile; in this way they decreased the conductivities of sample solutions. Besides, they used an electro-injection mode which led to a significant improvement in detection sensitivity due to a field-amplified sample stacking effect and values of LOD/LOQ were expressed in nanograms per milliliter [24]. In comparison, the average values of LOQ within the majority of reported studies are in the $\mu g\ mL^{-1}$ level (See Tables 1–5). For the analysis of inorganic and organic compounds (together with medicinal products) in an acidic or basic form, contactless conductivity detection can be implemented to overcome the limitations of optical detectors with low sensitivities. This method of detection can be comparable with CE–UV in some applications [7]. At present, the use of electrochemical detection is restricted mainly to conductivity detection, which is mainly employed for compounds that are difficult to detect by UV absorption. Moreover, the use of potentiometric and amperometric detectors is relatively rare [25]. A low limit of detection (LOD) may be also achieved because of the high sensitivity rendered by laser-induced fluorescence (LIF) detection. Reported studies with applications of CE–LIF are characterized by a limit of detection/quantification at the $ng\ mL^{-1}$ level) for more information, see Table 1). The main disadvantage of fluorescence detection is its necessity for derivatization of the analyte [25]. A microfluidic approach overcomes such inconveniences as poor resolution and poor LOD or LOQ values which herein are reported as microchip capillary electrophoresis coupled with the laser-induced fluorescence (MCE–LIF). This method is characterized by a very small sample and solvent consumption, a short operating time and a high mass sensitivity, which makes it favorable for the determination of minor compounds with fluorescence in complex samples. In a reported paper by Xiao at al., a developed and carefully applied MCE–LIF method for the fast quantification of aloin A and B present in seven aloe plant species and pharmaceutical formulations was presented. In this instance, the LOQ is expressed in $ng\ mL^{-1}$ [22]. Table 1 shows that in many cases, UV detectors are sufficient for the analysis of active constituents in pharmaceutical formulation or herbal raw material. However, when it comes to analysis of trace analytes in a complex biological matrix, an introduction of extremely sensitive detectors, such as mass spectrometry or laser-induced fluorescence is recommended [26].

Some reports describe the fabrication and subsequent application of novel detection electrodes for determining the bioactive ingredients by CE, for instance in *Belamcandae rhizome* [27], in *Bergeniae rhizoma* [28] and in *Cacumen platycladi* [29]. In those cases, the values of LOD/LOQ were also as low as ng per mL^{-1}. This approach gives better sensitivity, a considerably lower operating potential, an agreeable resistance to surface fouling, lower operating costs and enhanced stability. Not only does amperometric detection give impressive results, but also, a combination of high separation power of capillary electrophoresis with a high sensitivity of chemiluminescence is becoming very desirable. Wang et al. achieved ultrasensitive determination of epicatechin, rutin, and quercetin by CE with chemiluminescence detection with limits of detection expressed even in $pg\ mL^{-1}$ [30].

2.2.3. Sample Pretreatment Techniques in CZE

Despite its numerous advantages, CE is still considered a niche technique in separation sciences and the use of CE may be limited due to low sensitivity, which is on account of its short optical path and the small capillary dimensions as well as its small sample volume. To remove this inconvenience, sample pretreatment techniques are introduced to the CE system in order to achieve a lower LOD for many analytes, shorten the analysis time, reduce sample consumption, and decrease overall analysis cost. Sample pretreatment is essential for complex matrices, and especially for biological samples. Sample pretreatment may be either attached to CE through a dedicated interface (in-line mode) or online, i.e., unified with the CE separation space during or after sample injection. Liquid phase microextraction and solid phase microextraction are most frequently used as pretreatment techniques prior to sample injection. Among electrophoretic preconcentration techniques during/after sample injection one can distinguish: field-amplified/enhanced sample stacking, large volume sample stacking, field amplified/enhanced sample injection, sweeping, micelle to solvent stacking, isotachophoresis, transient isotachophoresis, and more [26].

For the utilization of pre-concentration techniques, Deng et al. elaborated a rapid and simple CE method for the separation and determination of two alkaloids in *Ephedra* herbs. They used a background electrolyte composed of 80 mM of NaH_2PO_4 (pH 3.0) with an addition of 15 mM of β–cyclodextrin and 0.3% of hydroxypropyl methyl–cellulose. In this study, the authors took advantage of the field-amplified sample injection and, in the presence of a low conductivity solvent plug, they achieved an approximately 1000-fold improvement in detection sensitivity in comparison to conventional sample injection without any negative impact on resolution [31]. On the other hand, Honegr et al. used large volume sample stacking with polarity switching in order to enhance sensitivity. In this study, sample injection represented 50% of capillary volume and polarity was switched at 1.6 min of analysis, under optimized conditions an average 90-fold heightening of absorbance signal of the analytes was accomplished [32]. The abovementioned authors, Honegr and Pospíšilova, also found a method for the determination of phenolic acids in plant extracts using capillary zone electrophoresis with on-line transient isotachophoretic preconcentration (tITP). The application of preconcentration techniques in this case enabled the injection of large plugs of low concentration samples without overloading the column capacity of the electrophoretic system and consequently led to low detection limits without any decline in separation efficiency [20].

The implementation of extraction techniques prior to separation and detection by capillary electrophoresis is the right approach to obtain an exceedingly sensitive determination. In Table 1 one can find that Zhang et al. employed solid-phase microextraction (SPME) for CE determination of three protoberberines. This group of researchers fabricated a pipette-based device for their new imprinted monolith-based SPME–CE method, which was used for loading, subsequent extraction and final elution of a sample. The positive influence of the addition of methanol to BGE on separation was also noticed. The study confirmed that three protoberberines can be well enriched by the use of imprinted SPME. The limits of detection obtained were lower than in previously reported methods, i.e., 0.1 µg mL^{-1} [32]. Wang et al. described the application of subcritical water extraction (SWE) for the determination of alkaloids in *Sophora*. This relatively new extraction technique was beneficial in terms of operation time, efficiency and lack of organic solvent consumption. Moreover, the electro-injection boosted reproducibility in capillary electrophoresis with field-amplified stacking through the addition of acid to the sample [33].

2.2.4. Time of Analysis

An important issue for CE methods and in general for the establishment of a drug quality control method is analysis time. Literature analysis shows that it can be reduced even to 4 min. This impressively short time of analysis is reported by Du and Wang, who applied CE for the determination of berberine in herbal medications [34].

2.2.5. CE or HPLC?

Analytical methods elaborated for quality control of herbal preparations based on CE techniques may be an attractive alternative, because of the short analysis time, good separation efficiency, minimal sample, and solvent requirements. However, there is a question of whether CE is able to give comparable results with high performance liquid chromatography (HPLC). In some cases, one can find an interesting answer, for instance, there was no significant difference between the two methods established by Chen et al.

using HPLC and CE to determine nine marker components in "samgiumgagambang" (SGMX, herbal medicinal preparation containing 14 herbs) on the basis of the results for the five main constituents in SGMX. What is more, CE stood out for its shorter time of operation (14 min vs. 50 min) and its higher separation efficiency [10]. Dresler et al. in turn verified that capillary electrophoresis may be an alternative to HPLC for assessing the content of metabolites in *Hypericum perforatum* and *H. annulatum* and likewise non-significant differences between those two elaborated methods were found (the difference less than 10%). However, a comparison between LOD and LOQ values achieved with each method demonstrated the advantage of HPLC over CE with respect to detection sensitivity, but the observed difference between these methods can be significant in the analysis of dilute samples with very small amounts of components [35]. Gufler et al. presented capillary electrophoresis as a rapid and potent technique for the analysis of *Urecola rosea* leaf extracts. On the one hand, in terms of qualitative and validation parameters, it was equivalent to HPLC. On the other hand, with respect to operation time and environmental sustainability, CE is definitely beneficial and may be an attractive alternative to HPLC [36]. Table 1 confirms the reader's opinion that low concentration sensitivity remains a challenge and is the subject of the continuous development of capillary electrophoresis.

2.2.6. Interactions between Analytes and Additives to the BGE

CZE separation is based on the differences between the electrophoretic mobilities of separated compounds. The development of the technique in the form of affinity electrophoresis allows us to obtain highly specific separation through the use of specific ligands (for instance, selective antibodies, proteins, metal ions, or lectins). It should be highlighted that specific interactions between analytes and ligands affect mobility; moreover, they provide possibilities for the isolation and detection of analytes from complex matrices [37]. It is well-established that two main factors influence the electrophoretic mobility, namely the intrinsic physical characteristics of the analyzed compound and chemical additives in BGE interacting with the analyte [38]. Secondary equilibria resulting from additive–analyte interactions are essential to accomplish good resolution. Despite the fact that borate buffer was the most frequently used as a background electrolyte in reported works, various organic solvents and compounds were also added to the BGE to optimize the separations; for instance, surfactants, neutral salts, organic amines, organic salts, and chiral selectors (see Table 1). On the one hand, those additives obviously have an impact on mobility. On the other hand, Table 1 reveals that additives and organic solvents, especially when they go hand in hand with sample pretreatment/preconcentration, have an influence on the detection performance, even when a conventional UV detection is used. Organic solvents and additives are totally different from water, but also from one another in terms of physical and chemical properties. Different solvent properties (in particular pH) strongly influence the acid-base behavior and generally increase the pKa values of analytes (significantly different in organic-water mixtures in comparison to aqueous media), electrophoretic mobility, and give more opportunities to control the overall separation processes, manipulate selectivity and to achieve separations unworkable in aqueous buffer [39–41]. During the optimization of CE conditions, it should be taken into consideration that the pKa values of acids and bases may be totally different in aqueous and nonaqueous media due to differences in dielectric constant between solvents, which also impacts on the mobilities of divalent ionic species, absorbance to the wall of the capillary, and finally affects the electroosmotic flow [42]. Apart from the effects of organic solvents on the acid-base properties of analyzed compounds, ion–ion interactions resulting from the presence of buffer electrolytes as well as other ionic modifiers in the background electrolyte and ion-solvent interactions could considerably impact the analyte's electrophoretic mobility [41]. An unwanted phenomenon of poor reproducibility in migration time occurs when analyte absorption onto the capillary wall changes its conditions and, as a result, affects the magnitude of electroosmotic flow. This happens especially when a bare fused silica capillary is employed for analysis and it is little wonder that the interaction between samples and the inner capillary's wall affects peak shape, resolution, and efficiency. This light capillary coating and surface modification establishes a good direction for future research and development of the technique [43].

2.2.7. Field of Application

Methods based on capillary zone electrophoresis may be successfully employed, even for a full-scale quality analysis of herbal formulations, as was proved in the study reported by Xu et al. In this study, a comprehensive, rapid, and accomplishable electrophoretic method for the simultaneous separation and determination of seven constituents in Guan–Xin–Ning injection was elaborated and subsequently employed for quality control purposes [44].

CZE was successfully employed for the quantification of different classes of secondary metabolites in plant extracts among others: phenolic compounds, coumarins, protoberberines, curcuminoids, inorganic cations, isoquinoline alkaloids, iridoid glycosides, benzoic acid compounds quinolizidine alkaloids, and triterpene acids. This technique was also used for the determination of various active constituents and adulterants in herbal formulations. This kind of utilization is extremely important for the quality control of herbal medicinal preparations (for details see Table 1). A detailed analysis of the column entitled "Remarks" confirms the abovementioned ways to increase separation performance and detection sensitivity of capillary zone electrophoresis.

Table 1. Application of capillary zone electrophoresis.

Sample	Analytes	BGE	Detection	LOQ ($\mu g\ mL^{-1}$)	Remarks	Ref.
"samgiumgagambang" (SGMX)	5-hydroxymethyl-furaldehyde, geniposidic acid, chlorogenic acid, paeoniflorin, 20-hydroxyecdysone, coptisine, berberine, luteolin and glycyrrhizic acid	70 mM borate buffer containing 10% methanol (pH 9.5)	UV (230 nm)	5.0–100.0	no significant difference between HPLC and CE results	[10]
12 herbal preparations used for the treatment of diabetes	metformin, chlorpropamide, glibenclamide and gliclazide	sodium acetate 20 mM L^{-1} (pH 10.0)	CM	3.21, 2.01, 4.46 and 5.77	determination of hypo-glycemics as adulterants	[7]
26 herbal formulations	furosemide, hydrochlorothiazide, chlorthalidone, amiloride, phenolphthalein, amfepramone, fluoxetine and paroxetine	phosphate buffer (pH 9.2)	CM	5.14–11.01 mg/kg	determination of adulterants in herbal formulations for weight loss	[9]
7 Aloe plant species, 10 Aloe pharmaceutical preparations	aloin A and B	20.0 mM borate buffer with 50 mM SDS and 10 mM β–CD (pH 9.3)	LIF	0.025	microchip capillary electrophoresis (MCE)	[22]
Abelia triflora extract	scutellarein and caffeic acid	40 mM borax buffer (pH 9.2)	UV (200 nm)	2.5		[45]

Table 1. Cont.

Sample	Analytes	BGE	Detection	LOQ (µg mL^{-1})	Remarks	Ref.
Aconite radix	aconitine, mesaconitine, hypaconitine, benzoylaconine, benzoylmesaconine and benzoylhypaconine	200 mM Tris, 150 mM perchloric acid and 40% 1,4–dioxane (pH 7.8)	UV (214 nm)	0.14, 0.13, 0.14, 0.14, 0.13 and 0.15	LOD/LOQ ng mL^{-1}	[24]
Aconitum carmichaeli (Aconiti radix: Chinese name: chuanwu)	aconitine, mesaconitine and hypaconitine	25 mM borax– 20 mM 1–ethyl–3– methylimidazo– lium tetrafluoroborate (pH 9.15)	ECL	5.62×10^{-8}, 2.78×10^{-8}, 3.50×10^{-9} mol L^{-1} 0.036, 0.018 and 0.002	LOD/LOQ ng mL^{-1}	[46]
Aesculus hippocastanum (dry, hydro-alcoholic and hydroglycolic extracts)	β–escin	25 mMol L^{-1} bicarbonate– carbonate buffer (pH 10.3)	UV (226 nm)	38760		[47]
Aesculus hippocastanum L., Cichorium intybus L., Melilotus officinalis L. and Juniperus communis L. "Pendula"	aesculin, aesculetin, umbelliferone, dihydrocoumarin	20 mM borax in 5% methanol (pH 10.1)	UV (194 and 206 nm)	0.4–2.5 ppm		[48]
Areca nut	arecoline (methyl–1,2,5,6– tetrahydro–1– methylnicotinate)	20 mMol L^{-1} phosphate with 10 mMol L^{-1} BMImBF$_4$ buffer (pH 7.50)	ECL	0.00077	LOD/LOQ pg mL^{-1}	[23]
Belamcandae rhizoma	tectoridin and irigenin	borate buffer (pH 9.8)	AM	nd, LOD: 0.111 and 0.076	detection electrode based on the composite of carbon nanotubes and polylactic acid	[27]
Bergeniae rhizoma	arbutin and bergenin	50 mM borate buffer (pH 9.2)	AM	0.057 and 0.076	carbon nanotube– epoxy composite electrode	[28]
Cacumen platycladi	rutin, quercitrin, kaempferol and quercetin	50 mM sodium borate buffer (pH 9.2)	AM	0.110, 0.085, 0.063, 0.070	a fabricated graphene/poly (ethylene–co– vinyl acetate) composite electrode	[29]

Table 1. Cont.

Sample	Analytes	BGE	Detection	LOQ (µg mL^{-1})	Remarks	Ref.
Camptotheca acuminata (Camptotheca bark and fruit)	camptothecin alkaloids (CPT, 9–ACPT 9–MCPT HCPT, 7–EHCPT)	25 mM borate buffer containing 20 mM Sulfobutylether–β–CD and 20 mM ionic liquid [EMIM] [L–Lac] (pH 9.0)	UV (254 nm)	0.00020–0.00078	Large-volume sample stacking	[49]
Cassia tora (*Cassiae semen* and Cassia seed tea)	physcion, aloe–emodin, chrysophanol, emodin, aurantio–obtusin, rhein	10 mM Na$_2$HPO$_4$ and 6 mM Na$_3$PO$_4$ 15% methanol (*v/v*) (pH 11.8)	UV (254 nm)	1.11–4.67	an accelerated solvent extraction procedure	[50]
Catha edulis	cathinone, cathine, and phenyl-propanolamine	25 mM TRIS phosphate buffer (pH 2.5)	UV (210 nm)	0.4		[51]
Chamomile and linden flower extracts	apigetrin, naringin, naringenin, catechin, galangin, apigenin, luteolin, quercetin, myricetin, kaempferol and kaempferide	40 mM borate buffer (pH 8.9)	UV (210 nm)	0.252–2.142		[52]
Chelidonium majus L	protopine, chelidonine, coptisine, sanguinarine, allocryptopine, chelerythrine, and stylopine	20 mM phosphate buffer (pH 3.1)	UV–LEDIF	0.06–5.5		[53]
Chuanxiong rhizoma (*Ligusticum wallichii*)	vanillin, ferulic acid, vanillic acid, caffeic acid and protocatechuic acid	50 mM borate buffer (pH 9.2)	AM	nd	carbon nanotube (CNT)–polydimethyl-siloxane (PDMS) composite electrode	[54]
Combretum aculeatum extracts	punicalagin	25 mM, phosphate buffer (pH 7.4)	UV (280 nm)	60 ppm		[55]
Connarus perrottetii var. *angustifolius* (aqueous infusions, ethanolic extracts and butanolic extracts)	catechin and rutin	20 mM borate buffer containing 15% methanol (*v/v*), (pH 9.2)	UV (230 nm)	0.97 and 2.46		[56]
Coptidis rhizoma and berberine hydrochloride tablets	berberine	10 mM L^{-1} PBS (pH 7.81)	ECL	0.005	LOD/LOQ ng mL^{-1}	[34]

Table 1. Cont.

Sample	Analytes	BGE	Detection	LOQ (µg mL^{-1})	Remarks	Ref.
Coreopsis tinctoria Nutt.	taxifolin–7–O–glucoside, flavanomarein, quercetagetin–7–O–glucoside, okanin 4′–O–glucoside, okanin and chlorogenic acid	50 mM borate buffer containing 15% acetonitrile (pH 9.0)	UV (280)	2.34–12.94		[57]
Daturae flos	atropine, scopolamine, and anisodamine	40 mM phosphate buffer containing 20% v/v methanol and 30% v/v acetonitrile (pH 7.0)	UV (196 nm)	0.50 (LOD)	capillary coated by graphene oxide	[58]
Duyiwei capsule and dried crude drug of Lamiophlomis rotata	8–O–acetylshanzhiside methylester and 8–deoxyshanzhiside, apigenin, quercetin and luteolin	10 mM sodium tetraborate–20 mM NaH$_2$PO$_4$–15% (v/v)methanol (pH 8.5)	UV (238 nm)	nd, nd, LOD: 2.6–9.2		[59]
Echium vulgare L. and Echium russicum L. radix	shikonin and rosmarinic acid	50 mM borate buffer (pH 9.5)	UV (218 and 202 nm)	nd, LOD: 0.603 and 0.270 ppm		[60]
Ephedra sinica herba	ephedrine and pseudoephedrine	80 mM of NaH$_2$PO$_4$, 15 mM of β–CD and 0.3% of hydroxypropyl methyl–cellulose (pH 3.0)	UV (214 nm)	nd, LOD: 0.7 and 0.6	Field-Amplified Sample Injection	[31]
Epilobium parviflorum extracts	caffeic acid, cinnamic acid, p–coumaric acid, ferulic acid, protocatechuic acid, syringic acid and vanilic acid	200 mM borate buffer with 37.5% methanol, 0.001% hexadimethrine bromide, and 15 mM 2–hydroxypropyl–β–CD (pH 9.2)	UV (214 nm)	0.032–0.094	On-line transient iso-tachophoretic preconcentration	[20]
Epimedii herba (Yin–Yang–Huo)	epimedin C, icariin, diphylloside A, epimedoside A and icarisoside A	30 mM borate buffer containing 40% methanol (pH 9.5)	UV (270 nm)	3.0, 2.0, 4.0, 2.0 and 3.0	coupled with SPE	[61]
Fengshi Maqian tablets and Yaotongning capsules	strychnine and brucine	75 mM phosphate buffer with 30% methanol (v/v) (pH 2.5)	UV (203 nm)	0.01	sample pre-concentration method by two-step stacking	[62]

Table 1. Cont.

Sample	Analytes	BGE	Detection	LOQ (µg mL^{-1})	Remarks	Ref.
Forsythia suspensa	galacturonic acid and glucuronic acid	130 mM sodium hydroxide, 36 mM disodium hydrogen phosphate dihydrate and 0.5 mM cetyltrimethylammonium bromide (pH 12.28)	UV (270 nm)	10.68 and 12.64	reversed electroosmotic flow (EOF) to improve separation of neutral sugars	[63]
Forsythia suspensa fructus and commercial extracts	phillyrin, phillygenin, epipinoresinol–4–O–β–glucoside, pinoresinol–4–O–β–glucoside, lariciresinol, pinoresinol, isolariciresinol and vladinol D	40 mM borate buffer containing 2 mM β–CD and 5% methanol (v/v) (pH 10.30)	UV (234 nm)	3.00–4.38		[21]
Forsythiae suspensae fructus	oleanolic acid, ursolic acid and betulinic acid	50.0 mM L^{-1} borax and 0.5 mM L^{-1} β–cyclodextrin (β–CD) (pH 9.5)	UV (200 nm)	4.8, 4.6 and 5.9		[30]
Fritillariae Thunbergii bulbus (chloroform extracts)	peimine and peiminine	66% MeOH–ACN (1:1, v/v), 34% aqueous buffer containing 15 mM NaH$_2$PO$_4$, 2.5 mM NED, 4 mM H$_3$PO$_4$ (pH 3.0)	UV (214 nm)	nd., LOD: 3.9 and 4.1	NED as the UV absorbing probe	[64]
Garcinia cambogia (fruit rinds) and *Hibiscus sabdariffa* (calyx)	sodium salts of (1S,2R)–hydroxycitric and (1S,2S)–hydroxycitric acids	50 mM sodium phosphate buffer (pH 7.0)	UV (193 nm)	32.89–68.52		[65]
Geranii herba	rutin, hyperin, kaempferol, corilagin, geraniin, gallic acid, and protocatechuic acid	50 mM borate buffer (pH 9.2)	AM	nd, LOD: 30.9–682.8	graphene/poly(methyl methacrylate) composite electrode as a sensitive amperometric detector	[66]
Ginkgo biloba extract and rutin tablet,	epicatechin, rutin, and quercetin	10.0 mM borate and 0.5 mM luminol (pH 8.5)	CL	6×10^{-7}, 5×10^{-7} and 1×10^{-6}	ultrasensitive determination	[67]
Glycyrrhiza uralensis Fisch *radix*	glycyrrhetinic acid and glycyrrhizic acid	10 mM borate buffer (pH 8.8)	UV (268 nm)	6.2 and 6.9	On-line extraction coupled with flow injection and CE	[68]

Table 1. Cont.

Sample	Analytes	BGE	Detection	LOQ ($\mu g\ mL^{-1}$)	Remarks	Ref.
Guan–Xin–Ning (GXN) injection	caffeic acid, danshensu, ferulic acid, isoferulic acid, salvianolic acid A, salvianolic acid B, tertamethylpyrazine	35 mM SDS and 45 mM borate solution (pH 9.3)	UV (212 nm)	1.5–4.90		[44]
Hippophae rhamnoides extract and Cerutin® tablets	quercetin and rutin	40 mM L^{-1} borax (pH 9.2)	EC	0.475 and 0.726	hot platinum microelectrodes, flow injection analysis	[33]
Houttuyniae herba	rutin, isoquercitrin, quercitrin, and chlorogenic	50 mM borate buffer (pH 9.2)	AM	41.4, 31.8, 38.2 and 65.6	graphene/polystyrene composite electrode for amperometric detection	[69]
Hypericum perforatum and *Hypericum annulatum*	chlorogenic acid, epicatechin, hyperoside, rutin, quercitrin and quercetin	40 mM borate buffer, 50 mM SDS and 12% acetonitrile	UV (348, 208, 370, 370 and 318)	4.960–9.458 ppm	Non-significant differences between CE and HPLC	[35]
Isatidis radix	benzoic acid, salicylic acid and ortho–aminobenzoic acid	20 mM borate and 30 mM sodium dodecyl sulfate buffer containing 2 mM b–CD and 4%methanol (v/v), (pH 9.8)	UV (250 nm)	nd, LOD–800		[70]
Komplex Kurkumin® (curcumin 375 mg, demethoxycurcumin 100 mg and bis-demethoxycurcumin 25 mg)	curcumin, demethoxycurcumin and bisdemethoxycurcumin	50 mM/ L CAPS, 100 mg mL^{-1} of HP–β–CD and 2 gL^{-1} of HEC	UV–VIS (480 nm)	5.30, 4.57 and 6.20	unconventional hydrodynamically closed CE systems	[71]
Lam– iophlomis rotate and Cistanche	homovanillyl alcohol, hydroxytyrosol, 3,4–dimethoxycinnamic acid, and caffeic acid	50 mM borate–100 mM phosphate buffer in addition to 5.0 mM L^{-1} β–CD (pH 9.48)	UV (290 nm)	nd, LOD: 0.0051–0.029		[72]
Lycoridis radiatae bulbus	galanthamine	18 mMol L^{-1} phosphate buffer (pH 9.0)	ECL	nd, LOD: 0.00025		[73]
Lycoris radiata	galanthamine, homolycorine, lycorenine and tazetteine	10.0 mMol L^{-1} PBS (pH 8.0)	ECL	nd, LOD: 0.014, 0.011, 0.0018 and 0.0031	Ultrasonic-assisted extraction	[74]
Lysium chinensis folium	mannitol, sucrose, glucose, and fructose	50 mM NaOH	AM	0.120, 0.394, 0.126 and 0.155	Far-infrared-assisted extraction	[75]

Table 1. Cont.

Sample	Analytes	BGE	Detection	LOQ (µg mL^{-1})	Remarks	Ref.
Macleaya cordata and *Chelidonium majus* extracts	chelerythrine and sanguinarine	40 mM ammonium acetate–acetic acid–water buffer containing 50% (v/v) formamide (pH 2.90)	LIF	nd, LOD: 5.0 and 0.002	microchip electrophoresis	[76]
Magnolia officinalis and Huoxiang Zhengqi Liquid.	honokiol and magnolol	16 mMol L^{-1} sodium tetraborate, 11% methanol (pH 10.0)	UV (210 nm)	1670 and 830		[77]
Origanum vulgare and Romanian propolis	resveratrol, pinostrobin, acacetin, chrysin, rutin, naringenin, isoquercitrin, umbelliferone, cinnamic acid, chlorogenic acid, galangin, sinapic acid, syringic acid, ferulic acid, kaempferol, luteolin, coumaric acid, quercetin, rosmarinic acid and caffeic acid	45 mM borate buffer with 0.9 mM sodium dodecyl sulfate (pH = 9.35)	UV (280 nm)	0.07–5.77		[78]
Orthosiphon stamineus Benth.	rutin, carnosolic acid, caffeic acid, rosmarinic acid, quercetin, luteolin, apigenin and cinnamnic acid	50 mM borate buffer (pH 9.0)	UV (200 nm)	0.053, 0.053, 0.046, 0.040, 0.040, 0.030, 0.023 and 0.020	large volume sample stacking with polarity switching	[32]
Peganum harmala semen infusions	harmine, harmaline, harmol, harmalol, harmane, and norharmane	50 mM tris–HCl (pH 7.8) with 20% (v/v) of methanol	UV (254 nm)	0.1–8.3		[79]
Penicillium glaucum, P. tenuifolium, P. dubium and *P. fugax* fruits	morphine, codeine and thebaine	100 mM sodium phosphate buffer, containing 5 mM α–CD (pH 3.0)	UV (214)	2.0	Ultrasound-assisted extraction	[80]
Phellodendri chinensis cortex	berberine, palmatine and jatrorrhizine	20 mM phosphate buffer with methanol 10% (v/v), (pH 7.0)	UV	0.3	imprinted solid-phase microextraction	[81]
Pholia magra (*Cordia ecalyculata vell*, 500 mg/capsule), *Persea americana* and *Cyperus rotundus*	NH_4^+, K^+, Ca^{2+}, Na^+, Mg^{2+}, Mn^{2+}, Tl^{3+}, Cr^{3+}, Pb^{2+}, Cd^{2+}, Zn^{2+}, Cu^{2+}, Co^{2+}, and Ni^{2+}	30 mM 2–N–MES /histidine, 1.5 mM 18–crown–6 ether, and 1 mM citric acid (pH 6.0)	C^{4}D	0.093, 0.182, 0.405, 0.475, 0.077, 0.170, 1.478, 0.988, 2.008, 1.749, 0.454, 1.193, 0.817 and 0.632		[82]

Table 1. Cont.

Sample	Analytes	BGE	Detection	LOQ (µg mL^{-1})	Remarks	Ref.
Phyllanthus urinaria	rutin, quercetin, ferulic acid, caffeic acid, and gallic acid	10 mM borate buffer (pH 9.0)	AM	nd, LOD–3.36, 0.45, 0.097, 0.072 and 1.00		[83]
Plumula nelumbinis	neferine, liensinine, isoliensinine, rutin and hyperoside	50 mM borate buffer (pH 9.2)	AM	0.42, 0.31, 0.38, 0.35 and 0.39	far infrared-assisted solvent removal	[84]
Portulaca oleracea L., *Crataegus pinnatifida* and *Aloe vera* L.	linolenic acid, lauric acid, p–coumaric acid, ascorbic acid, benzoic acid, caffeic acid, succinic acid, and fumaric acid	40 mM H_3BO_3–40 mM $Na_2B_4O_7$ (pH 8.70)	UV (200 nm)	nd, LOD: 0.02–3.44	field enhancement sample stacking for	[85]
propolis	pinocembrine; ferulic acid; p–coumaric acid; quercetin; and caffeic acid	100 mM borate buffer (pH = 8.7)	EC	nd, LOD: 0.1–0.5		[13]
Puerariae radix	3′-methoxypuerarin, puerarin, 3′-hydroxypuerarin, ononin, daidzin, daid–zein and genistin	35 mM sodium tetraborate, 9.0 mM sulfobutylether-β-CD α-cyclodextrin (pH9.34)	UV (254 nm)	2.5–9.5		[86]
Reduning injection	caffeic acid, isochlorogenic acid A, isochlorogenic acid B, isochlorogenic acid C, chlorogenic acid, neochlorogenic acid and cryptochlorogenic acid	20 mM NaH_2PO_4, 10 mM β–CD and 5% ACN (pH 4.2)	UV (325 nm)	0.8–1.5	DPPH–CE–DAD	[87]
Rhodiola	salidroside and tyrosol	50 mM borate buffer (pH 9.8)	AM	LOD: 0.72 and 0.39	a novel graphene/poly (urea–formalde-hyde) composite modified electrode as a sensitive am-perometric detector	[88]
Rourea minor stems	bergenin derivatives and catechins (new natural products)	30 mM borax solution with (pH 10.5)	UV (205 nm)	6.2–18.8		[89]

39

Table 1. Cont.

Sample	Analytes	BGE	Detection	LOQ (µg mL^{-1})	Remarks	Ref.
Salvia miltiorrhiza, S. przewalskii, S. castanea and Danshen	protocatechuic aldehyde, salvianolic acid C, rosmarinic acid, salvianolic acid A, danshensu, salvianolic acid B and protocatechuic acid	20 mM sodium tetraborate (pH 9.0)	UV (280 nm)	0.47–1.19		[90]
Sappan Lignum (the dried heartwood of *Caesalpinia sappan* L, methanolic extract)	brazilin and protosappanin B	20 mM borate buffer containing 6% v/v of methanol (pH 9.2)	UV (254 nm)	0.28 and 0.15	online concentration with acid barrage stacking	[91]
Scutellariae barbata extract	baicalein, baicalin, and quercetin	0.1 M borate buffer (pH 9.0)	EC	< 0.22		[92]
Shuxuening Injection	clitorin, rutin, isoquercitrin, quercetin–3–*O*–D–glucosyl]–(1–2)–L–rhamnoside, kaempferol–3–*O*–rutinoside, kaempferol–7–*O*–β–D–glucopyranoside, apigenin–7–*O*–Glucoside, quercetin–3–*O*–[2–*O*–(6–*O*–p–hydroxyl–E–coumaroyl)–D–glucosyl]–(1–2)–L–rhamnoside, 3–*O*–[2–*O*–[6–*O*–(p–hydroxyl–E–coumaroyl)–glucosyl]]–(1–2) rhamnosyl kaempfero	20 mM phosphate 5 mM β-cyclodextrin (β–CD), 40 mM sodium dodecyl sulfate and 7.5% ACN (pH 7.0)	UV–VIS (360 and 405 nm)	0.04–0.09	On-line 2,2′–Azinobis–(3–ethylbenzthiazoline–6–sulphonate)–ccapillary electrophoresis–diode array detector	[93]
Sophora flavescens	cytisine, sophocarpine, matrine, sophoridine, and oxymatrine	110 mM monosodium phosphate isopropanol (85:15, v/v) (pH 3.0)	UV (214 nm)	nd, LOD: 0.0004–0.0013	subcritical water extraction and field amplified sample stacking	[94]
Sophora flavescens (extract from the dried root)	matrine, oxymatrine, and sophoridine	50 mM sodium tetraborate solution, 500 mM boric acid and 1.2 mM citric acid (pH 7.98)	UV (210 nm)	60–100		[36]

Table 1. Cont.

Sample	Analytes	BGE	Detection	LOQ (µg mL^{-1})	Remarks	Ref.
Swertia mussotii Franch and preparations (herbs, granular, capsules)	oleanolic acid, ursolic acid, quercetin, and apigenin	50 mM borate–phosphate buffer with 5.0×10^{-3} mol L^{-1} β–cyclodextrin (pH 9.5)	UV (250 nm)	0.6829, 0.4007, 0.0124 and 0.5076		[95]
thyme and parsley extracts	luteolin and apigenin	20 mM borate buffer and methanol (90: 10, v/v), (pH 10.0)	UV (210 nm)	2.98 and 1.41		[96]
traditional Chinese medicines, *Hippophae rhamnoides*, *Hypericum perforatum*, and *Cacumen platycladi*	rutin, quercetrin, quercetin, kaempferol, kaempferide, catechin, apigenin and luteolin	18 mM borate buffer (pH 10.2)	AM	0.28, 0.22, 0.26, 0.24, 0.24, 0.22, 0.15 and 0.17		[97]
Trichilia catigua	epicatechin and procyanidin B2	80 mM borate buffer with 2–hydroxypropyl–β–cyclodextrin 10 mMol L^{-1}, (pH 8.80)	UV (214 nm)	17.16 and 15.26	CE method faster, more efficient, less expensive, less polluting than previously developed HPLC method	[98]
Trifolium alexandrinum seed	soyasaponin I, azukisaponin V, bersimoside I and bersimoside	80 mM borate buffer containing 24 mM β–CD (pH 10)	UV (195 nm)	23.33, 21.64, 23.30 and 22.94	diastereomeric separation in	[99]
Urceola rosea leaf extracts	five phenolic compounds	25 mM sodium tetraborate decahydrate solution with (pH 8.5)	UV (254 nm)	10.9–20.8	CE method was well comparable to HPLC	[100]
Valeriana officinalis extracts	acacetin, diosmetin, chlorogenic acid, kaempferol, apienin, luteolin, p–hydrox–benzoic acid and caffeic acid	60 mM borate buffer (pH 9.2)	AM	0.033–0.4		[101]
Yansuan Xiaobojian Pian (berberine tablets), and plant samples: Goldthread, Amor Cork Tree, Goldenseal, Plantain, Tree Tumeric, Yellow Root, Bupleurum and Oregon Grape	berberine	20 mM acetic acid, 35 mM 2–HP–β–CD, and 20% methanol (pH 5.0)	LIF	nd, LOD: 0.016		[102]

Table 1. Cont.

Sample	Analytes	BGE	Detection	LOQ (µg mL^{-1})	Remarks	Ref.
Yinqiaojiedu tablet	liquiritin, chlorogenic acid, and glycyrrhizic acid	103.1 mM boric acid, 51.6 mM sodium borate, 9.8 mM disodium hydrogen phosphate, and 15.6 mM sodium dihydrogen phosphate (pH 7.86)	UV (254 nm)	0.41, 0.79 and 0.68		[103]

nd—no data, LOQ—limit of quantification, LOD—limit of detection, CM—conductometric, β—CD–β–cyclodextrin, LIF—Laser Induced Fluorescence, Tris—tris(hydroxymethyl)aminomethane, ECL—electrochemiluminescence, AM—amperometric, UV—LEDIF–ultraviolet light-emitting diode-induced native fluorescence, PBS—sodium phosphate buffer solution, NED—N-(1–naphthyl)ethylenediamine dihydrochloride, SDS—sodium dodecyl sulfate, CL—chemiluminescence, SPE—solid phase extraction, MES—morpholinoethanesulfonic acid, ACN—acetonitrile, capacitively coupled, C^{4}D—contactless conductivity, DPPH—1,1-diphenyl-2-picryl-hydrazyl, DAD—diode array detector, 2-HP-β-CD-(2–hydroxypropyl)-β-cyclodextrin.

2.3. Micellar Electrokinetic Chromatography (MEKC)

MEKC is a powerful electrophoresis-driven separation technique, which offers good selectivity, high efficiency, optimization flexibility, and significantly reduces organic solvent consumption during its operation. However, it is not possible to avoid organic solvent consumption when MEKC is applied to the analysis of medicinal plant materials or pharmaceutical formulations. This technique allows for the resolution of both neutral and charged compounds and may be applied for the analysis of a broad selection of active constituents; for instance, flavonoids in herbal raw material. The running buffer in MEKC is fortified with surfactants at a concentration exceeding their critical micelle concentration, that leads to forming micelles. The micelles arrange for a pseudostationary phase that enables the differential separating of analyzed compounds as a result of the influence of dispersed surfactants [101]. In reported studies, various pseudostationary phases were introduced. There are four major classes of surfactants: anionic, cationic, zwitterionic, and nonionic [104]. However, anionic surfactant, i.e., sodium dodecyl sulfate, was most frequently used in reported analyses. For more details see Table 2.

In the past decade, MEKC was employed for the separation and quantification of different classes of secondary metabolites in plant extracts among others: coumarins, tanshinones, phenolic acids, terpenoids, iridoids, phenylethanoid glycosydes, phenylpropanoids, and flavonoids, saponin (see Table 2).

The compounds were detected and quantification was achieved mainly by UV absorption, but amperometric detection was also applied (see Table 2).

Recent studies confirm the high separation efficiency of MEKC and indicate tremendous potential for a wide range of analytical applications. Yang et al. employed polyvinylpyrrolidone-stabilized graphene-modified MEKC for the separation of tanshinones. The established method was successfully employed for the quality assessment of Danshentong capsules [105]. Cao et al. in turn used MEKC to resolve a mixture of flavonoids, phenolic acids, and saponin. In order to alter the electrophoretic behavior of analytes and to develop the resolution, they added ionic liquids-coated multi-walled carbon nanotubes to the running buffer, which influenced the partitioning of the analytes. Their results give real hope for the future analysis of complex samples based on considerable advantages in overcoming the effects of matrix-induced interferences exhibited in the study [106].

In the case of the use of large amounts of solvents, Chang and coworkers, in their paper, exhibited the elaboration of surfactant-assisted pressurized liquid extraction (PLE) for the effective extraction of flavonoids in *Costus speciosus* flowers prior to MEKC analysis. The reported work confirmed numerous advantages of PLE, i.e., short extraction time, simplicity, efficiency, automation, and environmental friendliness (organic-free). The PLC–MEKC approach enabled fast, eco-friendly, and effective extraction and assay of flavonoids in the abovementioned raw material [107].

In terms of improving the detection sensitivity in MEKC, a study by Chang et al. is reported, where the authors elaborated a magnetic iron oxide nanoparticle-based solid-phase extraction process in conjunction with the online concentration and separation of salicylic acid in in tobacco leaves through micellar electrokinetic chromatography–UV detection. The authors observed an approximately 1026-fold improvement in the detection sensitivity of the elaborated method in comparison to a single MEKC method without an online concentration [108].

Qualitative and quantitative methods based on MEKC are rapid, efficient, and eco-friendly, and are successfully employed for the routine quality control of herbal drugs and raw plant material (see Table 2).

Table 2. Application of micellar electrokinetic chromatography and microemulsion electrokinetic chromatography.

Sample	Analytes	BGE	Detection	LOQ ($\mu g\ mL^{-1}$)	Remarks	Ref.
Calendula officinalis, Hypericum perforatum, Galium verum and Origanum vulgare extracts	(+)–catechin, (−)–epigallocatechin, (−)–epigallocatechin gallate, (−)–epicatechin gallate and (−)–epicatechin	10 mM KH_2PO_4 and 8.3 mM sodium tetraborate buffer with 66.7 mM SDS, (pH 7.0)	UV (210 nm)	0.010–0/047	LOD/LOQ ng mL^{-1}	[109]
clove oil, litsea cubeba oil, and citronella oil	citronellal, citral (Z; E), a–pinene, limonene, linalool, and eugenol	20 m borate buffer, 50 mMSDS, 20% (v:v), (pH 9.5)	UV (210 nm)	0.8–5.9		[110]
Costus speciosus flos extract	rutin, quercitrin, and quercetin	10 mM phosphate, 10 mM borate, 50 mM SDS (pH 8.5)	UV (370 nm)	2.30, 1.57 and 1.07	surfactant–assisted pressurized liquid extraction	[107]
Curcuma wenyujin origin's Chinese herbal medicines	curdine, curcumenol, germacrone, furanodiene, and β–elemene	1.3% SDS, 5.0% 1–butanol, 0.5% ethyl acetate and 10% acetonitrile in 10 mM borate buffer (pH 9.0)	UV (215 nm)	16.0–78.0	MSPD extraction coupled with MEEKC	[111]
Danshentong capsule (Salvia miltiorrhiza)	tanshinone IIB, dihydrotanshinone I, tanshinone I, cryptotanshinone, 1,2–dihydrotanshinone I, miltirone, and tanshinone IIA	10 mM borate buffer (pH 9.3) containing 30 mM SDS, 10% v/v 2–propanol and 6 μg mL^{-1} graphene	UV (260 nm)	8.73–19.10		[105]
Hemidesmus indicus radix	2–hydroxy–4–methoxybenzaldehyde, 2–hydroxy–4–methoxybenzoic acid, and 3–hydroxy–4–methoxybenzaldehyde	50 mM phosphate buffer with 65 mM of sodium taurodeoxycholate (pH 2.5)	UV (254 nm)	0.40, 2.5, and 0.7	MEKC results confirmed by HPLC–MS	[112]

Table 2. Cont.

Sample	Analytes	BGE	Detection	LOQ (µg mL^{-1})	Remarks	Ref.
Heracleum sphondylium herb and *Aesculus hippocastanum* cortex	coumarin, scoparone, isoscopoletin, esculin, esculetin, umbelliferone, xanthotoxin, byakangelicin, isopimpinellin, bergapten, phellopterin, xanthotoxol	50 mM sodium tetraborate, 45 mM SC, and 20% of methanol (v/v) (pH 9.00)	UV (214 nm)	1.70–4.772		[113]
He–Shou–Wu	hypohorine, THSG, epicatechin, proanthocyanidin B2, proantocyanidin B1, catechin and gallic acid	50 mM phosphate buffer containing 90 mM SDS and 2% (m/v) HP–β–CD (pH 2.5)	UV (210 nm)	<5.5	pressurized liquid extraction and short-end injection MEKC	[114]
Larrea divaricata Cav. extracts	nordihydroguaiaretic acid	20 mM phosphate buffer 10 mM SDS and 10% acetonitrile, (pH 7.5),	UV (283 nm)	1.06		[115]
Lianqiao Baidu pill	genistein, caffeic acid, glycyrrhizic acid ammonium salt, wogonoside	30 mMol L^{-1} SB, 95 mMol L^{-1} SDS, and 100 mMol L^{-1} boric acid (pH 9.30)	UV (214 nm)	0.77–1.85		[116]
Ligaria cuneifolia extracts	catechin, epicatechin, procyanidin B2, rutin, quercetin–3–O–glucoside, quercetin–3–O–xyloside, quercetin–3–O–rhamnoside, quercetin–3–O–arabinofuranoside, quercetin–3–O–arabinopyranoside and quercetin	20 mM borate buffer, 50 mM SDS mM β–CD and 2% w/v S–β–CD and 10% v/v methanol (pH 8.3)	UV (255 and 280 nm)	0.26–1.33		[117]
Lippia alba leaves	genoposidic acid, 8-epi–loganin, mussaenoside, chrysoeriol–7–O–diglucuronide, triclin–7–O–diglucuronide, acetoside	50 mM borax buffer containing 75 mM SDS and 5% isopropanol		38.0–119.0	no statistically significant differences beetween CE and HPLC	[118]

Table 2. Cont.

Sample	Analytes	BGE	Detection	LOQ (μg mL^{-1})	Remarks	Ref.
Nicotiana tabacum L. leaves	salicylic acid	TB buffer containing 100 mM SDS and 15% (v/v) acetonitrile (pH 10.0)	UV (205 nm)	nd, LOD: 0.0005	magnetic iron oxide nanoparticle-based solid–phase extraction procedure followed by an online concentration technique	[108]
Petroselinum crispum, Rosmarinus officinalis, Thymus vulgaris L., *Origanum vulgare, Origanum majorana* L., *Salvia officinalis* L.,and *Levisticum officinale*	apigenin	30 mMol L^{-1} sodium borate 10% acetonitrile, and 10 mMol L^{-1} sodium dodecyl sulfate (pH 10.2)	UV (390 nm)	0.28		[119]
*Plantago lanceolata, Plantago major,*and *Plantago asiatica* leaf extracts and biotechnological product, plant tissue cultures (calli) of *P. lanceolata*.	aucubin, catalpol, verbascoside and plantamajoside	15 mM sodium tetraborate, 20 mM TAPS and 250 mM DOC (pH 8.50)	UV (200 and 350 nm)	1360, 1630, 2350 and 2720		[120]
Qishenyiqi dropping pills	calycosin–7–O–β–D–glucoside, formononetin, dihydroquercetin, rosmarinic acid, danshensu, salvianolic acid B, protocatechuic acid, ginsenoside Rg$_1$, ginsenoside Rb$_1$	10 mM borate buffer (pH 9.0) containing 100 mM SDS, 6% propanol and 4 μg mL^{-1} ILs–MWNTs	UV (200 nm)	nd, LOD: 1.01–76.32	ionic liquids coated multi–walled carbon nanotubes as pseudo-stationary phase	[106]
Salvia chionantha and *Salvia kronenburgii* acetone extracts	horminone and 7–O–acetylhorminone	50 mM SDS, 25% metanol (pH:11.5)	UV (230 nm)	nd, LOD: 3.269 and 4.518		[121]
Salvia miltiorrhiza, S. przewalskii, and *S. castanea*	dihydrotanshinone I, cryptotanshinone, protocatechuic aldehyde, tanshinone I, tanshinone IIa, salvianolic acid C, rosmarinic acid, 9′-methyl lithospermate b, danshensu, salvianolic acid B and protocatechuic acid	15 mM sodium tetraborate with 10 mM SDS, 5 mM β–CD, 10 mM [bmim]BF$_4$ and 15% ACN (v/v), (pH 9.8)	UV (254 nm)	0.90–4.63		[122]

Table 2. Cont.

Sample	Analytes	BGE	Detection	LOQ (µg mL^{-1})	Remarks	Ref.
Schisandra chinensis	schizandrin, schisandrol B, schisantherin B, schisantherin A, schisanhenol, deoxyschizandrin, schisandrin B	35 mM phosphate with 10 mM β–cyclodextrin (β–CD), 30 mM sodium dodecyl sulfate (SDS) and 10% ACN (pH 8.0)	UV (222 nm)	0.02–0.12	2,2–azinobis–(3–ethylbenzoth-iazoline–6–sulfonic acid)–sweeping micellar electroki-netic chromato-graphy–diode array detector	[123]

nd—no data, LOQ—limit of quantification, LOD—limit of detection, SDS—sodium dodecyl sulfate, SB—sodium borate, MSPD—micro matrix solid phase dispersion, TAPS—N-[(1S,2S,3R)-2,3-bis(acetyloxy)-1-[(acetyloxy)methyl]heptadecyl]-acetamide, DOC—anionic detergents sodium deoxycholate, ILs—MWNTs—ionic liquids coated multi-walled carbon nanotubes, β–CD—β–cyclodextrin, HP–β–CD—hydroxypropyl–β–cyclodextrin, THSG-2,3,5,4′—tetrahydroxystilbene 2-O-β-D-glucoside, SC—sodium cholate.

2.4. Non-Aqueous Capillary Electrophoresis

Non-aqueous capillary electrophoresis (NACE) is a potent alternative to aqueous electrophoretic techniques, especially when it is difficult to separate lipophilic compounds. Separation of analytes is achieved using non-aqueous background electrolytes and the principle is based on the diverse physical and chemical properties of organic solvents. This great variety of solvents broadens the scope of separation selectivity. The low generation of current in nonaqueous media allows the use of high electric field strengths and wide bore capillaries and subsequentially allows a larger volume of the sample. Other advantages of NACE include better solubility of analytes in organic solvents, MS compatibility, and finally enhanced detection selectivity in many cases [124]. Analogically to the CZE sample, preconcentration techniques are applied in order to develop detection sensitivity in NACE. Field-amplified sample stacking, large-volume stacking using the electroosmotic flow pump, and transient isotachophoresis proved to be suitable for the abovementioned purpose and not only in aqueous CE. What is more, employment of LVSEP for NACE allowed the sensitive determination of organic anions at the nanomolar range using conventional UV detection and the introduction of ITP shortened the time of analysis [41].

The non-aqueous approach is not as prevalent as CZE or MEKC, but literature analysis indicates some applications for the analysis of herbal drugs and plant material. This technique was used in the study published by Hou et al. for the efficient separation and determination of five alkaloids in *Coptidis rhizoma*. In this work, surfactant-coated multi-walled carbon nanotubes provided a pseudostationary phase. Numerous parameters affecting NACE separation were studied, and in consequence, the authors noticed an important enhancement in the resolution due to the π–π interactions between the analyzed compounds and the surface of the carbon nanotubes in comparison to conventional NACE [125]. Meanwhile, Yuan et al. proposed a fast and uncomplicated method for the analysis of atropine, anisodamine, and scopolamine in *Deturae flos* extract by NACE coupled with electrochemiluminescence and electrochemistry dual detection. The running buffer was composed of acetonitrile and 2–propanol containing 1 M acetic acid, 20 mM sodium acetate, and 2.5 mM tetrabutylammonium perchlorate. Despite using a short capillary of 18 cm, the decoupler was not necessary and the separation performance was respectable [126]. Dresler et at. analyzed lipophilic compounds (hypericin and hyperforin) in *Hypericum* extracts with the non-aqueous capillary electrophoresis. The separation of the abovementioned constituents was conducted using bare fused silica 75 µm i.d. capillaries with an effective total length of 80.0 cm. The running buffer was a mixture of methanol, dimethylsulfoxide, and N–methyl formamide (3:2:1 $v/v/v$) as a solvent, with 50 mM ammonium acetate, 150 mM sodium acetate, and 0.02% (w/v) of cationic polymer hexadimethrine bromide to reverse the flow. At the same time, flavonoids and chlorogenic were evaluated with traditional CE as described above (Section 2.2). Only non-significant statistical differences were observed between the HPLC and CE results, namely the average differences between the particular metabolite ranged, e.g., from less than 10% for rutin and hypericin to ca. 1% for quercitrin [35]. The NACE method was also optimized for the simultaneous determination of major bioactive curcuminoids and some of the degradation products in turmeric milk and herbal commercial products. Non-aqueous BGE for separation of analytes was composed of sodium tetraborate, sodium hydroxide, methanol, and 1–propanol. Moreover, an innovative ultrasonication-assisted phase separation method was optimized and employed for extraction of the analytes in turmeric milk and subsequent direct injection of the extract into the capillary without any pretreatment [12]. The abovementioned NACE methods are simple, fast, convenient, and economical and applicable to analysis of herb extracts and commercial products (see Table 3).

Table 3. Application of nonaqueous capillary electrophoresis.

Sample	Analytes	BGE	Detection	LOQ ($\mu g\ mL^{-1}$)	Remarks	Ref.
Coptidis rhizoma	coptisin, berberine, epiberberine, palmatine, jatrorrizine	20 mM sodium acetate in methanol–acetonitrile (80:20, v/v), 20% acetonitrile and 6 μg mL^{-1} SC–MWNTs	UV (254 nm)	0.31–0.34		[125]
Daturae flos extract	atropine, anisodamine, and scopolamine	acetonitrile and 2–propanol containing 1 M acetic acid, 20 mM sodium acetate and 2.5 mM tetrabutylammonium perchlorate	ECL and EC dual detection	0.5–50.0		[126]
Hypericum perforatum and Hypericum annulatum	hypericin and hyperfolin	methanol, dimethylsulfoxide, N–methyl formamide (3:2:1 $v/v/v$) with 50 mM ammonium acetate, 150 mM sodium acetate and 0.02% (w/v) of cationic polymer hexadimethrine bromide	UV (294 and 594)	2.191–2.948 ppm		[35]
Turmeric milk (Curcuma longa) and herbal products	curcumin, desmethoxycurcumin and bisdesmethoxycurcumin, vanillin, vanillic acid, ferulic acid, and 4–hydroxybenzaldehyde	a mixture of sodium tetraborate, sodium hydroxide, methanol and 1–propanol	UV-VIS (300 and 498 nm)	10.1–26.5	a novel ultrasonication–assisted phase separation method (US–PS) was used for extraction and subsequently the extract was directly injected into the capillary	[12]

nd—no data, LOQ—limit of quantification, LOD—limit of detection, SC-MWNTs—surfactant-coated multi-walled carbon nanotubes, ECL—electrochemiluminescence, EC—electrochemistry.

2.5. Capillary Electrochromatography (CEC)

Capillary electrochromatography is a hybrid technique because it merges features of both high performance liquid chromatography and capillary electrophoresis and may be applied for the determination of charged and neutral analytes. In CE, analytes are separated in a capillary column with electroosmotic flow as the driving force for bulk fluid movement. However, in capillary electrochromatography, the capillary contains a stationary phase as in HPLC. Hence, there is a capability to take advantage of different mechanisms to provide additional selectivity beyond that possible through HPLC or CE alone. This combination of CE has advantages, i.e., high-efficiency, low-solvent and sample consumption, and reverse-phase mechanism of HPLC makes this technique reliable and flexible and, what is more, it can be fully suitable for pharmaceutical analysis and can replace other more demanding techniques in terms of time and expenses [127,128]. On the other hand, in comparison to CZE, the optimization of CEC is more challenging, the efficiency is lower due to peak broadening and the reproducibility of retention times is poorer [129]. However, a continuous fulfilling CEC with nanoparticles as a pseudostationary phase coupled with MS detection demonstrates high separation efficiency, as well as high performance confirmed by such parameters as limit of detection, peak asymmetry, repeatability, and reproducibility [130]. Except for employing a detector with high sensitivity, other approaches to achieve good detection sensitivity, as well as resolution and separation efficiency, include bubble or Z-type cells to extend the optical path, and obviously sample preconcentration techniques. For instance, FASS and in-column detection [131,132]. CEC was reported for the fast separation and quantification of coumarins in *Angelica dahurica* extract. A methacrylate ester-based monolithic column was used as a stationary phase. In order to gain a significant raise in the selectivity, surfactant sodium desoxycholate was added to the mobile phase as the pseudostationary, so there was no need to increase the hydrophobicity of the stationary phase. The devised method was characterized not only by satisfactory separation and a running time of 6 min, but also by LOQs lower than 0.30 µg/mL^{-1} [133]. The second reported study describes a CEC method for the quality control of *Cnidii fructus* extracts. This method, taking advantage of the methacrylate ester-based monolithic column, was characterized by an acceptable resolution of LOQs between 1.0 and 2.8 µg/mL^{-1} and the time of operation was shortened to 5 min [134]. For more details see Table 4.

Table 4. Application of capillary electrochromatography.

Sample	Analytes	BGE	Detection	LOQ (µg mL^{-1})	Remarks	Ref.
Angelica dahurica extract	byakangelicin, oxypeucedanin hydrate, xanthotoxol, 5–hydroxy–8–methoxypsoralen and bergapten	30:70 v/v ACN–buffer containing 20 mM sodium dihydrogen phosphate (NaH$_2$PO$_4$) and 0.25 mM SDC (pH 2.51)	UV (210 nm)	<0.30	methacrylate ester–based monolithic column	[133]
Cnidii fructus extracts	isopimpinelline, bergapten, imperatorin and osthole	50% ACN and 50% of a 10 mM sodium dihydrogen phosphate (pH 4.95)	UV (210 nm)	1.0–2.8	poly(butyl methacrylate–co–ethylene dimethacrylate–co–[2–(methacryloyloxy)ethyl] trimethylammonium chloride) monolithic column	[134]

LOQ—limit of quantification, SDC—surfactant sodium desoxycholate, ACN—acetonitrile.

2.6. Capillary Electrophoresis–Mass Spectrometry (CE-MS)

Capillary electrophoresis has many advantages in HPLC (low solvent consumption, using inexpensive capillaries, short time of operation, high efficiency without sample retreatment) and can support complementary or supplementary information about the constitution of a sample. One of the limitations of CE techniques is the relatively poor sensitivity as a result of the injection of small sample volumes, which might be improved by the implementation of pre-concentration techniques. The other way is to take advantage of CE–MS hyphenation, which not only enhances LOD thanks to MS detection, but also allows for the measurement of the particular mass of analytes and offers structural information, including the opportunity to identify and determine co-migrating species in overlapping peaks [135]. In reported papers, authors have described the quantitative analysis mainly of alkaloid compounds in plant extracts/pharmaceutical formulations.

In the study of Liu et al., the CEC–MS method, fully applicable for the quality evaluation of *Evodiae fructus*, was elaborated. It should be underlined that 4–16 fold improvement of detection limits was achieved in comparison to the CEC method with conventional UV detection [136]. Wang et al. proposed matrix solid-phase dispersion microextraction combined with CE in conjunction with quadrupole time-of-flight mass spectrometry for the quantification of three alkaloids in *Fritillariae Thunbergii bulbus*. It is noteworthy that in this method the reported LOQ value is in the ng mL^{-1} level [137]. All reported CE–MS methods were effectively employed for qualitative and quantitative analysis of bioactive components in plant extract and pharmaceutical preparations (see Table 5).

Table 5. Application of capillary electrophoresis with MS detection.

Sample	Analytes	BGE	Method	LOQ (µg mL^{-1})	Remarks	Ref.
Catharanthus roseus	vinblastine, vindoline, and catharanthine	20 mM ammonium acetate and 1.5% acetic acid	CE–MS	nd, LOD: 0.1–0.8		[135]
Evodiae fructus	limonin, evodiamine, and rutaecarpine	30% acetonitrile (ACN) in 1% ammonia aqueous solution	CEC–MS	3.1, 0.63 and 0.15	provided 4–16 folds improvement of LODs when compared with CEC–UV method	[136]
Fritillariae Thunbergii bulbus	peimine, peiminine, and peimisine	20 mM ammonium acetate with MS–grade water	CE–Q–TOF–MS	0.004–0.005	solid acids assisted matrix solid-phase dispersion micro-extraction	[137]
Lycoris radiata roots	lycorine, lycoramine, lycoremine, lycobetaine, and dihydrolycorine	ACN and methanol (1:2, *v*/*v*), which 40 mM ammonium acetate and 0.5% acetic acid	NACE ESI-IT-MS	0.04–0.24		[138]
Psoralae fructus and pharmaceutical preparations	bavachin and isobavachalcone	20 mM aqueous solution of ammonium acetate (pH 10.0)	CE–ESI–MS	nd, LOD: 0.06		[139]
Banisteriopsis caapi, Datura stramonium, Mimosa tenuiflora, Peganum harmala, Voacanga africana, Ayahuasca	harmaline, harmine, harmalol, norharmane, harmane, harmol, tetrahydro-harmine, and tryptamine	58 mMol L^{-1} ammonium formate and 1.01 mol L^{-1} acetic acid in acetonitrile	NACE–MS	0.01, 0.01, 0.015, 0.012, 0.018, 0.019, 0.022 and 0.024		[140]

Table 5. Cont.

Sample	Analytes	BGE	Method	LOQ (µg mL^{-1})	Remarks	Ref.
Rheum (Rhubarb, Dahuang) extracts	physcion, chrysophanol, and aloe–emodin	80% methanol and 20% acetonitrile with 20 mM ammonium acetate	NACE–ESI–MS/MS	nd, LOD:84, 180 and 210 ppb		[141]
Stephaniae tetrandrae radix and *Menispermum dauricum rhizoma*	tetrandrine, fangchinoline, and sinomenine	80 mM solution of ammonium acetate with mixture of 70% methanol, 20% ACN, and 10% water, which also contained 1% acetic acid	NACE–IT–MS	nd, LOD: 0.05, 0.08, and 0.15		[142]
Tinosporae radix	palmatine, cepharanthine, menisperine, magnoflorine, columbin and 20–hydroxy-ecdysone	methanol and acetonitrile (4:1; v/v), which contained 40 or 50 mM ammonium acetate and 0.5% acetic acid	NACE–ESI–MS	0.06–4.0		[2]

nd—no data, CE–Q–TOF–MS—capillary electrophoresis coupled with quadrupole time-of-flight mass spectrometry, ESI–IT–MS—electrospray ionization ion trap mass spectrometry, ESI–MS—electrospray ionization mass spectrometry, IT–MS—ion trap mass spectrometry.

3. Materials and Methods

The present literature review is based on PRISMA guidelines. The selection criteria for the articles for the review were formulated on the basis of the PICOS process (see Table 6). For the purpose of this review, articles from 2010 to 2019 were taken into consideration. Searching of the literature for this publication was performed between January 2021 and March 2021 using the PubMed and Web of Science databases. The search strategy took place with the use of the following keywords:

1. "capillary electrophoresis" AND
2. "pharmaceutical analysis" OR "determination" OR "quantification" AND
3. "herbal drugs" OR "medicinal plants" OR "plant extracts" OR "plant metabolites".

Table 6. PICOS (Population, Intervention, Comparison, Outcome, Study type).

	Inclusion Criteria	Exclusion Criteria
Population	herbal drugs and medicinal plants	garden and ornamental plants, vegetables and fruits, edible products, beverages
Intervention	use of CE method	other methods
Comparison	capillary electrophoresis vs. other methods	not applicable
Outcome	analysis of active constituents	different outcomes
Study type	original research articles, full articles, English language	review articles, reports, abstracts, articles with no quantitative information or details

In the PubMed database, a combination of terms 'All fields' and in Web of Science base terms 'Topic' was used, which searches titles, abstracts, author keywords, keywords, and more. Only articles in English, available full texts and articles delineating the quantitative analysis of bioactive components in medicinal plants and pharmaceutical formulations by CE are included in this review. The exclusion criteria were opinion letters, conferences, abstracts, and papers not written in English (for example, in Chinese). Publications restricted only to fingerprinting or separation without quantitative analysis were rejected. Additionally, articles with urine, human plasma and blood serum, and edible products such as the matrix were eliminated. Studies in which amino acids in plant tissues, enzyme inhibition or alternations of secondary metabolites in plants under different factors analyzed using CE were also not taken into account. Duplicates were removed and found articles were sorted by title, abstract, and then main text. The articles were excluded if they did not meet the inclusion criteria. Selection of appropriate works taking into account the inclusion and exclusion criteria were controlled by the three authors of this paper (M. G., A. P, M. K.) Selection of the publications by them was made on the basis of a qualitative and quantitative evaluation of articles from the PubMed and Web of Science databases, especially by title of paper, first name of the author, and year publication.

4. Conclusions

The present review summarized the state of the art applications of capillary electrophoresis over a past decade. The versatile application of CE-based methods was recorded due to the possibility of using different techniques of CE adapted to the substance to be determined and their numerous modifications.

The present scrutiny reveals a large number of applications, including different formulations, various plant extracts, simultaneous identification, and quantification of even several active constituents in a complex matrix. In the reported works, CZE, MEKC, and NACE were successfully used for the assay of different classes of secondary metabolites, whereas NACE was employed for the analysis of lipophilic compounds, CEC for the analysis of coumarins and CE-MS mainly for alkaloid compounds. Due to its many advantages, such as little solvent and sample consumption, short time of operation, and high efficiency, CE is an attractive and eco-friendly approach in current pharmaceutical analysis and its continuous development gives hope for well-established, validated, and increasingly accurate and precise methods of quality control of pharmaceutical formulations and herbal raw material.

Among all reported methods, the most common is the CE–UV technique; however, in some cases, resolution and sensitivity are limited. For this reason, other methods of detection, such as conductometry, electrochemiluminescence, laser-induced fluorescence, and hyphenation of CE–MS, have successfully been applied. Moreover, among other electrophoretic techniques, MEKC and NACE are well established. Other ways to solve this problem are through the addition of some modifiers to the BGE, i.e., cyclodextrins are added as a chiral selectors during enantiomeric separation, or through introducing sophisticated extraction and/or pre-concentration techniques. The flexibility of CE is a great advantage, i.e., it includes many opportunities for optimizing the parameters of analysis, additives to BGE, introducing in-line and online preconcentration techniques and different methods of detection, and makes every electrophoretic technique capable of being used for the routine qualitative and quantitative analysis of active constituents in plant material or herbal formulation even at the ng mL^{-1} level. In some cases, a comparison of the results obtained with CE to HPLC methods exhibited no statistically significant differences. Moreover, differences in sensitivity are relevant only in the analysis of samples with very low analyte concentrations, which does not directly relate to pharmaceutical analysis, where the content of the active ingredient, for instance in tablets, is at the milligram level. This suggests the CE method may be better where it does not influence the quality of the analysis, because of its shorter time of execution, lower costs, and eco-friendly approach. It should also be noted that the future of CE is strongly connected to hyphenation with the MS technique because of its ability for both measuring molecular weight and for offering structural information. On the one hand, detection sensitivities of the reported methods based on CE-MS were relatively low, but in some cases, they were comparable to results achieved even with UV detection. On the other hand, more and more utilizations of CE-MS, as well as a constant development, indicate that this hyphenation is heading in the right direction.

In conclusion, CE is a powerful analytical tool, and after adequate optimization, it could be an auspicious alternative to more expensive methods in the pharmaceutical quality control of herbal drugs and herbal raw material.

Author Contributions: The article was prepared by all authors. Conceptualization: M.G., A.P.; methodology: M.G., S.K., M.K.; investigation: M.G., A.P., K.M.-G., A.B.; supervision: S.K., M.K.; visualization: A.P., K.M.-G.; writing—original draft: M.G.; writing—review and editing: A.P., K.M.-G., A.B. All authors have read and agreed to the published version of the manuscript.

Funding: This research received no external funding.

Data Availability Statement: All data obtained during the research appears in the submitted article.

Conflicts of Interest: The authors declare no conflict of interest.

References

1. Barnes, J.; Anderson, L.A.; Phillipson, J.D. *Herbal Medicine*, 3rd ed.; Pharmaceutical Press: London, UK, 2007.
2. Liu, Y.; Zhou, W.; Mao, Z.; Liao, X.; Chen, Z. Analysis of six active components in Radix tinosporae by nonaqueous capillary electrophoresis with mass spectrometry. *J. Sep. Sci.* **2017**, *40*, 4628–4635. [CrossRef] [PubMed]
3. Li, F.; Weng, J. Demystifying traditional herbal medicine with. *Nat. Publ. Gr.* **2017**, *3*, 1–7. [CrossRef]
4. Folashade, O.; Omoregie, H.; Ochogu, P. Standardization of herbal medicines-A review. *Int. J. Biodivers. Conserv.* **2012**, *4*, 101–112. [CrossRef]
5. European Medicines Agency. *The EU regulatory system for medicines. A consistent approach to medicines regulation across the European Union*; EMA: London, UK, 2016; pp. 1–6.
6. The FDA's Drug Review Process: Ensuring Drugs Are Safe and Effective. Available online: https://www.fda.gov/drugs/information-consumers-and-patients-drugs/fdas-drug-review-process-ensuring-drugs-are-safe-and-effective (accessed on 19 March 2021).
7. Viana, C.; Ferreira, M.; Romero, C.S.; Bortoluzzi, M.R.; Lima, F.O.; Rolim, C.M.B.; De Carvalho, L.M. A capillary zone electrophoretic method for the determination of hypoglycemics as adulterants in herbal formulations used for the treatment of diabetes. *Anal. Methods* **2013**, *5*, 2126–2133. [CrossRef]
8. Johnson, R.T.; Lunte, C.E. A capillary electrophoresis electrospray ionization-mass spectrometry method using a borate background electrolyte for the fingerprinting analysis of flavonoids in Ginkgo biloba herbal supplements. *Anal. Methods* **2016**, *8*, 3325–3332. [CrossRef]
9. Moreira, A.P.L.; Motta, M.J.; Dal Molin, T.R.; Viana, C.; de Carvalho, L.M. Determination of diuretics and laxatives as adulterants in herbal formulations for weight loss. *Food Addit. Contam. Part A Chem. Anal. Control. Expo. Risk Assess.* **2013**, *30*, 1230–1237. [CrossRef] [PubMed]
10. Chen, J.; Zhu, H.; Chu, V.M.; Jang, Y.S.; Son, J.Y.; Kim, Y.H.; Son, C.G.; Seol, I.C.; Kang, J.S. Quality control of a herbal medicinal preparation using high-performance liquid chromatographic and capillary electrophoretic methods. *J. Pharm. Biomed. Anal.* **2011**, *55*, 206–210. [CrossRef]
11. Feng, A.; Tian, B.; Hu, J.; Zhou, P. Recent Applications of Capillary Electrophoresis in the Analysis of Traditional Chinese Medicines. *Comb. Chem. High. Throughput Screen.* **2010**, *13*, 954–965. [CrossRef] [PubMed]
12. Anubala, S.; Sekar, R.; Nagaiah, K. Determination of Curcuminoids and Their Degradation Products in Turmeric (*Curcuma longa*) Rhizome Herbal Products by Non-aqueous Capillary Electrophoresis with Photodiode Array Detection. *Food Anal. Methods* **2016**, *9*, 2567–2578. [CrossRef]
13. Peng, Y.Y. Study on capillary electrophoresis-amperometric detection profiles from propolis and its medicinal preparations. *Adv. Mater. Res.* **2013**, *750–752*, 1617–1620. [CrossRef]
14. Tubaon, R.M.S.; Rabanes, H.; Haddad, P.R.; Quirino, J.P. Capillary electrophoresis of natural products: 2011–2012. *Electrophoresis* **2014**, *35*, 190–204. [CrossRef]
15. Zalewska, M.; Wilk, K.; Milnerowicz, H. Review capillary electrophoresis application in the analysis of the anti-cancer drugs impurities. *Acta Pol. Pharm. Drug Res. Drug Res.* **2013**, *70*, 171–180.
16. Jorgenson, J.W.; DeArman Lukacs, K. Zone electrophoresis in open-tubular glass capillaries: Preliminary data on performance. *J. High. Resolut. Chromatogr.* **1981**, *4*, 230–231. [CrossRef]
17. Hurtado-Fernández, E.; Gómez-Romero, M.; Carrasco-Pancorbo, A.; Fernández-Gutiérrez Alberto, A. Application and potential of capillary electroseparation methods to determine antioxidant phenolic compounds from plant food material. *J. Pharm. Biomed. Anal.* **2010**, *53*, 1130–1160. [CrossRef] [PubMed]
18. Li, S.F.Y. Chapter 1 Introduction. In *Capillary Electrophoresis—Principles, Practice and Applications*; Elsevier: Amsterdam, The Netherlands; London, UK; New York, NY, USA; Tokyo, Japan, 1992; pp. 1–30.
19. Altria, K.D. *Capillary Electrophoresis Guidebook*; Humana Press: Totowa, NJ, USA, 1995; Volume 52, ISBN 0-89603-315-5.

20. Honegr, J.; Pospíšilová, M. Determination of phenolic acids in plant extracts using CZE with on-line transient isotachophoretic preconcentration. *J. Sep. Sci.* **2013**, *36*, 729–735. [CrossRef]
21. Liang, J.; Gong, F.Q.; Sun, H.M. Simultaneous separation of eight lignans in Forsythia suspensa by β-cyclodextrin-modified capillary zone electrophoresis. *Molecules* **2018**, *23*, 514. [CrossRef] [PubMed]
22. Xiao, M.W.; Bai, X.L.; Liu, Y.M.; Yang, L.; Hu, Y.D.; Liao, X. Rapid quantification of aloin A and B in aloe plants and aloe-containing beverages, and pharmaceutical preparations by microchip capillary electrophoresis with laser induced fluorescence detection. *J. Sep. Sci.* **2018**, *41*, 3772–3781. [CrossRef]
23. Xiang, Q.; Gao, Y.; Han, B.; Li, J.; Xu, Y.; Yin, J. Determination of arecoline in areca nut based on field amplification in capillary electrophoresis coupled with electrochemiluminescence detection. *Luminescence* **2013**, *28*, 50–55. [CrossRef] [PubMed]
24. Song, J.Z.; Han, Q.B.; Qiao, C.F.; But, P.P.H.; Xu, H.X. Development and validation of a rapid capillary zone electrophoresis method for the determination of aconite alkaloids in aconite roots. *Phytochem. Anal.* **2010**, *21*, 137–143. [CrossRef]
25. de Jong, G. Detection in Capillary Electrophoresis-An Introduction. In *Capillary Electrophoresis-Mass Spectrometry (CE-MS): Principles and Applications*; Wiley-VCH Verlag GmbH & Co. KGaA: Weinheim, Germany, 2016; pp. 1–5.
26. Jarvas, G.; Guttman, A.; Miękus, N.; Bączek, T.; Jeong, S.; Chung, D.S.; Pátoprstý, V.; Masár, M.; Hutta, M.; Datinská, V.; et al. Practical sample pretreatment techniques coupled with capillary electrophoresis for real samples in complex matrices. *TrAC-Trends Anal. Chem.* **2020**, *122*, 115702. [CrossRef]
27. Mao, H.; Ye, X.; Chen, W.; Geng, W.; Chen, G. Fabrication of carbon nanotube-polylactic acid composite electrode by melt compounding for capillary electrophoretic determination of tectoridin and irigenin in Belamcandae Rhizoma. *J. Pharm. Biomed. Anal.* **2019**, *175*, 112769. [CrossRef]
28. Zhang, L.; Zhang, W.; Chen, G. Determination of arbutin and bergenin in Bergeniae Rhizoma by capillary electrophoresis with a carbon nanotube-epoxy composite electrode. *J. Pharm. Biomed. Anal.* **2015**, *115*, 323–329. [CrossRef]
29. Sheng, S.; Liu, S.; Zhang, L.; Chen, G. Graphene/poly(ethylene-co-vinyl acetate) composite electrode fabricated by melt compounding for capillary electrophoretic determination of flavones in *Cacumen platycladi*. *J. Sep. Sci.* **2013**, *36*, 721–728. [CrossRef]
30. Ren, T.; Xu, Z. Study of isomeric pentacyclic triterpene acids in traditional Chinese medicine of Forsythiae Fructus and their binding constants with β-cyclodextrin by capillary electrophoresis. *Electrophoresis* **2018**, *39*, 1006–1013. [CrossRef]
31. Deng, D.; Deng, H.; Zhang, L.; Su, Y. Determination of ephedrine and pseudoephedrine by field-amplified sample injection capillary electrophoresis. *J. Chromatogr. Sci.* **2014**, *52*, 357–362. [CrossRef]
32. Honegr, J.; Šafra, J.; Polášek, M.; Pospíšilová, M. Large-volume sample stacking with polarity switching in CE for determination of natural polyphenols in plant extracts. *Chromatographia* **2010**, *72*, 885–891. [CrossRef]
33. Magnuszewska, J.; Krogulec, T. Application of hot platinum microelectrodes for determination of flavonoids in flow injection analysis and capillary electrophoresis. *Anal. Chim. Acta* **2013**, *786*, 39–46. [CrossRef]
34. Du, J.-X.; Wang, M. Capillary Electrophoresis Determination of Berberine in Pharmaceuticals with End-Column Electrochemiluminescence Detection. *J. Chin. Chem. Soc.* **2010**, *57*, 696–700. [CrossRef]
35. Dresler, S.; Kováčik, J.; Strzemski, M.; Sowa, I.; Wójciak-Kosior, M. Methodological aspects of biologically active compounds quantification in the genus Hypericum. *J. Pharm. Biomed. Anal.* **2018**, *155*, 82–90. [CrossRef] [PubMed]
36. Hou, Z.; Sun, G.; Guo, Y.; Yang, F.; Gong, D. *Capillary Electrophoresis Fingerprints Combined with Linear Quantitative Profiling Method to Monitor the Quality Consistency and Predict the Antioxidant Activity of Alkaloids of Sophora flavescens*; Elsevier Ltd.: Amsterdam, The Netherland, 2019; Volume 1133, ISBN 8602423986286.
37. Olabi, M.; Stein, M.; Wätzig, H. Affinity capillary electrophoresis for studying interactions in life sciences. *Methods* **2018**, *146*, 76–92. [CrossRef]
38. He, X.; Ding, Y.; Li, D.; Lin, B. Recent advances in the study of biomolecular interactions by capillary electrophoresis. *Electrophoresis* **2004**, *25*, 697–711. [CrossRef] [PubMed]
39. Wehr, T. Capillary Zone Electrophoresis. In *Encyclopedia of Physical Science and Technology*; Elsevier: Amsterdam, The Netherland, 2003; pp. 355–368.
40. Steiner, F.; Hassel, M. Nonaqueous capillary electrophoresis: A versatile completion of electrophoretic separation techniques. *Electrophoresis* **2000**, *21*, 3994–4016. [CrossRef]
41. Huie, C.W. Effects of organic solvents on sample pretreatment and separation performances in capillary electrophoresis. *Electrophoresis* **2003**, *24*, 1508–1529. [CrossRef]
42. Beckers, J.L.; Ackermans, M.T.; Boček, P. Capillary zone electrophoresis in methanol: Migration behavior and background electrolytes. *Electrophoresis* **2003**, *24*, 1544–1552. [CrossRef]
43. Gao, Z.; Zhong, W. Recent (2018–2020) development in capillary electrophoresis. *Anal. Bioanal. Chem.* **2021**. [CrossRef]
44. Xu, L.; Chang, R.; Chen, M.; Li, L.; Huang, Y.; Zhang, H.; Chen, A. Quality evaluation of Guan-Xin-Ning injection based on fingerprint analysis and simultaneous separation and determination of seven bioactive constituents by capillary electrophoresis. *Electrophoresis* **2017**, *38*, 3168–3176. [CrossRef] [PubMed]
45. Alzoman, N.Z.; Maher, H.M.; Al-Showiman, H.; Fawzy, G.A.; Al-Taweel, A.M.; Perveen, S.; Tareen, R.B.; Al-Sabbagh, R.M. CE-DAD determination of scutellarein and caffeic acid in *Abelia triflora* crude extract. *J. Chromatogr. Sci.* **2018**, *56*, 746–752. [CrossRef] [PubMed]

46. Bao, Y.; Yang, F.; Yang, X. CE-electrochemiluminescence with ionic liquid for the facile separation and determination of diester-diterpenoid aconitum alkaloids in traditional Chinese herbal medicine. *Electrophoresis* **2011**, *32*, 1515–1521. [CrossRef] [PubMed]
47. Dutra, L.S.; Leite, M.N.; Brandão, M.A.F.; De Almeida, P.A.; Vaz, F.A.S.; De Oliveira, M.A.L. A rapid method for total β-escin analysis in dry, hydroalcoholic and hydroglycolic extracts of *Aesculus hippocastanum* L. by capillary zone electrophoresis. *Phytochem. Anal.* **2013**, *24*, 513–519. [CrossRef] [PubMed]
48. Kubrak, T.; Dresler, S.; Szymczak, G.; Bogucka-Kocka, A. Rapid Determination of Coumarins in Plants by Capillary Electrophoresis. *Anal. Lett.* **2015**, *48*, 2819–2832. [CrossRef]
49. Chen, M.; Huang, Y.; Xu, L.; Zhang, H.; Zhang, G.; Chen, A. Simultaneous separation and analysis of camptothecin alkaloids in real samples by large-volume sample stacking in capillary electrophoresis. *Biomed. Chromatogr.* **2018**, *32*. [CrossRef] [PubMed]
50. Wang, N.; Su, M.; Liang, S.; Sun, H. Investigation of six bioactive anthraquinones in slimming tea by accelerated solvent extraction and high performance capillary electrophoresis with diode-array detection. *Food Chem.* **2016**, *199*, 1–7. [CrossRef]
51. Roda, G.; Liberti, V.; Arnoldi, S.; Argo, A.; Rusconi, C.; Suardi, S.; Gambaro, V. Capillary electrophoretic and extraction conditions for the analysis of Catha edulis FORKS active principles. *Forensic Sci. Int.* **2013**, *228*, 154–159. [CrossRef]
52. Şanli, S.; Lunte, C. Determination of eleven flavonoids in chamomile and linden extracts by capillary electrophoresis. *Anal. Methods* **2014**, *6*, 3858–3864. [CrossRef]
53. Kulp, M.; Bragina, O.; Kogerman, P.; Kaljurand, M. Capillary electrophoresis with led-induced native fluorescence detection for determination of isoquinoline alkaloids and their cytotoxicity in extracts of *Chelidonium majus* L. *J. Chromatogr. A* **2011**, *1218*, 5298–5304. [CrossRef] [PubMed]
54. Zhang, L.; Zhang, W.; Chen, W.; Chen, G. Simultaneous determination of five bioactive constituents in Rhizoma Chuanxiong by capillary electrophoresis with a carbon nanotube-polydimethylsiloxane composite electrode. *J. Pharm. Biomed. Anal.* **2016**, *131*, 107–112. [CrossRef]
55. Diop, E.H.A.; Jacquat, J.; Drouin, N.; Queiroz, E.F.; Wolfender, J.L.; Diop, T.; Schappler, J.; Rudaz, S. Quantitative CE analysis of punicalagin in Combretum aculeatum extracts traditionally used in Senegal for the treatment of tuberculosis. *Electrophoresis* **2019**, *40*, 2820–2827. [CrossRef]
56. Müller, L.S.; Da Silveira, G.D.; Dal Prá, V.; Lameira, O.; Viana, C.; De Carvalho, L.M. Investigation of phenolic antioxidants as chemical markers in extracts of *Connarus perrottetii* var. *Angustifolius radlk* by capillary zone electrophoresis. *J. Liq. Chromatogr. Relat. Technol.* **2016**, *39*, 13–20. [CrossRef]
57. Deng, Y.; Lam, S.C.; Zhao, J.; Li, S.P. Quantitative analysis of flavonoids and phenolic acid in *Coreopsis tinctoria* Nutt. by capillary zone electrophoresis. *Electrophoresis* **2017**, *38*, 2654–2661. [CrossRef] [PubMed]
58. Ye, N.; Li, J.; Gao, C.; Xie, Y. Simultaneous determination of atropine, scopolamine, and anisodamine in Flos daturae by capillary electrophoresis using a capillary coated by graphene oxide. *J. Sep. Sci.* **2013**, *36*, 2698–2702. [CrossRef]
59. Lü, W.; Li, M.; Chen, Y.; Chen, H.; Chen, X. Simultaneous determination of iridoid glycosides and flavanoids in *Lamionphlomis rotate* and its herbal preparation by a simple and rapid capillary zone electrophoresis method. *Drug Test. Anal.* **2012**, *4*, 123–128. [CrossRef]
60. Dresler, S.; Kubrak, T.; Bogucka-Kocka, A.; Szymczak, G. Determination of shikonin and rosmarinic acid in Echium vulgare L. and Echium russicum J.F. Gmel. by capillary electrophoresis. *J. Liq. Chromatogr. Relat. Technol.* **2015**, *38*, 698–701. [CrossRef]
61. Xie, J.P.; Xiang, J.M.; Zhu, Z.L. Determination of Five Major 8-Prenylflavones in Leaves of Epimedium by Solid-Phase Extraction Coupled with Capillary Electrophoresis. *J. Chromatogr. Sci.* **2016**, *54*, 664–669. [CrossRef] [PubMed]
62. Yang, X.; Zhang, S.; Wang, J.; Wang, C.; Wang, Z. On-line two-step stacking in capillary zone electrophoresis for the preconcentration of strychnine and brucine. *Anal. Chim. Acta* **2014**, *814*, 63–68. [CrossRef]
63. Xia, Y.G.; Liang, J.; Yang, B.Y.; Wang, Q.H.; Kuang, H.X. A new method for quantitative determination of two uronic acids by CZE with direct UV detection. *Biomed. Chromatogr.* **2011**, *25*, 1030–1037. [CrossRef] [PubMed]
64. Liu, L.; You, W.; Zheng, L.; Chen, F.; Jia, Z. Determination of peimine and peiminine in Bulbus Fritillariae Thunbergii by capillary electrophoresis by indirect UV detection using N-(1-naphthyl)ethylenediamine dihydrochloride as probe. *Electrophoresis* **2012**, *33*, 2152–2158. [CrossRef] [PubMed]
65. Abhijith, B.L.; Mohan, M.; Joseph, D.; Haleema, S.; Aboul-Enein, H.Y.; Ibnusaud, I. Capillary zone electrophorsis for the analysis of naturally occurring 2-hydroxycitric acids and their lactones. *J. Sep. Sci.* **2017**, *40*, 3351–3357. [CrossRef]
66. Wang, X.; Li, J.; Qu, W.; Chen, G. Fabrication of graphene/poly(methyl methacrylate) composite electrode for capillary electrophoretic determination of bioactive constituents in *Herba geranii*. *J. Chromatogr. A* **2011**, *1218*, 5542–5548. [CrossRef]
67. Wang, J.; Wang, H.; Han, S. Ultrasensitive determination of epicatechin, rutin, and quercetin by capillary electrophoresis chemiluminescence. *Acta Chromatogr.* **2012**, *24*, 679–688. [CrossRef]
68. Chen, H.; Ding, X.; Wang, M.; Chen, X. An automated method of on-line extraction coupled with flow injection and capillary electrophoresis for phytochemical analysis. *J. Chromatogr. Sci.* **2010**, *48*, 866–870. [CrossRef] [PubMed]
69. Lu, Y.; Wang, X.; Chen, D.; Chen, G. Polystyrene/graphene composite electrode fabricated by in situ polymerization for capillary electrophoretic determination of bioactive constituents in *Herba houttuyniae*. *Electrophoresis* **2011**, *32*, 1906–1912. [CrossRef]
70. Gao, S.Y.; Li, H.; Wang, L.; Yang, L.N. Simultaneous separation and determination of benzoic acid compounds in the plant medicine by high performance capillary electrophoresis. *J. Chin. Chem. Soc.* **2010**, *57*, 1374–1380. [CrossRef]
71. Maráková, K.; Mikuš, P.; Piešťanský, J.; Havránek, E. Determination of curcuminoids in substances and dosage forms by cyclodextrin-mediated capillary electrophoresis with diode array detection. *Chem. Pap.* **2011**, *65*, 398–405. [CrossRef]

72. Dong, S.; Gao, R.; Yang, Y.; Guo, M.; Ni, J.; Zhao, L. Simultaneous determination of phenylethanoid glycosides and aglycones by capillary zone electrophoresis with running buffer modifier. *Anal. Biochem.* **2014**, *449*, 158–163. [CrossRef]
73. Deng, B.; Xie, F.; Li, L.; Shi, A.; Liu, Y.; Yin, H. Determination of galanthamine in Bulbus Lycoridis Radiatae by coupling capillary electrophoresis with end-column electrochemiluminescence detection. *J. Sep. Sci.* **2010**, *33*, 2356–2360. [CrossRef]
74. Sun, S.; Wei, Y.; Cao, Y.; Deng, B. Simultaneous electrochemiluminescence determination of galanthamine, homolycorine, lycorenine, and tazettine in *Lycoris radiata* by capillary electrophoresis with ultrasonic-assisted extraction. *J. Chromatogr. B Anal. Technol. Biomed. Life Sci.* **2017**, *1055–1056*, 15–19. [CrossRef] [PubMed]
75. Fu, Y.; Zhang, L.; Chen, G. Determination of carbohydrates in Folium Lysium Chinensis using capillary electrophoresis combined with far-infrared light irradiation-assisted extraction. *J. Sep. Sci.* **2011**, *34*, 3272–3278. [CrossRef] [PubMed]
76. Sun, Y.; Li, Y.; Zeng, J.; Lu, Q.; Li, P.C.H. Microchip electrophoretic separation and fluorescence detection of chelerythrine and sanguinarine in medicinal plants. *Talanta* **2015**, *142*, 90–96. [CrossRef] [PubMed]
77. Han, P.; Luan, F.; Yan, X.; Gao, Y.; Liu, H. Separation and determination of honokiol and magnolol in Chinese traditional medicines by capillary electrophoresis with the application of response surface methodology and radial basis function neural network. *J. Chromatogr. Sci.* **2012**, *50*, 71–75. [CrossRef]
78. Gatea, F.; Teodor, E.D.; Matei, A.O.; Badea, G.I.; Radu, G.L. Capillary Electrophoresis Method for 20 Polyphenols Separation in Propolis and Plant Extracts. *Food Anal. Methods* **2015**, *8*, 1197–1206 [CrossRef]
79. Tascón, M.; Benavente, F.; Vizioli, N.M.; Gagliardi, L.G. A rapid and simple method for the determination of psychoactive alkaloids by CE-UV: Application to Peganum Harmala seed infusions. *Drug Test. Anal.* **2017**, *9*, 596–602. [CrossRef] [PubMed]
80. Fakhari, A.R.; Nojavan, S.; Ebrahimi, S.N.; Evenhuis, C.J. Optimized ultrasound-assisted extraction procedure for the analysis of opium alkaloids in Papaver plants by cyclodextrin-modified capillary electrophoresis. *J. Sep. Sci.* **2010**, *33*, 2153–2159. [CrossRef]
81. Zhang, Y.; Li, Y.; Chen, Z. Selective and sensitive determination of protoberberines by capillary electrophoresis coupled with molecularly imprinted microextraction. *J. Sep. Sci.* **2015**, *38*, 3969–3975. [CrossRef]
82. de Carvalho, L.M.; Raabe, A.; Martini, M.; Sant'anna, C.S.; da Silveira, G.D.; do Nascimento, P.C.; Bohrer, D. Contactless Conductivity detection of 14 inorganic cations in mineral and phytotherapeutic formulations after capillary electrophoretic separation. *Electroanalysis* **2011**, *23*, 2574–2581. [CrossRef]
83. Deng, G.H.; Chen, S.; Wang, H.; Gao, J.; Luo, X.; Huang, H. Determination of active ingredients of *Phyllanthus urinaria* by capillary electrophoresis with amperometric detection. *J. Liq. Chromatogr. Relat. Technol.* **2012**, *35*, 2370–2380. [CrossRef]
84. Wan, D.; Han, Y.; Li, F.; Mao, H.; Chen, G. Far infrared-assisted removal of extraction solvent for capillary electrophoretic determination of the bioactive constituents in *Plumula nelumbinis*. *Electrophoresis* **2019**, *40*, 582–586. [CrossRef] [PubMed]
85. Zhu, Q.; Xu, X.; Huang, Y.; Xu, L.; Chen, G. Field enhancement sample stacking for analysis of organic acids in traditional Chinese medicine by capillary electrophoresis. *J. Chromatogr. A* **2012**, *1246*, 35–39. [CrossRef]
86. Guo, J.; Wang, M.; Guo, H.; Chang, R.; Yu, H.; Zhang, G.; Chen, A. Simultaneous separation and determination of seven isoflavones in *Radix puerariae* by capillary electrophoresis with a dual cyclodextrin system. *Biomed. Chromatogr.* **2019**, *33*. [CrossRef]
87. Chang, Y.X.; Liu, J.; Bai, Y.; Li, J.; Liu, E.W.; He, J.; Jiao, X.C.; Wang, Z.Z.; Gao, X.M.; Zhang, B.L.; et al. The activity-integrated method for quality assessment of Reduning injection by on-line DPPH-CE-DAD. *PLoS ONE* **2014**, *9*, e106254. [CrossRef]
88. Chen, B.; Zhang, L.; Chen, G. Determination of salidroside and tyrosol in Rhodiola by capillary electrophoresis with graphene/poly(urea-formaldehyde) composite modified electrode. *Electrophoresis* **2011**, *32*, 870–876. [CrossRef]
89. Ngoc, H.N.; Löffler, S.; Nghiem, D.T.; Pham, T.L.G.; Stuppner, H.; Ganzera, M. Phytochemical study of Rourea minor stems and the analysis of therein contained Bergenin and Catechin derivatives by capillary electrophoresis. *Microchem. J.* **2019**, *149*, 104063. [CrossRef]
90. Cao, J.; We, J.; Tian, K.; Su, H.; Wan, J.; Li, P. Simultaneous determination of seven phenolic acids in three Salvia species by capillary zone electrophoresis with β-cyclodextrin as modifier. *J. Sep. Sci.* **2014**, *37*, 3738–3744. [CrossRef] [PubMed]
91. Lu, Y.; Bai, H.; Kong, C.; Zhong, H.; Breadmore, M.C. Analysis of brazilin and protosappanin B in sappan lignum by capillary zone electrophoresis with acid barrage stacking. *Electrophoresis* **2013**, *34*, 3326–3332. [CrossRef] [PubMed]
92. Wang, Y.; Wei, Z.; Zhang, J.; Wang, X. Electrochemical determination of baicalein, baicalin and quercetin in Scutellaria barbata. *Int. J. Electrochem. Sci.* **2016**, *11*, 8323–8331. [CrossRef]
93. Ma, H.; Li, J.; An, M.; Gao, X.M.; Chang, Y.X. A powerful on line ABTS+-CE-DAD method to screen and quantify major antioxidants for quality control of Shuxuening Injection. *Sci. Rep.* **2018**, *8*, 1–10. [CrossRef] [PubMed]
94. Wang, H.; Lu, Y.; Chen, J.; Li, J.; Liu, S. Subcritical water extraction of alkaloids in *Sophora flavescens* Ait. and determination by capillary electrophoresis with field-amplified sample stacking. *J. Pharm. Biomed. Anal.* **2012**, *58*, 146–151. [CrossRef] [PubMed]
95. Gao, R.; Wang, L.; Yang, Y.; Ni, J.; Zhao, L.; Dong, S.; Guo, M. Simultaneous determination of oleanolic acid, ursolic acid, quercetin and apigenin in *Swertia mussotii* Franch by capillary zone electrophoresis with running buffer modifier. *Biomed. Chromatogr.* **2015**, *29*, 402–409. [CrossRef]
96. Maher, H.M.; Al-Zoman, N.Z.; Al-Shehri, M.M.; Al-Showiman, H.; Al-Taweel, A.M.; Fawzy, G.A.; Perveen, S. Determination of Luteolin and Apigenin in Herbs by Capillary Electrophoresis with Diode Array Detection. *Instrum. Sci. Technol.* **2015**, *43*, 611–625. [CrossRef]
97. Wang, W.; Lin, P.; Ma, L.; Xu, K.; Lin, X. Separation and determination of flavonoids in three traditional Chinese medicines by capillary electrophoresis with amperometric detection. *J. Sep. Sci.* **2016**, *39*, 1357–1362. [CrossRef]

98. Sereia, A.L.; Longhini, R.; Lopes, G.C.; de Mello, J.C.P. Capillary Electrophoresis as Tool for Diastereomeric Separation in a *Trichilia catigua* Fraction. *Phytochem. Anal.* **2017**, *28*, 144–150. [CrossRef]
99. Emara, S.; Masujima, T.; Zarad, W.; Mohamed, K.; Kamal, M.; Fouad, M.; EL-Bagary, R. Field-amplified sample stacking β-cyclodextrin modified capillary electrophoresis for quantitative determination of diastereomeric saponins. *J. Chromatogr. Sci.* **2014**, *52*, 1308–1316. [CrossRef]
100. Gufler, V.; Ngoc, H.N.; Stuppner, H.; Ganzera, M. Capillary electrophoresis as a fast and efficient alternative for the analysis of *Urceola rosea* leaf extracts. *Fitoterapia* **2018**, *125*, 1–5. [CrossRef]
101. Li, W.L.; Li, M.J.; Pan, Y.L.; Huang, B.K.; Chu, Q.C.; Ye, J.N. Study on electrochemical profiles of *Valeriana medicinal* plants by capillary electrophoresis. *J. Anal. Chem.* **2014**, *69*, 179–186. [CrossRef]
102. Uzaşçi, S.; Erim, F.B. Enhancement of native fluorescence intensity of berberine by (2-hydroxypropyl)-β-cyclodextrin in capillary electrophoresis coupled by laser-induced fluorescence detection: Application to quality control of medicinal plants. *J. Chromatogr. A* **2014**, *1338*, 184–187. [CrossRef]
103. Ma, D.; Yang, L.; Yan, B.; Sun, G. Capillary electrophoresis fingerprints combined with chemometric methods to evaluate the quality consistency and predict the antioxidant activity of Yinqiaojiedu tablet. *J. Sep. Sci.* **2017**, *40*, 1796–1804. [CrossRef]
104. Hancu, G.; Simon, B.; Rusu, A.; Mircia, E.; Gyéresi, Á. Principles of micellar electrokinetic capillary chromatography applied in pharmaceutical analysis. *Adv. Pharm. Bull.* **2013**, *3*, 1–8. [CrossRef] [PubMed]
105. Yang, H.; Ding, Y.; Gao, W.; Qi, L.W.; Cao, J.; Li, P. Efficient separation of tanshinones by polyvinylpyrrolidone-stabilized graphene-modified micellar electrokinetic chromatography. *Electrophoresis* **2015**, *36*, 2874–2880. [CrossRef]
106. Cao, J.; Li, P.; Yi, L. Ionic liquids coated multi-walled carbon nanotubes as a novel pseudostationary phase in electrokinetic chromatography. *J. Chromatogr. A* **2011**, *1218*, 9428–9434. [CrossRef] [PubMed]
107. Chang, Y.Q.; Tan, S.N.; Yong, J.W.H.; Ge, L. Surfactant-assisted pressurized liquid extraction for determination of flavonoids from Costus speciosus by micellar electrokinetic chromatography. *J. Sep. Sci.* **2011**, *34*, 462–468. [CrossRef] [PubMed]
108. Chang, Y.H.; Huang, C.W.; Fu, S.F.; Wu, M.Y.; Wu, T.; Lin, Y.W. Determination of salicylic acid using a magnetic iron oxide nanoparticle-based solid-phase extraction procedure followed by an online concentration technique through micellar electrokinetic capillary chromatography. *J. Chromatogr. A* **2017**, *1479*, 62–70. [CrossRef] [PubMed]
109. Matei, A.O.; Gatea, F.; Teodor, E.D.; Radu, G.L. Tannins analysis from different medicinal plants extracts using MALDI-TOF and MEKC. *Chem. Pap.* **2016**, *70*, 515–522. [CrossRef]
110. Huang, X.; Yi, L.; Gao, Z.; Li, H. Determination of Seven Active Ingredients in Three Plant Essential Oils by Using Micellar Electrokinetic Chromatography. *Anal. Lett.* **2012**, *45*, 2014–2025. [CrossRef]
111. Wei, M.; Chu, C.; Wang, S.; Yan, J. Quantitative analysis of sesquiterpenes and comparison of three Curcuma wenyujin herbal medicines by micro matrix solid phase dispersion coupled with MEEKC. *Electrophoresis* **2018**, *39*, 1119–1128. [CrossRef] [PubMed]
112. Fiori, J.; Leoni, A.; Fimognari, C.; Turrini, E.; Hrelia, P.; Mandrone, M.; Iannello, C.; Antognoni, F.; Poli, F.; Gotti, R. Determination of Phytomarkers in Pharmaceutical Preparations of *Hemidesmus indicus* Roots by Micellar Electrokinetic Chromatography and High-Performance Liquid Chromatography–Mass Spectrometry. *Anal. Lett.* **2014**, *47*, 2629–2642. [CrossRef]
113. Dresler, S.; Bogucka-Kocka, A.; Kováčik, J.; Kubrak, T.; Strzemski, M.; Wójciak-Kosior, M.; Rysiak, A.; Sowa, I. Separation and determination of coumarins including furanocoumarins using micellar electrokinetic capillary chromatography. *Talanta* **2018**, *187*, 120–124. [CrossRef]
114. Lao, K.M.; Han, D.Q.; Chen, X.J.; Zhao, J.; Wang, T.J.; Li, S. ping Simultaneous determination of seven hydrophilic bioactive compounds in water extract of *Polygonum multiflorum* using pressurized liquid extraction and short-end injection micellar electrokinetic chromatography. *Chem. Cent. J.* **2013**, *7*, 1–8. [CrossRef] [PubMed]
115. Stege, P.W.; Sombra, L.L.; Davicino, R.C.; Olsina, R.A. Analysis of nordihydroguaiaretic acid in *Larrea divaricata* Cav. extracts by micellar electrokinetic chromatography. *Phytochem. Anal.* **2011**, *22*, 74–79. [CrossRef]
116. Chen, S.; Sun, G.; Yang, L.; Zhang, J. Micellar electrokinetic chromatography fingerprinting combined with chemometrics as an efficient strategy for evaluating the quality consistency and predicting the antioxidant activity of *Lianqiao baidu* pills. *J. Sep. Sci.* **2017**, *40*, 2838–2848. [CrossRef]
117. Dobrecky, C.B.; Flor, S.A.; López, P.G.; Wagner, M.L.; Lucangioli, S.E. Development of a novel dual CD-MEKC system for the systematic flavonoid fingerprinting of *Ligaria cuneifolia* (R. et P.) Tiegh.—Loranthaceae—extracts. *Electrophoresis* **2017**, *38*, 1292–1300. [CrossRef]
118. Gomes, A.F.; Ganzera, M.; Schwaiger, S.; Stuppner, H.; Halabalaki, M.; Almeida, M.P.; Leite, M.F.; Amaral, J.G.; David, J.M. Simultaneous determination of iridoids, phenylpropanoids and flavonoids in *Lippia alba* extracts by micellar electrokinetic capillary chromatography. *Microchem. J.* **2018**, *138*, 494–500. [CrossRef]
119. Głowacki, R.; Furmaniak, P.; Kubalczyk, P.; Borowczyk, K. Determination of Total Apigenin in Herbs by Micellar Electrokinetic Chromatography with UV Detection. *J. Anal. Methods Chem.* **2016**, *2016*. [CrossRef] [PubMed]
120. Gonda, S.; Nguyen, N.M.; Batta, G.; Gyémánt, G.; Máthé, C.; Vasas, G. Determination of phenylethanoid glycosides and iridoid glycosides from therapeutically used *Plantago species* by CE-MEKC. *Electrophoresis* **2013**, *34*, 2577–2584. [CrossRef] [PubMed]
121. Öztekin, N.; Başkan, S.; Evrim Kepekçi, S.; Erim, F.B.; Topçu, G. Isolation and analysis of bioactive diterpenoids in Salvia species (*Salvia chionantha* and *Salvia kronenburgiii*) by micellar electrokinetic capillary chromatography. *J. Pharm. Biomed. Anal.* **2010**, *51*, 439–442. [CrossRef] [PubMed]

122. Cao, J.; Hu, J.; Wei, J.; Li, B.; Zhang, M.; Xiang, C.; Li, P. Optimization of micellar electrokinetic chromatography method for the simultaneous determination of seven hydrophilic and four lipophilic bioactive components in three salvia species. *Molecules* **2015**, *20*, 15304–15318. [CrossRef] [PubMed]
123. Ma, H.; Liu, T.; Li, J.; Ding, M.; Gao, X.M.; Chang, Y. xu the in-capillary- 2,2-azinobis-(3-ethylbenzothiazoline-6-sulfonic acid)-sweeping micellar electrokinetic chromatography-Diode array detector method for screening and quantifying trace natural antioxidants from *Schisandra chinensis*. *J. Chromatogr. A* **2019**, *1593*, 147–155. [CrossRef] [PubMed]
124. Riekkola, M.L.; Jussila, M.; Porras, S.P.; Valkó, I.E. Non-aqueous capillary electrophoresis. *J. Chromatogr. A* **2000**, *892*, 155–170. [CrossRef]
125. Hou, J.; Li, G.; Wei, Y.; Lu, H.; Jiang, C.; Zhou, X.; Meng, F.; Cao, J.; Liu, J. Analysis of five alkaloids using surfactant-coated multi-walled carbon nanotubes as the pseudostationary phase in nonaqueous capillary electrophoresis. *J. Chromatogr. A* **2014**, *1343*, 174–181. [CrossRef]
126. Yuan, B.; Zheng, C.; Teng, H.; You, T. Simultaneous determination of atropine, anisodamine, and scopolamine in plant extract by nonaqueous capillary electrophoresis coupled with electrochemiluminescence and electrochemistry dual detection. *J. Chromatogr. A* **2010**, *1217*, 171–174. [CrossRef]
127. Mistry, K.; Krull, I.; Grinberg, N. Capillary electrochromatography: An alternative to HPLC and CE. *J. Sep. Sci.* **2002**, *25*, 935–958. [CrossRef]
128. Svec, F. Recent developments in the field of monolithic stationary phases for capillary electrochromatography. *J. Sep. Sci.* **2005**, *28*, 729–745. [CrossRef]
129. de Jong, G. *Milestones in the Development of Capillary Electromigration Techniques*; Elsevier Inc.: Amsterdam, The Netherland, 2018; ISBN 9780128093757.
130. Viberg, P.; Spégel, P.; Carlstedt, J.; Jörntén-Karlsson, M.; Petersson, P. Continuous full filling capillary electrochromatography: Chromatographic performance and reproducibility. *J. Chromatogr. A* **2007**, *1154*, 386–389. [CrossRef] [PubMed]
131. Ping, G.; Schmitt-Kopplin, P.; Zhang, Y.; Baba, Y. Capillary electrochromatography and on-line concentration. *Methods Mol. Biol.* **2008**, *384*, 751–769. [CrossRef]
132. Yan, C.; Xue, Y.; Wang, Y. *Capillary Electrochromatography*; Elsevier Inc.: Amsterdam, The Netherland, 2018; ISBN 9780128093757.
133. Chen, Z.; Wang, J.; Chen, D.; Fan, G.; Wu, Y. Sodium desoxycholate-assisted capillary electrochromatography with methacrylate ester-based monolithic column on fast separation and determination of coumarin analogs in *Angelica dahurica* extract. *Electrophoresis* **2012**, *33*, 2884–2891. [CrossRef]
134. Wang, J.; Chen, D.; Chen, Z.; Fan, G.; Wu, Y. Fast separation and determination of coumarins in *Fructus cnidii* extracts by CEC using poly(butyl methacrylate-co-ethylene dimethacrylate-co-[2-(methacryloyloxy) ethyl] trimethylammonium chloride) monolithic columns. *J. Sep. Sci.* **2010**, *33*, 1099–1108. [CrossRef] [PubMed]
135. Chen, Q.; Li, N.; Zhang, W.; Chen, J.; Chen, Z. Simultaneous determination of vinblastine and its monomeric precursors vindoline and catharanthine in *Catharanthus roseus* by capillary electrophoresis-mass spectrometry. *J. Sep. Sci.* **2011**, *34*, 2885–2892. [CrossRef] [PubMed]
136. Liu, Y.; Zhou, W.; Mao, Z.; Chen, Z. Analysis of Evodiae Fructus by capillary electrochromatography-mass spectrometry with methyl-vinylimidazole functionalized organic polymer monolith as stationary phases. *J. Chromatogr. A* **2019**, *1602*, 474–480. [CrossRef] [PubMed]
137. Wang, Q.Y.; Dong, X.; Yang, J.; Zhen, X.T.; Ye, L.H.; Chu, C.; Wang, B.; Hu, Y.H.; Zheng, H.; Cao, J. Solid acids assisted matrix solid-phase dispersion microextraction of alkaloids by capillary electrophoresis coupled with quadrupole time-of-flight mass spectrometry. *J. Sep. Sci.* **2019**, *42*, 3579–3588. [CrossRef]
138. Zhang, Y.; Chen, Z. Nonaqueous CE ESI-IT-MS analysis of Amaryllidaceae alkaloids. *J. Sep. Sci.* **2013**, *36*, 1078–1084. [CrossRef]
139. Zhang, Y.; Chen, Z. Separation of isomeric bavachin and isobavachalcone in the Fructus Psoraleae by capillary electrophoresis-mass spectrometry. *J. Sep. Sci.* **2012**, *35*, 1644–1650. [CrossRef]
140. Posch, T.N.; Martin, N.; Pütz, M.; Huhn, C. Nonaqueous capillary electrophoresis-mass spectrometry: A versatile, straightforward tool for the analysis of alkaloids from psychoactive plant extracts. *Electrophoresis* **2012**, *33*, 1557–1566. [CrossRef]
141. Cheng, J.; Wang, L.; Liu, W.; Chen, D.D.Y. Quantitative Nonaqueous Capillary Electrophoresis-Mass Spectrometry Method for Determining Active Ingredients in Plant Extracts. *Anal. Chem.* **2017**, *89*, 1411–1415. [CrossRef] [PubMed]
142. Chen, Q.; Zhang, J.; Zhang, W.; Chen, Z. Analysis of active alkaloids in the Menispermaceae family by nonaqueous capillary electrophoresis-ion trap mass spectrometry. *J. Sep. Sci.* **2013**, *36*, 341–349. [CrossRef] [PubMed]

Article

Quantification and Metabolite Identification of Sulfasalazine in Mouse Brain and Plasma Using Quadrupole-Time-of-Flight Mass Spectrometry

Jangmi Choi, Min-Ho Park, Seok-Ho Shin, Jin-Ju Byeon, Byeong ill Lee, Yuri Park and Young G. Shin *

Institute of Drug Research and Development, College of Pharmacy, Chungnam National University, Daejeon 34134, Korea; jangmi.choi.cnu@gmail.com (J.C.); minho.park.cnu@gmail.com (M.-H.P.); seokho.shin.cnu@gmail.com (S.-H.S.); jinju.byeon.cnu@gmail.com (J.-J.B.); byungill.lee.cnu@gmail.com (B.i.L.); yuri.park.cnu@gmail.com (Y.P.)
* Correspondence: yshin@cnu.ac.kr; Tel.: +82-42-821-5931

Abstract: Sulfasalazine (SAS), an anti-inflammatory drug with potent cysteine/glutamate antiporter system xc-(SXC) inhibition has recently shown beneficial effects in brain-related diseases. Despite many reports related to central nervous system (CNS) effect of SAS, pharmacokinetics (PK) and metabolite identification studies in the brain for SAS were quite limited. The aim of this study was to investigate the pharmacokinetics and metabolite identification of SAS and their distributions in mouse brain. Using in vivo brain exposure studies (neuro PK), the PK parameters of SAS was calculated for plasma as well as brain following intravenous and oral administration at 10 mg/kg and 50 mg/kg in mouse, respectively. In addition, in vivo metabolite identification (MetID) studies of SAS in plasma and brain were also conducted. The concentration of SAS in brain was much lower than that in plasma and only 1.26% of SAS was detected in mouse brain when compared to the SAS concentration in plasma (brain to plasma ratio (%): 1.26). In the MetID study, sulfapyridine (SP), hydroxy-sulfapyridine (SP-OH), and N-acetyl sulfapyridine (Ac-SP) were identified in plasma, whereas only SP and Ac-SP were identified as significant metabolites in brain. As a conclusion, our results suggest that the metabolites of SAS such as SP and Ac-SP might be responsible for the pharmacological effect in brain, not the SAS itself.

Keywords: CNS; sulfasalazine; brain to plasma ratio; LC-ESI-TOF-MS

Citation: Choi, J.; Park, M.-H.; Shin, S.-H.; Byeon, J.-J.; Lee, B.i.; Park, Y.; Shin, Y.G. Quantification and Metabolite Identification of Sulfasalazine in Mouse Brain and Plasma Using Quadrupole-Time-of-Flight Mass Spectrometry. *Molecules* **2021**, *26*, 1179. https://doi.org/10.3390/molecules26041179

Academic Editors: Franciszek Główka and Marta Karaźniewicz-Łada

Received: 21 January 2021
Accepted: 18 February 2021
Published: 22 February 2021

Publisher's Note: MDPI stays neutral with regard to jurisdictional claims in published maps and institutional affiliations.

Copyright: © 2021 by the authors. Licensee MDPI, Basel, Switzerland. This article is an open access article distributed under the terms and conditions of the Creative Commons Attribution (CC BY) license (https://creativecommons.org/licenses/by/4.0/).

1. Introduction

Drug development for central nervous system (CNS) disorders has many hurdles in terms of brain penetrability due to the inherent physicochemical properties of drugs and the complex environment such as brain-blood barrier (BBB) [1,2]. Despite the challenges of discovery for new targets and mechanisms of CNS diseases, the current treatment strategies for CNS diseases are mostly limited to the modulation of disease-derived symptoms. One of the recent approaches was that inhibition of cysteine/glutamate antiporter system xc-(SXC) could be a therapeutic potential for CNS disorder in respect of regulating glutamate concentration in glial cells [3–5].

Sulfasalazine (SAS, Figure 1) is broadly used to treat chronic inflammation in the gut, joints, and retina [6]. The mechanisms that explain anti-inflammatory activity may involve the inhibition of NF-κB signaling pathway [7]. In addition, SAS has also shown clinical potential as an effective inhibitor of xc-transporter, which has a potential as a new therapeutic approach for CNS diseases [8–10]. Recently, some reports suggested that the effectiveness of SAS in vitro and in vivo model of brain tumor was likely due to the inhibition of the NF-κB signaling pathway by SAS [7]. Other studies have shown that SAS may contribute to the antidepressant-like effect because of its SXC inhibitory effects [10]. Another recent study explored that SAS was able to decrease the duration of epileptiform events associated with the modulation of glutamate system, which decrease the release of excitatory glutamate in the brain [11]. Therefore, due to its wide range of effects, SAS has been investigated in a number of preclinical and clinical studies [9,12–17]. Unfortunately, despite the fact that SAS has been studied recently in the areas of CNS-related research field, no/little information regarding drug metabolism and pharmacokinetics (DMPK) of SAS itself nor its metabolites in the brain have been reported.

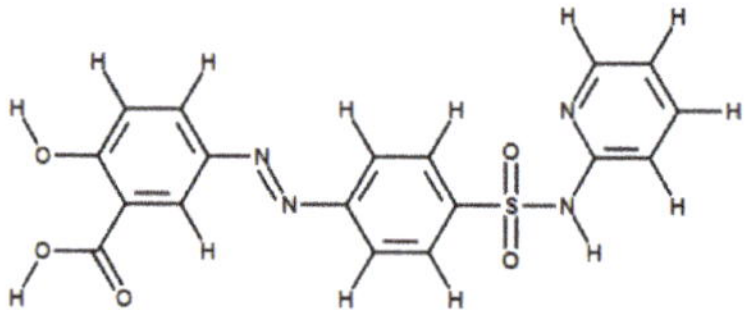

Figure 1. Chemical structure of sulfasalazine.

In this study, the quantification and the metabolite identification for SAS in mouse brain and plasma have been simultaneously explored to link the effects of SAS in the CNS-related disease models. Generally, in vivo brain exposure studies (a.k.a.; neuroPK) are involved with assessing the extent (Kp_{brain}) of partitioning into the brain from the blood [2]. Therefore, Kp_{brain} was calculated to evaluate the permeability of SAS passing through the BBB from its CNS effect perspectives [18–22]. In addition, in vivo metabolite identification (MetID) was conducted to investigate SAS and its metabolites in mouse plasma as well as in mouse brain. To our best knowledge, this is the first report regarding SAS exposure and metabolite identification in mouse brain. Hydrogen/deuterium exchange (H/D exchange) for the product ion spectra by the electrospray ionization (ESI) was also performed to confirm the fragment pattern of SAS [23,24].

2. Results

2.1. Method Development and Qualification

The liquid chromatography-electrospray-time-of-flight mass spectrometry (LC-ESI-TOF-MS) method was newly developed for SAS in mouse plasma over concentration ranges of 9.15~6670 ng/mL, as shown in Figure 1. Successive linear calibration curves were obtained over these concentration ranges with average correlation coefficients >0.99. The lower limit of quantification (LLOQ) was 9.15 ng/mL using the simple protein precipitation method.

Accuracy was evaluated at three different levels of quality control (QC) samples over three run times. There were three replicates of 15.0, 165, and 1820 ng/mL evaluated for the accuracy calculations. The inter-run accuracy ranged from 94.2% to 114% and the intra-run accuracy ranged from 98.7% to 109%, respectively. These results fall within the acceptance criteria internally made for this type of the fit-for-purpose research (±25% of the nominal value for the accuracy and precision). Precision was also evaluated at three different levels of QC samples over three run times. The intra-run precision ranged from 1.10% to 10.0% and the inter-run precision ranged from 4.09% to 8.71%, respectively. The precision data for both intra and inter-run did not exceed 10.01% from the nominal values and were less than the accepted limit of 25% CV. The repeated injection of LLOQ was also performed to assess the assay performance.

All QC samples met the acceptance criteria for this fit-for-purpose research study within ± 25% of the nominal value and the results are summarized in Tables 1 and 2. The accuracy and precision of LLOQ were also 111% and 13.0%, respectively, as shown in Table 3. In addition, Table 4 shows that the accuracy and precision of dilution QC were also 88.4% and 13.9%, respectively.

Table 1. The inter-run assays results for sulfasalazine ($n = 3$).

Run	Nominal Concentration (ng/mL)	Calculated Concentration (ng/mL)	Mean Accuracy (%)	Precision (% CV)
Run 1	QC low [15.0]	14.3	95.1	10.0
	QC medium [165]	167	101	6.90
	QC high [1820]	2070	114	3.79
Run 2	QC low [15.0]	14.2	94.2	5.98
	QC medium [165]	160	97.1	1.69
	QC high [1820]	1950	107	1.10
Run 3	QC low [15.0]	16.3	108	4.95
	QC medium [165]	161	97.6	7.00
	QC high [1820]	1930	106	4.56

Table 2. The intra-run assays and results for sulfasalazine ($n = 9$).

Nominal Concentration (ng/mL)	Calculated Concentration (ng/mL)	Mean Accuracy (%)	Precision (% CV)
QC low [15.0]	14.9	99.2	8.71
QC medium [165]	163	98.7	5.11
QC high [1820]	1980	109	4.09

Table 3. The repeat injection of lower limit of quantification (LLOQ) results for sulfasalazine ($n = 3$).

Nominal Concentration (ng/mL)	Calculated Concentration (ng/mL)	Mean Accuracy (%)	Precision (% CV)
9.15	10.1	111	13.0

Table 4. The dilution quality control (QC) results for sulfasalazine ($n = 3$).

Dilution Factor	Nominal Concentration (ng/mL)	Calculated Concentration (ng/mL)	Mean Accuracy (%)	Precision (% CV)
×10	20,000	17,700	88.4	13.9

2.2. Pharmacokinetic Study–Kp_{brain}

The developed LC-ESI-TOF-MS method was successfully applied to determine the PK parameters after intravenous (IV) administration at singly dose of 10 mg/kg for SAS in mice (Figure 2). Both plasma and brain homogenates were used to determine brain to plasma ratio (%) of SAS in assessing the kinetics of drug penetration across the BBB. While most of the plasma samples were within the range of the qualified calibration curve (range: 9.15–6670 ng/mL), some samples were diluted with blank mouse plasma for covering the higher concentration over the upper limit of quantification (ULOQ), particularly for the early time point samples. Final plasma concentrations were calculated by considering the dilution factors for 5 × and 10 × of the measured concentrations for those samples above the ULOQ.

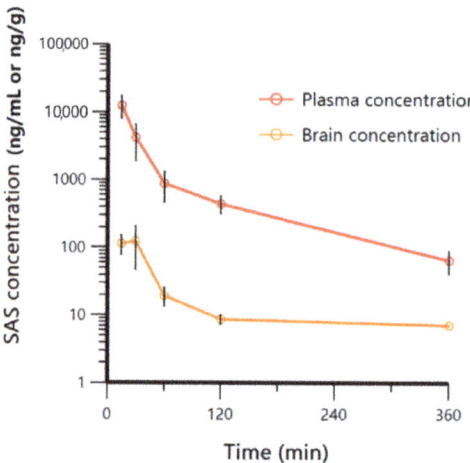

Figure 2. Time-concentration profiles of sulfasalazine for pharmacokinetics (PK) study (IV: 10 mg/kg) sample (plasma (red line) and brain (yellow line)).

For the brain homogenate sample analysis, the evaporation and reconstitution process was used to increase the sensitivity due to relatively poor SAS exposure in brain. Furthermore, the surrogate-matrix mixing with blank mouse plasma to the brain homogenates was used to reduce the unknown non-specific binding, as well as matrix effect during electrospray ionization for LC-ESI-TOF-MS analysis [25]. Although the intra/inter run assays for SAS in brain homogenates have not been fully performed, the calibration curve samples and QC samples for the brain homogenates were prepared freshly to minimize any stability issues during the brain homogenate sample analysis. The regression of brain homogenate standard curve was also acceptable with the regression coefficient (0.998) in the range of 3.05–2200 ng/mL and the accuracies of brain homogenate QC samples were 92.2~106%. It would be ideal to evaluate several stability tests and matrix effects in each matrix to demonstrate the robustness of this assay but the stability issue was not significant due to the fresh sample preparation in this study for the brain homogenate sample analysis. Furthermore, no significant internal standard peak response variation was observed during the entire sample analysis, which would be able to explain to some extent that the matrix effect was probably not significant. In addition, the same matrix used for each calibration curve (e.g., plasma calibration curve for the plasma samples and the brain homogenate calibration curve for the brain homogenate samples) would be also able to help to normalize the matrix specific peak response, if any. Final brain concentrations were calculated by considering the dilution factors 4 × of the measured concentrations. Upon obtaining concentration-time profile data of plasma (ng/mL) and brain (ng/g), the non-compartmental analysis (NCA) using WinNonlin (version 8.0.0) was performed to estimate the PK parameters for SAS, as presented in Table 5.

PK results have demonstrated the volume of distribution (Vd) and clearance (CL) calculated from plasma concentration of 1740 mL/kg and 14.8 mL/min/kg, respectively, showing that SAS has a low clearance in mice. The area under the curve up to last time point (AUC_{last}) for plasma and brain were 674,000 and 8460 min × ng/mL, respectively. The brain to plasma AUC ratios (Kp_{brain}) were then calculated to evaluate the efficiency of SAS passing through the brain. Kp_{brain} of SAS was calculated to be 1.26%. These results suggested that SAS, as an intact form, would hardly penetrate the BBB, which is challenging in correlating the role of SAS for the CNS-related disease studies. Instead, this result indicated that the SAS efficacy against CNS-related studies in vivo might be likely due to its metabolites in brain or other indirect modes of actions in vivo. To understand its clinical efficacy of SAS, a metabolite identification (MetID) study was designed and conducted in mice to investigate the metabolism of SAS in mouse brain.

Table 5. Pharmacokinetic parameters of sulfasalazine from IV administration PK study (10 mg/kg).

Dosing Route	Matrix	Dose (mg/kg)	$T_{1/2}$ (min)	C_{max} (ng/mL or ng/g)	AUC_{last} (min × ng/mL or min × ng/g)	Vd (mL/kg)	CL (mL/min/kg)	Brain to Plasma Ratio (%)
IV administration	Plasma	10	82	12,500	674,000	1740	14.8	1.26
	Brain		269	126	8460	—	—	

Cmax: observed maximum plasma concentration; Vd: Volume of distribution; CL: clearance.

2.3. In Vivo Metabolite Identification

2.3.1. Collision-Induced Dissociation (CID) Analysis

The collision-induced dissociation (CID) patterns of SAS were accomplished by LC-ESI-TOF-MS analysis. High resolution-electrospray ionization (ESI) mass spectra were obtained to determine exact masses of ions and to identify the mass fragment patterns of the compounds for the MetID study. H_2O and D_2O solvents were also used for the H/D exchange study to better understand unknown fragments. The fragmentation patterns of SAS obtained with both H_2O and D_2O solvent infusion methods were similar, however, some differences after the H/D exchange were also observed. Particularly the fragmentation using D_2O solvent showed some different fragment ions, which seem to be related to the neutral losses of SAS [23,24]. SAS has a molecular ion at m/z 399.0772 and 403.0993 with H_2O and D_2O solvent, respectively, and this result matches well with the number of exchangeable protons (n = 3) in SAS. The product ion scan of m/z 399.0772 with H_2O leads to the formation of fragment ions at m/z 94.0533, 119.0132, 147.0193, 165.0298, 223.0511, 315.0889, 317.1047, 333.0996, and 381.0663. However, the product ion scan of m/z 403.0993 with D_2O leads to the formation of fragment ions at m/z 119.0138, 167.0435, 315.0901, 319.1169, 335.1117, and 383.0789. The presence of common fragment ions at m/z 119.0132 (119.0138 in D_2O) and 315.0889 (315.0901 in D_2O) suggests that these fragments were not subjected to H/D exchange. For fragment ions of m/z 165.0298, 317.1047, 333.0996, and 381.0663 from the product of ions of 399.0772, two amu-increased fragment ions were observed at m/z 167.0435, 319.1169, 335.1117, and 383.0789 from the product of ions of 403.0993. Especially, the observation of a fragment ion at m/z 333.0996 from the product ion of m/z 399.0772 was quite interesting because it matched with the direct loss of sulfoxylic acid (H_2SO_2) from SAS based on the calculation of its neutral loss by high resolution mass. This was also confirmed by monitoring a unique fragment ion at m/z 335.1117 from the product ion spectrum of SAS with D2O, which indicates the loss of equivalent neutral ion with deuterium (D_2SO_2).

2.3.2. Brain Distribution

Representative extracted ion chromatogram (XIC) and the product ion spectra of metabolites from SAS from mouse plasma and brain are shown in Figures 3–5. Under our experimental conditions, at least three significant metabolites (SP, SP-OH and Ac-SP) were detected and characterized in vivo from mouse plasma and two significant metabolites SP and Ac-SP were detected and characterized in vivo from mouse brain after 10 mg/kg IV and 50 mg/kg PO administration. Additionally, as shown in Figure 4, the metabolite peak intensity from SAS following the IV and PO administration confirmed that no/little SAS as intact was present in the brain, whereas another two metabolites were detected to significant levels based on their intensities. There seemed no significant difference in terms of metabolite profiles between IV and PO samples as well as plasma and brain samples, except the significantly low level of SAS in brain regardless of administration. Sulfasalazine is normally administered orally, therefore the CNS-related efficacy of SAS in preclinical or clinical studies might be likely due to its metabolites, not the intact SAS based on this experiment.

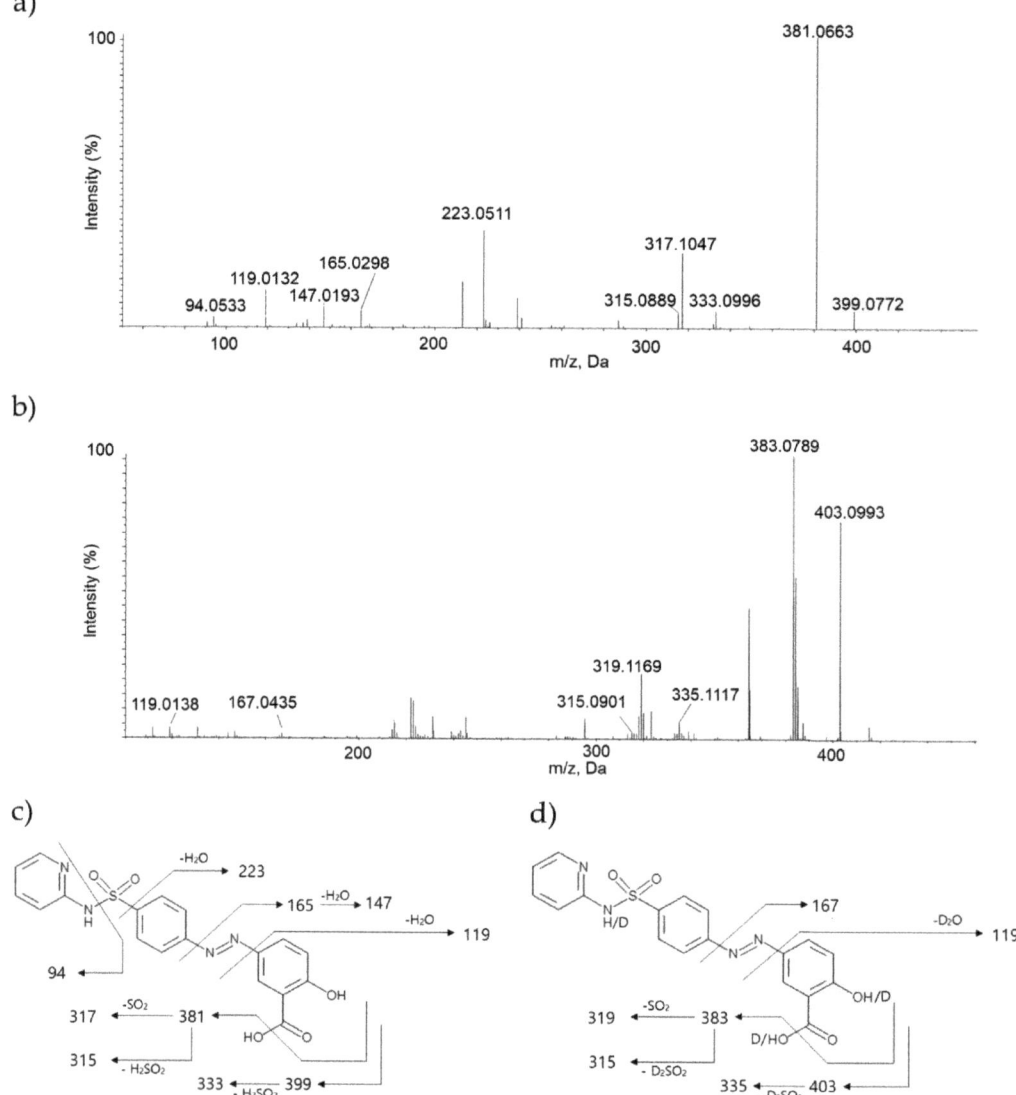

Figure 3. Comparison of the product ion spectra of sulfasalazine (SAS) infused in (**a**) H_2O solvent, (**b**) D_2O solvent, fragment pattern of SAS in (**c**) H_2O solvent and (**d**) D_2O solvent (representative m/z of each fragment was rounded off to zero decimal place).

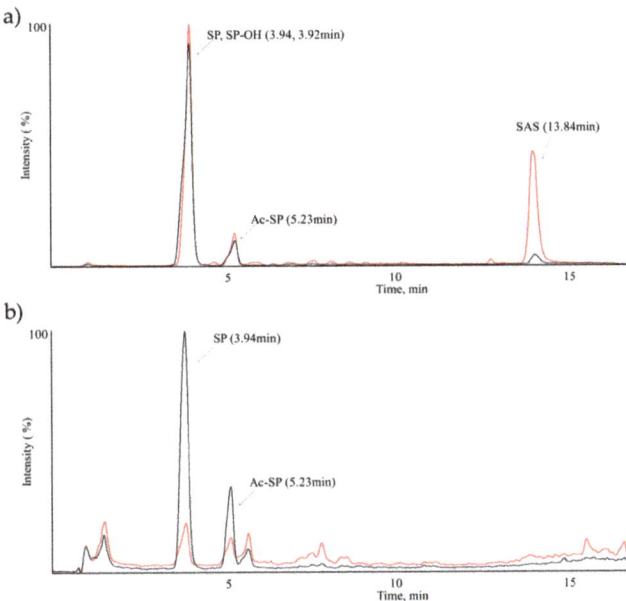

Figure 4. Extracted ion chromatograms (XIC) of SAS MetID samples in mouse (**a**) plasma and (**b**) brain (IV 10 mg/kg (Red line) and PO 50 mg/kg (Black line)).

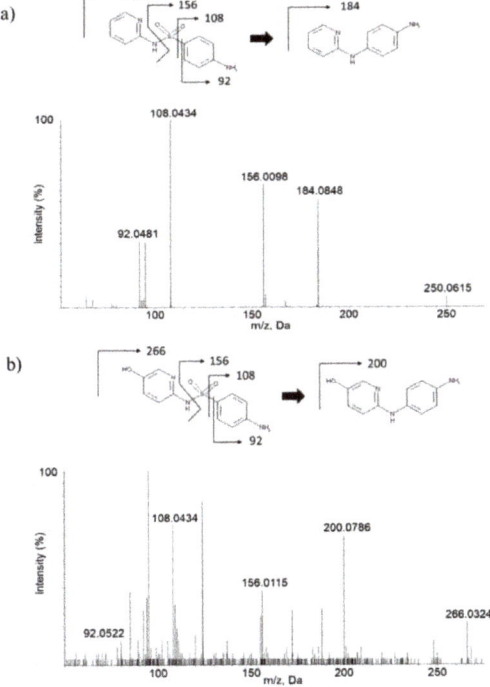

Figure 5. *Cont.*

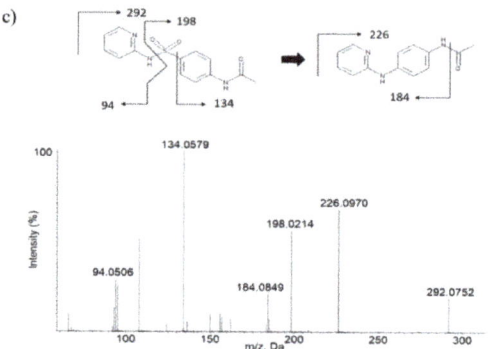

Figure 5. Product ion spectra of in vivo metabolites of SAS; (**a**) Sulfapyridine (SP), (**b**) Hydroxy-sulfapyridine (SP-OH), and (**c**) N-acetyl sulfapyridine (Ac-SP) in mouse plasma and brain (representative m/z of each fragment was rounded off to zero decimal place).

2.3.3. In Vivo Metabolites Identification

For LC-ESI-TOF-MS analysis of SP, the precursor ion of m/z 250.1 leads to the formation of fragment ions at m/z 92.0481, 108.0434, 156.0098, and 184.0848. For LC-ESI-TOF-MS analysis of SP-OH, the precursor ion of m/z 266.1 leads to the formation of fragment ions at m/z 92.0522, 108.0434, 156.0115, and 200.0786. The unchanged fragment ion m/z 156.0115 suggests that metabolism has occurred in the pyridine ring moiety of SAS. The product ion at m/z 200.0786, which is 16 amu higher than SP, suggests the metabolism of hydroxylation to SP has occurred. For LC-ESI-TOF-MS analysis of Ac-SP, precursor ion of m/z 292.1 leads to the formation of fragment ions at m/z 94.0506, 134.0579, 184.0849, 198.0214, and 226.0970. The unchanged fragment ion m/z 184.0849 suggests that metabolism has occurred in the amide group in phenyl ring moiety of SAS. The product ions at m/z 134.0579, 198.0214, and 226.0970 were 42 amu higher than the SAS fragments at m/z 92, 156, and 184. These results explained that Ac-SP is a N-acetylation metabolite of SP. All metabolites were also compared and confirmed with the authentic standards after the LC-ESI-TOF-MS analysis (Table 6).

Table 6. In vivo MetID results of sulfasalazine after PK study (IV: 10 mg/kg PO: 50 mg/kg).

Name	Formula	Exact m/z	Error (ppm)	Nominal Mass Change (Da)	RT (min)
SAS	$C_{18}H_{15}N_4O_5S^+$	399.0763	—	—	13.84
SP	$C_{11}H_{12}N_3O_2S^+$	250.0650	−7.2	−149	3.94
SP-OH	$C_{11}H_{12}N_3O_3S^+$	266.0599	−6.4	−133	3.92
Ac-SP	$C_{13}H_{14}N_3O_3S^+$	292.0756	−4.5	−107	5.23

RT: retention time (min).

3. Discussion and Conclusions

SAS is an anti-inflammatory and immune-modulating drug that has been used for rheumatology and inflammatory bowel disease. In addition to the commonly known inflammation related efficacy, research on the CNS-related effect of SAS has recently been carried out [9,10,18,21,22]. Previous studies revealed that orally administered SAS exerted the antidepressant-like effects that were at least as effective as fluoxetine [10]. In addition, concomitant of its major metabolites, 5-ASA and SP also showed a similar tendency [10]. Nevertheless, whether SAS and its metabolites can penetrate brain or not has not been studied so far. Thus, in this paper, in vivo brain exposure studies and brain MetID study were conducted to investigate the pharmacokinetics and brain distribution of SAS in a mouse model.

As a result, the finding of poor brain-penetration of SAS as 1.26% suggests that in vivo CNS activity of SAS is most likely not directly correlated with SAS itself but with other opportunities of metabolites activity or other unknown indirect mode of actions. According to previous studies of SAS metabolism and distribution, SP is relatively well absorbed from the intestine and mainly acetylated and conjugated as a glucuronide in the liver before excretion in the urine, whereas another metabolite 5-ASA is minimally absorbed and passes out in the feces [26]. Furthermore, highly bound to albumin (>99%) of SAS is reported whereas SP is only 70% bound to albumin. In conclusion, these results suggested that the probability of SAS in alleviating CNS symptoms, depressant, and seizures is possibly due to its presence of metabolites, not SAS itself in the brain.

Although the metabolism of SAS has been studied for a long time, the BBB penetration of SAS as well as the metabolism of SAS in brain has not been reported so far. In this MetID study, SP, SP-OH and Ac-SP were identified in plasma, and SP and Ac-SP were identified in brain, regardless of drug administration route. After PO administration, SAS is metabolized by gut bacterial azo-reduction in the colon, while liver-azoreductases serve cleavage of SAS after IV administration. The metabolites of SP, SP-OH and Ac-SP following IV administration of SAS are evidence that human azoreductases, which play a crucial role in the metabolism of SAS [27,28]. In addition, SP and its secondary metabolites, Ac-SP, were only observed in brain among the significant metabolites generated by azo-reduction. These results imply that there might be some differences in terms of gastrointestinal absorption of metabolites after azo-reduction in the gut.

SAS is still being studied in the CNS-related diseases by SXC inhibitory effects. Despite the limitation that the quantitative aspect and tissue protein binding of metabolites were not considered in this paper, the results of very low brain penetration of SAS as well as the significant levels of two metabolites (SP and Ac-SP) would be helpful to understand the new role of the metabolites in the brain. Further studies for these metabolites would be warranted for their CNS-related effects in vitro and in vivo.

4. Materials and Methods

4.1. Reagents and Chemicals

Sulfasalazine, verapamil, and deuterium oxide (D_2O) were obtained from Merck and Sigma-Aldrich (Yong-in, Gyeonggi, Korea). Dimethyl sulfoxide (DMSO) and formic acid were all obtained from Daejung reagents (Siheung, Gyeonggi, Korea). HPLC grade methanol (MeOH) was acquired from Duksan reagents (Ansan, Gyeonggi, Korea). HPLC grade acetonitrile (ACN) and distilled water (DW) were all obtained from Samchun reagents (Gangnam, Seoul, Korea). All other chemicals were commercial products of either analytical grade or reagent grade, and no further purification was used

4.2. Preparation of Analytical Stock and Standards Solutions

Stock solutions of SAS was prepared in DMSO at 1 mg/mL concentrations and stored at 4 °C when not in use. Stock standards were freshly prepared in DMSO to 1.02, 3.05, 9.15, 27.4, 82.3, 247, 741, 2220, and 6670 ng/mL for SAS. The internal standards working solution using verapamil was prepared with ACN to a concentration of 100 ng/mL for pharmacokinetic study and 10 ng/mL for MetID study.

4.3. Sample Preparation

For pharmacokinetic study sample preparation, 20 μL of plasma samples were transferred to cluster tubes and mixed with 4 μL of stock standards for plasma matrix standard curve. There were 100 μL of internal standards solution in ACN added to each sample for extraction. The resulting solutions were vortex-mixed for 30 s and centrifuged at 13,000 rpm for 5 min to precipitate proteins in the matrix. The supernatant was three times diluted by distilled water and transferred to LC vial for LC-ESI-TOF-MS analysis.

There were 150 μL of brain homogenates transferred to 1.7 mL Eppendorf tubes including 150 μL of blank plasma and mixed with 30 μL of stock standards for brain surrogate-matrix standard curve. Then, 1 mL of internal standards solution in ACN was added to each sample for extraction. The result solutions were vortex-mixed for 30 s and centrifuged at 13,000 rpm for 5 min to precipitate protein in the matrix. The supernatant was transferred to clean 1.7 mL Eppendorf tubes and evaporated to dryness. The dried extract was reconstituted in 200 µL of 30% ACN in DW and transferred to LC vial for LC-ESI-TOF-MS analysis.

For MetID study sample preparation, plasma samples collected at 15, 30, 60, 120, and 240 min were pooled according to the Hamilton pooling method and the same Hamilton pooling method was applied to the brain homogenate samples collected at the same time and were transferred to 15 mL tubes [29–31]. Total volume of pooled plasma and brain homogenate samples were 1 mL and 3mL, respectively. After pooling, 3 and 9 mL of 50% MeOH in ACN was added to each plasma and brain surrogate-matrix for extraction, respectively. After vortexed for 1 min, the extraction solutions were centrifuged at 13000 rpm for 5 min. The supernatant was transferred to another tube and the evaporation and reconstitution processes using the 50% MeOH were conducted for MetID study.

4.4. LC-ESI-TOF-MS Condition

The LC-ESI-TOF-MS system for this experiment consisted of a chromatographic pump system (Shimadzu CBM-20A/LC-20AD, Shimadzu Scientific Instruments, Riverwood Dr, Columbia, SC, USA) and an auto-sampler system (Eksigent CTC HTS PAL, Sciex, Redwood City, CA, USA) equipped with a mass spectrometer (TripleTOF^{TM} 5600, Sciex, Redwood City, CA, USA). Chromatographic separation was performed on a reversed-phase C_{18} column (Phenomenex® Kinetex XB-C18 column; 2.1 × 50 mm for bioanalytical sample quantification and 2.1 × 100 mm for MetID). A guard column was placed upstream of the analytical column. Mobile phase A (0.1% formic acid in distilled water) and mobile phase B (0.1% formic acid in acetonitrile) were used following an optimized gradient profile for the best separation of the analytes. The LC-gradient was optimized as follows: 3 min for the quantification (0–0.5 min, 10% B; 0.5–1.8 min, 10–95% B; 1.8–2.0 min, 95% B; 2.0–2.1 min, 95–10% B; and 2.1–3.0 min, 10% B with a flow rate of 0.4 mL/min), and 20 min for MetID (0–0.5 min, 5% B; 0.5–15 min, 5–40% B; 15–15.5 min, 40–95% B; 15.5–16.5 min, 95% B; 16.5–16.6 min, 95–5% B; and 16.6–20 min, 5% B with a flow rate of 0.3 mL/min).

In pharmacokinetic study, the curtain gas was 33 L/min, the gas source 1 and 2 were 50 psi, the ion spray voltage (ISVF) was set at 5500 V and the source temperature was 500 °C. The high resolution TOF full scan and two product ion scan for SAS and verapamil using single reaction monitoring at high resolution option mode was used for the PK sample analysis. Quantification was performed using the transitions m/z 399.1 > 381.1 (DP = 168 and CE = 22) for sulfasalazine and the transitions m/z 455.3 > 165.1 (DP = 125 and CE = 30) for verapamil, respectively.

In the MetID study, the curtain gas, ISVF and the source temperature were performed under identical experimental conditions. The high resolution TOF full scan and nine product ion scans for SAS and its known metabolites were performed. The following conditions were used to identify the SAS and its metabolites; TOF-MS scan (mass range: m/z 100–700, DP: 100 and CE: 10), product ion scan for SAS (DP: 168 and CE: 25). Other mass spectrometric conditions are summarized in Table 7.

Table 7. The mass spectrometric conditions for three sulfasalazine metabolites in high resolution product ion mode.

Metabolite	SP	SP-OH	Ac-SP
Mass Range (m/z)		50~500	
Product of (m/z)	250.1	266.1	292.1
DP (V)	100	100	100
CE (V)	30	30	30

SP: Sulfa pyridine; SP-OH: Hydroxy-sulfapyridine; Ac-SP: N-acetyl sulfapyridine.

4.5. Animal Study Design

All experimental protocols performed on mice were approved by the animal care institute from Chungnam National University (protocol no. CNU-01104). Male ICR mice were purchased from the Samtako Biokorea Co. (Gyeonggi, Korea) and housed in groups of 4~5 per cage with free access to standard rodent chow (labdiet 5L79, Orientbio, Korea).

SAS was administered to male ICR mice to evaluate the pharmacokinetics, brain-to-plasma coefficient (Kp_{brain}), and the MetID of SAS after single dose of 10 mg/kg IV and 50 mg/kg PO administration. Animals were randomly assigned to two groups for five time points (n = 3), a total fifteen mice per each administration type. Body weights for the mice assigned to the study ranged from 28 to 30 g. SAS was dissolved in 20% DMSO, 20% PEG400 in DW for IV administration, and 30% DMSO, 20% PEG400 in DW for PO administration. Animals were sacrificed at 15, 30, 60, 120, and 240 min after dosing. At each time point, blood samples were first obtained using the heparinized tubes and centrifuged at 13,000 rpm for 5 min. After the 20~30 mL systemic perfusion with phosphate buffered saline (PBS), the brain was removed from the skull and then washed and homogenized using PBS in a ratio of 1:3 for tissue to buffer. The plasma and brain homogenates were placed in clean Eppendorf tubes and frozen at $-20\ °C$ until analysis.

4.6. H/D Exchange Study

H/D exchange is a well-established technique for studying structure, stability, folding dynamics, and intermolecular interactions in proteins in solution. The use of LC-ESI-TOF-MS equipped with an ESI source and deuterium oxide (D_2O) as infusion solvent allows H/D exchange for compounds. Comparison of infusion method using 500 ng/mL of SAS dissolved in 50% H_2O in ACN with 0.1% formic acid and 50% D_2O in ACN with 0.1% formic acid were performed to elucidate CID pattern of SAS.

4.7. Data Analysis for Pharmacokinetic and MetID Study

The plasma and brain PK parameters (terminal half-life ($T_{1/2}$), time to reach maximum concentration (T_{max}), the area under the curve up to the last time point (AUC_{last})) were estimated from the mean concentrations at each time by non-compartmental analysis (NCA) using Phoenix WinNonlin® Version 8.0.0 (Certara, Princeton, NJ, USA). The $T_{1/2}$ and AUC_{last} were calculated using a linear trapezoidal linear interpolation method; T_{max} were observed values. The brain to plasma AUC ratios of the compounds (Kp_{brain}) were calculated as follows:

$$\text{brain to plasma AUC ratios } (Kp_{brain}) = AUC_{last\text{-}brain} / AUC_{last\text{-}plasma} \quad (1)$$

The $AUC_{last\text{-}brain}$ and $AUC_{last\text{-}plasma}$ were each AUC_{last} of the brain and plasma, respectively.

Data acquisition and LC-ESI-TOF-MS operation were conducted using Analyst® TF Version 1.6 (Sciex). MultiQuant® Version 2.1.1 (Sciex) was used for the peak integration of SAS for quantification. PeakView® Version 2.2 and MetabolitePilot™ Version 2.0.2 were used for the structural elucidation of SAS metabolites.

Author Contributions: Conceptualization, J.C. and Y.G.S.; methodology, J.C.; M.-H.P.; S.-H.S. and Y.G.S.; software, J.C.; formal analysis, J.C.; investigation, J.C. and M.-H.P.; data curation, J.C.; writing—original draft preparation, J.C.; writing—review and editing, S.-H.S.; B.i.L.; Y.P.; J.-J.B. and Y.G.S.; visualization, J.C.; supervision, Y.G.S.; project administration, Y.G.S. All authors have read and agreed to the published version of the manuscript.

Funding: This research was supported by a research fund of the Ministry of Food and Drug Safety (19172MFDS163).

Data Availability Statement: The data presented in this study are available on request from the corresponding author.

Conflicts of Interest: The authors declare no conflict of interest.

Sample Availability: Samples of the compounds are not available from the authors.

References

1. Upadhyay, R.K. Drug delivery systems, CNS protection, and the blood brain barrier. *BioMed Res. Int.* **2014**, *2014*, 869269. [CrossRef]
2. Di, L.; Kerns, E.H. *Blood-Brain Barrier in Drug Discovery: Optimizing Brain Exposure of CNS Drugs and Minimizing Brain Side Effects for Peripheral Drugs*; John Wiley & Sons: Hoboken, NJ, USA, 2015.
3. Huang, Y.; Dai, Z.; Barbacioru, C.; Sadée, W. Cystine-glutamate transporter SLC7A11 in cancer chemosensitivity and chemoresistance. *Cancer Res.* **2005**, *65*, 7446–7454. [CrossRef]
4. Lo, M.; Wang, Y.Z.; Gout, P.W. The x cystine/glutamate antiporter: A potential target for therapy of cancer and other diseases. *J. Cell Physiol.* **2008**, *215*, 593–602. [CrossRef]
5. Robert, S.M.; Sontheimer, H. Glutamate transporters in the biology of malignant gliomas. *Cell Mol. Life Sci.* **2014**, *71*, 1839–1854. [CrossRef] [PubMed]
6. Plosker, G.L.; Croom, K.F. Sulfasalazine: A review of its use in the management of rheumatoid arthritis. *Drugs* **2005**, *65*, 2591. [CrossRef]
7. Robe, P.A.; Bentires-Alj, M.; Bonif, M.; Rogister, B.; Deprez, M.; Haddada, H.; Khac, M.-T.N.; Jolois, O.; Erkmen, K.; Merville, M.-P. In vitro and in vivo activity of the nuclear factor-κB inhibitor sulfasalazine in human glioblastomas. *Clin. Cancer Res.* **2004**, *10*, 5595–5603. [CrossRef]
8. Gout, P.; Buckley, A.; Simms, C.; Bruchovsky, N. Sulfasalazine, a potent suppressor of lymphoma growth by inhibition of the xc−cystine transporter: A new action for an old drug. *Leukemia* **2001**, *15*, 1633–1640. [CrossRef] [PubMed]
9. Sontheimer, H.; Bridges, R.J. Sulfasalazine for brain cancer fits. *Expert Opin Investig Drugs* **2012**, *21*, 575–578. [CrossRef] [PubMed]
10. Nashed, M.G.; Ungard, R.G.; Young, K.; Zacal, N.J.; Seidlitz, E.P.; Fazzari, J.; Frey, B.N.; Singh, G. Behavioural effects of using sulfasalazine to inhibit glutamate released by cancer cells: A novel target for cancer-induced depression. *Sci. Rep.* **2017**, *7*, 1–11. [CrossRef]
11. Alcoreza, O.; Tewari, B.P.; Bouslog, A.; Savoia, A.; Sontheimer, H.; Campbell, S.L. Sulfasalazine decreases mouse cortical hyperexcitability. *Epilepsia* **2019**, *60*, 1365–1377. [CrossRef]
12. Chungi, V.S.; Dittert, L.W.; Shargel, L. Pharmacokinetics of sulfasalazine metabolites in rats following concomitant oral administration of riboflavin. *Pharm. Res.* **1989**, *6*, 1067–1072. [CrossRef] [PubMed]
13. Sjöquist, B.; Ahnfelt, N.-O.; Andersson, S.; Fjellner, G.; Hatsuoka, M.; Olsson, L.-I.; Ljungstedt-Påhlman, I. Pharmacokinetics of Salazosulfapyridine (Sulfasalazine, SASP)(I): Plasma kinetics and plasma metabolites in the rat after a single intravenous or oral administration. *Drug Metab. Pharmacokinet.* **1991**, *6*, 425–437. [CrossRef]
14. Zaher, H.; Khan, A.A.; Palandra, J.; Brayman, T.G.; Yu, L.; Ware, J.A. Breast cancer resistance protein (Bcrp/abcg2) is a major determinant of sulfasalazine absorption and elimination in the mouse. *Mol. Pharm.* **2006**, *3*, 55–61. [CrossRef] [PubMed]
15. Hanngren, Å.; Hansson, E.; Svartz, N.; Ullberg, S. Distribution and Metabolism of Salicyl-azo-sulfapyridine: I. A Study with C14-salicyl-azo-sulfapyridine and C14-5-amino-salicylic Acid 1. *Acta Med. Scand.* **1963**, *173*, 61–72. [CrossRef] [PubMed]
16. Zamek-Gliszczynski, M.J.; Bedwell, D.W.; Bao, J.Q.; Higgins, J.W. Characterization of SAGE Mdr1a (P-gp), Bcrp, and Mrp2 knockout rats using loperamide, paclitaxel, sulfasalazine, and carboxydichlorofluorescein pharmacokinetics. *Drug Metab. Dispos.* **2012**, *40*, 1825–1833. [CrossRef]
17. Salvamoser, J.D.; Avemary, J.; Luna-Munguia, H.; Pascher, B.; Getzinger, T.; Pieper, T.; Kudernatsch, M.; Kluger, G.; Potschka, H. Glutamate-mediated down-regulation of the multidrug-resistance protein BCRP/ABCG2 in porcine and human brain capillaries. *Mol. Pharm.* **2015**, *12*, 2049–2060. [CrossRef] [PubMed]
18. Mut, S.E.; Kutlu, G.; Ucler, S.; Erdal, A.; Inan, L.E. Reversible encephalopathy due to sulfasalazine. *Clin. Neuropharmacol.* **2008**, *31*, 368–371. [CrossRef]
19. Agarwal, V.; Kommaddi, R.P.; Valli, K.; Ryder, D.; Hyde, T.M.; Kleinman, J.E.; Strobel, H.W.; Ravindranath, V. Drug metabolism in human brain: High levels of cytochrome P4503A43 in brain and metabolism of anti-anxiety drug alprazolam to its active metabolite. *PLoS ONE* **2008**, *3*, e2337. [CrossRef]
20. Hong, L.; Jiang, W.; Pan, H.; Jiang, Y.; Zeng, S.; Zheng, W. Brain regional pharmacokinetics of p-aminosalicylic acid and its N-acetylated metabolite: Effectiveness in chelating brain manganese. *Drug Metab. Dispos.* **2011**, *39*, 1904–1909. [CrossRef] [PubMed]
21. Yerokun, T.; Winfield, L.L. Celecoxib and LLW-3-6 reduce survival of human glioma cells independently and synergistically with sulfasalazine. *Anticancer Res.* **2015**, *35*, 6419–6424. [PubMed]
22. Sehm, T.; Fan, Z.; Ghoochani, A.; Rauh, M.; Engelhorn, T.; Minakaki, G.; Dörfler, A.; Klucken, J.; Buchfelder, M.; Eyüpoglu, I.Y. Sulfasalazine impacts on ferroptotic cell death and alleviates the tumor microenvironment and glioma-induced brain edema. *Oncotarget* **2016**, *7*, 36021. [CrossRef] [PubMed]
23. Zalipsky, J.; Patel, D.; Reavey-Cantwell, N. Characterization of impurities in sulfasalazine. *J. Pharm. Sci.* **1978**, *67*, 387–391. [CrossRef] [PubMed]
24. Lam, W.; Ramanathan, R. In electrospray ionization source hydrogen/deuterium exchange LC-MS and LC-MS/MS for characterization of metabolites. *J. Am. Soc. Mass Spectrom.* **2002**, *13*, 345–353. [CrossRef]

25. Chen, S.; Wu, J.T.; Huang, R. Evaluation of surrogate matrices for standard curve preparation in tissue bioanalysis. *Bioanalysis* **2012**, *4*, 2579–2587. [CrossRef]
26. Das, K.M.; Dubin, R. Clinical Pharmacokinetics of Sulphasalazine. *Clin. Pharmacokinet.* **1976**, *1*, 406–425. [CrossRef]
27. Ryan, A. Azoreductases in drug metabolism. *Br. J. Pharmacol.* **2017**, *174*, 2161–2173. [CrossRef]
28. Guo, Y.; Lee, H.; Jeong, H. Gut microbiota in reductive drug metabolism. *Prog. Mol. Biol. Transl. Sci.* **2020**, *171*, 61–93. [PubMed]
29. Hop, C.E.; Wang, Z.; Chen, Q.; Kwei, G. Plasma-pooling methods to increase throughput for in vivo pharmacokinetic screening. *J. Pharm. Sci.* **1998**, *87*, 901–903. [CrossRef] [PubMed]
30. Hamilton, R.A.; Garnett, W.R.; Kline, B.J. Determination of mean valproic acid serum level by assay of a single pooled sample. *Clin. Pharmacol. Ther.* **1981**, *29*, 408–413. [CrossRef] [PubMed]
31. Graham, R.A.; Lum, B.L.; Morrison, G.; Chang, I.; Jorga, K.; Dean, B.; Shin, Y.G.; Yue, Q.; Mulder, T.; Malhi, V. A single dose mass balance study of the Hedgehog pathway inhibitor vismodegib (GDC-0449) in humans using accelerator mass spectrometry. *Drug Metab. Dispos.* **2011**, *39*, 1460–1467. [CrossRef] [PubMed]

Article

Development, Validation, and Comparison of Two Mass Spectrometry Methods (LC-MS/HRMS and LC-MS/MS) for the Quantification of Rituximab in Human Plasma

Aurélien Millet [1,2], Nihel Khoudour [3], Dorothée Lebert [4], Christelle Machon [1,2,5], Benjamin Terrier [6,7], Benoit Blanchet [3,8] and Jérôme Guitton [1,2,9,*]

1. Biochemistry and Pharmacology-Toxicology Laboratory, Lyon-Sud Hospital, Hospices Civils de Lyon, F-69495 Pierre Bénite, France; aurelien.millet@chu-lyon.fr (A.M.); christelle.machon@univ-lyon1.fr (C.M.)
2. Inserm U1052, CNRS UMR5286 Cancer Research Center of Lyon, F-69000 Lyon, France
3. Department of Pharmacokinetics and Pharmacochemistry, Cochin Hospital, AP-HP, CARPEM 75014 Paris, France; nihel.khoudour@aphp.fr (N.K.); benoit.blanchet@aphp.fr (B.B.)
4. Promise Proteomics, 7 Parvis Louis Néel, F-38040 Grenoble, France; dorothee.lebert@promise-proteomics.com
5. Analytical Chemistry Laboratory, Faculty of Pharmacy ISPBL, University Lyon 1, F-69373 Lyon, France
6. Department of Internal Medicine, National Referral Center for Rare Systemic Autoimmune Diseases, Assistance Publique Hôpitaux de Paris-Centre (APHP-CUP), University of Paris, F-75014 Paris, France; benjamin.terrier@aphp.fr
7. INSERM U970, PARCC, Université de Paris, F-75006 Paris, France
8. UMR8038 CNRS, U1268 INSERM, Faculty of Pharmacy, University of Paris, PRES Sorbonne Paris Cité, CARPEM 75006 Paris, France
9. Toxicology Laboratory, Faculty of Pharmacy ISPBL, University of Lyon 1, F-69373 Lyon, France
* Correspondence: jerome.guitton@univ-lyon1.fr

Citation: Millet, A.; Khoudour, N.; Lebert, D.; Machon, C.; Terrier, B.; Blanchet, B.; Guitton, J. Development, Validation, and Comparison of Two Mass Spectrometry Methods (LC-MS/HRMS and LC-MS/MS) for the Quantification of Rituximab in Human Plasma. *Molecules* **2021**, *26*, 1383. https://doi.org/10.3390/molecules26051383

Academic Editor: Joselito P. Quirino

Received: 9 January 2021
Accepted: 25 February 2021
Published: 4 March 2021

Publisher's Note: MDPI stays neutral with regard to jurisdictional claims in published maps and institutional affiliations.

Copyright: © 2021 by the authors. Licensee MDPI, Basel, Switzerland. This article is an open access article distributed under the terms and conditions of the Creative Commons Attribution (CC BY) license (https://creativecommons.org/licenses/by/4.0/).

Abstract: Rituximab is a chimeric immunoglobulin G1-kappa (IgG1κ) antibody targeting the CD20 antigen on B-lymphocytes. Its applications are various, such as for the treatment of chronic lymphoid leukemia or non-Hodgkin's lymphoma in oncology, and it can also be used in the treatment of certain autoimmune diseases. Several studies support the interest in therapeutic drug monitoring to optimize dosing regimens of rituximab. Thus, two different laboratories have developed accurate and reproductive methods to quantify rituximab in human plasma: one using liquid chromatography quadripolar tandem mass spectrometer (LC-MS/MS) and the other, liquid chromatography orbitrap tandem mass spectrometer (LC-MS/HRMS). For both assays, quantification was based on albumin depletion or IgG-immunocapture, surrogate peptide analysis, and full-length stable isotope-labeled rituximab. With LC-MS/MS, the concentration range was from 5 to 500 µg/mL, the within- and between-run precisions were <8.5%, and the limit of quantitation was 5 µg/mL. With LC-MS/HRMS, the concentration range was from 10 to 200 µg/mL, the within- and between-run accuracy were <11.5%, and the limit of quantitation was 2 µg/mL. Rituximab plasma concentrations from 63 patients treated for vasculitis were compared. Bland–Altman analysis and Passing–Bablok regression showed the interchangeability between these two methods. Overall, these methods were robust and reliable and could be applied to routine clinical samples.

Keywords: rituximab; quadripolar mass spectrometer; albumin depletion; pharmacokinetics; orbitrap mass spectrometer; IgG-immunocapture

1. Introduction

CD20 is a glycosylated transmembrane phosphoprotein expressed on the surface of pre-B and mature B-lymphocytes, as well as many B-cell malignancies. Rituximab (RTX) is a chimeric IgG1κ therapeutic monoclonal antibody (mAb) that targets the CD20 antigen on B-lymphocytes [1]. It marks B-cells for destruction through direct induction of B-cell apoptosis, antibody-dependent cell-mediated cytotoxicity, or complement-mediated cytotoxicity. In the oncology field, it is used for the treatment of chronic lymphoid leukemia and diffuse and follicular non-Hodgkin's lymphoma [2]. By depleting normal B-cells, rituximab also reduces the adaptive immune response against self. It is approved in the treatment of autoimmune diseases such as rheumatoid arthritis, granulomatosis with polyangiitis, and moderate to severe pemphigus vulgaris. Finally, it has also shown benefit in treating some other autoimmune diseases such as vasculitis [3]. Different factors such as constant fragment gamma (Fcγ) and CR3 polymorphisms, gender, rituximab pharmacokinetics are known to contribute to highly variable clinical response in patients treated with rituximab [4]. The interindividual variability in rituximab systemic exposure is usually large because of differences in antigenic burden, which can influence rituximab clearance [5]. Besides, several studies reported a relationship between rituximab plasma concentrations and efficacy in patients treated for lymphoproliferative disorders [6]. These results support the use of therapeutic drug monitoring to optimize dosing regimens of rituximab in oncology. In contrast, pharmacokinetic/pharmacodynamic data are sparse in patients treated for autoimmune diseases [7,8], which deserves more investigation in the future.

For a few years, works on mAbs quantifying methods are in full swing. Regarding RTX, some ELISA methods [9–11], the Gyrolab™ assay [12], and mass spectrometry methods [13–17] have been described for their quantification in blood. The mass spectroscopy (MS) method developed by Mills et al. was based on the quantification of the light chain of RTX after disulfide bonds reduction and using quadrupole time of flight (Q-TOF) detection and vedolizumab as an internal standard (I.S.) [13]. Another method was developed with a sample preparation consisting of methanol precipitation, followed by reduction, peptide digestion, and solid-phase extraction [14]. Detection was achieved using a Triple TOF mass spectrometer, and the labeled peptide was used as an I.S. Recently a quadripolar tandem mass spectrometer method was published with a sample preparation by IgG-immunocapture and digestion and stable-isotope-labeled-adalimumab as an I.S. [15]. The analytical performances of the methods based on MS are quite similar: the ranges of the calibration curves were from 1 to 200 µg/mL [14,15], from 0.586 to 300 µg/mL [16], or from 5 to 100 µg/mL [17]. All the MS methods have an accuracy and precision < 15%. The Gyrolab™ assay was found to have a dynamic range from 0.09 to 60 µg/mL [12] and ELISA from 0.5 to 800 µg/mL [11], from 6.6 to 3400 µg/mL [9], or from 2 to 50 µg/mL for a commercial kit [18]. With immunological methods, the precision and accuracy were below 15% [12,18] or 25% [9,11].

This article describes the development and validation of two RTX quantification methods performed in two separate laboratories, one using a quadripolar tandem mass spectrometer and the other a liquid chromatography orbitrap tandem mass spectrometer (LC-MS/HRMS) (Orbitrap™). In both cases, a full-length stable isotope-labeled RTX was used as an I.S., and two samples preparation (albumin depletion and IgG-immunocapture) were performed. Several comparisons using plasma samples from 63 patients treated for vasculitis were carried out based on the surrogate peptide selected, mass analyzers, and sample preparations.

2. Results and Discussion
2.1. Selection of Proteotypic Peptides

Since RTX is a chimeric monoclonal antibody, several tryptic peptides are proteotypic and could be used as surrogate peptides for quantification. Thus, from the in-silico study, eight candidate peptides were found (Figure 1).

The selection of surrogate peptides was based on the abundance of the parent ions and the signal/noise ratio from spiked plasma, considering the matrix effect and the sample preparation recovery. Among the candidate, QVQLQQPGAELVK (QVQ or HC1) and QIVLSQSPAILSASPGEK (QIVL or LC1) peptides were pre-selected with both MS methods because the abundance of the parent ions was clearly higher (Figure 2).

Rituximab—heavy chain

pQVQLQQPGAELVKPGASVKMSCKASGYTFTSY³⁵NMHWVKQTPGRGLEWIGAIYPG⁵⁹GD⁶¹T⁶²SYNQKFKGKAT
LTADKSSSTAYMQLSSLTSEDSAVYYCAR⁹⁹STYYGGDW¹⁰⁶YFNVWGAGTTVTVSAASTKGPSVFPLAPSSKSTSGGTA
ALGCLVKDYFPEPVTVSWNSGALTSGVHTFPAVLQSSGLYSLSSVVTVPSSSLGTQTYICNVNHKPSNTKVDKKAEPK
SCDKTHTCPPCPAPELLGGPSVFLFPPKPKDTLMISRTPEVTCVVVDVSHEDPEVKFNWYVDGVEVHNAKTKPREEQ
YNSTYRVVSVLTVLHQDWLNGKEYKCKVSNKALPAPIEKTISKAKGQPREPQVYTLPPSRDELTKNQVSLTCLVKGF
YPSDIAVEWESNGQPENNYKTTPPVLDSDGSFFLYSKLTVDKSRWQQGNVFSCSVMHEALHNHYTQKSLSLSPGK

Rituximab—light chain

pQIVLSQSPAILSASPGEKVTMTCRASSSVSYIHWFQQKPGSSPKPWIYATSNLASGVPVRFSGSGSGTSYSLTISR
VEAEDAATYYCQQWTS⁹³NPPTFGGGTKLEIKRTVAAPSVFIFPPSDEQLKSGTASVVCLLNNFYPREAKVQWKVD
NALQSGNSQESVTEQDSKDSTYSLSSTLTLSKADYEKHKVYACEVTHQGLSSPVTKSFNRGEC

XXX: Non-proteotypic; XAAXX: amino-acid involved with CD20 binding; XXX: proteotypic peptides; XXX: proteotypic and surrogate peptides pre-selected for the quantification

Figure 1. Amino-acid sequences of the heavy-chain (HC) and light-chain (LC) of rituximab (RTX).

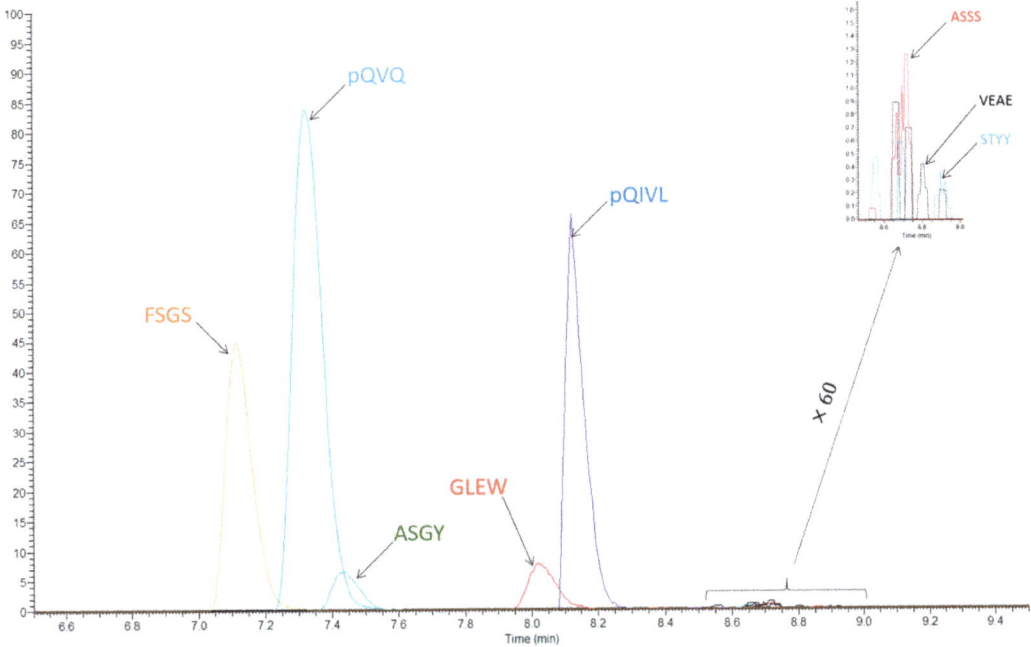

Figure 2. Liquid chromatography orbitrap tandem mass spectrometer (LC-MS/HRMS) chromatogram of proteotypic peptides obtained from a pure solution of RTX at 10 µg/mL analyzed in Full Scan Mode.

These two peptides exist in different forms and with two different charge states, as shown for QVQ. The two peptides present the transformation of N-terminal glutamine in pyroglutamate (loss of NH_3), and pQVQ also has a misscleavage before a proline generating the pQVQLQQPGAELVKPGASVK (pQVQ or HC1) peptide, which was the most abundant form (Figure 3). Ions with m/z +2 were the most abundant for pQVQ and pQIVLSQSPAILSASPGEK (pQIVL or LC1). The FSGSGSGTSYSLTISR (FSGS or LC6) was a third peptide only pre-selected for LC-MS/HRMS. Finally, pQILV was only selected as a qualifier peptide because an important ion suppression effect was measured for this compound (see below), pQVQ was used for the quantification with both MS methods and with LC-MS/HRMS, FSGS was also tested for the quantification.

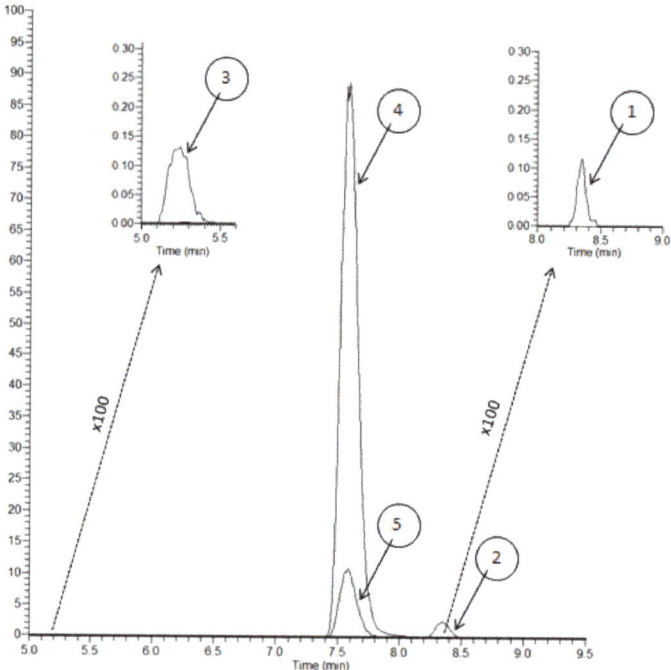

Figure 3. Chromatograms of the different forms of QVQLQQPGAELVK peptide after trypsin digestion of a pure solution of rituximab. ①: Wild type peptide: m/z = 719.4066 (charge: +2); ② N-terminal Pyroglutamate (loss of NH$_3$): m/z = 710.8934 (charge: +2); ③: Misscleavage (addition of PGASVK): m/z = 659.7091 (charge: +3); ④: Misscleavage (addition of PGASVK) and N-terminal pyroglutamate (loss of NH$_3$): m/z = 980.5467 (charge: +2); ⑤: Misscleavage (addition of PGASVK) and N-terminal pyroglutamate (loss of NH$_3$): m/z = 654.0336 (charge: +3).

None of the three peptides selected have amino-acids involved in the rituximab-CD20 bond [19]. The primary structure of three commercialized biosimilar peptides (PF-05280586, GP-2013, and ABP-798) were checked as being the same as RTX [20–22]. Thus, the two RTX princeps and the three biosimilar drugs could be quantified with the selected proteotypic peptides.

Previous studies also used pQVQ as a surrogate-peptide for quantification with quadripolar tandem mass spectrometer [15] or pQILV and GLEWIGAIYPGNGDTSYNQK, another proteotypic peptide, using a Triple TOF mass spectrometer [14].

2.2. Liquid Chromatography and Mass Spectrometry

Separation of the surrogate peptides was accomplished in 9.5 and 16 min with LC-MS/MS and LC-MS/HRMS, respectively. A step gradient was used with water and acetonitrile, each containing 0.1% formic acid. Representative parallel reaction monitoring (PRM) chromatograms obtained with LC-MS/HRMS are shown in Figure 4. The retention times of the peptides FSGS, pQVQ, and pQIVL were 4.8 min, 6.2 min, and 7.9 min, respectively. Figure 5 shown MRM chromatograms obtained with LC-MS/MS, and the retention times of the peptides pQVQ and pQIVL were 4.9 min and 6.0 min, respectively. All compounds were resolved completely with a good peak shape.

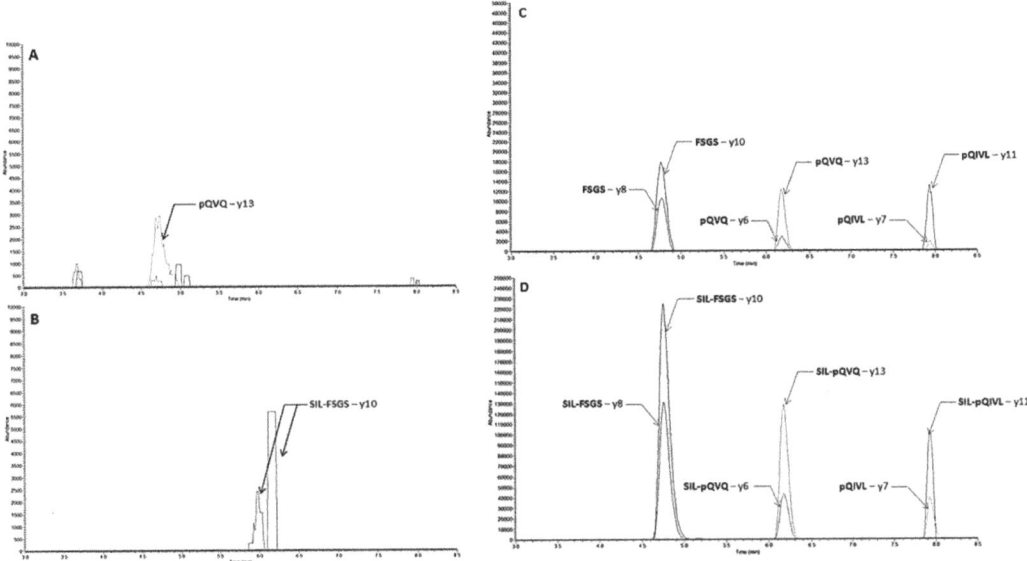

Figure 4. PRM chromatograms with LC-MS/HRMS assay of rituximab (**A**) and (**C**) and SIL-rituximab (**B**) and (**D**). **A** and **B**: blank sample; **C**: sample spiked at 2 µg/mL (lower limit of quantitation (LLOQ)); **D**: SIL-rituximab at 15 µg/mL.

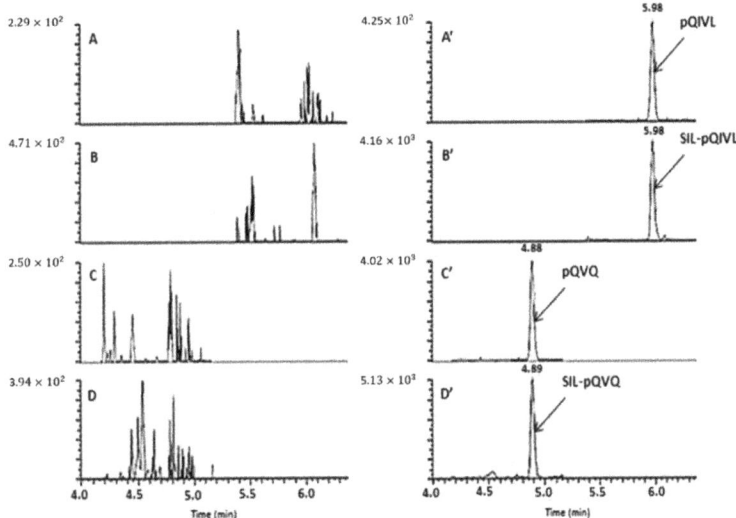

Figure 5. Multiple-reaction monitoring (MRM) chromatograms with LC-MS/MS assay of rituximab assay of rituximab (**A**), (**A'**) and (**C**), (**C'**) and SIL-rituximab (**B**), (**B'**) and (**D**), (**D'**). (**A**), (**B**), (**C**), (**D**): blank sample; **A'** and **C'**: sample spiked at 5 µg/mL (LLOQ); **B'** and **D'**: SIL-rituximab at 15 µg/mL.

Protonated molecules were predominantly formed, and the $[M + H]^{2+}$ ion of each peptide was selected as the precursor ion to find the most abundant product ion. For the quantification with LC-MS/HRMS, y8 + y10 and y6 + y13 were used as the product ion for FSGS and pQVQ, respectively (Table 1). For LC-MS/MS, the sum of the product ion y6 + y10 + y12 + y13 from pQVQ was used for quantification (Table 2).

Table 1. Surrogate peptides pre-selected for Rituximab (RTX) quantification by LC-MS/HRMS using the full-length stable isotope-labeled rituximab (SIL-RTX). Peptides in bold were used for the quantification.

Compound	Peptide	Precursor Ion (m/z)	Charge	Ion	Product Ion (m/z)	Charge
RTX	FSGS	803.8890	+2	y8	926.4942	+1
				y10	1084.5629	+1
	pQVQ	**980.5467**	**+2**	y13	1252.7260	+1
				y6	558.3246	+1
	pQIVL	904.4936	+2	y11	1069.5888	+1
				y7	675.3308	+1
SIL-RTX	FSGS	808.8931	+2	y8	936.5024	+1
				y10	1094.5716	+1
	pQVQ	**988.5609**	**+2**	y13	1268.7544	+1
				y6	566.3360	+1
	pQIVL	908.5007	+2	y11	1077.6030	+1
				y7	683.3450	+1

Table 2. Surrogate peptides pre-selected for Rituximab (RTX) quantification by LC-MS/MS using the full-length stable isotope-labeled rituximab (SIL-RTX). Peptide in bold was used for the quantification.

Compound	Peptide	Q1 [1] m/z (charge)	Q3 [2] Ion	m/z (charge)	CE [3] (eV)	CVt [4] (V)
RTX	**pQVQ**	**980.5 (+2)**	y6	558.3 (+1)	35	80
			y10	1027.6 (+1)		
			y12	1155.7 (+1)		
			y13	1252.7 (+1)		
	pQIVL	904.5 (+2)	y4	430.2 (+1)	20	90
			y7	675.3 (+1)		
			y9	788.4 (+1)		
			y11	1069.6 (+1)	30	
			y12	1156.6 (+1)		
SIL-RTX	**pQVQ**	**988.6 (+2)**	y6	566.3 (+1)	35	80
			y10	1043.6 (+1)		
			y12	1171.7 (+1)		
			y13	1268.7 (+1)		
	pQIVL	908.5 (+2)	y4	438.2 (+1)	20	90
			y7	683.3 (+1)		
			y9	796.4 (+1)		
			y11	1077.6 (+1)	30	
			y12	1164.6 (+1)		

[1] Q1: first quadrupole; [2] Q3: third quadrupole; [3] CE: collision energy; [4] CVt: cone voltage.

2.3. Result of Validation

2.3.1. Selectivity

With LC-MS/HRMS, no interference on chromatograms of RTX or isotope-labeled rituximab (SIL-RTX) was observed from ten blank plasmas (Figure 4). With LC-MS/MS, the signal observed due to endogenous components in comparison to the signal at the lower limit of quantitation (LLOQ) (5 µg/mL) was from 0.3% to 8.3% and 0.8 to 4.8% for RTX and SIL-RTX, respectively (Figure 5). These results show an excellent selectivity with both mass spectrometry methods.

2.3.2. Calibration, Accuracy, Precision, LLOQ, and Dilution

Quantification was based on the ratio of RTX and the SIL-RTX as an internal standard. Calibration curves were generated using a weighted (1/X) quadratic regression. With the LC-MS/MS, the LLOQ was set at 5 µg/mL (4.6 +/− 0.4 µg/mL), and the within-day, the between-day, and the accuracy were equal to 3.3%, 8.5%, and 91.1%, respectively. With the LC-MS/HRMS, the LLOQ was set at 2 µg/mL (2.1 +/− 0.2 µg/mL), and the within-day, the between-day, and the accuracy were equal to 7.5, 8.6, and 107.1%, respectively. The back-calculated calibration concentrations were within 91.8–103.1% (FSGS peptide) and within 88.4–102.1% (pQVQ peptide) of the theoretical value for LC-MS/HRMS (Table 3). With LC-MS/MS using (pQVQ peptide), the accuracy was within 94.1–108.2% (Table 4). The precision for standards was <12.6% (FSGS peptide) and <16.6 (pQVQ peptide) for LC-MS/HRMS and <11.3% for (pQVQ peptide) with LC-MS/MS.

Table 3. Inter-day validation of the determination of rituximab in plasma by LC-MS/HRMS using either FSGS or pQVQ proteotypic peptides. Data from seven calibration curves were prepared as a single replicate and analyzed on seven different days.

Spiked (µg/mL)	FSGS			pQVQ		
	Found (µg/mL) (mean ± s.d.)	Precision (%)	Accuracy (%)	Found (µg/mL) (mean ± s.d.)	Precision (%)	Accuracy (%)
10	9 ± 1	12.6	91.8	9 ± 2	16.6	88.4
25	26 ± 3	9.7	103.1	26 ± 3	11.8	102.8
50	51 ± 3	5.4	101.7	51 ± 5	9.1	102.0
100	98 ± 3	3.3	98.3	100 ± 6	6.0	100.0
150	150 ± 7	4.9	99.8	148 ± 8	5.6	98.6
200	200 ± 5	2.4	99.8	204 ± 5	2.6	102.1

s.d.: standard deviation.

Table 4. Inter-day validation of the determination of rituximab in plasma by LC-MS-MS using pQVQ as a surrogate peptide. Data from twelve calibration curves were prepared as a single replicate and analyzed on twelve different days.

Spiked (µg/mL)	pQVQ		
	Found (µg/mL) (mean ± s.d.)	Precision (%)	Accuracy (%)
5.0	5 ± 1	11.3	108.2
10.0	10 ± 1	6.2	103.1
20	20 ± 2	7.7	97.8
50	47 ± 4	9.2	94.1
100	96 ± 7	7.2	96.0
250	247 ± 16	6.7	98.6
500	510 ± 20	3.8	102.0

In Tables 5 and 6, a summary of the intra- and inter-day accuracy and precision assay performance is shown for quality control (QC) samples for both methods. All these parameters were <11.5% for LC-MS/HRMS and <10% for LC-MS/MS.

For LC-MS/HRMS, dilution integrity of RTX presented acceptable accuracy and precision after diluting 1/5 (v/v) in blank plasma with a mean concentration equal to 524 +/− 46 µg/mL (bias 4.8% and reproducibility 9.1%).

2.3.3. Matrix Effects, Carryover, and Sample Stability

Matrix effect due to endogenous compounds in a biological matrix is commonly encountered during the analysis by mass spectrometry in electrospray ionization (ESI) mode. The matrix effect varies greatly from one peptide to another among and according to the mass technology (Table 7). Since pQIVL exhibited the most important matrix effect with both MS methods, it was not selected for RTX quantification but only used as a qualifier peptide.

Table 5. Assessment of accuracy and precision for rituximab in plasma using the LC-MS/HRMS method and FSGS or pQVQ as surrogate peptides ($n = 6$, five days).

Peptide	Concentration (µg/mL)		Precision (%)		Accuracy (%)
	Spiked	Found (mean ± s.d.)	Within-Day	Between-Day	
FSGS	20	21 ± 2	8.4	7.3	106
	80	86 ± 7	5.3	6.9	107
	160	170 ± 13	7.3	0.5	106
pQVQ	20	22 ± 3	11.3	7.8	109
	80	84 ± 9	8.5	7.6	105
	160	168 ± 15	8.7	3.1	105

Table 6. Assessment of accuracy and precision for rituximab in plasma using the LC-MS/MS method and pQVQ as surrogate peptides ($n = 3$, ten days).

Peptide	Concentration (µg/mL)		Precision (%)		Accuracy (%)
	Spiked	Found (mean ± s.d.)	Within-Day	Between-Day	
pQVQ	15	16 ± 2	8.4	5.2	104
	75	76 ± 7	6.4	8.3	101
	300	291 ± 22	5.3	5.5	97

Table 7. Matrix effect calculated from six human plasmas spiked at 50 µg/mL (LC-MS/HRMS), or four human plasmas spiked at 15 and 300 µg/mL (LC-MS/MS).

Method	Peptide	Ion Suppression	Min; Max
LC-MS/HRMS	FSGS	−67%	−54%; −78%
	pQVQ	−11%	15%; −35%
	pQIVL	−93%	−88%; −95%
LC-MS/MS	pQVQ	−17%	0%; −34%
	pQIVL	−22%	1%; −44%

No peaks in the blank plasma samples were observed after three injections of the highest standard, indicating no carryover with both MS assays.

Pre-analytical stability tests did not show any degradation, either at −20 °C for quality controls (QCs) (Table 8) or after three freeze-thaw cycles from patient samples (Table 9). Furthermore, the post-analytical stability of RTX on autosampler (+4 °C) was checked (Table 8).

2.4. Application

As described, two MS methods for RTX quantification were validated in two different laboratories. In each laboratory, two sample preparations were performed (albumin depletion and IgG-immunocapture), and in all cases, a full-length stable isotope-labeled rituximab was used. From sixty-three plasmas from patients with vasculitis, several comparisons were conducted to evaluate the interchangeability between both MS assays.

Three intra-laboratory comparisons are shown in Figure 6 (Passing–Bablok and Bland–Altman). On LC-MS/HRMS system, we compared results obtained from the two surrogate peptides on the same extraction and injection (Figure 6A) and results obtained from two different peptides analyzed on two different days with a different sample preparation protocol (Figure 6B). For LC-MS/MS, we compared results obtained from albumin depletion and IgG-immunocapture (Figure 6C).

Table 8. Pre-analytical stability of rituximab at −20 °C over three months and autosampler stability (+4 °C, 72 h), calculated with the pQVQ peptide. Data obtained via LC-MS/MS assay.

Temperature	Concentration (µg/mL)		Precision (%)	Accuracy (%)
	Spiked	Found (mean ± s.d.)		
+4 °C	15	16 ± 1	7.9	107.5
	75	75 ± 2	3.2	101.0
	300	297 ± 6	2.0	98.8
−20 °C	15	14 ± 2	13.2	95.8
	75	72 ± 3	4.3	95.3
	300	319 ± 21	6.7	106.2

Table 9. Pre-analytical stability of rituximab at −20 °C over three months and autosampler stability (+4 °C, 72 h), calculated with the pQVQ peptide. Data obtained via LC-MS/MS assay.

Patient	FSGS			pQVQ		
	Found (µg/mL) (mean ± s.d.)	Repro. (%)	Diff. (%)	Found (µg/mL) (mean ± s.d.)	Repro. (%)	Diff. (%)
P1	100 ± 2	1.5	3.0	102 ± 7	6.5	9.5
P3	242 ± 25	10.3	−14.1	232 ± 15	6.4	−11.0
P5	244 ± 19	7.7	−12.2	229 ± 24	10.5	−6.8
P7	305 ± 8	2.7	4.6	286 ± 33	11.5	8.0
P9	<2	-	-	<2	-	-
P13	174 ± 13	7.7	−12.2	180 ± 7	4.1	−1.1
P15	139 ± 2	1.2	−1.8	147 ± 14	9.3	10.8

Repro.: Reproducibility; Diff.: Difference in concentration between the first and the forth dosage after three freezing/thawing cycles.

We also compared inter-laboratory results, as shown in Figure 7. In all cases, the Bland–Altman analysis did not show any significant bias between the two methods. The Passing–Bablock linear regression analysis also showed a good agreement between the methods whatever the surrogate peptide used, the sample preparation, or the mass spectrometer. However, we note that the comparison between results obtained from albumin depletion (Figure 7A) was less correlated than the results obtained from IgG-immunocapture (Figure 7B).

This could be explained by the fact that IgG-immunocapture produces a less complex extract than albumin depletion. The preparation of samples by albumin depletion is faster (approximately 2 h) but required many manual steps. Conversely, carrying out the IgG-immunocapture is longer (approximately 4 h), but several phases consist of incubation, agitation, and evaporation times. Regarding the cost, albumin depletion is less expensive, but the choice of the sample preparation must also take into account the production time and the selectivity of the preparation.

2.5. Comparison with Previously Published MS Methods

The analytical performances of MS methods previously published for the quantification of RTX were quite similar to the performance of the two methods described here. The LOQ (expressed in ng injected on column) of the methods were 8 ng (LC-MS/HRMS) and 20 ng (LC-MS/MS), in the same range previous MS assays using LOQ, from 1.2 to 21.1 ng [13–17]. These values, according to the sample preparation, corresponded to a LOQ from 0.586 to 5 µg/mL. However, obtaining a very low LOQ was not a major issue for us because it was reported, from a study of 166 patients treated by RTX for lymphoma, a median serum level around 6 µg/mL and 25 µg/mL in non-responders and in responders was detected, respectively [6]. Excepted for an assay based on a middle-down analysis [13], the precision of published MS methods was below 15%, as was observed in the present work. Thus, the analytical validation parameters of mass spectrometry methods meet the required criteria of the validation guidelines. In the present study, two-sample preparation protocol were tested. Albumin depletion and IgG immunocapture were simple and were quite quick to perform since proteolysis did not need human manual interventions. A previous study describes a less time-consuming sample preparation method based on nano-surface and molecular-orientation limited (nSMOL) technology and the use of a high trypsin concentration; however, the cost for each RTX quantification will be more expensive and should be taken into account for routine activity [16]. Other studies proposed sample preparation requiring more steps with human interventions such as reduction-alkylation [13] or solid-phase extraction [14]. Here we present the first comparison of two RTX quantification methods based on two different mass detector technologies performed in two separate laboratories. Thus, all the process from the patient sample to the result was separately performed in each lab. Moreover, more than 60 sample patients were used to show the applicability of both methods, whereas some assays did not present application on patient samples [14,16,17]. Finally, the robustness of both MS methods and the relevance to use a full-length stable isotope-labeled RTX-like as an internal standard was demonstrated, with the main focus being on the inter-laboratory transferability, which is an important point for the development of therapeutic drug monitoring of RTX.

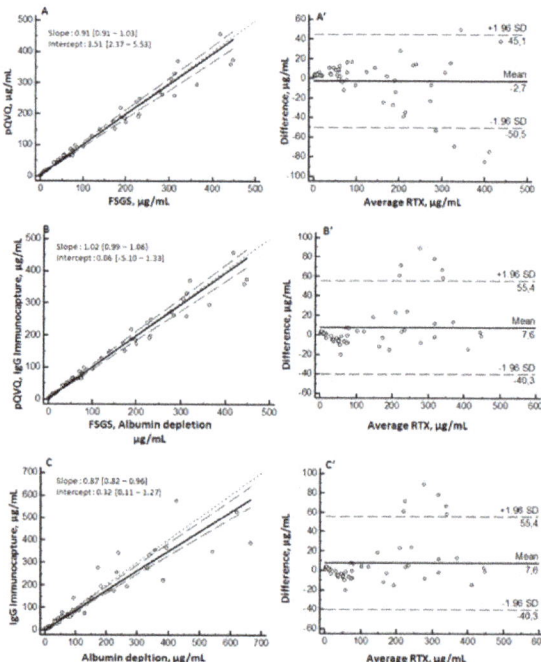

Figure 6. Intra-laboratory comparison of rituximab (RTX) quantification. Passing–Bablok regression (left), with solid line representing regression line and dashed line representing 95% confidence interval for regression line, and Bland–Altman difference-plot (right). Comparison between pQVQ and FSGS by albumin depletion protocol on the same extract sample (**A**), (**A′**); comparison between FSGS by albumin depletion and pQVQ with IgG-immunocapture (**B**), (**B′**) with LC-MS/HRMS and comparison between albumin depletion and immunocapture on pQVQ with LC-MS/MS (**C**), (**C′**). SD: standard deviation.

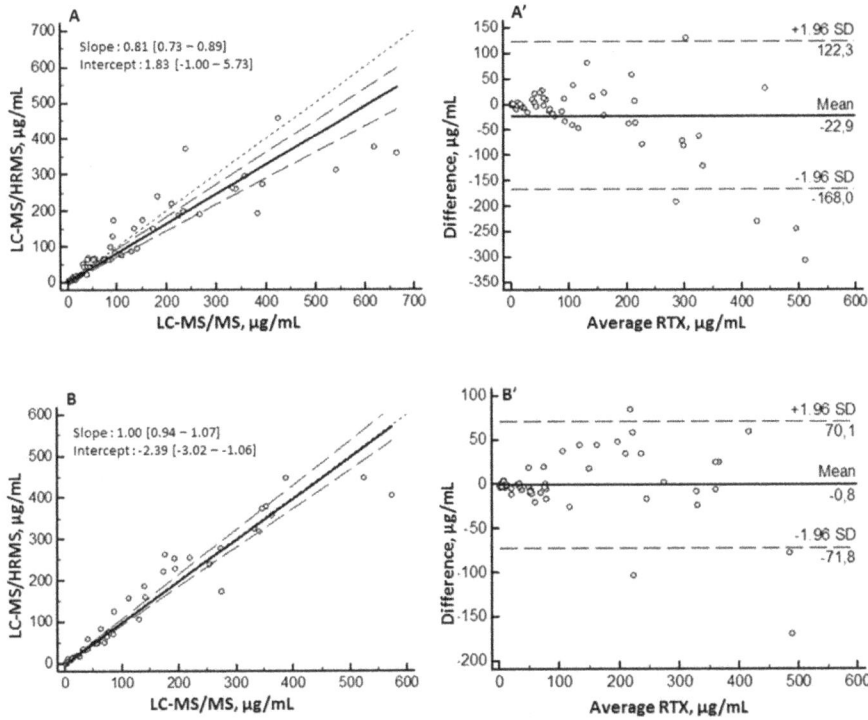

Figure 7. Inter-laboratory comparison of rituximab (RTX) quantification. Passing–Bablok regression (left), with solid line representing regression line and dashed line representing 95% confidence interval for regression line, and Bland–Altman difference-plot (right). Comparison between results obtained with the pQVQ and albumin depletion protocol on LC-MS/HRMS and LC-MS/MS (**A**), (**A'**) and comparison between results obtained with pQVQ and IgG-immunocapture protocol with LC-MS/HRMS and LC-MS/MS (**B**), (**B'**). SD: standard deviation.

3. Materials and Methods

3.1. Chemicals and Reagents

The pure solution of RTX was Truxima® (10 g/L, Celltrion Healthcare, Budapest, Hungary) for LC-MS/HRMS assay and Rixathon® (10 g/L, Sandoz, Kundl, Austria) for LC-MS/MS assay. Full-length stable isotope-labeled rituximab (Arginine 13C6-15N4 and Lysine 13C6-15N2) (SIL-RTX) was purchased from Promise Advanced Proteomics (Grenoble, France). SIL-RTX presented with a purity > 95% and labeling > 99%. Stock solutions of RTX and SIL-RTX were prepared in water at 1 g/L and at 100 mg/L, respectively, and stored at +4 °C.

ULC/MS grade acetonitrile was obtained from Biosolve (Dieuze, France) and formic acid (FA) from Fisher Chemicals (Illkirch, France). Ultrapure water (resistivity 18.2 mΩ.cm) was obtained using a Milli-Q Plus® system (Millipore, Molsheim, France). PBS buffer (pH 7.4, molarity 10X) was from Gibco (Thermo Fisher, Waltham, MA, USA). Trypsin Gold, Mass Spectrometry Grade were purchased from Promega (Madison, WI, USA). Propan-2-ol for analysis and trichloroacetic acid 20% for analysis were obtained from Carlo Erba Reagents (Val-de-Reuil, France). Ammonium bicarbonate for mass spectrometry was from Sigma-Aldrich (Saint-Quentin-Fallavier, France). Drug-free human plasma was provided by the regional blood service (EFS Rhône-Alpes and Ile-de-France, France). Low adsorption polypropylene microtubes from Dutsher (Brumath, France) were used throughout the study.

3.2. Chromatographic and Mass Spectrometric Conditions and Instrumentation

3.2.1. LC-MS/HRMS

The LC system was an UltiMate 3000 chromatographic system (ThermoScientific, USA) with two ternary pumps. An on-line SPE µ-Precolumn (Strata-X^{TM}; 20 × 2.0 mm, 25 µm, Phenomenex, Torrance, CA) was used to clean up extracted samples before the chromatographic separation, which was performed on a bioZen^{TM} Peptide PS-C18 chromatographic column (100 × 2.1 mm, 1.6µm, Phenomenex, Torrance, CA). The mobile phase was composed of water (A) and acetonitrile (B), each containing 0.1% formic acid. Pump 1 delivered an isocratic mobile phase composed of 0.1% formic acid (FA) aqueous solution and acetonitrile (95/5; v/v) through Strata-X^{TM} at 150 µL/min. Pump 2 performed the following gradient: 0–1min, 15% B; 1–8 min, 15–52.5% B; 8–10 min, 90% B; 10–16 min, 15% B (A: water, B: acetonitrile, each containing 0.1% formic acid). The chromatographic column was heated at 50 °C. A six-port valve was used to switch from precolumn to analytical column and to inject in backflush mode the 20 µL sample (Figure 8). The sample compartment in the autosampler was maintained at +4 °C.

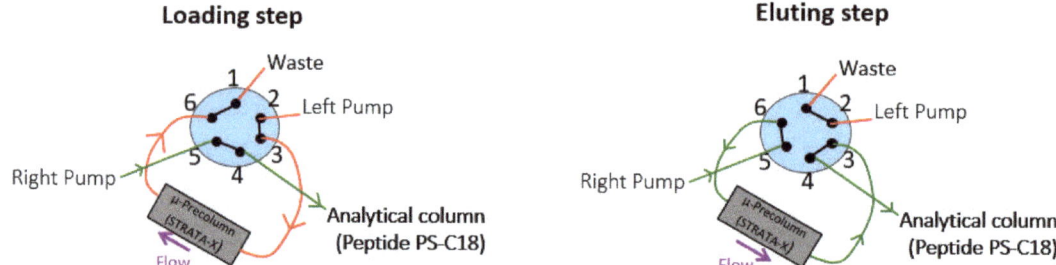

Figure 8. Switching valve used with the LC-MS/HRMS device to clean-up the sample.

Detection was performed with a Q-Exactive Plus Orbitrap mass spectrometer (Thermo Scientific, Bremen, Germany) via a heated electrospray ionization source (HESI) interface in positive ionization mode with a spray voltage of 4.0 kV and a capillary temperature of 300 °C. High-purity nitrogen gas was employed as the sheath gas (30 arbitrary units (au)) and auxiliary gas (10 au). Quantification of RTX was performed by using Parallel Reaction Monitoring (PRM) mode at a resolving power of 35,000 at m/z 200. The precursor ions filtered by the quadrupole in a 1 m/z isolation window were fragmented in a higher-energy collisional dissociation (HCD) collision cell with a normalized collision energy (NCE) of 25 au and a nitrogen collision pressure of 1.5 mTorr. Product ions were detected in the Orbitrap mass analyzer at an AGC value of 1 × 10^6 and an IT of 128 ms.

3.2.2. LC-MS/MS

LC device was a Vanquish system (ThermoScientific, USA) with a binary pump. The chromatographic conditions were close to those used for MS/HRMS method: the same column maintained at 50 °C and same mobile phase, but without the on-line extraction step. The gradient program, delivered at 400µL/min, was performed as follow: 0–1 min, 5% B; 1–6 min, 5–90% B; 6–7.5 min, 90% B; 7.5–9.5 min, 5% B (A: water, B: acetonitrile, each containing 0.1% formic acid). Extracted samples were maintained at +4 °C on the autosampler. The detector was a TSQ-Altis Triple Quadrupole mass spectrometer (Thermo Scientific, USA) with a HESI source. The instrument operated in positive ion mode, and the pressures for the nitrogen sheath gas, auxiliary gas, and sweep gas were maintained at 40, 10, and 1 au, respectively. Spray voltage and capillary temperature were set at 3.0 kV and 235 °C, respectively. The first (Q1) and the third (Q3) quadrupole were set with full-width at half maximum height of 0.7 Th. RTX was quantified in selected reaction monitoring (SRM) mode with a 20 ms dwell time. Argon was used as collision gas at 1.5 mTorr.

3.3. Selection of Peptides for Quantification

As described by El Amrani et al. [23], the first step to develop a quantification method of mAbs consisted of selecting signature peptides. For this purpose, an in silico digestion with Skyline® software (https://skyline.ms/project/home/begin.view) (accessed on 23 December 2020) followed by the analysis of the uniqueness of generated peptides with BLAST® software (http://blast.ncbi.nlm.nih.gov/Blast.cgi) (accessed on 23 December 2020) was completed. To define potential proteotypic peptides, we took into account in one hand that some peptides may undergo post-translational modifications, either on an amino-acid (methionine (di)oxidation, asparagine or glutamine deamidation, cysteine carbamylation) or such as a tryptic misscleavage, especially when an arginine/lysine was followed by a proline [24]. On the other hand, we also took into account that a transformation of N-terminal glutamic acid or glutamine in pyroglutamic acid (pE) or pyroglutamate (pQ) respectively may occur [25,26]. Then, to determine the most relevant peptides from an analytical point of view, samples obtain after trypsin digestion of pure solution of RTX and blank plasma were analyzed with LC-MS/HRMS and LC-MS/MS. The final selection of the peptides was based on the highest ratio signal/noise and the selectivity.

3.4. Sample Preparation

Validation of both methods and analysis of samples from patients were performed using a sample preparation based on albumin depletion adapted from the method described by Liu et al. [27]. To 10 µL of plasma (standard, quality control (QC) and patient samples), were added 20 µL of SIL-RTX at 15 µg/mL with LC-MS/HRMS and 25 µg/mL with LC-MS/MS in PBS and 300 µL of a mixture of isopropanol with 1% of trichloroacetic acid, in a low adsorption Eppendorf tube. After a brief vortexing step, eppendorfs were centrifuged at $2000\times g$ for 5 min. After removing the supernatant, which contains albumin, a washing step with 200 µL of methanol was carried out to remove trichloroacetic acid residues, followed by a second quick centrifugation at $2000\times g$ for 2 min. The supernatant was then removed, and the pellet was resuspended in 45 µL of ammonium bicarbonate (100 mM). Proteolysis was performed with 5 µL of Trypsin Gold at 0.2 µg/µL, and Eppendorfs were stored at 37 °C overnight. After centrifugation ($13,000\times g$, 5 min), the clear supernatant was transferred in a vial to inject 20 µL into the liquid chromatography system.

Sample preparation based on IgG immunocapture (Pierce™ Protein G Spin plate) was also tested for sample patients. The plate was washed twice with 200 µL PBS (1×), and the buffer was discarded, 20 µL of sample was mixed with 80µL of PBS (1×) containing the SIL-RTX (30 µg/mL at final concentration). Incubation for 1 h was performed with an orbital shaker at room temperature. Then the resin was washed three times with 200 µL PBS (1×), and IgG elution was obtained by applying two times 150 µL of a mixture containing water/acetonitrile (50/50, v/v, and 0.1% FA). Centrifugation ($1000\times g$, 1min) was performed, and fractions were combined, dried at room temperature, and the residue was resuspended with 45 µL of ammonium bicarbonate (100 mM). Trypsin Gold was added (1 µg/sample), and the mixture was incubated at 37 °C for 16 h. The digestion reaction was stopped by adding FA (final concentration at 1%), samples were centrifuged (5 min, $13,000\times g$), and 20 µL of supernatant were injected into the LC apparatus.

3.5. Method Validation

The selectivity of the methods was tested by analysis of six (LC-MS/MS) and ten (LC-MS/HRMS) blank plasma from patients not treated by RTX.

For the LC-MS/MS method, the calibration standard samples were prepared by spiking the blank plasma into concentrations 5–500 µg/mL (5, 10, 20, 50, 100, 250, and 500 µg/mL, three QCs were also prepared at 15, 75, and 300 µg/mL and SIL-RTX was added at a final concentration of 25 µg/mL. A total of 12 calibration curves (prepared as a single replicate and analyzed on 12 different days) were generated during the entire validation process. Ten runs included a calibration curve and QCs n three replicates.

For the LC-MS/HRMS method, the calibration standard samples were prepared by spiking the blank plasma into concentrations 10–200 µg/mL (10, 25, 50, 100, 150, and 200 µg/mL), three QCs were also prepared at 20, 80, and 160 µg/mL and SIL-RTX was added at a final concentration of 30 µg/mL. A total of 7 calibration curves (prepared as a single replicate and analyzed on 7 different days) were generated during the entire validation process. Five runs included a calibration curve and QC in six replicates.

In both cases, the accuracy and precision of the assays were assessed by the mean relative percentage deviation from the nominal concentrations and the within-run precision and between-run precision, respectively. The lower limit of quantitation (LLOQ) was tested at 2 µg/mL and 5 µg/mL for LC-MS/HRMS and LC-MS/MS, respectively. In both cases it was verified that the variance was within 20% for both precision and accuracy. For LC-MS/HRMS, the LLOQ was tested in triplicate on three different days, whereas for LC-MS/MS, the LLOQ was tested in duplicate on twelve different days. The upper limit of quantitation (ULOQ) was set as the concentration of the higher calibration standard. A dilution procedure was validated with LC-MS/HRMS method if the concentration from patient samples would be over the ULOQ. Thus, blank plasma was spiked with RTX at 500 mg/L and then diluted by 1/5 with plasma and analyzed in duplicate on three different days. The accuracy and precision should be <20%.

Carry-over was assessed by injection of three blank plasma samples after the highest calibration samples were also injected three times. This cycle was repeated twice. Peak area responses in the blank matrix samples were compared with the analyte area responses of the LLOQ of the method, and values ≤20% of the corresponding analyte response of the LLOQ level were considered acceptable.

The matrix effect was evaluated with both methods by analyzing extracted plasma spiked with RTX and compared to water samples spiked at the same final concentration. For LC-MS/HRMS, six different samples with 50 µg/mL of RTX were analyzed, and with LC-MS/MS four different plasmas at 15µg/mL (C1) and 300µg/mL (QC3) were processed. This experiment was performed on three different days.

Pre-analytical and autosampler stabilities were assessed with the LC-MS/MS method using the three levels of plasma QCs (low, medium, and high). The QCs were stored at −20 °C for 3 months and then re-analyzed. Stability on autosampler (+4 °C) was evaluated by re-analyzing the QCs samples 72 h after the first injection. The freeze-thaw stability was assessed with the LC-MS/HRMS method by re-analyzing seven samples of treated patients in triplicate following three cycles at −20 °C. In all cases, stability was confirmed if calculated bias was < +/−15% from the initial value and a precision of <15%.

3.6. Application and Method Comparison

Sixty-three patients with vasculitis were treated once monthly with RTX (Mabthera® or Rixathon®). The RTX dose depended on the period treatment: 375 mg/m^2 during induction treatment and 500 mg/m^2 during maintenance treatment. Samples were collected during the Mainritsan trial and in the context of routine clinical care. Blood samples were collected at a steady-state into tubes containing lithium heparin just prior to the next infusion (trough concentration) or at the end of infusion (peak concentration). The samples were centrifuged for 10 min at 3000× g; then plasma was frozen (−20 °C) until assay.

Medcalc software (version 7.2.1.0) was used to perform statistical analysis. The relationship between the different conditions of RTX quantification was performed by a nonparametric regression. Passing–Bablok regression analysis [28] was performed to investigate any linear relationship between the methods. The regression equation (slope and intercept) was expressed with a 95% confidence interval. Method agreement was evaluated using Bland–Altman analysis [29]. The scatter of the result from the patient samples between the two methods was also shown. The numerical results were reported as mean ± 1.96 SD. The samples below the LLOQ were excluded, and $p < 0.05$ was considered statistically significant.

4. Conclusions

We described two completely validated MS methods of quantification of RTX in plasma. To our best knowledge, these methods are the first using full-length stable-isotope-labeled RTX as an internal standard. This approach may be considered as the most robust since the same labeled mAb undergoes all the analytical steps and mimic the mAb quantified. Both MS methods were robust and reliable in terms of analytical performances and could be applied to routine clinical samples.

Author Contributions: Conceptualization, A.M., N.K., D.L., C.M., B.B., and J.G.; methodology, A.M., N.K., and D.L.; writing—original draft preparation, A.M., N.K., B.B., and J.G.; supervision, B.B. and J.G.; Resources, B.T.; All authors have read and agreed to the published version of the manuscript.

Funding: This work was supported in part by grant from Région Auvergne Rhône-alpes–France (Madmas project in Proof of concept-Cancéropole Lyon Auvergne Rhône-Alpes).

Institutional Review Board Statement: The study was conducted according to the guidelines of the Declaration of Helsinki, and approved by Ethics Committee (Comité de Protection des Personnes Île-de-France) (protocol MAINRITSAN 2). ClinicalTrials.gov Identifier: NCT01731561.

Informed Consent Statement: The authors state that they have obtained an appropriate institutional review board for the human experimental investigations. In addition, informed consent has been obtained from the participants involved.

Data Availability Statement: The data presented in this study are available on request from the corresponding author.

Acknowledgments: Authors thank ML Brandely (Department of pharmacy, Hôpital Cochin) for providing pure solutions of rituximab.

Conflicts of Interest: D. Lebert works for the company Promise Proteomics. B. Blanchet has received consultancy fees from Promise Proteomics.

Sample Availability: Samples of the compounds are not available from the authors.

Abbreviations

IgG	immunoglobulin G
CR3	complement receptor 3
Fcγ	constant fragment gamma
HC	heavy chain
LC	light chain
QVQ	peptide QVQLQQPGAELVKPGASVK
pQVQ	peptide QVQLQQPGAELVKPGASVK with pyroglutamination of N-terminal glutamine
ASGY	peptide ASGYTFTSYNMHWVK
pQIVL	peptide QIVLSQSPAILSASPGEK with pyroglutamination of N-terminal glutamine
FSGS	peptide FSGSGSGTSYSLTISR
GLEW	peptide GLEWIGAIYPGNGDTSYNQK
ASSS	peptide ASSSVSYIHWFQQKPGSSPKPWIYATSNLASGVPVR
VEAE	peptide VEAEDAATYYCQQWTSNPPTFGGGTK
STYY	peptide STYYGGDWYFNVWGAGTTVTVSAASTK
LLOQ	lower limit of quantification
PRM	parallel reaction monitoring
MRM	multiple-reaction monitoring
SD	standard deviation
QCs	quality controls
nSMOL	nano-surface and molecular-orientation limited
FA	formic acid

References

1. Kazkaz, H.; Isenberg, D. Anti B Cell Therapy (Rituximab) in the Treatment of Autoimmune Diseases. *Curr. Opin. Pharmacol* **2004**, *4*, 398–402. [CrossRef]
2. Salles, G.; Barrett, M.; Foà, R.; Maurer, J.; O'Brien, S.; Valente, N.; Wenger, M.; Maloney, D.G. Rituximab in B-Cell Hematologic Malignancies: A Review of 20 Years of Clinical Experience. *Adv. Ther.* **2017**, *34*, 2232–2273. [CrossRef]
3. Guillevin, L.; Pagnoux, C.; Karras, A.; Khouatra, C.; Aumaître, O.; Cohen, P.; Maurier, F.; Decaux, O.; Ninet, J.; Gobert, P.; et al. Rituximab versus Azathioprine for Maintenance in ANCA-Associated Vasculitis. *N Engl. J. Med.* **2014**, *371*, 1771–1780. [CrossRef]
4. Cartron, G.; Trappe, R.U.; Solal-Céligny, P.; Hallek, M. Interindividual Variability of Response to Rituximab: From Biological Origins to Individualized Therapies. *Clin. Cancer Res.* **2011**, *17*, 19–30. [CrossRef]
5. Harrold, J.M.; Straubinger, R.M.; Mager, D.E. Combinatorial Chemotherapeutic Efficacy in Non-Hodgkin Lymphoma Can Be Predicted by a Signaling Model of CD20 Pharmacodynamics. *Cancer Res.* **2012**, *72*, 1632–1641. [CrossRef] [PubMed]
6. Paci, A.; Desnoyer, A.; Delahousse, J.; Blondel, L.; Maritaz, C.; Chaput, N.; Mir, O.; Broutin, S. Pharmacokinetic/Pharmacodynamic Relationship of Therapeutic Monoclonal Antibodies Used in Oncology: Part 1, Monoclonal Antibodies, Antibody-Drug Conjugates and Bispecific T-Cell Engagers. *Eur. J. Cancer* **2020**, *128*, 107–118. [CrossRef] [PubMed]
7. Reddy, V.; Croca, S.; Gerona, D.; De La Torre, I.; Isenberg, D.; McDonald, V.; Leandro, M.; Cambridge, G. Serum Rituximab Levels and Efficiency of B Cell Depletion: Differences between Patients with Rheumatoid Arthritis and Systemic Lupus Erythematosus. *Rheumatology (Oxford)* **2013**, *52*, 951–952. [CrossRef] [PubMed]
8. Rekeland, I.G.; Fluge, Ø.; Alme, K.; Risa, K.; Sørland, K.; Mella, O.; de Vries, A.; Schjøtt, J. Rituximab Serum Concentrations and Anti-Rituximab Antibodies During B-Cell Depletion Therapy for Myalgic Encephalopathy/Chronic Fatigue Syndrome. *Clin. Ther.* **2019**, *41*, 806–814. [CrossRef]
9. Iacona, I.; Lazzarino, M.; Avanzini, M.A.; Rupolo, M.; Arcaini, L.; Astori, C.; Lunghi, F.; Orlandi, E.; Morra, E.; Zagonel, V.; et al. Rituximab (IDEC-C2B8): Validation of a Sensitive Enzyme-Linked Immunoassay Applied to a Clinical Pharmacokinetic Study. *Ther. Drug Monit.* **2000**, *22*, 295–301. [CrossRef] [PubMed]
10. Blasco, H.; Lalmanach, G.; Godat, E.; Maurel, M.C.; Canepa, S.; Belghazi, M.; Paintaud, G.; Degenne, D.; Chatelut, E.; Cartron, G.; et al. Evaluation of a Peptide ELISA for the Detection of Rituximab in Serum. *J. Immunol. Methods* **2007**, *325*, 127–139. [CrossRef]
11. Hampson, G.; Ward, T.H.; Cummings, J.; Bayne, M.; Tutt, A.L.; Cragg, M.S.; Dive, C.; Illidge, T.M. Validation of an ELISA for the Determination of Rituximab Pharmacokinetics in Clinical Trials Subjects. *J. Immunol. Methods* **2010**, *360*, 30–38. [CrossRef]
12. Liu, X.F.; Wang, X.; Weaver, R.J.; Calliste, L.; Xia, C.; He, Y.J.; Chen, L. Validation of a Gyrolab™ Assay for Quantification of Rituximab in Human Serum. *J. Pharmacol. Toxicol. Methods* **2012**, *65*, 107–114. [CrossRef] [PubMed]
13. Mills, J.R.; Cornec, D.; Dasari, S.; Ladwig, P.M.; Hummel, A.M.; Cheu, M.; Murray, D.L.; Willrich, M.A.; Snyder, M.R.; Hoffman, G.S.; et al. Using Mass Spectrometry to Quantify Rituximab and Perform Individualized Immunoglobulin Phenotyping in ANCA-Associated Vasculitis. *Anal. Chem.* **2016**, *88*, 6317–6325. [CrossRef]
14. Mekhssian, K.; Mess, J.-N.; Garofolo, F. Application of High-Resolution MS in the Quantification of a Therapeutic Monoclonal Antibody in Human Plasma. *Bioanalysis* **2014**, *6*, 1767–1779. [CrossRef] [PubMed]
15. Truffot, A.; Jourdil, J.-F.; Seitz-Polski, B.; Malvezzi, P.; Brglez, V.; Stanke-Labesque, F.; Gautier-Veyret, E. Simultaneous Quantification of Rituximab and Eculizumab in Human Plasma by Liquid Chromatography-Tandem Mass Spectrometry and Comparison with Rituximab ELISA Kits. *Clin. Biochem.* **2020**. [CrossRef]
16. Iwamoto, N.; Takanashi, M.; Hamada, A.; Shimada, T. Validated LC/MS Bioanalysis of Rituximab CDR Peptides Using Nano-Surface and Molecular-Orientation Limited (NSMOL) Proteolysis. *Biol. Pharm. Bull.* **2016**, *39*, 1187–1194. [CrossRef] [PubMed]
17. Willeman, T.; Jourdil, J.-F.; Gautier-Veyret, E.; Bonaz, B.; Stanke-Labesque, F. A Multiplex Liquid Chromatography Tandem Mass Spectrometry Method for the Quantification of Seven Therapeutic Monoclonal Antibodies: Application for Adalimumab Therapeutic Drug Monitoring in Patients with Crohn's Disease. *Anal. Chim. Acta* **2019**, *1067*, 63–70. [CrossRef]
18. LISA TRACKER–Theradiag. Available online: https://www.theradiag.com/lisa-tracker/ (accessed on 1 February 2021).
19. Du, J.; Wang, H.; Zhong, C.; Peng, B.; Zhang, M.; Li, B.; Huo, S.; Guo, Y.; Ding, J. Structural Basis for Recognition of CD20 by Therapeutic Antibody Rituximab. *J. Biol. Chem.* **2007**, *282*, 15073–15080. [CrossRef]
20. Ryan, A.M.; Sokolowski, S.A.; Ng, C.-K.; Shirai, N.; Collinge, M.; Shen, A.C.; Arrington, J.; Radi, Z.; Cummings, T.R.; Ploch, S.A.; et al. Comparative Nonclinical Assessments of the Proposed Biosimilar PF-05280586 and Rituximab (MabThera®). *Toxicol. Pathol.* **2014**, *42*, 1069–1081. [CrossRef]
21. Visser, J.; Feuerstein, I.; Stangler, T.; Schmiederer, T.; Fritsch, C.; Schiestl, M. Physicochemical and Functional Comparability between the Proposed Biosimilar Rituximab GP2013 and Originator Rituximab. *BioDrugs* **2013**, *27*, 495–507. [CrossRef]
22. Seo, N.; Huang, Z.; Kuhns, S.; Sweet, H.; Cao, S.; Wikström, M.; Liu, J. Analytical and Functional Similarity of Biosimilar ABP 798 with Rituximab Reference Product. *Biologicals* **2020**, *68*, 79–91. [CrossRef] [PubMed]
23. El Amrani, M.; Donners, A.A.M.; Hack, C.E.; Huitema, A.D.R.; van Maarseveen, E.M. Six-Step Workflow for the Quantification of Therapeutic Monoclonal Antibodies in Biological Matrices with Liquid Chromatography Mass Spectrometry-A Tutorial. *Anal. Chim. Acta* **2019**, *1080*, 22–34. [CrossRef] [PubMed]
24. Siepen, J.A.; Keevil, E.-J.; Knight, D.; Hubbard, S.J. Prediction of Missed Cleavage Sites in Tryptic Peptides Aids Protein Identification in Proteomics. *J. Proteome Res.* **2007**, *6*, 399–408. [CrossRef] [PubMed]

25. Chelius, D.; Jing, K.; Lueras, A.; Rehder, D.S.; Dillon, T.M.; Vizel, A.; Rajan, R.S.; Li, T.; Treuheit, M.J.; Bondarenko, P.V. Formation of Pyroglutamic Acid from N-Terminal Glutamic Acid in Immunoglobulin Gamma Antibodies. *Anal. Chem.* **2006**, *78*, 2370–2376. [CrossRef] [PubMed]
26. Liu, Y.D.; Goetze, A.M.; Bass, R.B.; Flynn, G.C. N-Terminal Glutamate to Pyroglutamate Conversion in Vivo for Human IgG2 Antibodies. *J. Biol. Chem.* **2011**, *286*, 11211–11217. [CrossRef]
27. Liu, G.; Zhao, Y.; Angeles, A.; Hamuro, L.L.; Arnold, M.E.; Shen, J.X. A Novel and Cost Effective Method of Removing Excess Albumin from Plasma/Serum Samples and Its Impacts on LC-MS/MS Bioanalysis of Therapeutic Proteins. *Anal. Chem.* **2014**, *86*, 8336–8343. [CrossRef]
28. Passing, H.; Bablok, W. Comparison of Several Regression Procedures for Method Comparison Studies and Determination of Sample Sizes. Application of Linear Regression Procedures for Method Comparison Studies in Clinical Chemistry, Part II. *J. Clin. Chem. Clin. Biochem.* **1984**, *22*, 431–445. [CrossRef]
29. Bland, J.M.; Altman, D.J. Regression Analysis. *Lancet* **1986**, *1*, 908–909. [CrossRef]

Article

Fast Screening of Biomembrane-Permeable Compounds in Herbal Medicines Using Bubble-Generating Magnetic Liposomes Coupled with LC–MS

Xiaoting Gu, Dongwu Wang, Xin Wang, Youping Liu and Xin Di *

Laboratory of Drug Metabolism and Pharmacokinetics, Shenyang Pharmaceutical University, 103 Wenhua Road, Shenyang 110016, China; guxiaoting320@126.com (X.G.); sydww2021@163.com (D.W.); wangxin68k@163.com (X.W.); yp-liu@163.com (Y.L.)
* Correspondence: 101040401@syphu.edu.cn; Tel.: +86-24-4352-0568

Abstract: A novel strategy based on the use of bionic membrane camouflaged magnetic particles and LC–MS was developed to quickly screen the biomembrane-permeable compounds in herbal medicines. The bionic membrane was constructed by bubble-generating magnetic liposomes loaded with NH_4HCO_3 (BMLs). The lipid bilayer structure of the liposomes enabled BMLs to capture biomembrane-permeable compounds from a herbal extract. The BMLs carrying the compounds were then separated from the extract by a magnetic field. Upon heat treatment, NH_4HCO_3 rapidly decomposed to form CO_2 bubbles within the liposomal bilayer, and the captured compounds were released from BMLs and analyzed by LC–MS. Jinlingzi San (JLZS), which contains various natural ingredients, was chosen to assess the feasibility of the proposed method. As a result, nine potential permeable compounds captured by BMLs were identified for the first time. Moreover, an in vivo animal study found that most of the compounds screened out by the proposed method were absorbed into the blood. The study provides a powerful tool for rapid and simultaneous prediction of multiple biomembrane-permeable components.

Keywords: bubble-generating magnetic liposomes; bionic membrane; permeable compounds; herbal medicines; LC–MS

1. Introduction

Chinese herbal medicines have been deemed as a momentous combinatorial chemical library of therapeutic agents [1]. Screening out bioactive compounds from complex herbal medicines is an important way to find new lead compounds in early drug discovery. Until now, the routine procedure of screening bioactive compounds from herbal medicines is to extract and isolate complex mixtures and then to evaluate the pharmacological activities of the individual compounds in vivo or in vitro [2]. However, this screening method is expensive and time consuming, with no consideration of the synergistic action of the multiple components in herbal medicines. Developing a more suitable method for rapid screening of active ingredients is helpful for the application of herbal medicines and the discovery of lead compounds. This study presents a novel screening strategy based on bubble-generating magnetic liposomes (BMLs) to screen the potential biomembrane-permeable components from herbal medicines.

The permeability of compounds across the cell membrane is an important prerequisite for screening the compounds that could exert their effects in the body. Liposomes are artificially prepared vesicles made of a phospholipid bilayer which are commonly used as oral delivery carriers for drugs and nutrients to increase absorption [3,4]. Due to their structural similarity to cell membranes, liposomes can mimic the cell membrane and have been used to predict the permeability of the compounds into cells. Recently, liposomes, as the simplest bionic membrane, have seen increasing use in screening biomembrane-permeable compounds or indicating the permeability of compounds through cell membranes. For instance, immobilized liposome chromatography [5,6] and liposome electrokinetic chromatography [7–9] are deemed to be promising approaches to investigate drug–membrane interactions, and have been employed in screening biomembrane-permeable compounds in herbal medicines. However, the brief lifespan and cumbersome preparation of the immobilized bionic membranes restrict their application. In the last few years, the strategy of screening membrane-permeable compounds in herbal medicines with liposomes by equilibrium dialysis combined with offline HPLC was developed, with the merits of easy operation and low cost [10–12]. However, the time-consuming equilibrium dialysis of the strategy has much room for improvement through burgeoning separation technologies such as magnetic nanoparticles (MNPs).

Magnetic separation technology has shown the benefits of avoiding the time-consuming steps of centrifugation, precipitation, and filtration [13]. MNPs have been widely used to separate various biomolecules or natural compounds from samples through a magnetic field [14,15]. The integration of MNPs and liposomes provides a rapid screening platform for biomembrane-permeable compounds [16–18]. In recent years, some creative strategies for controlled drug release from liposomes have been developed. Notably, bubble-generating liposomes, originally proposed by Sung [19], are produced by encapsulating NH_4HCO_3 in the aqueous cores of liposomes and used for controlled drug release in drug delivery. NH_4HCO_3 can produce CO_2 bubbles upon heat treatment, which bestows liposomes with good temperature-triggered release properties [20]. By encapsulating NH_4HCO_3 and MNPs into liposomes directly, BMLs were prepared for targeted drug delivery [21,22]. Inspired by the above research, we developed a strategy based on BMLs that integrated features of capturing fat-soluble compounds into the lipid layer, magnetic separation from impurities by a magnetic field, and thermo-responsive controlled release of captured compounds to quickly screen the biomembrane-permeable compounds in herbal medicines.

As a case study, Jinlingzi San (JLZS), composed of Fructus Toosendan (mature fruits of *MeLia toosendan* Sieb.et Zucc) and Rhizoma Corydalis (dry tubers of *Corydalis yanhusuo* W. T. Wang), was selected as a model sample. As a classical traditional Chinese medicine prescription for promoting Qi circulation to relieve pain, JLZS has been commonly used in the treatment of gastric ulcers [23]. JLZS contains various natural ingredients that provide a good resource for therapeutic agents. The experimental process was as follows (Figure 1): amino-modified magnetic nanoparticles were synthesized first, then encapsulated into liposomes loaded with NH_4HCO_3 to form BMLs. After incubating BMLs with JLZS extract, the potential biomembrane-permeable components in JLZS were captured into the lipid bilayer of BMLs. Then, the BMLs carrying the components were separated from the extract by a magnetic field. Upon heat treatment, CO_2, born by NH_4HCO_3, converged into bubbles to destroy the structure of the lipid bilayer, and the entrapped components were released and determined by LC–MS. Finally, an in vivo animal study was performed to verify whether the biomembrane-permeable components screened out by the proposed method could be absorbed into the blood of rats. The proposed screening strategy can provide a powerful tool to quickly and simultaneously predict multiple biomembrane-permeable components in herbal medicines.

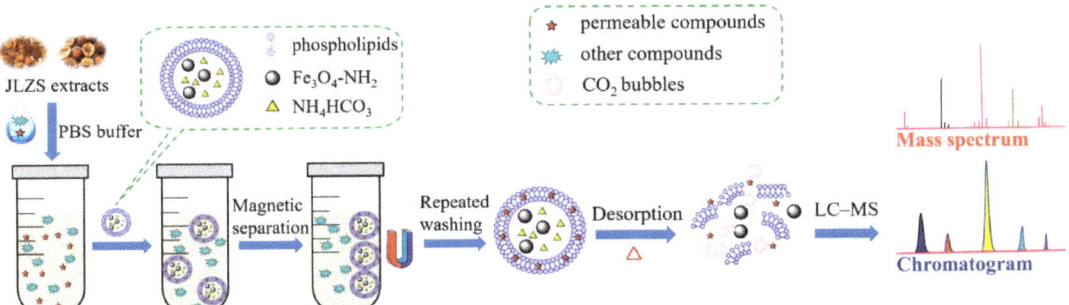

Figure 1. Schematic illustration of the proposed screening strategy using bubble-generating magnetic liposomes coupled with LC–MS.

2. Results and Discussion

2.1. Characterization of BMLs

FTIR spectra of samples are shown in Figure 2A. Amino-modified magnetic nanoparticles (AMNPs) showed a strong characteristic absorption peak at 578 cm^{-1} (asymmetric stretching vibration peak of Fe–O) [24], while transmissions around 1629 cm^{-1} and 3425 cm^{-1} could be assigned to the N–H stretching and bending vibrations of free –NH$_2$, respectively [25], confirming the successful synthesis of AMNPs. The FTIR spectrum of blank liposomes displayed absorption peaks at 1090 and 1236 cm^{-1}, attributed to symmetric and asymmetric –PO$_2$ stretching vibrations; at 1467 cm^{-1}, assigned to –CH$_2$ bending vibration; and at around 2853 and 2924 cm^{-1}, attributed to symmetric and asymmetric –CH$_2$ stretching vibrations in the acyl chain [21], respectively. It should be noted that all the characteristic peaks of blank liposomes and AMNPs were observed in the spectrum of BMLs, which confirmed the successful encapsulation of AMNPs in BMLs.

The particle size distribution of AMNPs and BMLs tested by dynamic light scattering (DLS) was recorded, as shown in Figure 2B,C. The mean particle sizes of AMNPs and BMLs were 39.5 ± 0.8 nm (n = 3) and 235.9 ± 5.2 nm (n = 3). The PDI (polydispersity index) value is an indicator to evaluate the uniformity of particle sizes present in the suspension [26]. The results showed that the PDI values of AMNPs and BMLs were 0.179 ± 0.003 (n = 3) and 0.184 ± 0.002 (n = 3), respectively, reflecting the homogenous size distribution of AMNPs and BMLs. The zeta potential value indicates the stability of nanoparticles [26], the higher the absolute value of zeta potential, the higher the stability of nanoparticles. The average zeta potentials of AMNPs and BMLs were −16.03 ± 1.01 (n = 3), and −29.17 ± 1.27 (n = 3), respectively, indicating that AMNPs and BMLs were stable.

The magnetic hysteresis loops of AMNPs and BMLs, measured by vibrating sample magnetometer (VSM) at room temperature, are shown in Figure 2D. No magnetic hysteresis was observed in magnetization curves, suggesting that AMNPs and BMLs exhibited superparamagnetic properties. Generally, the covering of phospholipid on the surface of AMNPs would decrease the saturation magnetization [21]. As expected, the saturation magnetization value of BMLs (9 emu/g) was lower than that of AMNPs (27 emu/g). Although a downturn of magnetization value was observed after encapsulating AMNPs into liposomes, the saturation magnetization of BMLs could still be used to separate the BMLs from sample solution by a magnetic field.

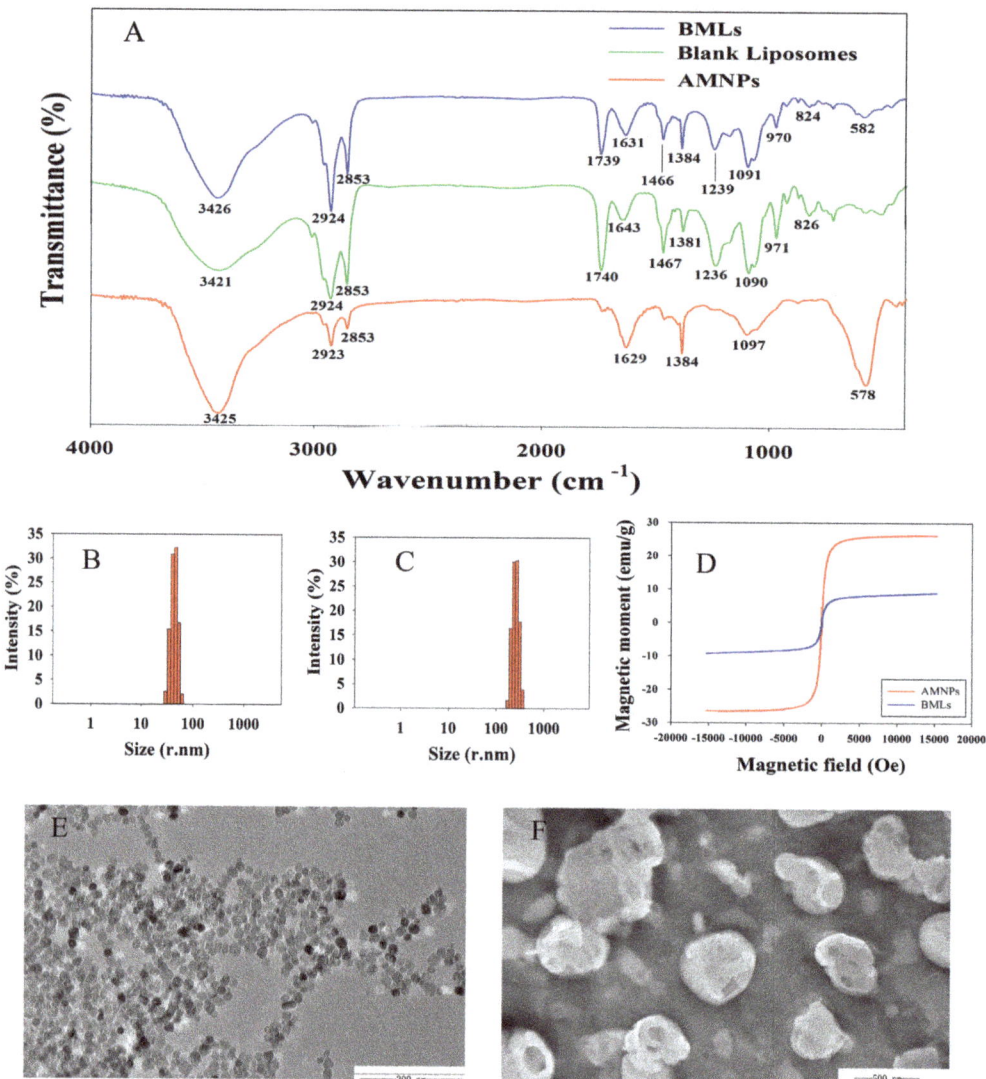

Figure 2. Characterization of bubble-generating magnetic liposomes (BMLs): FTIR spectroscopy of BMLs, blank liposomes, and amino-modified magnetic nanoparticles (AMNPs) (**A**). Particle size distribution of AMNPs (**B**) and BMLs (**C**). Magnetic hysteresis loops of BMLs and AMNPs (**D**). TEM images of AMNPs (**E**) and BMLs (**F**).

The morphology of AMNPs and BMLs deduced from transmission electron microscopy (TEM) are shown in Figure 2E,F. The discrete AMNPs observed by TEM had a spherical shape with an average size below 50 nm, and were in the range of superparamagnetic particles. From the TEM images of BMLs, it can be seen that most of the BMLs were smooth and spherical or ellipsoidal in shape. An outer bright area of lipid film and an inside dark area of magnetic nanoparticles with high electron density suggest that the AMNPs were successfully inserted in liposomes.

2.2. Optimization of Rlelease Condition

The capability of hyperthermia-induced drug release in BMLs was investigated using tetrahydropalmatine (THP), the main bioactive compoment in JLZS that possesses high permeability, as a model drug. A unique temperature-dependent drug release curve is displayed in Figure 3A. It can be seen that the drug release increased with increasing temperature. The reason of thermo-responsive drug release was that the hyperthermic temperature led to the decomposition of NH_4HCO_3 and the generation of a large number of CO_2 bubbles, triggering a rapid drug release [21]. The burst release of THP under hyperthermic conditions at 42 °C was caused by the cavitation force of exploding CO_2 bubbles inside the liposomes. Peak areas of THP released from BMLs with different heating times were recorded, as shown in Figure 3B. An increasing trend of drug release from BMLs was observed with the increase of heating time, and a release equilibrium was achieved after heating for 4 min. By placing THP-loaded BMLs in the water bath at 42 °C for 4 min, THP could be released from BMLs satisfactorily.

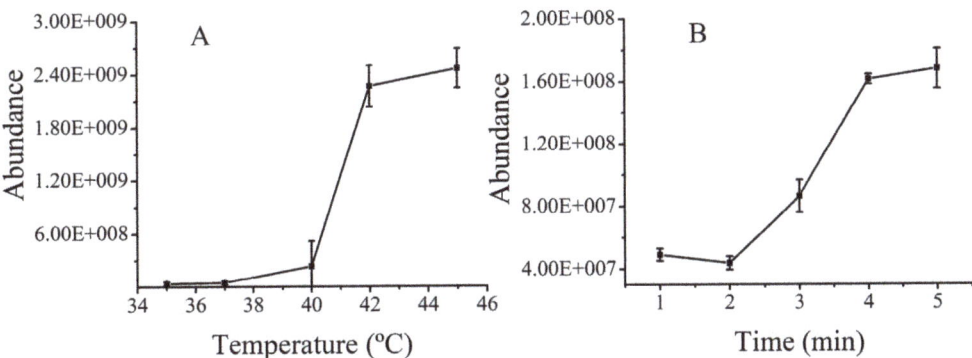

Figure 3. Optimization of the temperature and heating time for the destruction of liposomes: Peak areas of tetrahydropalmatine (THP) released from BMLs at different temperatures for 4 min (**A**). Peak areas of THP released from BMLs at 42 °C with different heating times (**B**).

2.3. Screening of Biomembrane-Permeable Compounds in JLZS

Until now, biomembrane-permeable compounds in JLZS had not been elaborated systematically. In this study, the chemical constituents of JLZS were initially identified by LC–MS. By comparing the MS splitting decomposition law and retention time with those acquired from the reference substances, 13 constituents were identified from JLZS: rutin (RUT), tetrahydropalmatine (THP), protopine (PRO), allocryptopine (ALL), jatrorrhizine (JAT), coptisine (COP), tetrahydrocoptisine (THC), tetrahydroberberine (THB), corydaline (COR), palmatine (PAL), berberine (BER), dehydrocorydaline (DHC), and toosendanin (TSN). The retention time and MS^2 fragment ions of 13 constituents are listed in Table 1. Screening and identification of the potential biomembrane-permeable compounds from JLZS were conducted using as-prepared BMLs coupled with LC–MS. Typical full-scan product ion chromatograms of JLZS extract and reconstituted solution of decomposed BMLs are shown in Figure 4A, B. Compounds C2, C6, C7, C8, C9, C10, C11, C12, and C13 were screened out and identified as THP, COP, THC, THB, COR, PAL, BER, DHC, and TSN. The chemical structures and product ion mass spectra of the nine compounds in JLZS extract are shown in Figure 5. Among them, eight compounds (THP, COP, THC, THB, COR, PAL, BER, and DHC) were attributed to protoberberine alkaloids from Rhizoma Corydalis, while TSN was the representative limonoid from Fructus Toosendan. This result indicated that protoberberine alkaloids from Rhizoma Corydalis and limonoid from Fructus Toosendan could possess stronger liposolubility and would be more likely to penetrate biofilm to reach the target and exert the potential pharmacological effects.

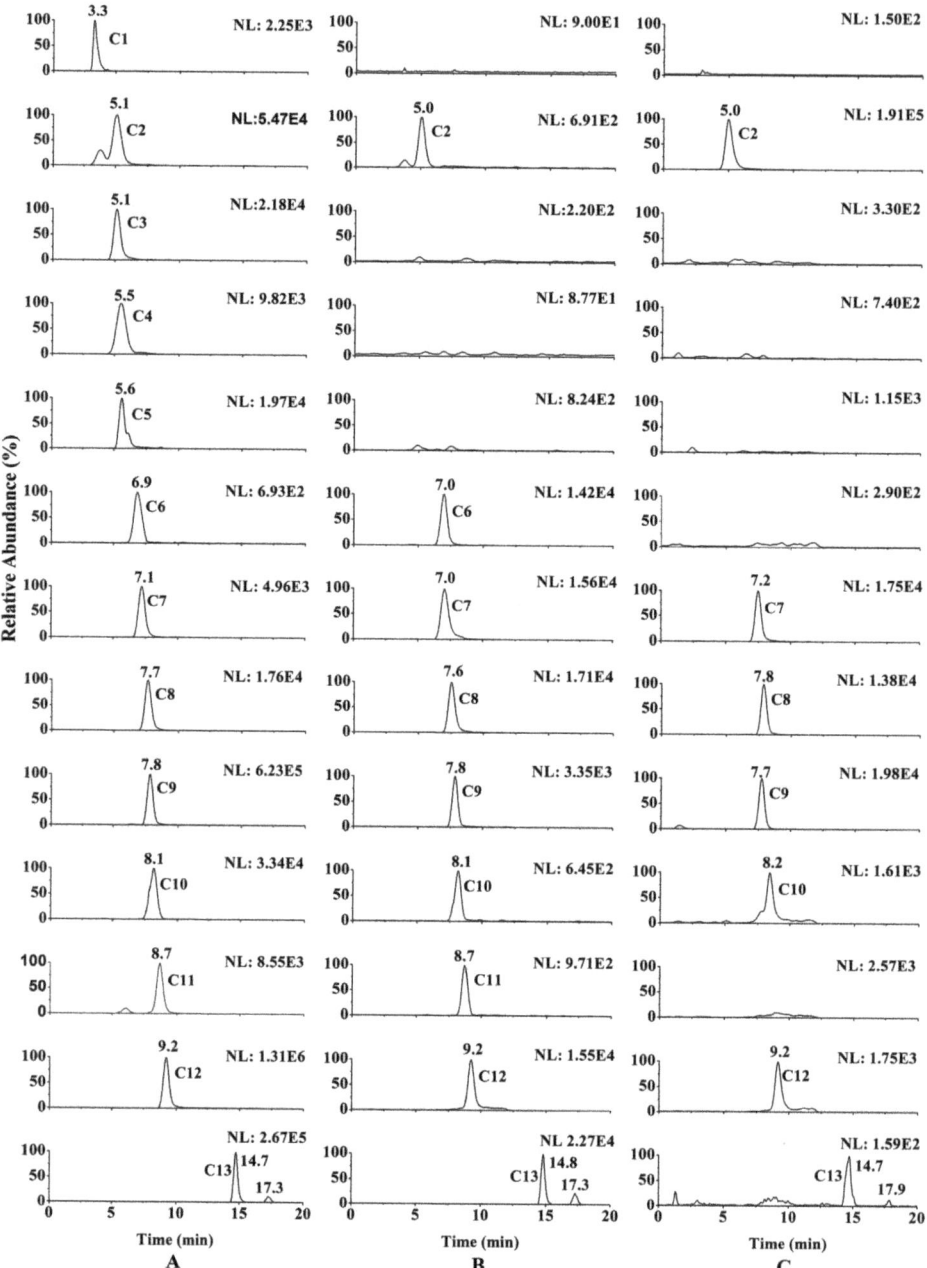

Figure 4. Typical full-scan product ion chromatograms of Jinlingzi San (JLZS) extract (**A**), reconstituted solution of decomposed BMLs (**B**), and real plasma sample at 1 h after JLZS administration (**C**).

Table 1. Retention time (t_R) and MS^2 fragment ions of thirteen components identified in JLZS.

Peak No.	Compound	Formula	t_R (min)	MS	MS^2
C1	RUT	$C_{27}H_{30}O_{16}$	3.3	609	300, 271, 178
C2	THP	$C_{21}H_{25}NO_4$	5.1	356	192, 174, 165
C3	PRO	$C_{20}H_{20}NO_5$	5.1	354	336, 275, 189, 188
C4	ALL	$C_{21}H_{24}NO_5$	5.5	370	352, 290, 206, 188
C5	JAT	$C_{20}H_{20}NO_4$	5.6	338	321, 303, 237
C6	COP	$C_{19}H_{14}NO_4$	6.9	320	292, 277, 262, 233
C7	THC	$C_{19}H_{17}NO_4$	7.1	324	176, 174, 149
C8	THB	$C_{20}H_{21}NO_4$	7.7	340	176, 149, 114
C9	COR	$C_{22}H_{27}NO_4$	7.8	370	295, 192
C10	PAL	$C_{21}H_{22}NO_4$	8.1	352	336, 321, 308
C11	BER	$C_{20}H_{18}NO_4$	8.7	336	321, 320, 292
C12	DHC	$C_{22}H_{24}NO_4$	9.2	366	350, 336, 322, 308
C13	TSN	$C_{30}H_{38}O_{11}$	14.7, 17.3	573	531

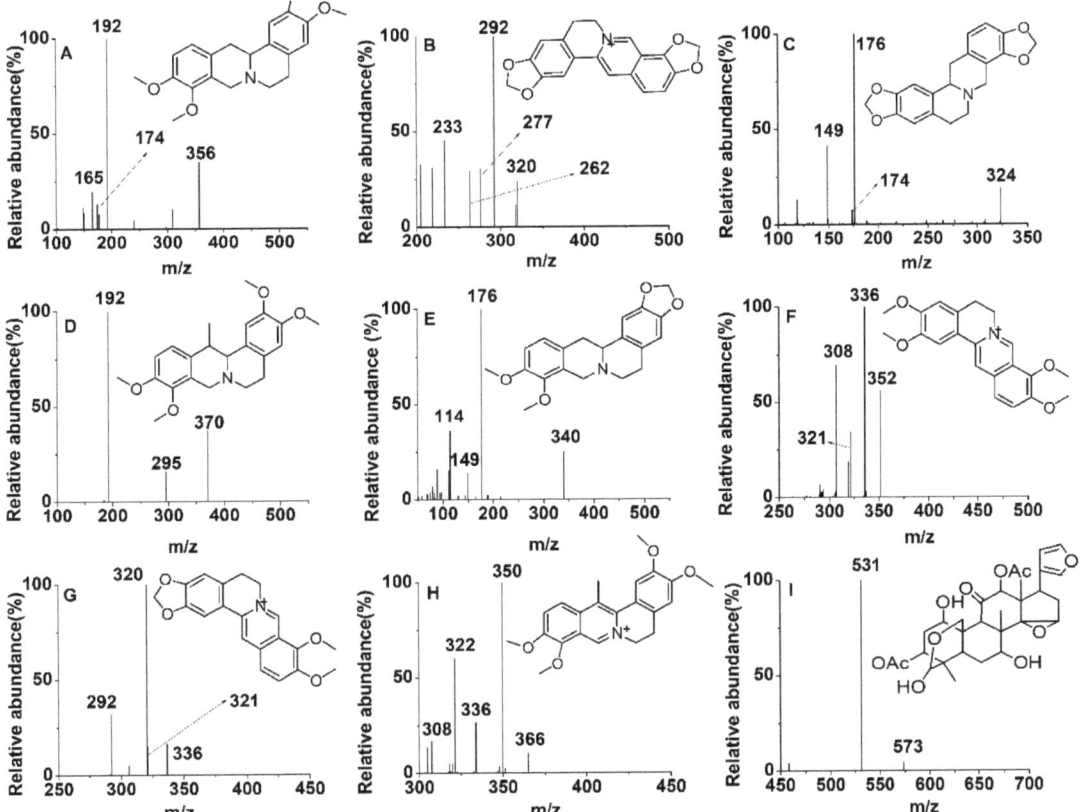

Figure 5. Product ion (MS^2) spectra and chemical structures of tetrahydropalmatine (THP) (**A**), coptisine (COP) (**B**), tetrahydrocoptisine (THC) (**C**), corydaline (COR) (**D**), tetrahydroberberine (THB) (**E**), palmatine (PAL) (**F**), berberine (BER) (**G**), dehydrocorydaline (DHC) (**H**), and toosendanin (TSN) (**I**).

2.4. In Vivo Animal Study

Biomembrane-permeable compounds in JLZS screened out by the proposed method are theoretically more likely to be absorbed into the blood through cells of the small intestine after oral administration of JLZS extract. To verify the practicability and effectiveness of the proposed method, the compounds of JLZS ingested into the blood of rats were identified by LC–MS and compared with the captured drugs screened out by the proposed method. Full-scan product ion chromatogram of a real plasma sample at 1 h after JLZS administration is shown in Figure 4C. It can be seen that seven components (THP, THC, THB, COR, PAL, DHC, and TSN) were absorbed into the blood of rats, while other compounds with weaker membrane permeability were not absorbed. This result suggests that the proposed strategy is effective and workable for the screening of biomembrane-permeable compound in herbal medicines or functional foods.

3. Materials and Methods

3.1. Chemicals and Materials

Egg yolk lecithin was provided by Lipoid Gmbh (Ludwigshafen, Germany). Cholesterol was purchased from Beijing Bailingwei Technology Co., Ltd. (Beijing, China). Ethylenediamine and ammonium bicarbonate (NH_4HCO_3) were purchased from Damao chemical reagent factory (Tianjin, China). Anhydrous sodium acetate and $FeCl_3 \cdot 6H_2O$ were obtained from Tianjin Hengxing Chemical Preparation CO., Ltd. (Tianjin, China). Ethylene glycol was provided by Tianjin Fuyu Fine Chemical CO., Ltd. (Tianjin, China). Fructus Toosendan and Rhizoma Corydalis were provided by GuoDa Pharmacy (Shenyang, China). COR, TSN, and PRO standards were obtained from Chendu Pufei De Biotech CO., Ltd. (Sichuan, China). The reference standards of THB, THC, DHC, and JAT were purchased from Chendu Herbpurify CO., Ltd. (Sichuan, China). THP, ALL, COP, BER, and palmatine chloride standards were provided by Chengdu MUST Bio-tech Co., Ltd. (Sichuan, China). The reference standard of RUT was provided by the National Institute for the Control of Pharmaceutical and Biological Products (Beijing, China). HPLC-grade acetonitrile was provided by Concord Technology CO., Ltd. (Tianjin, China). Formic acid (HPLC grade) was purchased from DIKMA Technologies, Inc. (Beijing, China). Deionized water used throughout the study was provided by Wahaha Corporation (Hangzhou, China).

3.2. Sample Preparation

The reference standards of THP, COR, THC, THB, COP, JAT, DHC, BER, PAL, PRO, ALL, RUT, and TSN were weighed accurately and dissolved in methanol/water (50:50, *v/v*) to prepare individual standard solutions with appropriate concentrations.

The smashed powder of JLZS (Fructus Toosendan 20 g + Rhizoma Corydalis 20 g) was evenly blended, then refluxed twice with 200 mL 50% ethanol for 2 h. The filtrates were pooled and condensed by rotary evaporation and then vacuum-dried. The residue was dissolved into 40 mL PBS solution and centrifuged at 4000 r/min for 30 min to obtain the supernatant JLZS extract.

3.3. Synthesis of Amino-Modified Magnetic Nanoparticles

AMNPs were synthesized according to the literature with slight modifications [25]. Briefly, 0.5 g of $FeCl_3 \cdot 6H_2O$, 1.0 g of anhydrous sodium acetate, and 5 mL of ethylenediamine were added to 15 mL ethylene glycol and quickly stirred at $50 \pm 2\ °C$, then transferred into the polytetrafluoroethylene autoclave when the solution was transparent. Reaction was carried out at $198 \pm 2\ °C$ for 6 h to obtain crude samples of AMNPs. Then, washing the crude samples with methanol and water twice, respectively, to remove residual solvent. Finally, the samples of AMNPs were dried at $50 \pm 2\ °C$. The resulting AMNPs were utilized for characterization and application.

3.4. Preparation of Bubble-Generating Magnetic Liposomes

BMLs were prepared by the thin-film-dispersion method [21]. In brief, egg yolk lecithin and cholesterol (mass ratio 3:1) were dissolved in a 3:1 (*v/v*) mixture of dichloromethane-ethanol solution assisted with ultrasound for 5 min to obtain a phospholipid solution. Next, 10 mL of this solution was transferred to a 250 mL round-bottom flask. Then, the organic solvents were removed by rotary evaporation at 35 °C to form a uniform lipid film on the bottle wall, and vacuum-dried for 12 h to remove the residual organic solvents completely. The lipid film on the bottle wall was sufficiently hydrated by rotating with 10 mL of PBS (pH 7.0) buffer, which contained 3 M NH_4HCO_3 and 0.02 g dispersed AMNPs at 20 °C for 0.5 h, sonicated using a sonicator (SCIENTZ-II D) for 10 min (200 W × 4 min, 400 W × 6 min), then passed through a microfiltration membrane of 0.8 μm to obtain a liposome preliminary product. To remove the unencapsulated free AMNPs, the liposome suspension was centrifuged at 1000× *g* for 10 min. As a comparison, PBS without AMNPs and NH_4HCO_3 were used instead of hydration solution to prepare blank liposomes in the same way.

3.5. Characterization

Fourier transform infrared (FTIR) spectroscopy of the samples was performed using a Bruker IFS-55 spectrometer (Saarbrucken, Germany), with blending and further compressing of the samples with KBr to form a pellet. A vibrating sample magnetometer (VSM) (BKT-4500Z, China) was used to measure the magnetic hysteresis loops of samples with a maximum magnetic field of 20 kOe. The particle size distribution, PDI, and zeta potential of the samples were evaluated by dynamic light scattering (DLS) using a Zetasizer (Nano ZS, Malvern, UK) by suspending the sample in deionized water. The suspensions of BMLs or AMNPs were diluted with deionized water and dropped on the copper-coated grid separately, followed by staining with 2% phosphowolframic acid (*w/v*) for 30 s. Then, the samples were dried and examined under a high-resolution transmission electron microscope (TEM) (JEM2100, JEOL, Japan). Finally, the morphology of the prepared samples was deduced from the TEM.

3.6. LC–MS Instrumentation

The LC–MS system was a Thermo TSQ Quantum Ultra triple-quadrupole mass spectrometer equipped with an electrospray ionization source (San Jose, CA, USA). LCquan quantitation software (version 2.5.6, Thermo Fisher Scientific, Waltham, MA, USA) was used to control the LC–MS system for data acquisition and analysis. Chromatographic separation was conducted using a Waters XTerra® MS C18 column (3.0 mm × 50 mm, 5 μm) maintained at 20 °C. By injecting 5 μL samples with flow rate of 0.2 mL/min, chromatographic separation was achieved using a mobile phase of 0.1% formic acid water (A) and acetonitrile (B) via gradient elution, 0–2 min, 20% B; 2–3 min, 30% B; 3–20 min, 30% B. The analytes were detected in ESI source with positive and negative ion modes in individual runs with multiple reaction monitoring.

3.7. Optimization of Release Condition

The properties of hyperthermia-induced bubble generation and drug release from BMLs were evaluated using THP as a model drug to optimize the temperature and time required for drug release from thermosensitive BMLs. First, the suspension of BMLs was incubated with THP solution at 4 °C for 30 min. BMLs carrying THP were separated from the impurities by a magnetic field, discarding the upper suspension and eluting the BMLs with PBS buffer 6 times; an equal volume of fresh PBS buffer was used to suspend the BML sample. Then, the hyperthermia-induced drug release of BMLs was studied by immersing the samples in a water bath at various temperatures (35, 37, 40, 42, 45 °C) for different amounts of time (1, 2, 3, 4, 5 min). The peak areas of THP in PBS buffer at specific conditions were determined by an LC–MS system to obtain the optimal temperature and time required for burst drug release from BMLs.

3.8. Screening of Biomembrane-Permeable Compound in JLZS

1 mL of JLZS extract was incubated with 1 mL of BMLs in a glass test tube at 4 °C for 30 min, with occasional gentle shaking for ample reaction. BMLs carrying biomembrane-permeable compounds were separated from the suspension by magnetic separation; the upper suspension was discarded and the BMLs were further eluted with PBS 6 times. Then, the BMLs were placed in a water bath at 42 °C for 4 min to release the drugs. After separating the released drugs by magnetic attraction, the top layer was collected and centrifuged at 12,000 × g for 4 min. Finally, 5 µL of the as-obtained supernatant was used to screen the permeable compounds in JLZS by LC–MS.

3.9. In Vivo Animal Study

Six healthy male Sprague Dawley rats (about 220 g) were provided by Liaoning Changsheng Biotechnology Co., Ltd. (SYXK2018-0009, Shenyang, China) and kept in a controlled environment with a 12 h/12 h light/dark cycle. The animal experiment program was approved by the Animal Ethics Committee of Shenyang Pharmaceutical University (SYPU-IACUC-C2020-12-11-101). After 12 h of fasting, but with free access to water, rats received gavage administration of 1.35 g/kg for JLZS extract (calculated as raw herbs). At 15 min, 30 min, 1 h, and 2 h after oral administration, 0.3 mL blood samples were collected from the retro-orbital vein of each rat, placed in heparinized tubes, and further centrifuged immediately at 3500× g for 10 min at 4 °C to separate the plasma. Plasma samples were pretreated by precipitation protein with methanol. In brief, we mixed 50 µL plasma with 150 µL methanol, vortexed it for 1 min, and further centrifuged it at 12,000× g for 4 min. Finally, 5 µL of supernatant was injected into the LC–MS system.

4. Conclusions

In this study, BMLs were prepared and characterized successfully, and a screening strategy using BMLs coupled with LC–MS was developed for the first time and successfully used to quickly screen the biomembrane-permeable compounds from JLZS. As a result, nine compounds (THP, COP, THC, THB, COR, PAL, BER, DHC, and TSN) of JLZS were screened out which possessed stronger liposolubility and would be more likely to penetrate biofilms to reach their target and exert potential pharmacological effects. The as-prepared BMLs are promising for the rapid screening of biomembrane-permeable compounds from herbal medicines in vitro.

Author Contributions: Conceptualization, X.G. and X.D.; Data curation, X.G.; Formal analysis, X.G. and X.D.; Investigation, X.G.; Methodology, X.G. and X.D.; Supervision, Y.L. and X.D.; Validation, D.W. and X.W.; Visualization, Y.L. and X.D.; Writing—original draft, X.G.; Writing—review and editing, X.D. All authors have read and agreed to the published version of the manuscript.

Funding: This research received no external funding.

Institutional Review Board Statement: The authors are thankful to Mengmeng Zhang and Mengyang Liu for their technical help.

Informed Consent Statement: Not applicable.

Data Availability Statement: Data are contained within the article.

Conflicts of Interest: The authors declare no conflict of interest.

Sample Availability: Samples of the compounds are not available from the authors.

References

1. Cheung, F. TCM made in China. *Nature* **2011**, *480*, S82–S83. [CrossRef] [PubMed]
2. Luo, H.; Chen, L.; Li, Z.; Ding, Z.; Xu, X. Frontal immunoaffinity chromatography with mass spectrometric detection: A method for finding active compounds from traditional Chinese herbs. *Anal. Chem.* **2003**, *75*, 3994–3998. [CrossRef]
3. Yu, M.; Song, W.; Tian, F.; Dai, Z.; Zhu, Q.; Ahmad, E.; Guo, S.; Zhu, C.; Zhong, H.; Yuan, Y.; et al. Temperature- and rigidity-mediated rapid transport of lipid nanovesicles in hydrogels. *Proc. Natl. Acad. Sci. USA* **2019**, *116*, 5362–5369. [CrossRef]

4. Liu, W.; Hou, Y.; Jin, Y.; Wang, Y.; Xu, X.; Han, J. Research progress on liposomes: Application in food, digestion behavior and absorption mechanism. *Trends Food Sci. Tech.* **2020**, *104*, 177–189. [CrossRef]
5. Wang, Y.; Kong, L.; Lei, X.; Hu, L.; Zou, H.; Welbeck, E.; Bligh, S.W.; Wang, Z. Comprehensive two-dimensional high-performance liquid chromatography system with immobilized liposome chromatography column and reversed-phase column for separation of complex traditional Chinese medicine Longdan Xiegan Decoction. *J. Chromatogr. A* **2009**, *1216*, 2185–2191. [CrossRef] [PubMed]
6. Zhang, C.; Li, J.; Xu, L.; Shi, Z. Fast immobilized liposome chromatography based on penetrable silica microspheres for screening and analysis of permeable compounds. *J. Chromatogr. A* **2012**, *1233*, 78–84. [CrossRef] [PubMed]
7. Ruokonen, S.K.; Dusa, F.; Rantamaki, A.H.; Robciuc, A.; Holma, P.; Holopainen, J.M.; Abdel-Rehim, M.; Wiedmer, S.K. Distribution of local anesthetics between aqueous and liposome phases. *J. Chromatogr. A* **2017**, *1479*, 194–203. [CrossRef]
8. Wang, Y.; Sun, J.; Liu, H.; Wang, Y.; He, Z. Prediction of Human Drug Absorption Using Liposome Electrokinetic Chromatography. *Chromatographia* **2007**, *65*, 173–177. [CrossRef]
9. Wang, T.; Feng, Y.; Jin, X.; Fan, X.; Crommen, J.; Jiang, Z. Liposome electrokinetic chromatography based in vitro model for early screening of the drug-induced phospholipidosis risk. *J. Pharm. Biomed. Anal.* **2014**, *96*, 263–271. [CrossRef]
10. Chen, X.; Deng, Y.; Xue, Y.; Liang, J. Screening of bioactive compounds in Radix Salviae Miltiorrhizae with liposomes and cell membranes using HPLC. *J. Pharm. Biomed. Anal.* **2012**, *70*, 194–201. [CrossRef]
11. Chen, X.; Xia, Y.; Lu, Y.; Liang, J. Screening of permeable compounds in Flos Lonicerae Japonicae with liposome using ultrafiltration and HPLC. *J. Pharm. Biomed. Anal.* **2011**, *54*, 406–410. [CrossRef] [PubMed]
12. Hou, G.; Niu, J.; Song, F.; Liu, Z.; Liu, S. Studies on the interactions between ginsenosides and liposome by equilibrium dialysis combined with ultrahigh performance liquid chromatography-tandem mass spectrometry. *J. Chromatogr. B* **2013**, *923–924*, 1–7. [CrossRef]
13. Borlido, L.; Azevedo, A.M.; Roque, A.C.; Aires-Barros, M.R. Magnetic separations in biotechnology. *Biotechnol. Adv.* **2013**, *31*, 1374–1385. [CrossRef] [PubMed]
14. Liu, S.; Yu, B.; Wang, S.; Shen, Y.; Cong, H. Preparation, surface functionalization and application of Fe_3O_4 magnetic nanoparticles. *Adv. Colloid Interface Sci.* **2020**, *281*, 102165. [CrossRef]
15. Hu, Q.; Bu, Y.; Zhen, X.; Xu, K.; Ke, R.; Xie, X.; Wang, S. Magnetic carbon nanotubes camouflaged with cell membrane as a drug discovery platform for selective extraction of bioactive compounds from natural products. *Chem. Eng. J.* **2019**, *364*, 269–279. [CrossRef]
16. Zhang, Y.; Xue, Q.; Liu, J.; Wang, H. Magnetic bead-liposome hybrids enable sensitive and portable detection of DNA methyltransferase activity using personal glucose meter. *Biosens. Bioelectron.* **2017**, *87*, 537–544. [CrossRef]
17. Pei, J.; Li, X.; Han, H.; Tao, Y. Purification and characterization of plantaricin SLG1, a novel bacteriocin produced by Lb. plantarum isolated from yak cheese. *Food Control* **2018**, *84*, 111–117. [CrossRef]
18. Pei, J.; Feng, Z.; Ren, T.; Jin, W.; Li, X.; Chen, D.; Tao, Y.; Dang, J. Selectively screen the antibacterial peptide from the hydrolysates of highland barley. *Eng. Life Sci.* **2018**, *18*, 48–54. [CrossRef]
19. Chung, M.F.; Chen, K.J.; Liang, H.F.; Liao, Z.X.; Chia, W.T.; Xia, Y.; Sung, H.W. A Liposomal System Capable of Generating CO_2 Bubbles to Induce Transient Cavitation, Lysosomal Rupturing, and Cell Necrosis. *Angew. Chem.* **2012**, *124*, 10236–10240. [CrossRef]
20. Cheng, Y.; Zou, T.; Dai, M.; He, X.Y.; Peng, N.; Wu, K.; Wang, X.Q.; Liao, C.Y.; Liu, Y. Doxorubicin loaded tumor-triggered targeting ammonium bicarbonate liposomes for tumor-specific drug delivery. *Colloid Surface B* **2019**, *178*, 263–268. [CrossRef]
21. Jose, G.; Lu, Y.J.; Chen, H.A.; Hsu, H.L.; Hung, J.T.; Anilkumar, T.S.; Chen, J.P. Hyaluronic acid modified bubble-generating magnetic liposomes for targeted delivery of doxorubicin. *J. Magn. Magn. Mater.* **2019**, *474*, 355–364. [CrossRef]
22. Dai, M.; Wu, C.; Fang, H.; Li, L.; Yan, J.; Zeng, D.; Zou, T. Thermo-responsive magnetic liposomes for hyperthermia-triggered local drug delivery. *J. Microencapsul.* **2017**, *34*, 408–415. [CrossRef] [PubMed]
23. Zhao, X.; Li, J.; Meng, Y.; Cao, M.; Wang, J. Treatment Effects of Jinlingzi Powder and Its Extractive Components on Gastric Ulcer Induced by Acetic Acid in Rats. *Evid. Based Complement. Alternat. Med.* **2019**, *2019*, 7365841. [CrossRef]
24. Wang, J.; Zheng, S.; Shao, Y.; Liu, J.; Xu, Z.; Zhu, D. Amino-functionalized $Fe_3O_4@SiO_2$ core-shell magnetic nanomaterial as a novel adsorbent for aqueous heavy metals removal. *J. Colloid Interface Sci.* **2010**, *349*, 293–299. [CrossRef]
25. Wang, L.; Bao, J.; Wang, L.; Zhang, F.; Li, Y. One-pot synthesis and Bioapplication of Amine-Functionalized Magnetite Nanoparticles and Hollow Nanospheres. *Chem. Eur. J.* **2006**, *12*, 6341–6347. [CrossRef] [PubMed]
26. Rasti, B.; Jinap, S.; Mozafari, M.R.; Yazid, A.M. Comparative study of the oxidative and physical stability of liposomal and nanoliposomal polyunsaturated fatty acids prepared with conventional and Mozafari methods. *Food. Chem.* **2012**, *135*, 2761–2770. [CrossRef]

Article

HPLC-UV and GC-MS Methods for Determination of Chlorambucil and Valproic Acid in Plasma for Further Exploring a New Combined Therapy of Chronic Lymphocytic Leukemia

Katarzyna Lipska [1], Anna Gumieniczek [1,*], Rafał Pietraś [1] and Agata A. Filip [2]

1 Department of Medicinal Chemistry, Medical University of Lublin, 20-090 Lublin, Poland; katarzyna.lipska@gmail.com (K.L.); rafal.pietras@umlub.pl (R.P.)
2 Department of Cancer Genetics with Cytogenetics Laboratory, Medical University of Lublin, 20-080 Lublin, Poland; a.filip@umlub.pl
* Correspondence: anna.gumieniczek@umlub.pl; Tel.: +48-81448-7382

Citation: Lipska, K.; Gumieniczek, A.; Pietraś, R.; Filip, A.A. HPLC-UV and GC-MS Methods for Determination of Chlorambucil and Valproic Acid in Plasma for Further Exploring a New Combined Therapy of Chronic Lymphocytic Leukemia. *Molecules* **2021**, *26*, 2903. https://doi.org/10.3390/molecules26102903

Academic Editors: Franciszek Główka and Marta Karaźniewicz-Łada

Received: 8 April 2021
Accepted: 10 May 2021
Published: 13 May 2021

Publisher's Note: MDPI stays neutral with regard to jurisdictional claims in published maps and institutional affiliations.

Copyright: © 2021 by the authors. Licensee MDPI, Basel, Switzerland. This article is an open access article distributed under the terms and conditions of the Creative Commons Attribution (CC BY) license (https:// creativecommons.org/licenses/by/ 4.0/).

Abstract: High performance liquid chromatography with ultra-violet detection (HPLC-UV) and gas chromatography–mass spectrometry (GC-MS) methods were developed and validated for the determination of chlorambucil (CLB) and valproic acid (VPA) in plasma, as a part of experiments on their anticancer activity in chronic lymphocytic leukemia (CLL). CLB was extracted from 250 µL of plasma with methanol, using simple protein precipitation and filtration. Chromatography was carried out on a LiChrospher 100 RP-18 end-capped column using a mobile phase consisting of acetonitrile, water and formic acid, and detection at 258 nm. The lowest limit of detection LLOQ was found to be 0.075 µg/mL, showing sufficient sensitivity in relation to therapeutic concentrations of CLB in plasma. The accuracy was from 94.13% to 101.12%, while the intra- and inter-batch precision was ≤9.46%. For quantitation of VPA, a sensitive GC-MS method was developed involving simple pre-column esterification with methanol and extraction with hexane. Chromatography was achieved on an HP-5MSUI column and monitored by MS with an electron impact ionization and selective ion monitoring mode. Using 250 µL of plasma, the LLOQ was found to be 0.075 µg/mL. The accuracy was from 94.96% to 109.12%, while the intra- and inter-batch precision was ≤6.69%. Thus, both methods fulfilled the requirements of FDA guidelines for the determination of drugs in biological materials.

Keywords: chlorambucil and valproic acid; HPLC-UV and GC-MS methods; optimization and validation; determination in plasma; combined anticancer therapy

1. Introduction

Chlorambucil (CLB) is a nitrogen mustard agent, chemically known as 4-(4-(bis(2-chloroethyl)amino)phenyl)butanoic acid (Figure 1). For many decades, CLB has been used as a standard therapy for patients with chronic lymphocytic leukemia (CLL). It acts as a bifunctional alkylating compound and possesses an ability to bind to different cellular structures by adding alkyls to a variety of electronegative groups, e.g., DNA bases. Its main anticancer activity is shown to be through cross-linking of DNA and inhibition of DNA replication [1,2]. It results in DNA fragmentation and wrong mRNA transcription from the damaged DNA. As a consequence, the cancer cells are unable to further divide [3]. CLB when administered orally is easily absorbed through the gastrointestinal tract and enters cells by simple diffusion [2]. However, the optimal dose of CLB and duration of treatment have not been established so far, despite the presence of CLB as a therapy for over 50 years [4]. Accordingly, its therapeutic concentration range in plasma is difficult to conclude and should be evaluated in a wide spectrum. In a study of 12 patients administered 0.2 mg/kg of CLB orally, a mean plasma concentration of 492 ± 160 ng/mL was measured [5].

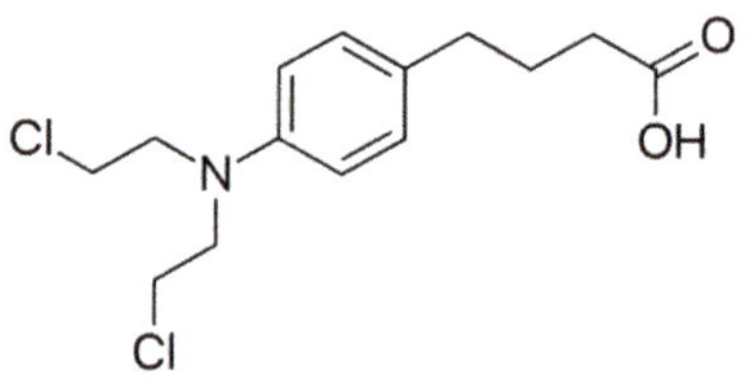

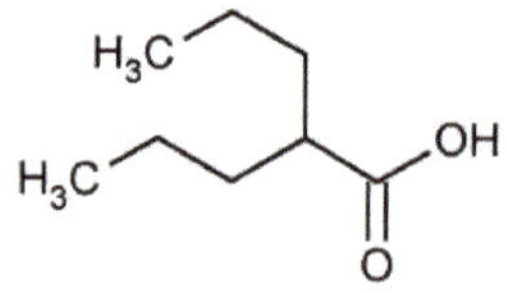

Figure 1. Chemical structures of (**a**) chlorambucil (CLB) and (**b**) valproic acid (VPA).

A review of the literature on the quantitative determination of CLB in biological materials revealed only a few methods from the 1980s to 1990s, i.e., liquid chromatographic methods with ultraviolet detection (HPLC-UV) [6–8] and one liquid chromatography–tandem mass spectrometry (LC-MS/MS) method [9]. In recent years, LC-MS/MS methods were elaborated to determine CLB in human and animal plasma, and successfully applied to pharmacokinetic studies [10,11].

Valproic acid (VPA), a chemical molecule composed of two propyl groups combined with an acetic acid moiety (Figure 1), can be classified into the group of short-chain fatty acids [12]. It is a well known anticonvulsant drug which nowadays is widely examined for new therapeutic indications [13]. Analytical methods, such as HPLC [14–18], LC-MS/MS [19–23], gas chromatography (GC) [24–26] and gas chromatography–mass spectrometry (GC-MS) [27] were reported for the determination of VPA in biological matrices. Most of the previously reported HPLC and GC methods required complex sample pretreatment or detection, e.g., multistep derivatization for UV detection [14,17], multistep derivatization for fluorimetric detection [15,16], or complex extraction [18,24–27].

Due to the results from many experimental studies showing the ability of VPA to inhibit proliferation of different cancer cells, it was proposed as a candidate for the therapy of various types of cancer. Its antitumor activity could be based on inhibiting specific enzymes, i.e., histone deacetylases, up-regulating the expression of some types of G protein-coupled receptors and affecting Notch signaling activity [28]. What is more, it could potentiate the anticancer activity of many classic cytostatic drugs used in conventional anticancer therapy [13]. That is why the present study was designed as part of a large project embracing in vitro experiments in which the impact of CLB and VPA on the viability and apoptosis of cells isolated from the blood of patients with CLL was examined. The obtained results suggested synergistic anticancer effects of CLB and VPA that might lead to the development of a new therapy for CLL [29]. Thus, the main goal of the present study was to develop, optimize, and validate the methods for quantitative measurement of CLB and VPA in biological samples for further pharmacological, and pharmacokinetic studies.

2. Results and Discussion

The main goal of the present study was to develop reliable quantitative methods for the determination of CLB and VPA in plasma, as a part of our large project focused on the potent anticancer activity of CLB and VPA in combination. From the literature survey and our preliminary experiments, it became clear that it would not be possible to develop a simple and sensitive LC-UV or GC-UV method for the simultaneous determination of these drugs. It was because VPA is a compound without a strong chromophore that does not allow detection by conventional UV spectrometry above 210 nm (Figure S1). Previously used HPLC methods required many steps for derivatization, or other sophisticated procedures. In other methods, 2-bromo-4´-nitroacetophenone or 2-bromo-2´-acetonaphtone and 18-crown-6 ether were needed [14,17]. Before this, many steps, i.e., liquid-liquid extraction, centrifugation, separation and evaporation were required. Other authors proposed precolumn derivatization with a non-commercially available reagent, i.e., nOePeS [15] or 2-(2-naphthoxy)ethyl 2-(piperidino)ethanesulfonate reagent [16], which both needed fluorimetric detection. In addition, one previously published LC-MS/MS method was based on a two step derivatization with 2-chloro-1-methylpyridinium and 4-dimethylaminobenzylamine, shaking for 1 h and using evaporation [23].

Bearing in mind the high volatility of VPA, the GC-MS method seemed to be a suitable approach to increase the sensitivity of the assay. Thus, it was taken into account and used for the determination of VPA and CLB. Unfortunately, these preliminary GC-MS experiments showed extremely low volatility of CLB and a limit of quantification above 5 µg/mL (Figure S2). Since we needed a simple and relatively cheap procedure, and having available the HPLC-UV and GC-MS apparatus in our laboratory, we decided to develop two simple independent methods, i.e., HPLC-UV for CLB, and GC-MS for VPA. As a result, the presented methods were developed, optimized and validated according to international guidelines [30,31], and finally applied for the determination of CLB and VPA in plasma, at concentrations adequate for further pharmacological and pharmacokinetic experiments.

2.1. Optimization of HPLC Conditions for CLB Assay

The chromatographic conditions were optimized over various experiments (different organic modifiers, different pH of the mobile phase, different columns) to achieve a short run time of analysis and symmetrical peaks for CLB and mefenamic acid (internal standard (I.S.)). It was found that the RP18 end-capped column and a simple mobile phase containing acetonitrile, water and formic acid at pH 2.6 were sufficiently effective for separation the analytes in a short time of below 3.5 min, as well as for the reduction of the peak tailing (Figure 2). At the same time, it was confirmed that VPA did not interfere with the assay of CLB (Figure S3).

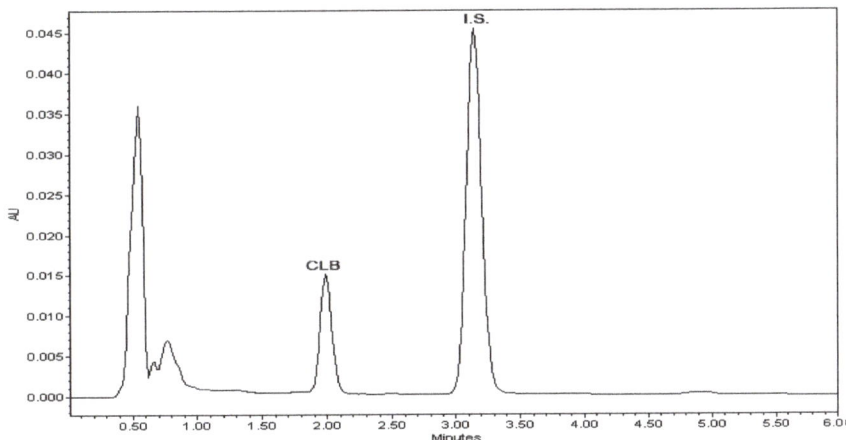

Figure 2. Representative chromatogram of plasma spiked with CLB at a concentration of 7.5 µg/mL; CLB and internal standard (I.S.) represent chlorambucil and mefenamic acid, respectively.

Six solutions containing CLB and I.S. at a concentration of 15 µg/mL were injected onto the column for estimation of the system suitability. Retention times, resolution and the peak area ratios (CLB to I.S.) were evaluated according to the obtained results. Finally, we obtained the relative standard deviation (RSD) of peak area ratios below 4.0%, the RSD of the retention times (both CLB and I.S.) below 1.0%, and resolution between the analytes above 4 (Table 1), confirming that the system suitability parameters were met [31].

Table 1. Parameters of HPLC-UV method for the determination of chlorambucil (CLB) in plasma ($n = 6$).

Parameter	Values
Retention time for CLB (mean ± SD) [min]	2.114 ± 0.019
Internal standard (I.S.)	mefenamic acid
Retention time for I.S. (mean ± SD) [min]	3.245 ± 0.026
Resolution (between CLB and I.S.)	4.2
System suitability (RSD for CLB/I.S. peak areas ratio) [%]	3.58
Linearity range [µg/mL]	0.075-15
Slope	0.15501
SD of the slope	0.00142
Intercept	0.00244
SD of the intercept	0.00123
r^2	0.9996
The limit of detection (LOD) [µg/mL]	0.024
The lower limit of quantification (LLOQ) [µg/mL]	0.075

SD = standard deviation.

What is more, the presented CLB retention time was much shorter than that of 12 min which was reported previously [7]. Mefenamic acid was selected as an optimal I.S., based on the peak shape and suitable retention time under the chromatographic conditions described above, as well as on its extraction efficiency from plasma. Using the proposed method, the centrifuged and filtered layers from deproteinized plasma samples were analyzed without evaporation and concentration steps. As a result, the time and steps required for sample preparation were visibly reduced.

2.2. Selectivity

The selectivity of the method was shown as the responses of the blank plasma (Figure 3a), zero plasma containing I.S. (Figure 3b) and the plasma sample containing CLB at the lower limit of quantification (LLOQ) level and I.S. (Figure 3c). These representative chromatograms indicated convincingly that no significant interferences from endogenous substances around the retention times of CLB and I.S. were observed.

2.3. Calibration and Sensitivity

The calibration curves showed linearity over the concentration range of 0.075–15 µg/mL with the mean correlation coefficient (r^2) equal to 0.9996. The weighed (1/x) linear regression equation y = (0.15501 ± 0.00123)x + (0.00244 ± 0.00141) was obtained (mean ± SD), where y was the peak area ratio of CLB to I.S. and x was the concentration of CLB in µg/mL.

The sensitivity of the method was determined as the lowest concentration of the standard samples within the range of quantification, with a signal-to-noise ratio of at least 10:1 (LLOQ), with an acceptable precision of less than 20%, and accuracy of 80–120% [30]. For the Quality Control (QC) samples at the LLOQ level, the mean intra-batch precision and accuracy were 9.09% and 94.13%, while the mean inter-batch values were 9.82% and 95.33%, respectively. All of these results were obtained using five replicated samples (Table 2).

While solid phase extraction (SPE) was used for the determination of CLB by the LC-MS/MS method, a limit of CLB quantification of 4 ng/mL was reported using 200 µL of plasma [9]. In turn, the HPLC-UV method offered an LLOQ of 30 ng/mL, but showed a much longer chromatographic run time (above 12 min) [7]. According to the literature, the mean plasma concentrations of CLB measured in plasma of patients after typical oral dosing were 492 ± 160 ng/mL [5]. Thus, the LLOQ proposed in the present study, which is equal to 75 ng/mL using 250 µL of plasma and a 10 µL of injection, could be acceptable for the determination of CLB in biological samples.

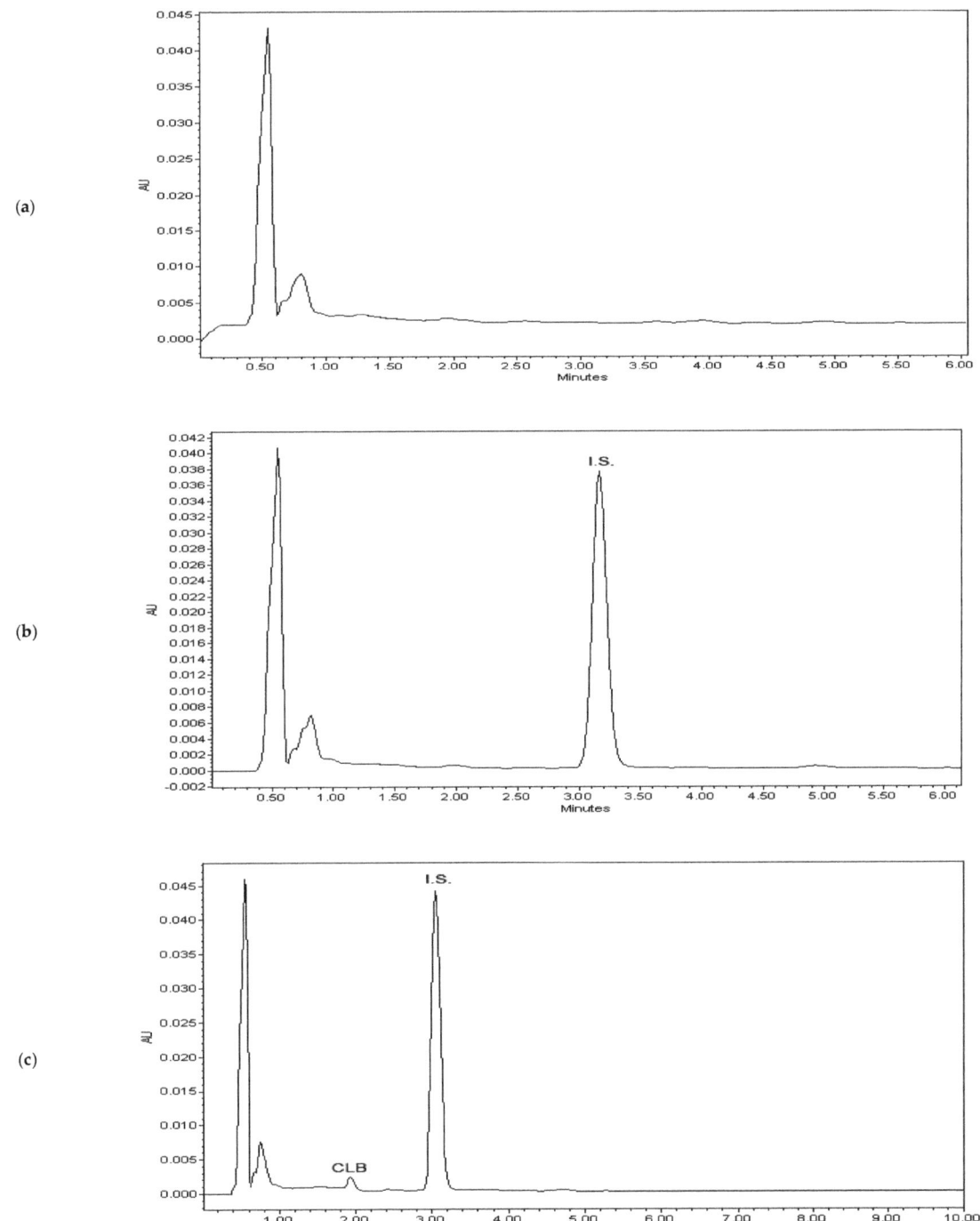

Figure 3. Representative chromatograms of (**a**) blank plasma, (**b**) zero sample with I.S. and (**c**) the lower limit of quantification (LLOQ) (0.075 µg/mL) for CLB.

Table 2. Intra-batch and inter-batch precision and accuracy of CLB assay ($n = 5$).

Spiked conc. [µg/mL]	Intra-Batch Precision and Accuracy			Inter-Batch Precision and Accuracy		
	Measured conc. Mean ± SD [µg/mL]	CV [%]	Accuracy [Recovery]	Measured conc. Mean ± SD [µg/mL]	CV [%]	Accuracy [Recovery]
LLOQ	0.0706 ± 0.0064	9.09	94.13	0.0715 ± 0.0071	9.82	95.33
QC 0.15	0.1428 ± 0.0135	9.46	95.21	0.1464 ± 0.0099	6.81	97.60
QC 3.0	3.0337 ± 0.1089	3.59	101.12	2.8991 ± 0.1166	4.02	96.64
QC 15.0	14.5761 ± 0.3694	2.53	97.14	14.2586 ± 0.4651	3.26	95.06

Conc. = concentration; CV = coefficient of variation; QC = Quality Control.

2.4. Precision and Accuracy

The acceptance criteria for the intra-batch and inter-batch precision, set at 15% for all QC samples, were achieved [30]. The intra-batch accuracy ranged from 94.13% to 101.12% with a precision (coefficient of variation, CV) of ≤9.46%. In addition, the inter-batch accuracy ranged from 95.06% to 97.60% with a precision of ≤9.82% (Table 2). These results indicated that our HPLC-UV method was sufficiently reproducible and accurate for quantification of CLB in plasma. Bearing in mind the results reported in the literature, i.e., the inter-day precision equal to 9.11% [10], the present method demonstrated a visibly comparable precision.

2.5. Recovery and Matrix Effect

The accuracy, expressed as recovery of CLB extracted from plasma in relation to nominal values, was in the range of 95.21–101.12% (Table 3). Thus, the mean recovery of the present method was higher than the 89% reported previously [9]. What is more, accurate and repeatable results were obtained using simple deproteinization with methanol followed by simple filtration. The present results also confirmed that the matrix components in plasma did not significantly affect the analytical responses. The matrix effect was evaluated by comparing the peak area ratios of CLB and I.S. after extraction from plasma, with the peak area ratios of the compounds analyzed directly. As a result, the levels from 95.45% to 100.33% were obtained for low, medium, and high concentrations of CLB (Table 3). Thus, it was suggested that significant matrix effects were not seen in the present method.

Table 3. Recovery and matrix effect of CLB in plasma ($n = 3$).

Chemical Standards [µg/mL]	Measured Conc. Mean ± SD [µg/mL]	Spiked Conc. in Plasma	Measured Conc. Mean ± SD [µg/mL]	Matrix Effect [%]
0.15	0.1497 ± 0.0071	QC 0.15	0.1428 ± 0.0135	95.45 ± 8.04
3.0	3.0471 ± 0.1315	QC 3.0	3.0337 ± 0.1089	100.33 ± 6.78
15.0	15.1031 ± 0.2509	QC 15.0	14.5761 ± 0.3694	96.55 ± 3.49

2.6. Stability

The stability of CLB was determined as the percentage ratio of the measured CLB concentration to its initial concentration in two QC samples, i.e., 0.15 µg/mL and 15 µg/mL. The results from the short-term, long-term and freeze-thaw stability tests are presented in Table 4. According to official guidelines the samples were considered stable when concentrations were within ± 15% of the nominal concentrations, and the precision was less than 15% [30]. CLB was stable in plasma at 23 °C for 24 h without visible degradation and with recovery in the range of 97.05–98.13%. CLB was also sufficiently stable in the long-term conditions with recovery in the range of 93.53–95.62%, and after three freeze-thaw cycles with recovery in the range of 93.67–94.99%. Furthermore, CLB was stable in the processed samples at 23 °C for 6 h with recovery ranging from 92.52% to 95.73%. In addition, the stock solutions for CLB and I.S. were stable for at least 3 weeks when stored at 4 °C.

Table 4. Stability data of CLB in plasma ($n = 3$).

Storage Conditions	Spiked Conc. [µg/mL]	Measured Conc. Mean ± SD [µg/mL]	RSD [%]	RE [%]
Short-term	QC 0.15	0.1472 ± 0.0054	3.67	1.87
	QC 15.0	14.5576 ± 0.3761	2.58	2.95
Long-term	QC 0.15	0.1403 ± 0.0082	5.84	2.39
	QC 15.0	14.3432 ± 0.3542	2.47	4.38
Freeze-thaw	QC 0.15	0.1405 ± 0.0089	6.33	6.33
	QC 15.0	14.2486 ± 0.3465	2.43	5.01
Post-preparative	QC 0.15	0.1423 ± 0.0071	4.99	5.13
	QC 15.0	14.1032 ± 0.2741	1.94	5.98

RE = relative error; RSD = relative standard deviation.

2.7. Carryover Test

A carryover test was conducted by injecting a blank sample after injecting the sample with the highest concentration of CLB in the standard curve (15 µg/mL). The acceptance criterion of the carryover was that the peak in the blank sample should be less than 20% of the peak in the LLOQ concentration [30]. As a result, no peak of CLB above 20% was shown in the blank sample. Thus, it was confirmed that the carryover had no effect on further analysis of CLB.

2.8. Optimization of GC-MS Method for the Determination of VPA

An important aspect of the present study was that a relevant internal standard was used, tracking VPA as a main analyte during derivatization and extraction steps, and compensating for the possible matrix effect. Benzoic acid was found to be the best option for all these purposes. Similarly to VPA, it belongs to a group of carboxylic acids, but possesses a sufficiently distinct fragmentation pattern that allowed us to obtain clean MS chromatograms without significant interferences in the selective ion monitoring (SIM) channel. To optimize our experiments, a simple derivatization and extraction method was chosen from the literature [32] and optimized for both VPA and I.S. After that, the optimum conditions were established, the compounds were mixed together, derivatized and extracted, in order to perform their simultaneous determination.

We realize that precolumn derivatization of the analytes can be considered as a difficulty by many researchers. However, it was shown to be an essential step in the presented method, visibly reducing the peak tailing and increasing sensitivity. What is more, the simple esterification with methanol followed by the simple extraction with hexane occurred sufficiently fast, and was reproducible and effective. In addition, our procedure was very cheap, not requiring any sophisticated reagents, and allowing sufficient sensitivity.

During the development of the method, the GC and MS parameters were optimized as well. The optimal choice of the SIM ions was based on the ion mass spectra of VPA and benzoic acid (I.S.). It provided three SIM ions for VPA (m/z 159, 116, and 87) and three ions for benzoic acid (m/z 136, 105 and 77), which were applied for quantitative determinations. It was found that the optimized GC-MS procedure in the SIM mode clarified the chromatograms and provided the single peaks of VPA and I.S. (Figure 4).

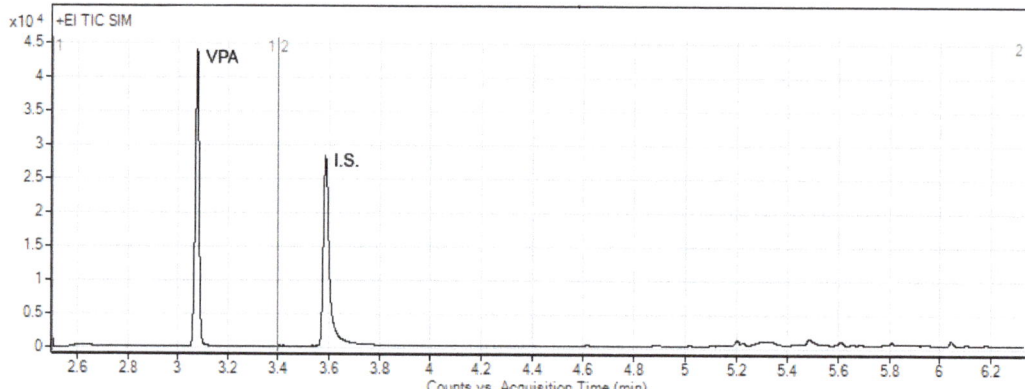

Figure 4. Representative chromatogram of plasma spiked with VPA at concentration of 7.5 μg/mL; VPA and I.S. represent valproic acid and benzoic acid, respectively.

2.9. Selectivity

The selectivity of the method was shown as the responses of (a) the blank plasma, (b) zero plasma containing I.S. and (c) the plasma sample containing VPA at an LLOQ level and I.S. Representative chromatograms demonstrating no significant interferences from endogenous substances around the acquisition times of the analytes were shown in Figure 5. At the same time, it was confirmed that CLB spiked with respective plasma samples did not interfere with the determination of VPA (Figure S2).

2.10. Calibration and Sensitivity

The calibration curves were obtained by plotting the ratios of peak areas (VPA to I.S.) versus concentrations of VPA, and were shown to be linear over the concentration range of 0.075–15.0 μg/mL, with the mean determination coefficients (r^2) equal to 0.9987 (n = 6). The weighed (1/x) linear regression equation y = (0.32003 ± 0.01209)x − (0.02029 ± 0.00279) (mean ± SD) was obtained, where y was the peak area ratio of VPA to I.S. and x was the concentration of VPA in μg/mL (Table 5).

Table 5. Parameters of GC-MS method for the determination of valproic acid (VPA) in plasma (n = 6).

Parameter	Values
Acquisition time for VPA (mean ± SD) [min]	3.076 ± 0.009
SIM ions for VPA (m/z)	159, 116, 87
I.S.	benzoic acid
Acquisition time for I.S. (mean ± SD) [min]	3.587 ± 0.008
SIM ions for I.S. (m/z)	136, 105, 77
Linearity range [μg/mL]	0.075–15.0
Slope	0.32003
SD of the slope	0.01209
Intercept	−0.02029
SD of the intercept	0.00279
r^2	0.9987
LOD [μg/mL]	0.026
LLOQ [μg/mL]	0.075

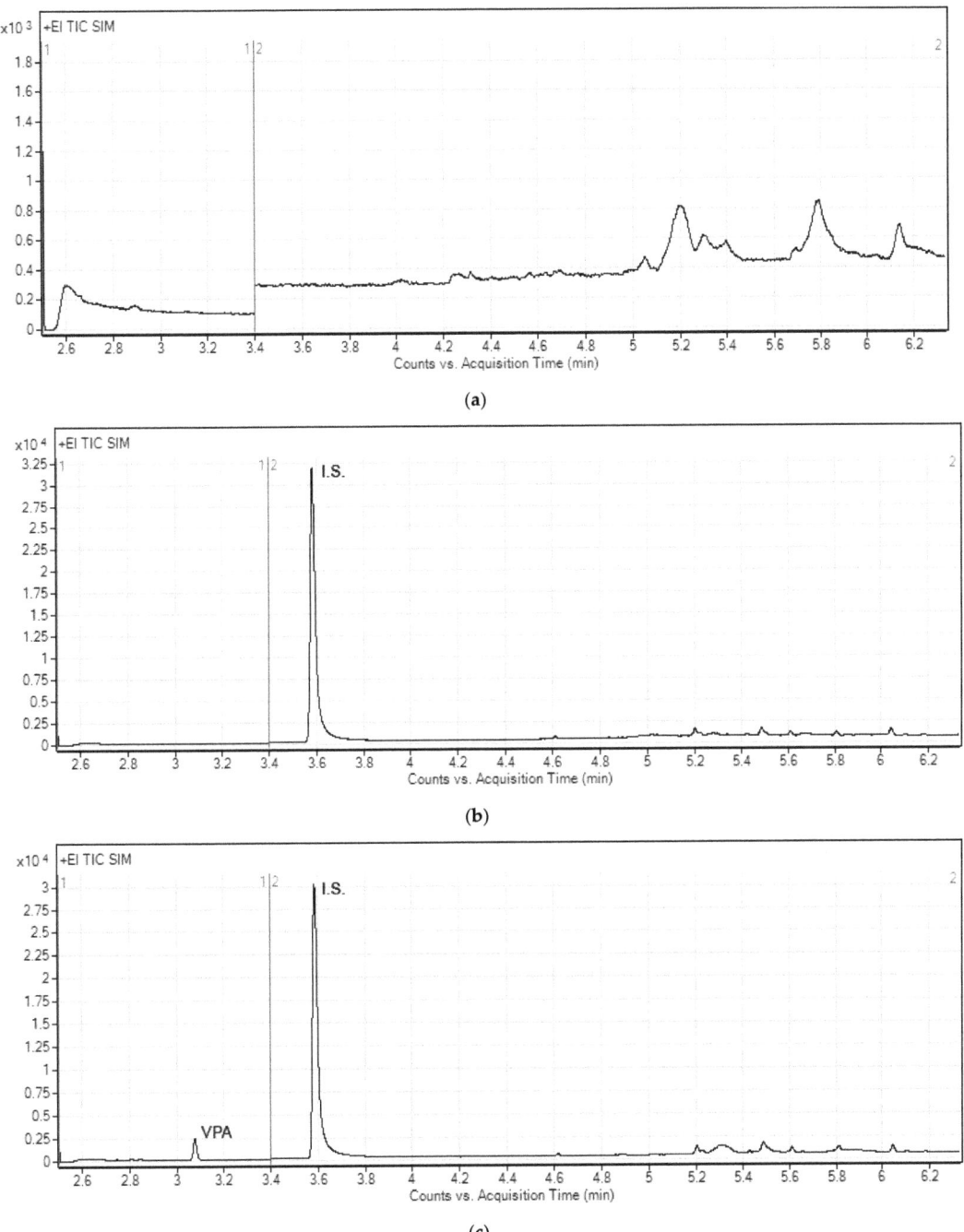

Figure 5. Representative chromatograms of (**a**) blank plasma, (**b**) zero sample with I.S. and (**c**) lower limit of quantification (0.075 µg/mL) for VPA. The selected ion monitoring (SIM) mode was used to monitor two subsets of fragments with their related mass values, the first (1) for VPA in the range 0–3.4 min, and the second (2) starting from 3.4 min for the internal standard (benzoic acid).

The sensitivity of the method (LLOQ) was determined as the lowest concentration of VPA, within the range of linearity, with a signal-to-noise ratio of at least 10:1, with an acceptable precision of less than 20%, and accuracy of 80–120% [30]. For the QC samples at the LLOQ level, the mean intra-batch precision and accuracy were 5.78% and 94.13%, while the mean inter-batch values were 6.69% and 94.96%, respectively (Table 6).

Table 6. Intra-batch and inter-batch precision and accuracy of VPA assay ($n = 5$).

Spiked Conc. [µg/mL]	Intra-Batch Precision and Accuracy			Inter-Batch Precision and Accuracy		
	Measured Conc. Mean ± SD [µg/mL]	CV [%]	Accuracy [%]	Measured Conc. Mean ± SD [µg/mL]	CV [%]	Accuracy [%]
LLOQ	0.0706 ± 0.0041	5.78	94.13	0.07122 ± 0.0048	6.69	94.96
QC 0.15	0.1426 ± 0.0042	2.96	95.07	0.1436 ± 0.0051	3.58	95.73
QC 3.0	3.2735 ± 0.1196	3.65	109.12	3.2427 ± 0.1863	5.75	108.09
QC 15.0	15.3429 ± 0.2282	1.45	102.29	15.3954 ± 0.3364	2.19	102.64

CV = coefficient of variation.

One previously described HPLC method, based on the extraction of VPA with hexane from acidified plasma, evaporation and derivatization with 2-bromo-2´-acetonaphtone, offered the LLOQ of 0.05 µg/mL using 160 µL of plasma [14]. When LC-MS/MS methods were applied, the LLOQ of 0.2–5 µg/mL using 200 µL of plasma were reported [20,21,23]. In one of these methods, MS/MS detection required the extraction of VPA from acidified plasma with methylene chloride, then two step derivatization with 2-chloro-1-methylpyridinium iodide and 4-dimethylaminobenzylamine dihydrochloride, shaking for 1 h and using evaporation [23]. In turn, the centrifugal ultrafiltration followed by a direct GC method offered the linearity range from 0.56 µg/mL [24]. Because the mean concentration of VPA in plasma of patients with epilepsy is usually between 40 and 120 µg/mL [14,20,21], these LLOQ are optimal for the determination of VPA in patients with epilepsy after typical dosing. However, effective VPA dosing in different types of cancer is yet to be discovered. On the one hand, it is expected to be similar to that in epilepsy [33]. On the other hand, it could be lower, in order to reduce the side effects like somnolence and disorientation [34]. Thus, we were focused on developing a method which could be as sensitive as possible, allowing the determination of VPA when the typical therapeutic levels of VPA were not yet achieved. As a result, we obtained a linearity range of 0.075–15 µg/mL, with the required precision and accuracy. A similar linearity range of 50–5000 ng/mL was proposed recently in the literature, using a GC-MS/MS method and the Quick, Easy, Cheap, Effective, Rugged and Safe (QuEChERS) extraction [27].

2.11. Precision and Accuracy

Precision and accuracy were determined at four concentrations of VPA (LLOQ, QC 0.15 µg/mL, QC 3 µg/mL and QC 15.0 µg/mL) on two different days, and the acceptance criteria for the intra-batch and inter-batch precision were achieved [30]. The intra-batch accuracy ranged from 94.13% to 109.12% with a precision of ≤5.78%. In addition, the inter-batch accuracy ranged from 94.96% to 108.09% with a precision of ≤6.69% (Table 6).

2.12. Recovery and Matrix Effect

The accuracy, expressed as recovery of VPA after derivatization and extraction from plasma in relation to the nominal values, was in the range 96.07–106.61%, proving the method was adequately reliable and reproducible within the required analytical range. The matrix effect was evaluated by comparing the concentrations of VPA after extraction from plasma with those analyzed directly from methanol. The obtained results of 98.39–101.31% suggested that significant matrix effects were not seen in the present method. All these results are summarized in Table 7.

Table 7. Recovery and matrix effect of VPA in plasma (n = 3).

Chemical Standards [µg/mL]	Measured Conc. Mean ± SD [µg/mL]	Spiked Conc. in Plasma [µg/mL]	Measured Conc. Mean ± SD [µg/mL]	Matrix Effect [%]
0.15	0.1441 ± 0.0014	QC 0.15	0.1427 ± 0.0042	99.03 ± 3.31
1.5	1.5991 ± 0.0979	QC 1.5	1.5735 ± 0.1196	98.39 ± 2.51
15.0	15.1453 ± 0.1891	QC 15.0	15.3429 ± 0.2282	101.31 ± 2.73

2.13. Carryover

A carryover test was conducted by injecting a blank sample after injecting the sample with the highest concentration of VPA in the standard curve (15.0 µg/mL). As a result, no peak of VPA more than 20% [30] of the peak at LLOQ concentration (0.075 µg/mL) was shown in the blank sample. Thus, it was confirmed that the carryover had no effect on further analysis of VPA.

2.14. Stability

The stability of VPA was determined as the percentage ratio of the measured VPA concentration to the initial VPA concentration in two QC samples, i.e., QC 0.15 µg/mL and QC 15.0 µg/mL. The results from stability studies are shown in Table 8. VPA was stable in plasma at 23 °C for 24 h without visible degradation, with recovery in the range of 95.06–96.87% and RSD in the range of 3.33–4.61%. VPA was also sufficiently stable in the long-term conditions with recovery in the range of 93.01–94.32% and precision in the range of 3.11–7.96%, and after three freeze-thaw cycles with recovery in the range of 86.86–96.41% and precision in the range of 2.55–8.64%. Furthermore, VPA was stable in the processed samples at 23 °C for 6 h with recovery ranging from 92.53% to 95.73%. Thus, all these ranges were within the limits recommended in the official guidelines [30]. In addition, the stock solutions for VPA and I.S. were stable for at least 3 weeks when stored at 4 °C.

Table 8. Stability data of VPA in plasma (n = 3).

Storage Conditions	Spiked Conc. [µg/mL]	Measured Conc. Mean ± SD [µg/mL]	RSD [%]	RE [%]
Short-term	QC 0.15	0.1453 ± 0.0067	4.61	3.13
	QC 15.0	14.2586 ± 0.4751	3.33	4.94
Long-term	QC 0.15	0.1395 ± 0.0111	7.96	7.01
	QC 15.0	14.1486 ± 0.4398	3.11	5.68
Freeze-thaw	QC 0.15	0.1446 ± 0.0125	8.64	3.60
	QC 15.0	13.0295 ± 0.3325	2.55	3.14
Post-preparative	QC 0.15	0.1436 ± 0.0116	8.08	4.27
	QC 15.0	13.8795 ± 0.3125	2.25	7.47

RE = relative error.

Finally, we could conclude that the elaborated GC-MS method with simple derivatization ensured the quantitative determination of VPA, with desired sensitivity, precision and accuracy. Consequently, it could be a good alternative to previously described LC-MS methods [20,21,23,24] when used for monitoring VPA levels in biological samples.

3. Materials and Methods

3.1. Chemicals and Apparatus

Pure substances, CLB, VPA, mefenamic acid and benzoic acid, were purchased from Sigma-Aldrich (St. Louis, MO, USA). Human pooled plasma (HCV, HIV, HBsAG free) was purchased from MP Biomedicals Inc. (Aurora, OH, USA). Acetonitrile, methanol, hexane and formic acid for HPLC and GC-MS were obtained from Sigma-Aldrich or Merck KGaA (Darmstadt, Germany). Sulfuric acid 98% and sodium bicarbonate for analysis were obtained from POCh (Gliwice, Poland). Deionized water was produced in our laboratory with a Simplicity UV Water Purification System from Merck-Millipore (Burlington, MA, USA). A model of 375 centrifuge from MPW Med. Instruments (Warsaw, Poland), a drying oven SL115 from Pol-Eko Aparatura Sp. J. (Wodzisław Śląski, Poland), and a pH-meter HI9024C from Hanna Instruments (Villafranca Padovana, PD, Italy) were used. Nylon membrane filters (0.45 µm) and hydrophilic PTFE syringe filters (0.2 µm) from Merck were also applied.

3.2. Statistical Analysis

Statistica version 13.3 from TIBCO Software Inc. (Palo Alto, CA, USA) and free GNU R computational environment version 3.4.0 were used for the statistical analysis.

3.3. Validation of the Methods

The elaborated methods were validated according to the international regulations recommended by the Food and Drug Administration Agency [30,31]. Validation was performed by assessing selectivity, linearity, sensitivity, precision and accuracy, recovery and matrix effect, carryover and stability.

3.4. Selectivity

The selectivity of the methods was shown as the response of (a) the blank plasma samples without both drugs and respective I.S., (b) zero plasma samples without drugs but containing respective I.S., and (c) the plasma samples at the LLOQ concentration of CLB or VPA and I.S.

3.5. Linearity Estimation

Six point calibration curves were constructed by plotting the peak area ratios of CLB or VPA to respective I.S. (y) versus CLB or VPA concentrations (x). The slope, intercept and correlation coefficient r^2 were calculated by weighted linear regression ($1/x$).

3.6. LOD and LLOQ

The limits of detection (LOD) were determined from the SD of the intercept and the slope of the mean regression lines. The lowest limits of quantification (LLOQ) were determined by analyzing progressively lower concentrations of CLB or VPA under the procedures described below, using a serial dilution method. The obtained LLOQ values were based on the signal-to-noise ratio of 10:1 which was determined by visual inspection.

3.7. Stability Tests

The stability of all stock solutions was examined after storage for 3 weeks at 4 °C. In addition, stability of CLB and VPA in plasma was assessed for short-term, long-term and freeze-thaw storages. Two different QC samples at low and high concentrations, i.e., 0.15 µg/mL and 15 µg/mL of CLB or VPA, were used for all stability tests. The samples were analyzed five times for every storage condition. The short-term stability test was conducted by maintaining the QC samples at 23 °C for 24 h, while the long-term stability was determined by freezing QC samples at −20 °C for 2 weeks. The QC samples were also stored at −20 °C for 24 h and then thawed at 23 °C. These steps were repeated three times for estimation of the freeze-thaw stability. In addition, the processed QC samples after extraction from plasma were checked after being placed on the table (CLB) or in an autosampler (VPA) at 23 °C for 6 h, for estimation of the post-preparative stability.

3.8. Carryover Tests

Carryover tests were conducted by injecting blank samples without the drugs after injecting the samples with the highest concentration of CLB or VPA in their calibration ranges (15 µg/mL).

3.9. Chromatographic Conditions for CLB Assay

The HPLC-UV method was performed using a model 515 pump, a Rheodyne 10 µL injector and a model UV 2487 detector, controlled by Empower 3 software, all from Waters UK (Elstree, Herts, England). Separation was performed on a LiChrospher®100 RP18 end-capped column (125 × 4.0 mm, 5 µm particle size) from Merck. The mobile phase was a freshly prepared mixture of acetonitrile and water (60:40, v/v) adjusted to pH 2.6 with formic acid, filtered using nylon membrane filters (0.45 µm) and degassed. The flow rate of the mobile phase was 1.0 mL/min, while the UV detection was set at 258 nm. All analyses were performed at room temperature 23 °C. Mefenamic acid was used as an I.S.

3.10. System Suitability

System suitability was determined before the sample analysis. Six solutions containing 15 µg/mL of CLB and 15 µg/mL of I.S. were prepared in methanol and injected onto the column. The similarity of peak area ratios of CLB to I.S. and similarity of retention times of both CLB and I.S. were taken into account.

3.11. Solutions for Calibration

The stock solution of CLB was prepared by dissolving the pure substance in methanol to obtain the concentration of 1.5 mg/mL. The stock solution of I.S. was prepared by dissolving the pure mefenamic acid in methanol to obtain the concentration of 1.0 mg/mL. The stock solution of CLB was diluted with methanol to obtain the working solutions at concentrations of 0.15 mg/mL (C1 solution), 0.075 mg/mL (C2 solution), 0.015 mg/mL (C3 solution), 0.0075 mg/mL (C4 solution), 0.0015 mg/mL (C5 solution) and 0.00075 mg/mL (C6 solution).

3.12. Calibration

A 100 µL volume of the C6-C1 solutions of CLB and 650 µL of the solution of I.S. (1.0 mg/mL) were added to 250 µL of plasma, to obtain CLB concentrations of 0.075 µg/mL, 0.15 µg/mL, 0.75 µg/mL, 1.5 µg/mL, 7.5 µg/mL and 15 µg/mL. The samples were thoroughly vortexed and centrifuged at 5000 rpm at 23 °C for 5 min. The supernatants were injected onto the HPLC system through hydrophilic PTFE syringe filters (0.2 µm).

3.13. Precision and Accuracy

The stock solution of CLB was obtained in methanol at a concentration of 1.5 mg/mL and used for the QC samples at concentrations of 0.15 mg/mL (QC1 solution), 0.03 mg/mL (QC2 solution), 0.0015 mg/mL (QC3 solutions) and 0.00075 mg/mL (QC4 solution). These QC solutions were stored in the freezer at −20 °C and brought to room temperature for precision and accuracy, as well as recovery and stability studies.

The QC samples in plasma were prepared by adding 100 µL volumes of QC4-QC1 solutions to 250 µL of plasma and spiking with 650 µL of the I.S. working solution (1.0 mg/mL). In this way, CLB concentrations of 0.075 µg/mL (LLOQ level), 0.15 µg/mL (a low level), 3 µg/mL (a medium level), and 15 µg/mL (a high level) were obtained. The samples were thoroughly vortexed and centrifuged at 5000 rpm at 23 °C for 5 min. The supernatants were filtered through hydrophilic PTFE syringe filters and injected onto the HPLC system. The intra-batch precision and accuracy were determined by analyzing the QC samples five times on the same day. The inter-batch precision and accuracy were determined by analyzing the QC samples on two different days.

3.14. Recovery and Matrix Effect

The recovery of CLB from plasma was assessed in the QC samples at low (0.15 µg/mL), medium (3 µg/mL) and high (15 µg/mL) levels in five replicates, and calculated by comparing the determined concentrations with the nominal values. The matrix effect was evaluated by comparing the peak area ratios of CLB to I.S. which were obtained for the samples extracted from plasma, with the peak area ratios of the analytes at the same concentrations analyzed directly.

3.15. Chromatographic Conditions for VPA Assay

A GC chromatograph 7890A equipped with a 7692AALS autosampler coupled to a 7000A triple quadrupole mass spectrometer from Agilent (Agilent Technologies Deutschland GmbH, Waldbronn, Deutschland) was used. Separation was done using an HP-5MS UI column (30 m × 0.25 mm, 0.25 μm film thickness) from Agilent. Ionization was carried out by an electron impact (EI) ionization with SIM mode. Three SIM ions for VPA at m/z 159, 116, and 87, and for benzoic acid (I.S.) at m/z 136, 105 and 77 were selected for the quantitative determinations. The mass spectrometer was tuned manually using perfluorotributylamine from Sigma-Aldrich, with m/z 69 and 502. Helium was used as the carrier gas at a constant flow rate of 1 mL/min. The operating parameters used were, the injector temperature 250 °C, source temperature 280 °C, MS transfer line temperature 300 °C and quadrupole temperature 150 °C. The starting oven temperature was 80 °C (1.2 min held) and it increased at a rate of 25 °C/min to 250 °C (2 min held). Injections were done in the split mode with the split ratio 10:1. Standard electron impact conditions (70 eV) were used. For collision induced dissociation (CID), ultra-high purity nitrogen (1.5 mL/min) was used and 25 V as a collision cell voltage was applied. Helium gas (2.25 mL/min) was also used as a quenching gas to eliminate the meta stable helium species. The chromatographic data were analyzed by MassHunter Data Analysis Reporting B.07.01 software from Agilent.

3.16. Solutions for Calibration

The stock solution of VPA was prepared by dissolving the substance in methanol to obtain the concentration of 1.5 mg/mL. It was diluted with methanol to obtain working solutions at the following concentrations: 0.15 mg/mL (C1 solution), 0.075 mg/mL (C2 solution), 0.015 mg/mL (C3 solution), 0.0075 mg/mL (Q4 solution), 0.0015 mg/mL (Q5 solution) and 0.00075 mg/mL (C6 solution). The stock solution of benzoic acid (I.S.) was prepared at a concentration of 1.0 mg/mL and diluted with methanol to obtain the working concentration of 0.05 mg/mL.

3.17. Derivatization and Calibration

To obtain VPA concentrations of 0.075 μg/mL, 0.15 μg/mL, 0.75 μg/mL, 1.50 μg/mL, 7.5 μg/mL and 15.0 μg/mL, aliquots of 100 μL of C6-C1 solutions and 650 μL of the I.S. working solution (0.05 mg/mL) were added to 250 μL aliquots of plasma. The samples were thoroughly vortexed and centrifuged at 5000 rpm at 4 °C for 5 min. Supernatant volumes of 200 μL were placed in chemically inert glass vials and subjected to derivatization and extraction procedures. Derivatization was carried out via simple esterification with methanol, according to the literature [32]. In short, volumes of 2 mL of sulphuric acid in methanol (10%, v/v) were added to each sample and thoroughly mixed. Then, the vials were heated at 60 °C for 30 min. After cooling to room temperature, 1 mL volumes of saturated sodium bicarbonate solution and 1 mL volumes of hexane were added to each sample for extraction. The mixtures were vortexed and centrifuged at 5000 rpm at 4 °C for 5 min. Then, aliquots of 500 μL of the upper (organic) layers were transferred to GC-MS vials, filled up with hexane to an equal volume of 1.5 mL and taken for GC-MS analysis.

3.18. Precision and Accuracy

The stock solution of VPA at a concentration of 1.5 mg/mL was prepared by independent weighing of the pure substance. Then, it was diluted to obtain the concentrations of 0.15 mg/mL (QC1 solution), 0.03 mg/mL (QC2 solution), 0.015 mg/mL (QC3 solution) and 0.0075 mg/mL (QC4 solution). These QC solutions were stored in the freezer at −20 °C and brought to room temperature before use.

The QC samples at concentrations of 0.075 μg/mL (LLOQ level), 0.15 μg/mL (a low level), 3.0 μg/mL (a medium level) and 15.0 μg/mL (a high level) were prepared by adding 100 µL volumes of QC4-QC1 solutions to 250 µL aliquots of plasma, and spiking with 650 µL of the I.S. working solution (0.05 mg/mL). The samples were mixed and centrifuged at 5000 rpm at 4 °C for 5 min. Then, 200 µL of the supernatants were subjected to derivatization and extraction procedures described above, and taken for GC-MS analysis.

3.19. Recovery and Matrix Effect

The recovery of VPA from plasma was assessed in the QC samples at low (0.15 µg/mL), medium (3 µg/mL) and high (15.0 µg/mL) levels in five replicates, and calculated by comparing the measured concentrations with the nominal values. The matrix effect was evaluated by comparing the peak area ratios of VPA to I.S. obtained after derivatization and extraction from plasma, with those obtained for the samples derivatized directly from methanol. In short, the solutions of VPA at three concentrations, i.e., 0.15 mg/mL, 0.03 mg/mL and 0.0015 mg/mL were prepared. Next, the 100 µL volumes of each VPA solution and 650 µL of I.S. working solution (0.05 mg/mL) were mixed and filled up with methanol to an equal volume of 1 mL. Then, 200 µL volumes were taken for esterification followed by heating at 60 °C, extraction with hexane and finally GC-MS analysis, as described above.

4. Conclusions

Accumulating data shows that VPA belongs to a group of epigenetic modifying agents and can effect many alterations in cancer cells, through modulation of multiple signaling pathways. Thus, VPA could be effective in some anticancer therapy, especially in combination with other anticancer drugs, e.g., CLB. Our in vitro study (sent for publication, under review) confirmed potent anticancer activity of VPA in CLL cells, especially due to the enhancement of CLB activity when used in combination. Based on these results, we can conclude that the combined treatment with VPA and CLB might be a new alternative in the therapy of CLL. Thus, two simple quantitative HPLC-UV and GC-MS methods were elaborated and validated for further studies of CLB and VPA in cultured cells and plasma, and to examine pharmacological aspects of such a combined therapy. As was stated above, in the literature there is no method for the simultaneous determination of CLB and VPA in biological materials. What is more, there are a limited number of procedures for the determination of CLB, as well as a limited number of simple methods for the determination of VPA. As a consequence, their determination in plasma is an ongoing challenge for researchers. On the one hand, it was not possible for us to develop one simple method for the simultaneous determination of CLB and VPA with the required sensitivity, accuracy and precision, mainly due to large differences in their physico-chemical properties. On the other hand, the presented methods allow the determination of CLB and VPA in the same plasma samples, without any mutual interference. As far as the required sensitivity, linearity, precision, accuracy, matrix effect, carryover and stability are concerned, they could be good alternatives to previously described HPLC or LC-MS methods. In addition, we think that the GS-MS method as such, should be promoted on a large scale in medical institutions at all levels as a suitable method for bioanalytical studies. What is more, both methods could be easily adopted for monitoring of CLB and VPA in a variety of in vitro as well as ex vivo scenarios, e.g., cell cultures.

Supplementary Materials: The following are available online, Figure S1: Representative LC chromatogram obtained for VPA at concentration of 30 µg/mL, using UV detection at 210 nm; Figure S2. Representative GC chromatograms of plasma samples spiked with VPA and CLB (both at concentration of 15 µg/mL) at two different scales (a,b) and the MS-SIM mode spectrum of CLB (c); Figure S3. HPLC chromatograms of: (a) CLB (30 µg/mL) with the internal standard (I.S.), (b) CLB (30 µg/mL), I.S. and VPA (30 µg/mL) and (c) VPA (30 µg/mL) and I.S. (VPA was not detected because of a lack of respective chromophores and absorptivity at 258 nm).

Author Contributions: Conceptualization, A.G., K.L. and R.P.; methodology, A.G., K.L. and R.P.; Software, A.G. and K.L.; validation, A.G. and K.L.; formal analysis, K.L. and R.P.; investigation,

K.L. and R.P.; resources, A.G. and A.A.F.; data Curation, A.G., K.L. and R.P; writing—original draft preparation, A.A.F., A.G. and K.L.; writing—review and editing, A.G.; supervision, A.A.F. All the authors approved the final manuscript for publication. All authors have read and agreed to the published version of the manuscript.

Funding: This research received no external funding.

Institutional Review Board Statement: Not applicable.

Informed Consent Statement: Not applicable.

Data Availability Statement: The data presented in this study are available on request from the corresponding author.

Conflicts of Interest: The authors declare no conflict of interest. The funders had no role in the design of the study; in the collection, analyses, or interpretation of data; in the writing of the manuscript, or in the decision to publish the results.

Sample Availability: Samples of chlorambucil and valproic acid are not available from the authors.

References

1. Zhang, W.; Zhu, W.; He, R.; Fang, S.; Zhang, Y.; Yao, C.; Ismail, M.; Li, X. Improvement of stability and anticancer activity of chlorambucil-tetrapeptide conjugate vesicles. *Chin. J. Chem.* **2016**, *34*, 609–616. [CrossRef]
2. Goede, V.; Eichhorst, B.; Fischer, K.; Wendtner, C.M.; Hallek, M. Past, present and future role of chlorambucil in the treatment of chronic lymphocytic leukemia. *Leuk. Lymphoma.* **2015**, *56*, 1585–1592. [CrossRef] [PubMed]
3. Dey, S.; Samal, H.B.; Monica, P.; Reddy, S.; Karthik, G.; Gujrati, R.; Geeta, B. Method development and validation for the estimation of chlorambucil in bulk and pharmaceutical dosage forms using UV-VIS spectrophotometric method. *J. Pharm. Res.* **2011**, *4*, 3244–3246.
4. Catovsky, D.; Else, M.; Richards, S. Chlorambucil-still not bad: A reappraisal. *Clin. Lymphoma Myeloma Leuk.* **2011**, *11*, 2–6.
5. Leukeran, (Chlorambucil Tablets). Available online: https://www.accessdata.fda.gov/drugsatfda_docs/label/2010/010669s030lbl.pdf (accessed on 12 May 2021).
6. Adair, C.G.; McElnay, J.C. Studies on the mechanism of gastrointestinal absorption of melphalan and chlorambucil. *Cancer Chemother. Pharmacol.* **1986**, *17*, 95–98. [CrossRef] [PubMed]
7. Oppitz, M.M.; Musch, E.; Malek, M.; Riib, H.P.; Gerd, E.U.; Loos, U.; Miihlenbrnch, B. Studies on the pharmacokinetics of chlorambucil and prednimustine in patients using a new high-performance liquid chromatographic assay. *Cancer Chemother. Pharmacol.* **1989**, *23*, 208–212. [CrossRef] [PubMed]
8. Lof, K.; Hovinen, J.; Reinikainen, P.; Vilpo, L.M.; Seppala, E.; Vilpo, J.A. Kinetics of chlorambucil in vitro: Effects of fluid matrix, human gastric juice, plasma proteins and red cells. *Chem. Biol. Interact.* **1997**, *103*, 187–198. [CrossRef]
9. Davies, I.D.; Allanson, J.P.; Causon, R.C. Rapid determination of the anti-cancer drug chlorambucil (Leukeran E) and its phenyl acetic acid mustard metabolite in human serum and plasma by automated solid-phase extraction and liquid chromatography-tandem mass spectrometry. *J. Chromatogr. B* **1999**, *732*, 173–184. [CrossRef]
10. Wang, X.Y.; Zhang, Q.; Lin, Q.; Zhang, Y.; Zhang, Z.R. Validated LC-MS/MS method for the simultaneous determination of chlorambucil and its prodrug in mouse plasma and brain, and application to pharmacokinetics. *J. Pharm. Biomed. Anal.* **2014**, *99*, 74–78. [CrossRef]
11. Reese, M.J.; Knapp, D.W.; Anderson, K.M.; Mund, J.A.; Case, J.; Jones, D.R.; Packer, R.A. In vitro effect of chlorambucil on human glioma cell lines (SF767 and U87-MG), and human microvascular endothelial cell (HMVEC) and endothelial progenitor cells (ECFCs), in the context of plasma chlorambucil concentrations in tumor-bearing dogs. *PLoS ONE* **2018**, *9*, e0203517. [CrossRef]
12. Mottamal, M.; Zheng, S.; Huang, T.L.; Wang, G. Histone deacetylase inhibitors in clinical studies as templates for new anticancer agents. *Molecules* **2015**, *20*, 3898–3941. [CrossRef]
13. Sun, L.; Coy, D.H. Anti-convulsant drug valproic acid in cancers and in combination anticancer therapeutics. *Mod. Chem. Appl.* **2014**, *2*, 118. [CrossRef]
14. Zhang, J.F.; Zhang, Z.Q.; Dong, W.C.; Jiang, Y. A new derivatization method to enhance sensitivity for the determination of low levels of valproic acid in human plasma. *J. Chromatogr. Sci.* **2014**, *52*, 1173–1180. [CrossRef] [PubMed]
15. Kulza, M.; Florek, E.; Grzybek, A.; Piekoszewski, W.; Hassan-Bartz, M.; Michalak, S.; Seńczuk-Przybyłowska, M. Oznaczanie kwasu walproinowego w surowicy pacjentów dla potrzeb terapii monitorowanej. *Przegląd Lekarski.* **2011**, *68*, 816–819.
16. Lin, M.C.; Kou, H.S.; Chen, C.C.; Wu, S.M.; Wu, H.L. Simple and sensitive fluorimetric liquid chromatography method for the determination of valproic acid in plasma. *J. Chromatogr. B* **2004**, *810*, 169–172. [CrossRef]
17. Rahayu, S.T.; Harahap, Y. Bioanalytical method validation of valproic acid in human plasma in-vitro after derivatization with 2,4-dibomoacetophenon by high performance liquid chromatography photo diode array and its application to in-vivo study. *IJPTP* **2013**, *4*, 644–648.
18. Yaripour, S.; Zaheri, M.; Mohammadi, A. An electromembrane extraction-HPLC-UV analysis for the determination of valproic acid in human plasma. *J. Chin. Chem. Soc.* **2018**, *65*, 989–994. [CrossRef]

19. Ramakrishna, N.V.S.; Vishwottam, K.N.; Manoj, S.; Koteshwara, M.; Santosh, M.; Chidambara, J.; Kumar, B.R. Liquid chromatography/electrospray ionization mass spectrometry method for the quantification of valproic acid in human plasma. *Rapid Commun. Mass Spectrom.* **2005**, *19*, 1970–1978. [CrossRef]
20. Jain, D.S.; Subbaiah, G.; Sanyal, M.; Shrivastav, P. A high throughput and selective method for the estimation of valproic acid an antiepileptic drug in human plasma by tandem LC-MS/MS. *Talanta* **2007**, *72*, 80–88. [CrossRef] [PubMed]
21. Vlase, L.; Popa, D.S.; Muntean, D.; Leucuta, S. A new high-throughput LC-MS/MS assay for therapeutic level monitoring of valproic acid in human plasma. *Sci. Pharm.* **2008**, *76*, 663–671. [CrossRef]
22. Zhao, M.; Zhang, T.; Li, G.; Qiu, F.; Sun, Y.; Zhao, L. Simultaneous determination of valproic acid and its major metabolites by UHPLC-MS/MS in Chinese patients: Application to therapeutic drug monitoring. *J. Chromatogr. Sci.* **2017**, *55*, 436–444. [CrossRef] [PubMed]
23. Cheng, H.; Liu, Z.; Blumb, W.; Byrd, J.C.; Klisovic, R.; Grever, M.R.; Marcucci, G.; Chan, K.K. Quantification of valproic acid and its metabolite 2-propyl-4-pentenoic acid in human plasma using HPLC-MS/MS. *J. Chromatogr. B* **2007**, *850*, 206–212. [CrossRef] [PubMed]
24. Gu, X.; Yu, S.; Peng, Q.; Ma, M.; Hu, Y.; Zhou, B. Determination of unbound valproic acid in plasma using centrifugal ultrafiltration and gas chromatography: Application in TDM. *Anal. Biochem.* **2020**, *588*, 113475. [CrossRef]
25. Feriduni, B.; Barzegar, M.; Sadeghvand, S.; Shiva, S.; Khoubnasabjafari, M.; Jouyban, A. Determination of valproic acid and 3-heptanone in plasma using air-assisted liquid-liquid microextraction with the assistance of vortex: Application in the real samples. *BioImpacts* **2019**, *9*, 105–113. [CrossRef] [PubMed]
26. Mostafa, M.S.; Elshafie, H.S.; Ghaleb, S. A rapid and simple procedure for monitoring valproic acid by gas chromatography. *J. Biol. Res.* **2017**, *90*, 6359. [CrossRef]
27. Mizuno, S.; Lee, X.P.; Fujishiro, M.; Matsuyama, T.; Yamada, M.; Sakamoto, Y.; Kusano, M.; Zaitsu, K.; Hasegawa, C.; Hasegawa, I.; et al. High-throughput determination of valproate in human samples by modified QuEChERS extraction and GC-MS/MS. *Legal Med.* **2018**, *31*, 66–73. [CrossRef]
28. Brodie, S.A.; Brandes, J.C. Could valproic acid be an effective anticancer agent? The evidence so far. *Expert Rev. Anticancer. Ther.* **2014**, *14*, 1097–1100. [CrossRef]
29. Lipska, K.; Filip, A.; Gumieniczek, A. The impact of chlorambucil and valproic acid on cell viability, apoptosis and expression of p21, HDM2, BCL2 and MCL1 genes in chronic lymphocytic leukemia. *Cells* **2021**, *10*, 1088. [CrossRef]
30. *Guidance for Industry. Bioanalytical Method Validation*; U.S. Department of Health and Human Services, Food and Drug Administration, Center for Drug Evaluation and Research (CDER), Center for Veterinary Medicine (CVM): Rockville, MD, USA, May 2018.
31. *Guidance for Industry. Analytical Procedures for Drugs and Biologics*; U.S. Department of Health and Human Services, Food and Drug Administration, Center for Drug Evaluation and Research (CDER), Center for Biologics Evaluation and Research (CBER): Silver Spring, MD, USA, July 2015.
32. Sigma-Aldrich. Available online: https://www.sigmaaldrich.com/content/dam/sigma-aldrich/docs/Aldrich/General_Information/methanol_h2so4.pdf (accessed on 12 May 2021).
33. Wang, D.; Jing, Y.; Ouyang, S.; Liu, B.; Zhu, T.; Niu, H.; Tian, Y. Inhibitory effect of valproic acid on bladder cancer in combination with chemotherapeutic agents in vitro and in vivo. *Oncol. Lett.* **2013**, *6*, 1492–1498. [CrossRef]
34. Atmaca, A.; Al Batran, S.E.; Maurer, A.; Neumann, A.; Heinzel, T.; Hentsch, B.; Schwarz, S.E.; Hövelmann, S.; Göttlicher, M.; Knuth, A.; et al. Valproic acid (VPA) in patients with refractory advanced cancer: A dose escalating phase I clinical trial. *Br. J. Cancer* **2007**, *97*, 177–182. [CrossRef]

Article

Development, Validation, and Application of the LC-MS/MS Method for Determination of 4-Acetamidobenzoic Acid in Pharmacokinetic Pilot Studies in Pigs

Paulina Markowska [1], Zbigniew Procajło [2], Joanna Wolska [3], Jerzy Jan Jaroszewski [1] and Hubert Ziółkowski [1,*]

[1] Department of Pharmacology and Toxicology, Faculty of Veterinary Medicine, University of Warmia and Mazury in Olsztyn, Oczapowskiego 13, 10-718 Olsztyn, Poland; paulina.markowska@uwm.edu.pl (P.M.); jerzyj@uwm.edu.pl (J.J.J.)
[2] Department of Epizootiology, Faculty of Veterinary Medicine, University of Warmia and Mazury in Olsztyn, Oczapowskiego 13, 10-718 Olsztyn, Poland; zbigniew.procajlo@uwm.edu.pl
[3] Department of Anaesthesiology and Intensive Care, Faculty of Medicine, Collegium Medicum, University of Warmia and Mazury in Olsztyn, Al. Warszawska 30, 11-082 Olsztyn, Poland; joanna.wolska@uwm.edu.pl
* Correspondence: hubert.ziolkowski@uwm.edu.pl; Tel.: +48-89-523-3758

Citation: Markowska, P.; Procajło, Z.; Wolska, J.; Jaroszewski, J.J.; Ziółkowski, H. Development, Validation, and Application of the LC-MS/MS Method for Determination of 4-Acetamidobenzoic Acid in Pharmacokinetic Pilot Studies in Pigs. *Molecules* **2021**, *26*, 4437. https://doi.org/10.3390/molecules26154437

Academic Editor: Young G. Shin

Received: 24 June 2021
Accepted: 20 July 2021
Published: 23 July 2021

Publisher's Note: MDPI stays neutral with regard to jurisdictional claims in published maps and institutional affiliations.

Copyright: © 2021 by the authors. Licensee MDPI, Basel, Switzerland. This article is an open access article distributed under the terms and conditions of the Creative Commons Attribution (CC BY) license (https://creativecommons.org/licenses/by/4.0/).

Abstract: Each drug has pharmacokinetics that must be defined for the substance to be used in humans and animals. Currently, one of the basic analytical tools for pharmacokinetics studies is high-performance liquid chromatography coupled with mass spectrometry. For this analytical method to be fully reliable, it must be properly validated. Therefore, the aims of this study were to develop and validate a novel analytical method for 4-acetamidobenzoic acid, a component of the antiviral and immunostimulatory drug Inosine Pranobex, and to apply the method in the first pharmacokinetics study of 4-acetamidobenzoic acid in pigs after oral administration. Inosine Pranobex was administered under farm conditions to pigs via drinking water 2 h after morning feeding at doses of 20, 40, and 80 mg/kg. For sample preparation, we used liquid–liquid extraction with only one step—protein precipitation with 1 mL of acetonitrile. As an internal standard, we used deuterium labeled 4-acetamidobenzoic acid. The results indicate that the described method is replicable, linear ($r^2 \geq 0.99$), precise (2.11% to 13.81%), accurate (89% to 98.57%), selective, and sensitive (limit of quantitation = 10 ng/mL). As sample preparation requires only one step, the method is simple, effective, cheap, and rapid. The results of the pilot pharmacokinetics study indicate that the compound is quickly eliminated (elimination half-life from 0.85 to 1.42 h) and rapidly absorbed (absorption half-life from 0.36 to 2.57 h), and that its absorption increases exponentially as the dose is increased.

Keywords: 4-acetamidobenzoic acid; validation; pharmacokinetic; pigs; LC-MS/MS

1. Introduction

4-acetamidobenzoic acid (PAcBA) is one of the naturally occurring, acetylated metabolites of 4-aminobenzoic acid [1] and is a component of the antiviral and immunostimulatory drug Inosine Pranobex (Inosiplex; Isoprinosine; Methisoprinol) [2–5]. Because of the increasing level of bacterial resistance to antimicrobials and the limitations in direct anti-viral therapy, an attempt has been made to improve immune defense mechanisms by searching for potential new drugs that stimulate the immune system [6–8]. For this reason, Inosine Pranobex is a potential candidate for preventive and/or therapeutic purposes, especially as an antimicrobial alternative in veterinary medicine.

However, although Inosine Pranobex shows antiviral and immunostimulatory effects under in vitro conditions, it is unclear how this translates to in vivo conditions in animals. This is because, as with every drug, it undergoes pharmacokinetic (PK) processes in the body such as absorption, distribution, metabolism, and elimination [2]. To test the real in vivo effect of this drug (its pharmacodynamic effect), it is crucial to know its PK based on the quantitation of concentrations using a validated analytical method, which has not been performed in animals.

Additionally, due to the fact that Inosine Pranobex consists of three different components, the PK has to be determined for each component. So far, the PK of PAcBA has only been examined in humans [9–11]. That study did not account for a potential matrix effect, and it used HPLC with UV light detection, which is less sensitive than tandem mass spectrometry. Moreover, tandem mass spectrometry allows the use of an isotopic standard to ensure more consistent results.

Therefore, the objectives of the present study were to develop a novel method that uses tandem mass spectrometry for PAcBA analysis and to validate it with plasma from humans and 12 animal species, including testing for potential matrix effects. The method was subsequently applied in the first PK study of PAcBA in pigs after oral administration of three doses of Inosine Pranobex under farm conditions.

2. Results and Discussion

No papers on PAcBA analyses conducted with HPLC-MS/MS could be found in the available databases. Therefore, a new method for analysis of this drug was developed via the following sequence of steps: first, the most suitable detector parameters were chosen; second, the appropriate chromatographic conditions were selected; third, a method of quick, short, effective, easy, and cheap analyte extraction was developed; and fourth, the method was validated with human and animal plasma. Once developed and validated, the method was subsequently applied in a pilot study to determine the PK of PAcBA following oral administration of Inosine Pranobex to pigs at three different doses.

2.1. LC-MS/MS Parameters

The molecular weight of PAcBA is 179.175 g/mol and that of the internal standard (IS) is 182.193 g/mol. On this basis, the parent ions of the molecules were sought, assuming that each substance is ionized only once ($m/z = 1/1$) within the nitrogen atom. A thorough analysis of the PAcBA and IS mass spectra obtained by MS/MS operating in positive electrospray mode gave an m/z ratio of 180.20 for PAcBA, and 183.20 for the IS. Subsequently, precursor ion fragmentation was conducted, and the product ions were identified, with the best results achieved for particles with m/z values of 94.0 and 95.0 for PAcBA and IS, respectively. For each analyte, one transition was measured. The detailed parameters of HPLC-MS/MS are summarized in Table 1.

2.2. Chromatographic Conditions

The first step toward establishing the chromatographic conditions was the choice of a suitable column. This choice was affected by the properties of PAcBA, which is a relatively polar compound (XLogP3 1.3, Hydrogen Bond Donors—2, Hydrogen Bond Acceptors—3). A 150 × 3 mm Atlantis T3 analytical column with a 3 μm particle size was used in the analyses; however, a 150 × 3 mm XBridge column with a 3.5 μm particle size was also suitable for studying these analytes (Supplementary Figure S1). The method developed by Chen et al. 2013 for PAcBA analysis employed a universal C18 column, with a diameter and particle size much larger than those in the column used in this experiment [11]. Moreover, the Atlantis T3 column retains polar compounds better than the C18 column, which is why the former was considered a better option. Additionally, the optimum LC chromatographic conditions were determined, such as the mobile phase composition, gradient, flow rate, and temperature (Table 1). The initial phase comprised 0.2% formic acid (FA) in water with 0.2% FA in acetonitrile (ACN) at a ratio of 9:1 v/v, and the temperature was set at 35 °C. However, in these conditions, PAcBA separation proved unsatisfactory because of interference from unidentified background noise, and the shape of the IS peak was asymmetrical at the top (Supplementary Figure S2). For this reason, the separation was modified by increasing the ratio of 0.2% FA in water to 0.2% FA in ACN. The best result was achieved with a ratio of 99:1 v/v when the gradient started with the column temperature set at 20 °C (Table 1). For comparison, Chen et al., 2013 used a mobile phase composed of methanol—0.2% ammonium acetate and a 0.2% acetic acid solution at a ratio of 15:85 v/v [11]. We rejected the use of methanol due to maximal reduction of the baseline during analysis.

Table 1. Selected liquid chromatography and mass spectrometry parameters.

MS/MS Parameters	Compound	
	PAcBA	PAcBA-d3
Precursor ions (m/z)	180.2	183.2
Product ions (m/z)	94.0	95.0
Desolvation gas	nitrogen	
Desolvation gas temperature (°C)	350	
Desolvation gas flow (L/h)	800	
Collision gas	argon	
Source temperature (°C)	120	
Gas cell pirani pressure (mbar)	3.24×10^{-3}	
Electrospray mode	Positive	
Cone voltage (V)	30	
Capillary voltage (kV)	3	
Collision energy (eV)	15	
Dwel (s)	0.200	
Delay (s)	0.010	
Retention time window (min)	6.82–7.35	

Time (min)	Mobile Phase (%)		Curve	Elution (mL/min)
	A	B		
0.00	99	1	1	0.40
6.00	30	70	6	0.40
7.00	0	100	6	0.40
10.00	0	100	6	0.40
11.50	99	1	6	0.40

PAcBA—4-acetamidobenzoic acid; PAcBA-d3—deuterium labeled (d3) 4-acetamidobenzoic acid (internal standard; IS). A—Phase A: 0.2% formic acid in water. B—Phase B: 0.2% formic acid in acetonitrile.

2.3. Development of Sample Preparation

The preparation of a sample for analysis is the main element in the development of a new analytical method for a given compound because, to ensure reproducibility, the validation of the whole method focuses mainly on the way the sample is prepared [12,13]. In the literature, only one publication by Chen et al., 2013 could be found regarding the preparation of the matrix, which was plasma (for the extraction of PAcBA) [11]; thus, it was decided to develop a new method for the preparation of plasma samples for the determination of PAcBA. As the main goal was to obtain a quick, easy, short, sensitive, and low-cost method of plasma purification, it was decided to use the liquid–liquid extraction (LLE) technique despite its disadvantages. As organic solutions for the purification of samples are most commonly used in LLE techniques, 1 mL of ACN was used for protein precipitation. For the extraction procedure, 1.5 mL of 1,2-dichloroethane or ethyl acetate was used. In contrast, Chen et al., 2013 used an LLE extraction procedure with hydrochloric acid for acidification and ethyl acetate as an extractant [11]. However, as they used UV detection in their analytical method, they might not have noticed potential problems with the matrix effect. The results of the present study were surprising; despite the fact that the samples were only treated by denaturation with ACN, the recovery was over 80%, and the matrix effect remained on an acceptable level, as shown by the validation protocol (Figure 1). However, in the samples extracted with 1,2-dichloroethane or ethyl acetate, even though the recovery was higher, the matrix effect was also higher (Supplementary Table S1). An advantage of the method presented here is that, because MS is used for detection, a stable isotopically labeled (SIL) analog of PAcBA can be used as an IS. In contrast, the method presented by Chen et al., 2013 used a UV detector, which meant that paracetamol was used as an IS instead of a SIL [11]. In such a situation, the extraction procedure may be much more difficult than simple protein precipitation. Thus, even though protein precipitation is not the most effective method of purifying the sample, based on the results obtained from the validation protocol in the present study and the fact that it allows more rapid utilization of the column, it was decided to use it for validation and subsequent analysis of pilot tests.

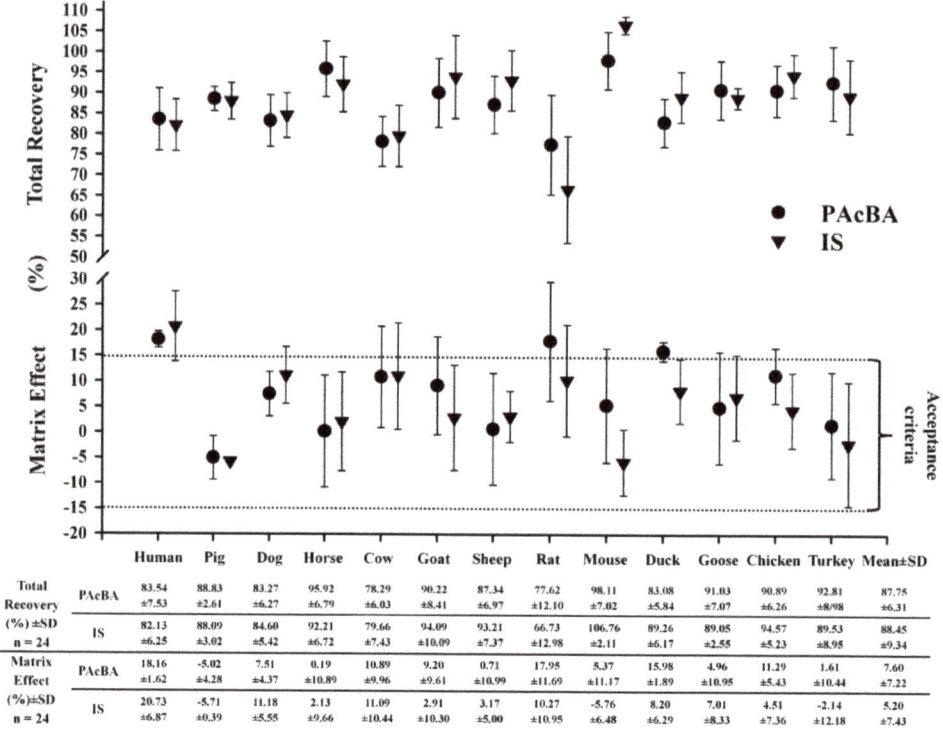

		Human	Pig	Dog	Horse	Cow	Goat	Sheep	Rat	Mouse	Duck	Goose	Chicken	Turkey	Mean±SD
Total Recovery (%) ±SD n = 24	PAcBA	83.54 ±7.53	88.83 ±2.61	83.27 ±6.27	95.92 ±6.79	78.29 ±6.03	90.22 ±8.41	87.34 ±6.97	77.62 ±12.10	98.11 ±7.02	83.08 ±5.84	91.03 ±7.07	90.89 ±6.26	92.81 ±8.98	87.75 ±6.31
	IS	82.13 ±6.25	88.09 ±3.02	84.60 ±5.42	92.21 ±6.72	79.66 ±7.43	94.09 ±10.09	93.21 ±7.37	66.73 ±12.98	106.76 ±2.11	89.26 ±6.17	89.05 ±2.55	94.57 ±5.23	89.53 ±8.95	88.45 ±9.34
Matrix Effect (%)±SD n = 24	PAcBA	18.16 ±1.62	-5.02 ±4.28	7.51 ±4.37	0.19 ±10.89	10.89 ±9.96	9.20 ±9.61	0.71 ±10.89	17.95 ±11.69	5.37 ±11.17	15.98 ±1.89	4.96 ±10.95	11.29 ±5.43	1.61 ±10.44	7.60 ±7.22
	IS	20.73 ±6.87	-5.71 ±0.39	11.18 ±5.55	2.13 ±9.66	11.09 ±10.44	2.91 ±10.30	3.17 ±5.00	10.27 ±10.95	-5.76 ±6.48	8.20 ±6.29	7.01 ±8.33	4.51 ±7.36	-2.14 ±12.18	5.20 ±7.43

Figure 1. Matrix effect and total recovery of 4-acetamidobenzoic acid (PAcBA) and internal standard (IS) in humans and twelve animal species. Each point represents the mean value (±SD) calculated from six replicates of four quality control points for each matrix and reference sample (n = 24 per point).

2.4. Validation

The method presented here was fully validated using pig plasma. Additionally, total recovery and matrix effect tests were also carried out with blood drawn from humans and eleven animal species.

The values of "r" and "r²" for the linear regressions with the data from calibration were above 0.99, which met the acceptance criteria for linearity (Table 2). Additionally, because the literature contains no information on expected concentrations of PAcBA in pig blood [12], the range of the curve was expanded as far as possible, with a 1000× difference between the concentrations at the first and the last points on the curve. Throughout this expanded range, all acceptance criteria were met (Table 2, Supplementary Table S2).

Table 2. Methods of calculation and acceptance criteria for validation parameters.

Parameter		Acceptance Criteria
Linearity	Calibration points	At least 75% of calibration points, but not less than 6, should have a deviation (residual) between nominal and back-calculated concentrations of ±15% or less
	Coefficient of determination (r^2)	≥ 0.99
	Relative residuals (Y_i)	$\left\|\frac{y_i - \hat{y}_i}{\hat{y}_i}\right\| \times 100\% \leq 20\%$
	SD of Relative residuals (S_{Yi})	$\sqrt{\frac{\sum(Y_i - \bar{Y})^2}{n-2}} \leq 0.1$
Stability	Stock and working standard	$\frac{S_t}{S_0} \times 100\% = \pm 15\%\ of\ S_0$
	Autosampler	
	Freeze and thaw	
	Sample processing temperature	
Precision (RSD or CV)		$\frac{SD}{C_{mean}} \times 100\% =$ $\pm 15\%\ within\ nominal\ concentration$
Accuracy (Deviation) (for at least 5 points per group/day)		$\frac{\|(C_t - C_n)\|}{C_n} \times 100\% =$ $\pm 15\%\ within\ nominal\ concentration$
Limit of detection (LOD)		$3 \times SD_{C fortified}$ where $S/N \geq 3:1$
The lowest limit of quantitation (LLOQ) with accuracy and precision		$6 \times C_{fortified}$ where $S/N \geq 10:1$
Matrix Effect		$100 - \left(\frac{X_i}{X} \times 100\%\right) =$ $\pm 15\%\ compare\ to\ sample\ without\ matrix$
Total Recovery		$\frac{X_z}{X_i} \times 100\%$ $= \pm 15\%\ RSD$
Selectivity/Specificity		No endogenous peaks in retention time of analyte
Carry Over		Area of carried peaks ≤20% of LLOQ area, and for IS, 5% of its area

y_i—experimental signal; $\hat{y}_i$—calculated signal; Y_i—relative residual; $\bar{Y}$—mean value of relative residuals; S_t—peak area obtained when analysis is carried out while making a pause with duration t in the analysis; S_0—initial peak area determined without introducing any extra pauses in the analysis process (freshly prepared standards); $SD_{Cfortified}$—standard deviation calculated from fortified samples with the lowest acceptable concentration; C_{mean}—mean concentration (ng/mL); C_n—nominal concentration (ng/mL); C_t—calculated individual concentration (ng/mL); $C_{fortified}$—minimal fortified concentration that meets requirements; X—peak area of analyte in final solvent; X_i—peak area of analyte added to matrix after extraction; X_z—peak area of analyte added to matrix before extraction.

The accuracy and precision were estimated for all quality controls (QC) and the lowest limit of quantitation (LLOQ), both for the preparation of a single sample and for the preparation of six samples over a period of three days. The intra-day precision for all QC and the LLOQ was 2.11% to 13.81%, and the accuracy was 1.43% to 11.0%. The inter-day precision for all QC and the LLOQ was 3.43% to 10.93%, and the accuracy was 2.7% to 8.78% (Table 3, Supplementary Table S3).

The limit of detection (LOD) was set at 3.27 ng/mL ± 1.48 (a signal-to-noise ratio S/N not lower than 3:1) and the LLOQ was 10.0 ng/mL ± 1.09 (S/N = 15.69 ± 3.95) (Table 3, Supplementary Table S4, Supplementary Figure S3). As the samples for analysis were diluted four times for sample preparation during protein precipitation, the use of this method in such conditions did not allow very low PAcBA levels (<3 ng/mL) to be determined, which may be a limitation when investigating endogenous PAcBA in animals.

Table 3. Selected validation tests results of 4-acetamidobenzoic acid (PAcBA) and deuterium-labeled (d3) 4-acetamidobenzoic acid (internal standard; IS).

Linearity [a]				I	II	III	IV	Mean
	r^2			0.9992	0.9992	0.9975	0.9988	0.9990
				LLOQ	LQC	IQC	MQC	HQC
Precision (%) and accuracy (%)	Intra-day $n = 6$; 3 repetitions	Precision	I	5.22	3.64	6.17	4.87	2.81
		Accuracy		4.33	2.3	3.91	3.93	2.31
		Precision	II	13.71	11.37	4.45	3.45	4.65
		Accuracy		11.0	8.57	3.61	2.74	3.24
		Precision	III	13.81	5.30	3.10	2.11	3.04
		Accuracy		11.0	4.47	2.38	1.43	2.49
	Inter-day $n = 18$	Precision		10.93	7.08	4.45	3.43	3.38
		Accuracy		8.78	5.11	3.3	2.7	2.68

			Concentration [b]	S/N
LLOQ and LOD [c]	LLOQ overall mean $n = 18$		10.00	15.69
	LLOQ overall SD $n = 18$		1.09	3.95
	LOD overall mean $n = 18$		3.27	10.89
	LOD overall SD $n = 18$		1.48	6.52

		Sample	Peak Area of PAcBA	Peak Area of Mobile Phase	Carry Over (%)
Carry over	Mean	PAcBA	193,810.4	6.27	4.69
		IS	24,601.28	0	0
	SD	PAcBA	2382.238	6.93	5.19
		IS	218.815	0	0

LQC—low-concentration quality control (50 ng mL^{-1}); IQC—intermediate-concentration quality control (500 ng mL^{-1}); MQC—medium-concentration quality control (5000 ng mL^{-1}); HQC—high-concentration quality control (10,000 ng mL^{-1}); LLOQ—the lowest limit of quantitation (nominal concentration 10 ng mL^{-1}, mean peak area 133.74); LOD—limit of detection; S/N—signal-to-noise ratio. [a] calibration curve range: 10, 25, 50, 100, 250, 500, 1000, 2500, 5000, 7500, and 10,000 ng mL^{-1}. [b] in ng mL^{-1}. [c] LOD = $3 \times SD_{LLOQ}$ or S/N_{mean}.

The selectivity/specificity of the method showed no significant peaks during PAcBA and IS retention in drug-free plasma obtained from blood collected from clinically healthy human and animals, and there was no significant carryover of PAcBA and IS during the analysis of high concentrations of these analytes (Table 3, Supplementary Table S5). However, it should be noted that there were several species (Supplementary Figure S4) for which small peaks appeared with an S/N lower than 10:1, which could indicate the presence of endogenous PAcBA. This phenomenon should be monitored by blank sample analyses in each test because of the risk that an endogenous analyte could appear that could be of particular importance in an attempt to optimize the method for the investigation of concentrations much lower than those possible with the method used in this experiment.

In the method presented here, the mean total recovery for all species was 87.75% ± 6.31% for PAcBA and 88.45% ± 9.34% for IS (Figure 1). Both analytes appeared to be stable after 3 h at the sample processing temperature, after 48 h in an autosampler at 4 °C, after 1680 h of thawing and freezing cycles, and for 120 h as a prepared working standard stored in a refrigerator (2 °C) (Table 4).

Table 4. Stability tests results.

Stability	Period (h)	Compound	Decrease/Increase of Quality Control Concentration (%)			
			LQC	IQC	MQC	HQC
Stock2 −75 °C	120	PAcBA	−8.42	−11.12	−14.59	−12.85
		IS	−8.74	−4.17	−4.99	−10.21
Working standard 2 °C	72	PAcBA	−7.64	7.81	3.72	1.71
		IS	8.31	−2.74	−4.28	−11.71
	120	PAcBA	−14.01	−0.14	0.65	2.80
		IS	−2.21	−8.27	−9.54	−14.88
Autosampler 4 °C	24	PAcBA	2.17	−0.825	0.19	1.05
		IS	1.44	6.01	3.14	1.76
	48	PAcBA	2.03	1.14	1.35	4.31
		IS	0.60	−7.85	−4.09	−4.53
Freeze and thaw −75 °C	24	PAcBA	4.60	−4.33	0.07	4.60
		IS	−5.17	2.04	−4.63	−3.09
	48	PAcBA	4.60	−3.43	−1.77	1.06
		IS	9.58	4.35	5.26	1.22
	96	PAcBA	3.68	−2.02	−2.84	1.54
		IS	0.34	−7.24	−10.10	−11.78
	1680	PAcBA	6.09	−4.80	−1.78	1.14
		IS	−13.06	−6.98	−10.23	−12.21
Sample processing temperature 21 °C	3	PAcBA	−3.36	0.28	−0.14	−0.48
		IS	−5.50	1.01	2.09	1.85

LQC—low-concentration quality control; IQC—intermediate-concentration quality control; MQC—medium-concentration quality control; HQC—high-concentration quality control; PAcBA—4-acetamidobenzoic acid; IS—internal standard.

Endogenous PAcBA signals were not found in the matrices of the examined species (Supplementary Figure S3). The mean matrix effect for all species was 7.60% ± 7.22% for PAcBA and 5.20% ± 7.43% for IS, although in humans, ducks, and rats, the matrix effect was above the acceptable limit (Figure 1). As this method of sample purification is the "dirtiest" sample preparation technique, attention should be paid to each matrix used for the analysis, as the presence of endogenous PAcBA probably varies from individual to individual and may change depending on the matrix origin (Supplementary Figure S4).

2.5. Pharmacokinetics

The experiment demonstrated that the drug was absorbed relatively quickly in pigs (Figure 2; Table 5), although the mean absorption time (MAT) and half-life in the absorption phase ($t_{1/2kab}$) differed somewhat between groups (Table 5). These differences, as well as the delayed absorption time, most likely resulted from the way the drug was administered (farm conditions). The animals received the drug in water 2 h after feeding. As they had probably drunk water when being fed, they most likely began to consume the drug some time after it was provided to them. The present results are consistent with those on the PK of Inosine Pranobex that are available in the literature; in studies of both Rhesus Monkeys [14] and healthy volunteers [11], PAcBA was also rapidly absorbed. Interestingly, in the present study, the $AUC_{(0 \to t)}$ increased around 3.65 times despite the drug being administered in water for several hours with each doubling of the dose. The best fit to this data was obtained with power regression; the equation was $y = 2.6707x^{1.9367}$, with $r^2 = 0.99995$ (Supplementary Figure S5). In contrast, the best fit to the change in the values of C_{max} was obtained with logarithmic regression ($y = 2788.3976 \ln(x) − 8033.1062$; $r^2 = 0.994$). Although the precise determination of these parameters is complicated by the limitations associated with administration of the drug and the number of animals, these results indicate that changes in PAcBA concentration in plasma clearly reflect changes in drug dose.

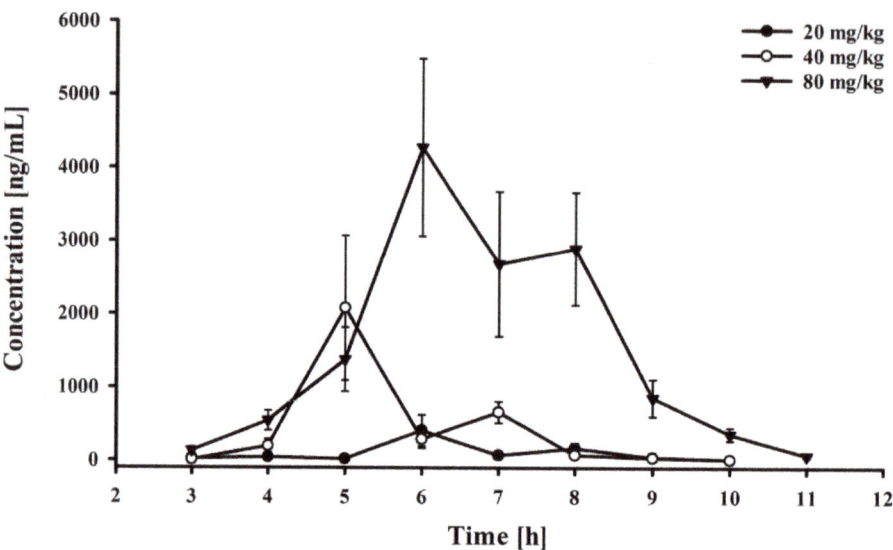

Figure 2. Plasma-concentration–time profile of 4-acetamidobenzoic acid after oral administration to pigs at a different doses.

Table 5. Pharmacokinetic parameters (noncompartmental analysis) of 4-acetamidobenzoic acid in pigs after oral administration at different doses.

Pharmacokinetic Parameters	Dose (mg/kg)		
	20	40	80
	Mean ± SD	Mean ± SD	Mean ± SD
$AUC_{(0 \to t)}$ (µg·h/L)	878.74 ± 372.3 [a]	3402.52 ± 1687.26 [a]	12,868.1 ± 4896.6 [b]
$AUMC_{(0 \to t)}$ (µg·h·h/L)	5935.97 ± 2453 [a]	19,015.92 ± 7157.79 [a]	88,002.9 ± 38,611.43 [b]
C_{max} (ng/mL)	406.73 ± 211.6 [a]	2079.87 ± 787.56 [a]	4272.27 ± 1713 [b]
t_{max} (h)	6 ± 1	5 ± 1.5	6 ± 1
C_{last} (ng/mL)	12.32 ± 3.89 [a]	10.21 ± 2.11 [a]	68.11 ± 25.1 [b]
t_{last} (h)	11	11	11
k_{el} (h^{-1})	0.49 ± 0.21	0.62 ± 0.31	0.81 ± 0.37
$t_{1/2kel}$ (h)	1.42 ± 0.87	1.12 ± 0.42	0.85 ± 0.29
$MRT_{(0 \to t)}$ (h)	6.76 ± 3.21	5.59 ± 2.78	6.84 ± 3.41
Cl_B/F (L·h)	682.8 ± 214.23 [a]	352.68 ± 131.2 [a,b]	186.51 ± 97.6 [b]
Vd_{area}/F (L)	1399.78 ± 731.6	568.33 ± 243.65	229.56 ± 112.38
k_{ab} (h^{-1})	0.27 ± 0.11 [a]	0.81 ± 0.37 [a]	1.96 ± 0.89 [b]
$t_{1/2kab}$ (h)	2.57 ± 1.21 [a]	0.86 ± 0.31 [a]	0.36 ± 0.13 [b]
MAT (h)	3.70 ± 1.7 [a]	1.23 ± 0.41 [a]	0.51 ± 0.19 [b]

$AUC_{0 \to t}$—area under the concentration vs. time curve from 0 to t; $AUMC_{0 \to t}$—area under the first moment of the curve; C_{max}—maximum plasma concentration; t_{max}—time of maximum concentration; C_{last}—last measured plasma concentration; t_{last}—time of last measured concentration; k_{el}—elimination rate constant; $t_{1/2kel}$—half-life in elimination phase; $MRT_{0 \to t}$—mean residence time; Cl_B/F—total body clearance without bioavailability correction; Vd_{area}/F—apparent volume of distribution without bioavailability correction; k_{ab}—absorption rate constant; $t_{1/2kab}$—half-life in absorption phase; MAT—mean absorption time. [a,b,c] Pharmacokinetic parameters differ significantly ($p < 0.05$) between the groups.

In the study presented here, the apparent volume of distribution without bioavailability correction (Vd_{area}/F) was high, similar to what was observed by Chen et al., 2013 [11]. This result can be explained by several phenomena: either the drug absorbed poorly, or it was subject to efficient, rapid metabolism and/or rapid elimination. Streeter and Pfadenhayer 1984 suggested that the bioavailability of PAcBA is high, as well as the metabolism of the compound [14]. However, in the present study, these phenomena were difficult to assess. Nevertheless, the half-life in the elimination phase ($t_{1/2kel}$) and MRT values of PAcBA (Table 5), as well as the results of other studies based on observations of concentration–time changes of PAcBA [11,14], indicate that it is quickly eliminated. Such rapid elimination is likely to increase the Vd_{area}/F value, but further studies using single intravenous administration are necessary to assess the actual value of total body clearance (Cl_B) and Vd_{area}.

3. Materials and Methods

3.1. Animals and Drugs

Nine Polish Large White × Polish Landrace piglets from one litter, with an initial body weight of 58.3 ± 4.09 kg, were used in the experiment. The animals were kept in an experimental animal facility at the Faculty of Veterinary Medicine of the University of Warmia and Mazury in Olsztyn, Poland, in three separate 4 × 4 m stalls (three animals in each) with automatic drinking bowls in which water was available ad libitum. The animal facility was equipped with a forced ventilation system protected by HEPA filters. The facility structure and the installed equipment allowed constant temperature (21 °C), relative humidity (65%) and air flow (0.2 m/s) to be maintained. Granular feed (WIPASZ, Wadąg, Poland) was given to the animals in an amount of 850 g/animal twice a day at 8.00 a.m. and 5.00 p.m., throughout the experiment. The content of nutrients in the feed, as declared by the manufacturer, is presented in Supplementary Table S6. The pigs did not receive any pharmacological treatment during the acclimatization period. The study was registered and approved by the Local Ethics Committee in Olsztyn (Ethics Committee Opinion No. 17/2014).

3.2. Chemicals and Reagents

For liquid chromatography–tandem mass spectrometry (LC-MS/MS), water, ACN, FA, and methanol (all LC/MS grade) were purchased from Sigma-Aldrich. Analytical standards for PAcBA and deuterium-labeled PAcBA (for use as an IS) were purchased from Sigma-Aldrich (Darmstadt, Germany) and Toronto Research Chemicals (Toronto, ON, Canada), respectively. The stock solution of 1 mg/mL of PAcBA and 0.2 mg/mL of IS was prepared in methanol in 5 mL volumetric borosilicate flasks supplied by the Duran Group (Mainz, Germany). Next, these solutions were taken to make working solutions that were used during the experiments and validation of the method. They were prepared in 5 mL volumetric borosilicate flasks by diluting stock solutions in methanol at the following concentrations: 0.25, 0.625, 1.25, 2.5, 6.25, 12.5, 25.0, 62.5, 125.0, 187.5, and 250.0 µg/mL. All solutions were refrigerated at 4 °C. The gases required for the LC-MS/MS system were nitrogen from a NitroGen N110R nitrogen generator, which was supplied by Peak Scientific (Inchanian, Scotland, UK), and argon, which was purchased from EUROGAZ-BOMBI (Olsztyn, Poland).

3.3. Experimental Design

After one week of acclimatization, the animals were divided into three equal groups (n = 3 in each group), in which the drug was administered via drinking water 2 h after morning feeding (water was given ad libitum) at doses of 20, 40, and 80 mg/kg in groups 1, 2, and 3, respectively. Blood samples (2 mL each) were collected into heparinized tubes from the right jugular vein through injection needles (1.2 × 80 mm) at 0, 1.0, 2.0, 3.0, 4.0, 5.0, 6.0, 7.0, 8.0, 9.0, 10.0, and 11.0 h after drug administration. Plasma was separated by centrifugation at 1650× g for 10 min at 4 °C and was stored at −81 °C until analysis.

3.4. Chromatography

The plasma concentration of PAcBA was determined using LC-MS/MS (Supplementary Figure S6). Drug levels were quantified with a Waters Alliance 2695 reversed-phase liquid chromatography system coupled with a tandem mass spectrometer (MS/MS) Quattro micro API MS (Milford, MA, USA). The analytical column was an Atlantis T3 (150 × 3 mm) with a 3 µm particle size, supplied by Waters. The optimal mobile phase was composed of phase A, water with 0.2% FA, and phase B, ACN with 0.2% FA. The gradient elution was based on the time set on the pump. The injection volume was 10 µL, the column temperature was set at 20 °C, and the flow rate was 0.40 mL/min. PAcBA was monitored from *m/z* 180.20 to 94.0 and IS was monitored from *m/z* 183.20 to 95.0 (Table 1).

3.5. Sample Preparation

Plasma obtained from humans and 12 different animals was thawed at room temperature, and 250 µL of each sample was combined with 10 µL of IS (1 µg/mL) and mixed in a vortex at 1000 rpm for 5 s. Next, 1 mL of ACN was added for protein precipitation, and the samples were vortexed at 3000 rpm for 10 s. After centrifugation at 40,000× *g* for 10 min at 4 °C, 150 µL of the supernatant was transferred through a 0.22 µm nylon syringe filter (13 mm in diameter) into chromatographic total recovery vials and injected into the chromatographic system.

3.6. Method Validation

The analytical method was fully validated using the analytical method validation protocol of the United States Food and Drug Administration (FDA), the European Medicines Agency (EMA) bioanalytical method validation requirements [12,13], and a tutorial review of liquid chromatography–mass spectrometry method validation [15,16].

During the validation procedure, the following parameters were determined: linearity, accuracy, precision (repeatability/intra-day precision and intermediate precision/inter-day precision), LOD, LLOQ, selectivity, recovery, matrix effect (ionization suppression/enhancement), carry-over, and stability (freeze–thaw stability, autosampler stability, working standard, stock stability, and sample processing temperature stability). The acceptance criteria were established based on [12,13,15,16], and they are summarized in Table 2. For validation, plasma from healthy pigs was used, and then an additional test was performed to determine the matrix effect and recovery from plasma obtained from healthy human volunteers and from 12 different animal species.

3.7. Linearity

The linearity of the method for assaying PAcBA in plasma by HPLC-MS/MS was determined using an 11-point standard curve (0.01, 0.025, 0.5, 0.1, 0.25, 0.5, 1.0, 2.5, 5.0, 7.5, and 10.0 µg/mL) that was prepared four times at one-day time intervals. Each curve was analyzed twice, and each analysis was preceded by a sample without any analytes (blank sample) and a sample containing only IS (zero sample). The values obtained from this test are summarized in Table 2, including the back-calculated concentration, the slope "a" and the intercept "b" in the equation y = bx + a, the Pearson correlation coefficient "r" (and the coefficient of determination, "r^2"), and the acceptance criteria.

3.8. Precision and Accuracy

Precision and accuracy were determined by preparing analyte concentrations at the four QC points and the LLOQ, which were all within the range of the standard curve; this was done three times (at specified time intervals) in six replicates together with IS, according to the method of drug determination previously established by experiment. The analysis yielded concentrations relative to the declared nominal concentration for each QC, obtained by back calculation, and the acceptance criteria are summarized in Table 2.

3.9. Limit of Detection

The limit of detection was determined based on the results for the LLOQ obtained in accordance with the parameters specified in Table 2.

3.10. Selectivity/Specificity

To identify endogenous matrix elements that may be present during PAcBA or IS retention (Table 2), an analysis was performed on plasma samples obtained from blood collected from pigs not exposed to drugs and, additionally, from 11 animal species and from humans free from exogenous substances. Each sample was prepared six times using the method developed here. The analysis of each matrix sample was separated by analysis of a sample with only a mobile phase.

3.11. Recovery

To estimate the degree/effectiveness of extraction, six replicates of each QC were prepared, in which the analytes were added to the plasma either before or after extraction. For samples that had analytes added after extraction, 10 µL of analyte and 10 µL of IS were added after the extraction of an "empty sample." The concentrations of the analyte and IS added following the extraction were regarded as 100% recovery (Table 2).

3.12. Matrix Effect

PAcBA and IS were added to the phase obtained following extraction of an empty matrix in six replicates for each QC. Next, the signal from each compound was compared to that of PAcBA and IS added to a mixture of water (which replaced plasma) and ACN, which was also prepared in six replicates for each QC. Subsequently, following the method presented here, all 48 samples for each of the tested matrices were analyzed. Signals from the analysis of PAcBA and IS in the mixture of water and ACN were considered to have values of 100% (Table 2).

3.13. Carry Over

This test was conducted to eliminate any possible PAcBA and IS carry over (ghost peaks) between samples when using the chromatographic system elements (e.g., injector, column, mobile phase). Six replicates of HQC (with IS) and six blank samples were prepared, and the blank sample was analyzed after each HQC sample analysis (Table 2).

3.14. Stability

The stability of PAcBA and IS in the matrix, stock solutions, and working solutions was determined at each stage of sample storage, sample preparation, and chromatographic analysis. The results (as peak areas) were compared with the results obtained with freshly prepared standards (Table 2). Moreover, an analysis of blank samples was always prepared and conducted in each stability test to verify the sample preparation procedure for further test stages.

This test was conducted for the QC working solutions and the stock solutions of PAcBA and IS that were stored in a refrigerator at 2 °C. The test was performed at 72 and 120 h for the working solutions and at 120 h for the stock solutions. These solutions were prepared on the first day and 30 samples (six replicates for each solution) were analyzed without extraction. After 72 h of storage of the working solutions, 24 samples were prepared and analyzed, and after 120 h, 30 samples were prepared and analyzed (working solutions + stock solutions).

This test was also performed to check the sample stability in the autosampler operating at 4 °C after 24 and 48 h. A set of samples of six replicates of each QC (24 samples) was prepared on the first day in accordance with the sample preparation protocol and was subsequently analyzed. A second analysis of the same samples was performed after 24 h, and a third was performed after 48 h.

A test was conducted to determine the stability of the drug and IS in the matrix at the sample storage temperature (−81 °C). For each QC, five sets of samples were prepared with six replicates each. The test was conducted at 0, 24, 48, 96, and 1680 h (the final time was the long-term freeze and thaw stability test). The first set of 24 samples was analyzed immediately after preparation (without freezing—day 0), and the remaining samples were frozen at −81 °C. All of the samples were thawed on the next day, the second set was analyzed, and the remaining samples were refrozen. The same procedure was followed with sets three (two days after day 0) and four (four days after day 0). Set five was stored for 70 days (−81 °C) and then analyzed.

Another test was conducted to determine the stability of the drug and IS in the matrix under the sample preparation conditions. Twenty-four samples (all QCs) were prepared from freshly prepared standard solutions and analyzed. At the same time, drug standards were added to another 24 samples (all QCs) and the sample preparation procedure was stopped for 3 h. After that time, the samples were prepared and analyzed.

3.15. Pharmacokinetics

The PK analysis was performed using noncompartmental analysis with ThothPro™ software (ThothPro, Gdańsk, Poland). The PK analysis determined the area under the curve ($AUC_{0\rightarrow t}$) according to the linear trapezoidal rule, the area under the first moment of the curve from 0 to t ($AUMC_{0\rightarrow t}$), the mean residence time from 0 to t ($MRT_{0\rightarrow t}$), the slope of the elimination phase (k_{el}), the half-life in the elimination phase ($t_{1/2kel}$), the apparent volume of distribution (Vd_{area}/F), and the total body clearance (Cl_B/F) without bioavailability correction. The mean absorption time (MAT) and the half-life in the absorption phase ($t_{1/2kab}$) were calculated using one-compartment analysis according to Gibaldi and Perrier 1982 [17]. The maximum (C_{max}) and the last (C_{last}) plasma concentrations and the time of C_{max} and C_{last} were determined individually for each animal and were expressed as mean values ± SD.

4. Conclusions

This is the first report on the development and validation of a method that uses HPLC-MS/MS for the quantification of PAcBA in the blood plasma of thirteen species. The results indicate that the method is replicable, precise, accurate, selective, and sensitive. The advantages of the method are its simplicity and effectiveness, as well as the rapidity of sample preparation, which make the method more economical and allow rapid and precise assays of PAcBA in plasma. Although the method uses only protein precipitation for plasma purification (it is relatively "dirty"), it has a high recovery rate, and the matrix effect is small enough to validate the method with proper accuracy and precision. Despite a few other limitations (namely, lower recovery in plasma samples of some species and an LOD that is probably not low enough to estimate potential endogenous PAcBA concentrations), the method is suitable for practical application, as shown by its successful application in a PK study of exogenous PAcBA in pigs after oral Inosiplex administration at three different dosages. The results of the PK study indicate that this compound is rapidly eliminated and that its absorption is not only fast but also increases exponentially depending on the dose.

Supplementary Materials: The following are available online: Figure S1: Chromatograms from an Atlantis T3 analytical column (150 × 3 mm with 3 μm particle size)—A, A', and from an XBridge column (150 × 3 mm with 3.5 μm particle size)—B, B' (A,B—4-acetamidobenzoic acid; A',B'—deuterium labeled 4-acetamidobenzoic acid as an internal standard). Figure S2: Chromatograms obtained from initial phase comprised of 0.2% formic acid in water with 0.2% formic acid in acetonitrile at a ratio of 9:1 v/v with the column temperature set at 35 °C (A—unidentified background noise; A'—asymmetrical peak at the top). Figure S3: Chromatograms obtained at the lowest limit of quantitation, 10 ng/mL (A—4-acetamidobenzoic acid; A'—deuterium labeled 4-acetamidobenzoic acid as an internal standard). Figure S4: Chromatograms from a selectivity/specificity test. The signal-to-noise ratios of all identified peaks are lower than 10:1. Figure S5: Power regression with the coefficient of determination (R^2) of the area under the concentration-time curve calculated from 0 to t ($AUC_{0\rightarrow t}$) and logarithmic regression with the coefficient of determination (R^2) of the maximum plasma concentration (Cmax) of 4-acetamidobenzoic acid after oral administration to pigs at doses of 20, 40 and 80 mg/kg BW (n = 3). Figure S6: Chromatograms obtained from the pharmacokinetics study: A—4-acetamidobenzoic acid; A'—deuterium labeled 4-acetamidobenzoic acid as an internal standard. Table S1: Matrix effect and total recovery results of PAcBA and IS using three different extractants. Each quality control point is the mean value calculated from six replicates. Table S2: Linearity results. Table S3: Precision and accuracy results. Table S4: The lowest limit of quantitation (LLOQ) and limit of detection (LOD) results. Table S5: Carry over test results of 4-acetamidobenzoic acid (PAcBA) and deuterium labeled (d3) 4-acetamidobenzoic acid (internal standard; IS). Table S6: Feed composition.

Author Contributions: P.M. conceived of the study and participated in its design, treatments, sample collection, and instrumental analysis, as well as data analysis, and was the primary author of the manuscript. Z.P. and J.W. participated in treatments, sample collection, and instrumental and statistical data analysis. J.J.J. participated in the study design and coordination. H.Z. significantly contributed to the conception and design of the study, and instrumental and pharmacokinetic analyses, and they revised the manuscript. All authors have read and agreed to the published version of the manuscript.

Funding: This research was funded primarily by National Centre for Research and Development, grant No. UDA-POIG.01.03.01-28-108/12-01. Moreover, the project was financially supported by the Ministry of Science and Higher Education in the range of the program entitled "Regional Initiative of Excellence" for the years 2019–2022, Project No. 010/RID/2018/19, amount of funding 12.000.000 PLN.

Institutional Review Board Statement: All the procedures were approved by the Local Ethics Commission (The Local Ethics Commission for Animal Experiments in Olsztyn; Ethical permission No. 17/2014).

Informed Consent Statement: Not applicable.

Data Availability Statement: The data presented in this study are available on reasonable request from the corresponding author.

Conflicts of Interest: The authors declare no conflict of interest.

Sample Availability: Not available.

References

1. Wan, H.; von Lehmann, B.; Riegelman, S. Renal Contribution to Overall Metabolism of Drugs III: Metabolism of p-Aminobenzoic Acid. *J. Pharm. Sci.* **1972**, *61*, 1288–1292. [CrossRef] [PubMed]
2. Campoli-Richards, D.M.; Sorkin, E.M.; Heel, R.C. Inosine pranobex. A preliminary review of its pharmacodynamic and pharmacokinetic properties, and therapeutic efficacy. *Drugs* **1986**, *32*, 383–424. [CrossRef] [PubMed]
3. Majewska, A.; Lasek, W.; Janyst, M.; Młynarczyk, G. Inhibition of adenovirus multiplication by inosine pranobex and interferon α in vitro. *Cent. Eur. J. Immunol.* **2015**, *40*, 395–399. [CrossRef] [PubMed]
4. Lasek, W.; Janyst, M.; Wolny, R.; Zapała, Ł.; Bocian, K.; Drela, N. Immunomodulatory effects of inosine pranobex on cytokine production by human lymphocytes. *Acta Pharm.* **2015**, *65*, 171–180. [CrossRef] [PubMed]
5. Rumel, A.S.; Newman, A.S.; O'Daly, J.; Duffy, S.; Grafton, G.; Brady, C.A.; Curnow, J.S.; Barnes, N.M.; Gordon, J. Inosine Acedoben Dimepranol promotes an early and sustained increase in the natural killer cell component of circulating lymphocytes: A clinical trial supporting anti-viral indications. *Int. Immunopharmacol.* **2017**, *42*, 108–114. [CrossRef] [PubMed]
6. Cheng, G.; Hao, H.; Xie, S.; Wang, X.; Dai, M.; Huang, L.; Yuan, Z. Antibiotic alternatives: The substitution of antibiotics in animal husbandry? *Front. Microbiol.* **2014**, *5*, 217. [CrossRef] [PubMed]
7. Vannucci, L.; Krizan, J.; Sima, P.; Stakheev, D.; Caja, F.; Rajsiglova, L.; Horak, V.; Saieh, M. Immunostimulatory properties and antitumor activities of glucans (Review). *Int. J. Oncol.* **2013**, *43*, 357–364. [CrossRef]
8. Kedl, R.M.; Griesgraber, G.W.; Zarraga, I.A.E.; Wightman, P.D. Immunostimulatory compositions and methods of stimulating an immune response. U.S. Patent 7,427,629 B2, 23 September 2008.
9. Nielsen, P.; Beckett, A.H. The metabolism and excretion in humans of NN-dimethylamino-isopropanol and p-acetamido-benzoic acid after administration of Isoprinosine. *J. Pharm. Pharmacol.* **1981**, *33*, 549–550. [CrossRef]
10. Chan, K.; Miners, J.O.; Birkett, D.J. Direct and simultaneous high-performance liquid chromatographic assay for the determination of p-aminobenzoic acid and its conjugates in human urine. *J. Chromatogr.* **1988**, *426*, 103–109. [CrossRef]
11. Chen, M.; Zhang, Y.; Que, X.T.; Ding, Y.; Yang, L.; Wen, A.; Hang, T. Pharmacokinetic study of inosiplex tablets in healthy Chinese volunteers by hyphenated HPLC and tandem MS techniques. *J. Pharm. Anal.* **2013**, *3*, 387–393. [CrossRef] [PubMed]
12. EMA. *Guideline on Bioanalytical Method Validation EMEA/CHMP/EWP/192217/2009*; EMA: Amsterdam, The Netherlands, 2011; pp. 1–25.
13. FDA. Bioanalytical Method Validation. In *Guidance for Industry*; FDA: Rockwell, MD, USA, 2013; pp. 1–25.
14. Streeter, D.G.; Pfadenhauer, E.H. Inosiplex: Metabolism and excretion of the dimethylaminoisopropanol and p-acetamidobenzoic acid components in rhesus monkeys. *Drug Metab. Dispos.* **1984**, *12*, 199–203. [PubMed]

15. Kruve, A.; Rebane, R.; Kipper, K.; Oldekop, M.L.; Evard, H.; Herodes, K.; Ravio, P.; Leito, I. Tutorial review on validation of liquid chromatography-mass spectrometry methods: Part II. *Anal. Chim. Acta* **2015**, *870*, 8–28. [CrossRef] [PubMed]
16. Kruve, A.; Rebane, R.; Kipper, K.; Oldekop, M.L.; Evard, H.; Herodes, K.; Ravio, P.; Leito, I. Tutorial review on validation of liquid chromatography-mass spectrometry methods: Part I. *Anal. Chim. Acta* **2015**, *870*, 29–44. [CrossRef] [PubMed]
17. Gibaldi, M.; Perrier, D. Noncompartmental Analysis Based on Statistical Moment Theory. In *Pharmacokinetics*, 2nd ed.; Gibaldi, M., Perrier, D., Eds.; Informa Healthcare: New York, NY, USA, 1982; pp. 409–417.

Article

Photometric Determination of Iron in Pharmaceutical Formulations Using Double-Beam Direct Injection Flow Detector

Stanislawa Koronkiewicz

Department of Chemistry, University of Warmia and Mazury in Olsztyn, 10-957 Olsztyn, Poland; stankor@uwm.edu.pl; Tel.: +48-89-5234137

Abstract: In this work, an innovative, flow-through, double-beam, photometric detector with direct injection of the reagents (double-DID) was used for the first time for the determination of iron in pharmaceuticals. For stable measurement of the absorbance, double paired emission-detection LED diodes and a log ratio precision amplifier have been applied. The detector was integrated with the system of solenoid micro-pumps. The micro-pumps helped to reduce the number of reagents used and are responsible for precise solution dispensing and propelling. The flow system is characterized by a high level of automation. The total iron was determined as a Fe(II) with photometric detection using 1,10-phenanthroline as a complexing agent. The optimum conditions of the propose analytical procedure were established and the method was validated. The calibration graph was linear in the range of 1 to 30 mg L^{-1}. The limit of detection (LOD) was 0.5 mg L^{-1}. The throughput of the method was 90 samples/hour. The repeatability of the method expressed as the relative standard deviation (R.S.D.) was 2% (n = 10). The method was characterized by very low consumption of reagents and samples (20 µL each) and a small amount of waste produced (about 540 µL per analysis). The proposed flow method was successfully applied for determination of iron in pharmaceutical products. The results were in good agreement with those obtained using the manual UV-Vis spectrophotometry and with values claimed by the manufacturers. The flow system worked very stably and was insensitive to bubbles appearing in the system.

Keywords: pharmaceutical analysis; iron determination; spectrophotometry; flow analysis; direct injection detector; multi-pumping flow system

Citation: Koronkiewicz, S. Photometric Determination of Iron in Pharmaceutical Formulations Using Double-Beam Direct Injection Flow Detector. *Molecules* **2021**, *26*, 4498. https://doi.org/10.3390/molecules26154498

Academic Editor: Constantinos K. Zacharis

Received: 21 June 2021
Accepted: 23 July 2021
Published: 26 July 2021

Publisher's Note: MDPI stays neutral with regard to jurisdictional claims in published maps and institutional affiliations.

Copyright: © 2021 by the author. Licensee MDPI, Basel, Switzerland. This article is an open access article distributed under the terms and conditions of the Creative Commons Attribution (CC BY) license (https://creativecommons.org/licenses/by/4.0/).

1. Introduction

Iron is an essential nutrient in the human diet. It plays an important role in cellular processes such as respiration, synthesis of DNA and RNA, electron transport, and regulation of gene expression. Iron deficiencies leading to anemia are one of the world's most common nutritional diseases. This is a global-scale public health problem affecting about 15% of human population (mainly children, teenagers, and woman at reproductive age) [1]. In order to avoid such deficiencies, an adequate supply of iron is needed. Iron preparations are one of the most commonly used medicaments.

Quality control of pharmaceuticals is a key factor at all stages of their development and production. The international pharmacopoeia requirements and standards related to quality control of pharmaceuticals have become increasingly strict. Most methods used for the assay of iron in pharmaceuticals are based on spectrophotometry [2–5]. Several other methods have also been reported including atomic absorption spectrometry [6], chemiluminescence [7], and electrochemical methods [8,9]. Most of these methods are easily applicable as detection methods in automated flow analysis systems. In such configuration, the most popular one is again spectrophotometric detection [3–5,10,11] due to its versatility, simplicity, rapidity, and relatively low cost.

For dedicated analytical applications, highly advanced and expensive spectrophotometers are unnecessary. Simple optical devices based on semiconductor components can be sufficient [12–15]. Spectrophotometric detection in such devices is achieved using light emission diodes (LEDs) as a source of inexpensive, stable, nearly monochromatic and high-intensity light. They are increasingly popular as a part of optical sensors since they offer several advantages including small size, low power consumption, robustness and long lifespan. In order to detect the light, the LEDs are usually coupled with a photodiode [16] or a phototransistor [10]. Optical detectors operating according to the paired emitter-detector diode (PEDD) principle are also commonly constructed [13,17–20]. PEDDs utilize two ordinary LEDs, the first of which is used as a source of the light and the second of which serves as a partially selective light detector. Wavelength selectors and optical fiber are unnecessary, allowing for reduction of optical complexity. Appropriately designed PEDDs have been successfully applied for photometric analytical measurements of iron in pharmaceuticals [21], as well as in wastewater and wine samples [11].

Application of flow analysis techniques has created new possibilities for automation of analytical procedures, often improving the precision, reducing the consumption of reagent, and increasing the sample throughput. Waste production and cost of analysis are reduced. Unfortunately, commercially available devices mostly rely on technology dating back whole decades, mainly continuous flow analysis (CFA) and flow injection analysis (FIA). More advanced techniques, such as sequential injection analysis (SIA) and systems commonly called "Lab-on-valve" (LOV, also called micro-SIA), are also available [4]. Modern solutions, better suited to the requirements of contemporary analytical chemistry, have been developed using electronically controlled actuators. These include, among others, micropumps and multi-way valves controlled using a solenoid. Solenoid micropumps are used in a technique called MPFS (multi pumping flow system) [22,23]. Solenoid micropumps can replace both the peristaltic pump and the system for precise dosing of the sample and reagents. The micropumps are characterized by their small size, which allows for miniaturization of the entire flow system. They have high precision and accuracy of dosing comparable with that of micropipettes. Operation of those elements can be programmed with a computer.

MPFS systems can be coupled with an original direct injection detector (DID). This type of detector is based on a PEDD system and allows for direct injection of reagents into the photometric detector's chamber without using a reaction coil. It utilizes the fact that reagents are effectively mixed when rapidly injected in counter-current [24]. As a result, the samples are mixed with the reagents within a fraction of a second, directly in the detector chamber. It allows the time of analysis to be considerably shortened and avoids unnecessary sample dilution. So far, two types of photometric PEDD-based DID detectors have been developed and described, namely a single-beam DID [18] and a double-beam DID [19]. The single-beam DID was successfully adapted to determine Fe(III) in ground water using the thiocyanate method [20] and to determine Fe(II) in wastewater and wine [11]. In the double-beam DID, a log ration amplifier to measure absorbance has been applied providing a voltage signal proportional to the absorbance (linear function). The double-beam DID is superior to the single-beam DID in terms of better reproducibility of the results due to its autozeroing function. The design of this system minimizes the Schlieren effect. The flow system is stable even over a long period of time and insusceptible to gas bubbles.

In pharmaceutical products, iron is in the Fe(II) oxidation state since this is the form of iron easily absorbed in human body. Moreover, the majority of pharmaceuticals contain antioxidants in excessive amounts compared to the concentration of Fe(II), and this prevents quantitative oxidation of Fe(II) to Fe(III). Sample preparation is therefore difficult when oxidation of Fe(II) is needed before its determination as Fe(III). Sometimes it takes several hours to prepare the sample by digestion and separation of organic parts, and the process is very reagent-consuming [5]. Due to this, a direct, simple, precise and accurate, robust, and inexpensive method for determination of Fe(II) in pharmaceutical products would be greatly desired.

This work describes a novel, simple, automated procedure for iron determination in pharmaceuticals. In this method total iron was determined as Fe(II) through photometric detection using 1,10-phenanthroline as a complexing agent. In order to achieve correct analytical parameters, an automated method using a double-beam DID system was used. This is the first time this detector was applied for analytical purposes in pharmaceutical analysis. The full analytical features of the method were investigated.

2. Results and Discussion

2.1. Optimization of the Parameters

2.1.1. Stop-Flow Time

For Fe(II) spectrophotometric determination of iron, the selective method based on the reaction between iron(II) and 1,10-phenanthroline is recommended (Scheme 1):

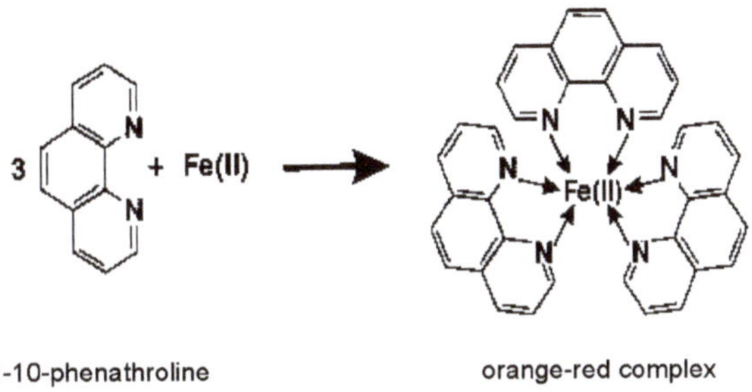

Scheme 1. Reaction of iron(II) with 1,10-phenanthroline.

This reaction is known to be relatively slow. The time required for complete formation of the reaction product (ferroin complex) and recommended for batch condition is about 5 min [25]. Application of direct injection detectors allows for the colored product of the reaction to be monitored from the moment of mixing appropriate reagents [18–20]. Once that time (i.e., 5 min) has passed, the signal of absorbance should be stable. In order to verify whether this was true in the described method and to determine the optimum stop-flow time, the change of absorbance during the 300 s was registered. It was found that the absorbance signal is stable after about 90 s (Figure 1). However, in reproducible flow conditions the time required to read the analytical signal can be significantly shortened. Therefore, as the minimum stop-flow time, 30 s can be selected. This is a compromise between the time of analysis and the height of the analytical signal. In order to obtain the highest possible signal, a stop-flow time of 90 s is recommended. Most of the experiments presented in this paper were conducted using a stop-flow time of 70 s The choice of a stop-flow time of 70 s allowed to obtain high analytical signal in a relatively short time.

2.1.2. Chemical Parameters

The effect of various chemical parameters was evaluated following the univariate method. At the beginning the pH of carrier solution was optimized. The studies were conducted in the range from 0 to 13 using appropriate concentration of NaOH or HCl (Figure 2). It was observed that the absorbance was approximately on the same level for pH in the range from 2 to 12. Therefore, the deionized water was chosen as a carrier. This facilitates saving the reagents. All standard solutions and samples were prepared in HCl of concentration 1.2 mol L^{-1} because the reduction of iron(III) to iron(II) using ascorbic acid has to be carried out under very acidic condition [25,26].

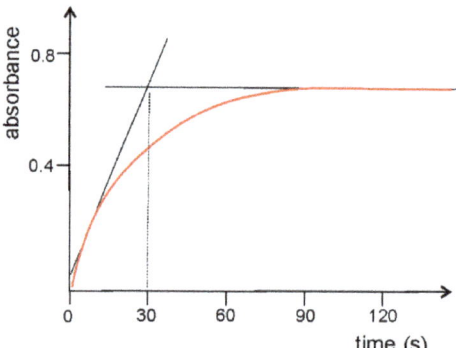

Figure 1. The typical analytical signal obtained for Fe(II) determination. Sample: Fe(II) 10 mg L^{-1}; reagent: 1% solution of 1,10-phenanthroline in sodium acetate (1 mol L^{-1}) and ethanol (25:5, v/v); carrier, water.

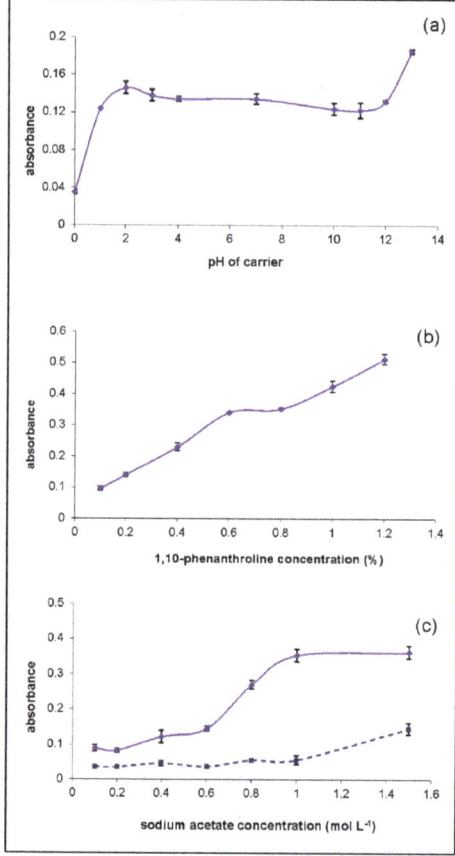

Figure 2. Effect of the most influential chemical parameters on the analytical signal. (**a**) Effect of carrier pH, (**b**) effect of 1,10-phenanthroline concentration, and (**c**) effect of sodium acetate concentration. Solid line, sample containing Fe(II) 10 mg L^{-1}; dashed line, blank. Each point represents the mean of at least four consecutive measurements, bars indicate the standard deviations.

The concentration of 1,10-phenanthroline was studied between 0.1 and 1.2%. It was observed that the analytical signal steadily increased with increasing concentration (Figure 2b). However, for the concentration of 0.6%, the precision (standard deviation) was found the best and this concentration was chosen as a compromise between quality of the analytical signal and cost of analysis.

When formation of ferroin complex was realized in flow conditions, a solution of 1,10-phenentroline in sodium acetate was applied [26,27]. The concentration of CH_3COONa in 1,10-phenanthroline solution was studied in the range of 0.1 to 1.5 mol L^{-1} (Figure 2c). It was observed that the absorbance was very low when the concentration of sodium acetate was below 0.6 mol L^{-1}. Then the absorbance increased and was stable for concentrations higher than 1 mol L^{-1}. Unfortunately, over the sodium acetate concentration of 1 mol L^{-1}, an increase in blank signal was observed. Therefore, the concentration of 1 mol L^{-1} was chosen as an optimum. The level of blank in all experiments conducted under optimized conditions was always stable and very low, below 0.01.

2.2. Method Evaluation

2.2.1. Analytical Parameters

The proposed flow system was evaluated while maintaining the optimum conditions described above. Linear calibration curves were attained for Fe(II) concentration between 1 and 30 mg L^{-1}. The typically obtained relationship between absorbance and Fe(II) concentration is shown in Figure 3.

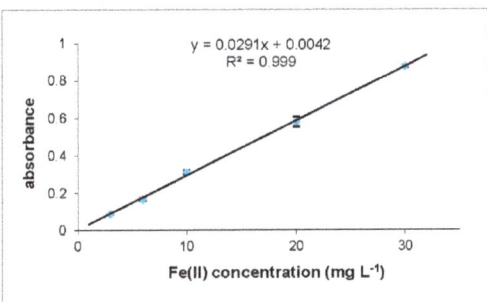

Figure 3. The calibration graph for Fe(II) determination. Each point represents the mean of four consecutive measurements. Bars indicate the standard deviations.

The limit of detection (LOD), calculated as $3sb/S$, where sb is the standard deviation for 10 measurements of the blank and S is the slope of the calibration graph, was in the range of 0.5 mg L^{-1}. Limit of quantification (LOQ), calculated as $10sb/S$, was equal to 1.7 mg L^{-1}. Taking into account the high content of iron(II) in the pharmaceuticals, the LOD value of the developed method is satisfactory.

The precision of the method evaluated as a relative standard deviation (R.S.D.) of 10 successive determinations of the standard solution 15 mg L^{-1} was found to be 2%. The estimated reagent consumption was very low, namely 20 µL of the sample and 20 µL of the 1,10-phenanthroline solution. The total volume of generated waste was 540 µL per analysis. Time of analysis applied for the calibration graph was 80 s (sample throughput of 45 determinations per hour). It can be shortened to about 40 s if very high sensitivity is unnecessary. In such condition the high throughput of 90 determinations per hour is possible.

2.2.2. Application to the Real Samples

Usefulness of the presented flow system was evaluated through determination of iron(II) in a set of pharmaceutical formulations. The obtained results were compared with the results received from a batch manual spectrophotometric procedure and with a value claimed by the producer (information given on the packaging). The samples were diluted to obtain the concentrations in the range of calibration graph and next, the iron content was appropriately recalculated. The results are shown in Table 1.

Table 1. Results of determination of Fe(II) in pharmaceutical preparations (iron content in mg per tablet, capsule or sachet). Results represent the average of at least four determinations ± S.D.

Sample	Proposed Flow System	Manual Procedure	Claimed Value
Sorbifer Durules	100.63 ± 0.25	99.25 ± 0.27	100
Hemofer Prolongtum	113.03 ± 0.28	108.41 ± 0.31	105
Chela-ferr biocomplex	13.95 ± 0.06	13.63 ± 0.03	14
Ascofer	23.66 ± 0.07	24.02 ± 0.08	23.2
Actiferol	28.78 ± 0.09	29.05 ± 0.08	30

To establish whether the proposed method produces reliable results and whether those results are in agreement with the traditional method of determination, Student's t-test was applied. It was found that the calculated t value (from 0.20 for Sorbifer Durules to 0.96 for Hemofer Prolongatum) was lower than the tabulated t value (t = 2.78, n = 4, $p = 0.05$). This suggested that, at 95% confidence level, the difference between the results obtained by the proposed method and the reference method was statistically insignificant.

2.2.3. Recovery Test

In order to check the accuracy of the proposed method, a recovery study was carried out. The samples of pharmaceutical formulations were spiked. The percentage of recoveries was calculated. It was in the range of 93 to 107%. The results shown in Table 2 confirm the validity of the method proposed.

Table 2. Study of recovery. Results of determination of Fe(II) in pharmaceutical preparations. Results represent the average of at least three determinations ± S.D.

Sample	Determined Concentration of Fe(II) (mg L^{-1})	Added Fe(II) (mg L^{-1})	Found Fe(II) (mg L^{-1})	Recovery (%)
Sorbifer Durules	10.57 ± 0.09	5.00	4.79	96
		10.00	9.74	97
		15.00	16.12	107
Hemofer Prolongatum	16.72 ± 0.28	5.00	4.90	98
		10.00	9.93	99
		15.00	13.98	93
Chela-ferr biocomplex	12.28 ± 0.25	5.00	5.01	100
		10.00	9.73	97
		15.00	14.61	97

2.3. Comparison of Proposed Methodology with Other Modern Metodologies

Comparison of the analytical features of the proposed approach with other systems, e.g., traditional, stationary spectrophotometry [2], flow methods applying spectrophotometers [3–5] and a commercially available flow system [4] was carried out (Table 3). The detection methods described in previously mentioned publications mainly concern conventional spectrophotometers [2–5] and LED-based techniques [10,11]. Various types of flow methods were also used for comparison: FIA [3], SIA [4,5], multicommutation based on solenoid valve application [10] and MPFS [11]. The detection process based mainly on the reaction of Fe(II) with 1,10-phenanthroline.

Table 3. Comparison of spectrophotometric methods of iron(II) determination. LOD, limit of detection; LOQ, limit of quantification; R.S.D., relative standard deviation.

Detection	Reagent	Flow System	Sample/Reagent Consumption (Waste) (µL)	LOD or LOQ (mg L^{-1})	Working Range (mg L^{-1})	Repeatability R.S.D. (%)	Type of Sample/Throughput (Injection h^{-1})	Reference
Conventional spectrophotometer	4-(2-pyridylazo) resorcinol	not applicable	no inf.	LOD = 0.008	0.025–0.2	2.9	pharmaceuticals/10	[2]
Conventional spectrophotometer	1,10-phenanthroline	FIA	1000/150 (no inf.)	LOD = 0.2	0.5–4.0	<3.0	pharmaceuticals/41	[3]
Conventional spectrophotometer FIAlab 3500 instrument	2,2-bipyridyl	SIA	50/200 (no inf.)	LOD = 1.0	5.0–40	5.0	pharmaceuticals/100	[4]
Conventional spectrophotometer	1,10-phenanthroline	SIA	187/140 (no inf.)	LOD = 0.02	0.25–5.0	3.0	pharmaceuticals/40	[5]
LED-phototransistor	1,10-phenanthroline	multicommutated FA	no inf. (2400)	LOD = 0.5	0.5–6.0	1.0	surface water/40	[10]
Single-DID	1,10-phenanthroline	MPFS	80/20 (500)	LOQ = 0.07	0.07–1.00 1.00–7.00	14.8 9.6	wine, wastewater/36	[11]
Double-DID	1,10-phenanthroline	MPFS	20/20 (540)	LOD = 0.5 LOQ = 1.7	1–30	2.0	pharmaceuticals/90	this work

It can be concluded that the discussed method is distinguished from others by its high repeatability (R.S.D. of about 2%) and wide working range (from 0.5 to 30 mg L^{-1}), which is satisfactory when one considers iron(II) content in pharmaceuticals. Limit od detection (LOD) and limit of quantification (LOQ) depend mainly on analytical reaction applied. For methods using 1,10-phenanthroline LOD (or LOQ) is similar. The differences may arise from different methods of determining this value.

The sampling rate (method's throughput) is also very good, comparable to the commercially available system FIAlab [4]. Presented methodology allows for very low sample and reagent consumption (both of 20 µL) and low waste generation (total volume of about 540 µL per analysis). Generally, all flow methods should be expected to use less reagents than stationary, classical methods [2]. However, typical flow systems (e.g., FIA, SIA) [3–5,10] require the use of peristaltic pumps and reaction coils. As a result, the volumes of the solutions used are still relatively large. Flow system with DID detector, described in this work, do not use such elements. This allows for an even greater reduction in sample/reagent consumption and waste generation than FIA and SIA.

Comparing the obtained results with those of the simplified, single-beam version of the DID [11], it can be concluded that the analytical parameters have been significantly improved.

3. Materials and Methods

3.1. Chemicals and Reagents

All solutions were prepared with analytical-grade chemicals and using deionized water obtained from a Milli-Q (Millipore) water purification system (resistivity > 18.2 MΩ cm). A 1 mg mL -1 stock solution of Fe(II) was prepared by dissolving 0.4982 g of iron(II) sulfate heptahydrate (Sigma-Aldrich, Darmstadt, Germany) in 100 mL of 1.2 mol L^{-1} HCl (Merck, Darmstadt, Germany). Working standard solutions were prepared by appropriate dilution of stock solution with 1.2 M of HCl to obtain the final concentration of Fe(II) ions from 1 to 30 mg L^{-1}. All working solutions additionally contained a reducing agent, namely ascorbic acid (Chempur, Poland) at a concentration of 10 mg L^{-1}. Solutions of ascorbic acid were prepared daily. The chromogenic reagent (solution of 1,10-phenanthroline) was prepared by dissolving an appropriate amount of 1,10-phenanthroline monohydrate (Sigma-Aldrich, Germany) in a mixture of ethanol and sodium acetate (5:25, v/v) (POCh, Poland). As a carrier, solutions of HCl, NaOH (POCh, Poland) or deionized water were used.

3.2. Samples

For the demonstration of practical utility of the developed flow system, samples of pharmaceutical formulations listed in Table 4 were used.

Table 4. Pharmaceutical formulations used for validation of developed method.

No.	Name/Supplier	Form	Active Substance
1	Sorbifer Durules/Egis Pharmaceuticals PLC (Budapest, Hungary)	tablet	FeSO$_4$
2	Hemofer Prolongatum/GlaxoSmithKline Pharmaceuticals (Poznan, Poland)	tablet	FeSO$_4$
3	Chela-Ferr/OlimpLabs (Debica, Poland)	capsule with powder	iron(II) diglycinate
4	Ascofer/Espefa (Cracow, Poland)	tablet	iron(II) gluconate
5	Actiferol/Sequoia (Warsaw, Poland)	sachet with powder	iron(II) pyrophosphate

All pharmaceuticals used also contained different additives (e.g., talc, stearic acid, beeswax, paraffin, microcrystalline cellulose) which were insoluble in water. To prepare the samples, the tablets or the contents of capsules/sachets were placed into the mortar and carefully grinded. Then, the resulting powder was dissolved in 100 mL of 1.2 mol L^{-1} HCl with addition of 10 mg L^{-1} of ascorbic acid to prevent oxidation of iron(II) to iron(III). Insoluble pharmaceutical ingredients were filtered. If necessary, the extracts were additionally diluted 10- to 100-fold using HCl and ascorbic acid before analysis to obtain the iron content compatible with the linear range of determination. The pharmaceuticals also contained some water-soluble excipients which were presented in the sample matrix. The most important and common were: ascorbic acid, lactose, sucrose, gelatin, acacia gum, and maltodextrin. Due to the highly selective nature of the reaction of Fe(II) ions with 1,10-phenanthroline, it was assumed that they should not significantly affect the analytical signal.

3.3. Apparatus

3.3.1. Flow System

The flow manifold consisted of three solenoid-operated micro-pumps, flow lines, and a double-beam direct-injection detector (double-DID). The micro-pumps were purchased from Cole-Parmer (Boonton, USA) and have a nominal volume of 20 µL (product no. P/N 73120-10) or 50 µL (product no. P/N 73120-22). The flow lines were made of a PTFE tube (ID 0.8 mm) and were also obtained from Cole-Parmer. The 20 µL volume pumps were used for injecting the sample and the chromogenic reagent into the reaction chamber of the DID. The third micro-pump was used for propelling the carrier. To decrease the time of analysis, the nominal volume of this pump was bigger (50 µL). The schematic diagram of the applying flow network is shown in Figure 4.

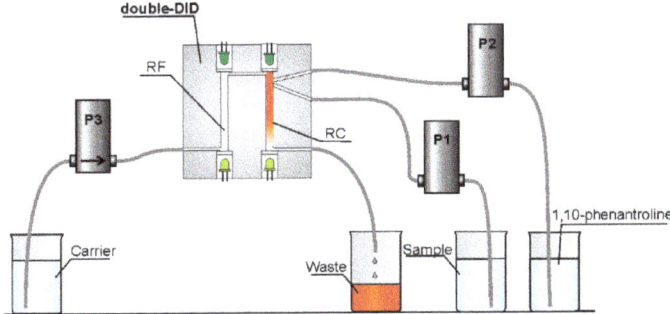

Figure 4. The schematic diagram of the applying flow network. P1, P2, P3, solenoid micro-pumps; RC, reaction chamber; RF, reference chamber.

The double-beam DID was made using one block of Teflon. Inside the detector there were two cells identical in shape and size, reaction (RC) and reference (RF). The optical path length was 20 mm, and the volume of each of the cells was 60 µL. All details regarding the principle of operation and characteristics of this detector were discussed in an earlier publication [19]. As emission diodes for iron(II) determination, two identical green LEDs were chosen. The maximum of the emission spectrum of these LEDs was consistent with the maximum of absorption of the Fe(II)-1,10-phenanthroline (ferroin) complex (λ_{max} = 512 nm [25]). Since the spectral sensitivity of the LEDs is usually shifted towards the shorter wavelength (Stokes shift) [13–15], two yellow-green LEDs with emission λ_{max} = 540 nm were selected as detection diodes, since they were well suited to detection of light emitted by the green diodes, which have lower λ_{max}. The LEDs were purchased at a local electronics parts shop.

The work of the DID and the micro-pumps were PC-controlled by the measurement system developed specifically for flow analysis in our laboratory [28]. The software enables the user to control the work of the solenoid micro-pumps and to calibrate the signal of absorbance. For absorbance measurements, an integrated logarithmic amplifier LOG101 (Burr-Brown, Tucson, AZ, USA) was used.

3.3.2. Procedure

All solutions were aspirated and then properly injected into the detector by three independent solenoid micro-pumps: P1, P2, and P3. The pumps were responsible not only for the propelling of all solutions, but also for precise dispensing. Rapid injection in counter-current facilitated efficient mixing of the solutions used. An example of the micro-pumps program used for the Fe(II) determination is shown in Figure 5.

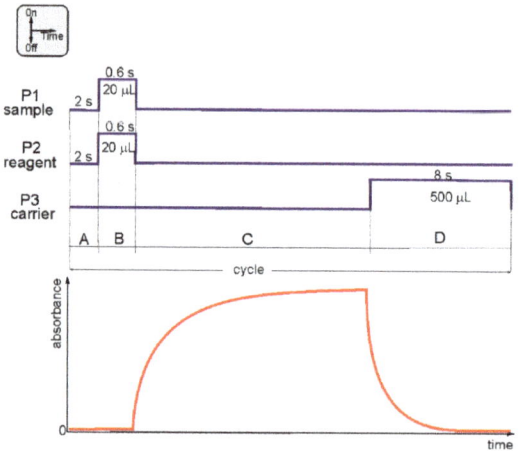

Figure 5. Example of the micro-pumps program (blue line) with absorbance signal (red line) measured during one cycle. A, zeroing the absorbance signal; B, injection of the sample and the reagent; C, formation of a colored reaction product; and D, cleaning of the reaction chamber.

Before the analytical cycle began, both cells of the detector (reaction (RC) and reference (RF)) were filled with the carrier solution using pump P3. The measuring cycle started by zeroing the absorbance signal (Figure 5, section A).

Next, simultaneous injection of the sample (pump P1) and the solution of 1,10-phenanthroline (pump P2) occurred (Figure 5, section B). Injection was rapid and conducted in counter-current. Such a method of introduction facilitated effective homogenization of the reagents. That way the reaction cell was filled with the total volume of sample and reagent of 40 µL. This volume was lower than the volume of the cell (60 µL) and the absorbing the light molecules should not escape the reaction cell.

From the moment of sample and reagent injection and mixing, the reaction started and absorbance was measured (Figure 5, section C). The stop-flow time for this section and, consequently, the measuring cycle time, depended mainly on the kinetics of the reaction used.

At the end of each cycle, the detector was cleaned using the carrier, by pump P3 (Figure 5, section D). To ensure the removal of reagents from reaction cell was thorough, several strokes of pump P3 were applied. The influence of the contaminants originating from previous cycle was not observed when the detector was cleaned using about 500 µL of the carrier.

3.3.3. Reference Method

The manual procedure for determination of Fe(II) used was identical to that recommended by Marczenko [25]. Absorbance was measured at 512 nm with a conventional spectrophotometer (Tomos Life Science Group, Singapore, model V-1100).

4. Conclusions

It can be concluded that the proposed novel flow procedure for iron(II) determination is a very valuable solution from the point of view of quality control analysis of pharmaceuticals. The chief advantages of the presented flow system are its very high stability and repeatability, high sampling frequency, simple and compact construction and high level of automation. Obtained results confirmed the analytical usefulness of the developed approach, and they are highly repeatable and accurate. Additionally, the method is inexpensive and environmentally friendly, since the reagent consumption and total waste production are very low (in the range of 20 µL and 540 µL per analysis, respectively).

The described system can work independently. However, it is possible to combine it with other flow systems, resulting in a more sophisticated overall system. It would be worthwhile to conduct research in the direction of developing a procedure for on-line (in a flow system) sample preparation. Obviously, this would increase the degree of automation of the entire determination, and would allow for a real, final minimization of sample consumption. In such configuration described flow system can be especially recommended for the Green Chemistry monitoring station working in a fully automatic mode.

5. Patents

The technical solutions and potential applications in analytical chemistry of a double-beam direct injection detector are described in Polish Patent [29].

Funding: The results presented in this paper were obtained as a part of comprehensive study financed by the University of Warmia and Mazury in Olsztyn, Faculty of Agriculture and Forestry, Department of Chemistry (grant No. 30.610.001-110).

Institutional Review Board Statement: Not applicable.

Informed Consent Statement: Not applicable.

Conflicts of Interest: The author declares no conflict of interest.

Sample Availability: Pharmaceutical samples were obtained from the local pharmacy.

References

1. Beard, J.L.; Dawson, H.; Pinero, D.J. Iron metabolism: A comprehensive review. *Nutr. Rev.* **1996**, *54*, 296–317. [CrossRef]
2. Karpinska, J.; Kulikowska, M. Simultaneous determination of zinc (II), manganese (II) and iron (II) in pharmaceutical preparations. *J. Pharm. Biomed.* **2002**, *29*, 153–158. [CrossRef]
3. Vakh, C.; Freze, E.; Pochivalov, A.; Evdokimova, E.; Kamencev, M.; Moskvin, L.; Bulatov, A. Simultaneous determination of iron (II) and ascorbic acid in pharmaceuticals based on flow sandwich technique. *J. Pharmacol. Toxicol. Methods* **2015**, *73*, 56–62. [CrossRef] [PubMed]
4. Oliveira, P.C.C.; Masini, J.C. Sequential injection of iron (II) in antianemic pharmaceutical formulations with spectrophotometric detection. *Anal. Lett.* **2001**, *34*, 389–397. [CrossRef]
5. Tesfaldet, Z.O.; van Staden, J.F.; Stefan, R.I. Sequential injection spectrophotometric determination of iron as Fe (II) in multi-vitamin preparations using 1,10-phenanthroline as complexing agent. *Talanta* **2004**, *64*, 1189–1195. [CrossRef] [PubMed]
6. Li, S.X.; Deng, N.S. Speciation analysis of iron in traditional Chinese medicine by flame atomic absorption spectrometry. *J. Pharm. Biomed. Anal.* **2003**, *32*, 51–57. [CrossRef]
7. Waseem, A.; Yaqoob, M.; Nabi, A. Analytical Applications of Flow Injection Chemiluminescence for the Determination of Pharmaceuticals—A Review. *Curr. Pharm. Anal.* **2013**, *9*, 363–395. [CrossRef]
8. Merli, D.; Profumo, A.; Dossi, C. An analytical method for Fe (II) and Fe (III) determination in pharmaceutical grade iron sucrose complex and sodium ferric gluconate complex. *J. Pharm. Anal.* **2012**, *2*, 450–453. [CrossRef] [PubMed]
9. Mahmoud, W.H. Iron ion-selective electrodes for direct potentiometry and potentiotitrimetry in pharmaceuticals. *Anal. Chim. Acta* **2001**, *436*, 199–206. [CrossRef]
10. Feres, M.A.; Reis, B.F. A downsized flow set up based on multicommutation for the sequential photometric determination of iron(II)/iron(III) and nitrite/nitrate in surface water. *Talanta* **2005**, *68*, 422–428. [CrossRef]
11. Paluch, J.; Kościelniak, P.; Moleda, I.; Machowski, K.; Kalinowski, S.; Koronkiewicz, S.; Kozak, J. Novel approach to determination of Fe(II) using a flow system with direct-injection detector. *Mon. Chem. Chem. Mon.* **2020**, *151*, 1305–1310. [CrossRef]
12. Dasgupta, P.K.; Eom, I.Y.; Morris, K.J.; Li, J. Light emitting diode-based detectors: Absorbance, fluorescence and spectroelectrochemical measurements in a planar flow-through cell. *Anal. Chim. Acta* **2003**, *500*, 337–364. [CrossRef]

13. O' Toole, M.; Lau, K.T.; Diamond, D. Photometric detection in flow analysis systems using integrated PEDDs. *Talanta* **2005**, *66*, 1340–1344. [CrossRef]
14. Li, S.; Pandharipande, A. LED-based color sensing and control. *IEEE Sens. J.* **2015**, *15*, 6116–6124. [CrossRef]
15. Bui, D.A.; Hauser, P.C. Absorbance measuring with light-emitting diodes as sources: Silicon photodiodes or light-emitting diodes as detector? *Talanta* **2013**, *116*, 1073–1078.
16. Pires, C.K.; Reis, B.F.; Morales-Rubio, A.; de la Guardia, M. Speciation of chromium in natural waters by micropumping multicommutated light emitting diode photometry. *Talanta* **2007**, *72*, 1370–1377. [CrossRef] [PubMed]
17. Tymecki, L.; Pokrzywnicka, M.; Koncki, R. Paired emitter detector diode (PEDD) based photometry—An alternative approach. *Analyst* **2008**, *133*, 1501–1504. [CrossRef]
18. Koronkiewicz, S.; Kalinowski, S. A novel direct-injection detector integrated with solenoid pulse-pump flow system. *Talanta* **2011**, *86*, 436–441. [CrossRef]
19. Kalinowski, S.; Koronkiewicz, S. Double-beam photometric direct-injection detector for multi-pumping flow system. *Sens. Actuators A* **2017**, *258*, 146–155. [CrossRef]
20. Koronkiewicz, S.; Kalinowski, S. Application of direct-injection detector integrated with the multi-pumping flow system to photometric, stop-flow determination of total iron. *Talanta* **2012**, *96*, 68–74. [CrossRef] [PubMed]
21. Rybkowska, N.; Koncki, R.; Strzelak, K. Optoelectronic iron detectors for pharmaceutical flow analysis. *J. Pharm. Biomed. Anal.* **2017**, *145*, 504–508. [CrossRef] [PubMed]
22. Lapa, R.A.S.; Lima, J.L.F.C.; Reis, B.F.; Santos, J.L.M.; Zagatto, E.A.G. Multi-pumping in flow analysis: Concepts, instrumentation, potentialities. *Anal. Chim. Acta* **2002**, *466*, 125–132. [CrossRef]
23. Santos, J.L.M.; Ribeiro, M.F.T.; Dias, A.C.B.; Lima, J.L.F.C.; Zagatto, E.E.A. Multi-pumping flow systems: The potential of simplicity. *Anal. Chim. Acta* **2007**, *600*, 21–28. [CrossRef] [PubMed]
24. Fortes, P.R.; Feres, M.A.; Sasaki, M.K.; Alves, E.R.; Zagatto, E.A.G.; Prior, J.A.V.; Santos, J.L.M.; Lima, J.L.F.C. Evidences for turbulent mixing in multi-pumping flow systems. *Talanta* **2009**, *79*, 978–983. [CrossRef]
25. Marczenko, Z. *Separation and Spectrophotometric Determination of Elements*; Wiley & Sons: New York, NY, USA, 1986.
26. Ferreira, A.M.R.; Lima, J.L.F.C.; Rangel, A.O.S.S. Colorimetric determination of available iron in soil by flow injection analysis. *Analusis* **1996**, *24*, 343–346.
27. Gomes, D.M.C.; Segundo, M.A.; Lima, J.L.F.C.; Rangel, A.O.S.S. Spectrophotometric determination of iron and boron in soil extracts using a multi-syringe flow injection system. *Talanta* **2005**, *66*, 703–711. [CrossRef] [PubMed]
28. Software for Flow Injection Analysis. Available online: http://www.uwm.edu.pl/kchem/el_team/software/fia.htm (accessed on 21 June 2021).
29. Kalinowski, S.; Koronkiewicz, S. Instantaneous Photometric Detector. Polish Patent No. Pat.222359, 11 September 2015.

Article

Analysis of IV Drugs in the Hospital Workflow by Raman Spectroscopy: The Case of Piperacillin and Tazobactam

Ioanna Chrisikou [1,2], Malvina Orkoula [1] and Christos Kontoyannis [1,2,*]

1. Department of Pharmacy, University of Patras, University Campus, GR-26504 Rio Achaias, Greece; ioannach94@gmail.com (I.C.); malbie@upatras.gr (M.O.)
2. Institute of Chemical Engineering Sciences, Foundation of Research and Technology-Hellas (ICE-HT/FORTH), GR-26504 Platani Achaias, Greece
* Correspondence: kontoyan@upatras.gr; Tel.: +30-2610-962328

Citation: Chrisikou, I.; Orkoula, M.; Kontoyannis, C. Analysis of IV Drugs in the Hospital Workflow by Raman Spectroscopy: The Case of Piperacillin and Tazobactam. *Molecules* 2021, 26, 5879. https://doi.org/10.3390/molecules26195879

Academic Editor: Franciszek Główka

Received: 24 August 2021
Accepted: 23 September 2021
Published: 28 September 2021

Publisher's Note: MDPI stays neutral with regard to jurisdictional claims in published maps and institutional affiliations.

Copyright: © 2021 by the authors. Licensee MDPI, Basel, Switzerland. This article is an open access article distributed under the terms and conditions of the Creative Commons Attribution (CC BY) license (https://creativecommons.org/licenses/by/4.0/).

Abstract: Medical errors associated with IV preparation and administration procedures in a hospital workflow can even cost human lives due to the direct effect they have on patients. A large number of such incidents, which have been reported in bibliography up to date, indicate the urgent need for their prevention. This study aims at proposing an analytical methodology for identifying and quantifying IV drugs before their administration, which has the potential to be fully harmonized with clinical practices. More specifically, it reports on the analysis of a piperacillin (PIP) and tazobactam (TAZ) IV formulation, using Raman spectroscopy. The simultaneous analysis of the two APIs in the same formulation was performed in three stages: before reconstitution in the form of powder without removing the substance out of the commercial glass bottle (non-invasively), directly after reconstitution in the same way, and just before administration, either the liquid drug is placed in the infusion set (on-line analysis) or a minimal amount of it is transferred from the IV bag to a Raman optic cell (at-line analysis). Except for the successful identification of the APIs in all cases, their quantification was also achieved through calibration curves with correlation coefficients ranging from 0.953 to 0.999 for PIP and from 0.965 to 0.997 for TAZ. In any case, the whole procedure does not need more than 10 min to be completed. The current methodology, based on Raman spectroscopy, outweighs other spectroscopic (UV/Vis, FT-IR/ATR) or chromatographic (HPLC, UHPLC) protocols, already applied, which are invasive, costly, time-consuming, not environmentally friendly, and require specialized staff and more complex sample preparation procedures, thus exposing the staff to hazardous materials, especially in cases of cytotoxic drugs. Such an approach has the potential to bridge the gap between experimental setup and clinical implementation through exploitation of already developed handheld devices, along with the presence of digital spectral libraries.

Keywords: medical errors; hospital workflow; patient safety; Raman spectroscopy; IV drugs; piperacillin; tazobactam; non-invasively

1. Introduction

In a hospital workflow routine, the intravenous administration of drugs to patients is a usual phenomenon. Some of the most important advantages of intravenous drug administration include the immediate and fast therapeutic result, which is of utmost importance in case of emergency and the treatment of patients who are unable to receive oral administration [1].

The errors that are associated with the preparation and administration of intravenous drugs constitute one of the most usual categories of medical errors, which can lead not only to morbidity but also to mortality due to the direct effect they have on patients. Up to date, a large number of studies have been conducted regarding the detection of intravenous medication errors. The majority of them have focused on the preparation and administration procedure and the frequency the errors occurred [2–13].

Two worth mentioning, typical examples of individual incidents of IV errors, include the unsuccessful identification of the drug administered and the administration of an incorrect drug dose. In the first case, epinephrine was administered instead of midazolam as anesthetic, due to the similarity in size and color of packaging between the two formulations, in a health unit in Egypt [14]. In the second case, an excessive paracetamol dose was administered to a three-month-old baby at a pediatric department of a university hospital in Turkey. The drug quantity finally administered was ten-fold larger than the desired, which caused transient hepatotoxicity to the infant [15].

Obviously, there is a need for measures to be taken in order for the above-mentioned medication errors, mainly those associated with the wrong drug or the wrong concentration–dose of it, to be prevented. The present study aims at proposing a suitable methodology, for identification and quantification of IV drugs. The proposed methodology should be non-invasive (without removal of the substance from the commercial glass bottle, the infusion bag, or the intravenous infusion set used in each stage of administration process), aseptic, non-destructive for the sample, and simple. Additionally, it should be fast and effective as it should be applied in a hospital workflow in cases of emergency, as well as of low cost so that it can be applied on a large scale. A possible candidate is Raman spectroscopy.

In the medical and biomedical field, Raman spectroscopy lists numerous applications, including oncologic (cancer diagnosis and monitoring), cardiovascular, and neurological (Alzheimer disease) developments [16–19]. In recent years, and regarding the hospital context, the potential of Raman spectroscopy to bridge the distance between bench and bedside has been emphasized. More specifically, Bourget and collaborators determined the concentration of three hospital antineoplastic drugs, 5-fluoruracil in elastomeric portable infusion pumps, so as to develop an analytical tool for geometrically complex therapeutic objects [20], and andriamycin and epirubicin in polyethylene syringes [21]. Le et al. used a handheld Raman device for the analysis of two isomeric anticancer drugs, doxorubicin and epirubicin, by direct measurement of the substances in containers [22], while Amin and his team validated a method for the determination of ganciclovir, a cytotoxic anticancer agent [23]. The analysis of the latter was performed directly, through sealed glass vials. In two other studies, five antineoplastic drugs (5-fluorouracil, gemcitabine, cyclophosphamide, ifosfamide, and doxorubicin), as well as three taxane drugs (cabazitaxel, docetaxel, and paclitaxel) were packaged in glass vials and then measured [24–26]. In the case of taxane drugs, a handheld spectrometer had been used. Furthermore, Makki et al. investigated the identification of four chemotherapeutic drugs (doxorubicin, daunorubicin, ifosfamide, and methotrexate) by means of Raman spectroscopy and the macroanalysis was performed from solutions placed in quartz cuvettes [27]. Lastly, Makki and his team analyzed solutions of three hardly distinguishable isomers, doxorubicin, daunorubicin, and epirubicin, which were placed in quartz cuvettes once more [28].

This paper reports on the employment of Raman spectroscopy for the development of a new methodology, aiming at the identification and quantification of a piperacillin (PIP) and tazobactam (TAZ), intravenously administered formulation, during the preparation process and before administration to the patient. The current methodology was developed with a view to being implemented in the clinical environment and applied to IV drugs, extensively, irrespective of the category to which they belong (antibiotic, chemotherapeutic, etc.). The drug used in the current study is an injectable formulation, in the form of powder for reconstitution, in which the mass ratio between the two APIs is equal to 8:1 (89% PIP/11% TAZ). Piperacillin is a broad-spectrum b-lactam antibiotic, which inhibits the synthesis of the bacterial cell wall, while tazobactam is a beta-lactamase inhibitor, which increases and expands the antimicrobial spectrum of piperacillin. In particular, it protects the antibiotic from degradation caused by beta-lactamase enzymes [29–31]. Up to date, in most cases, PIP and TAZ in formulations have been simultaneously analyzed through exploitation of time-consuming LC (liquid chromatography) protocols [32–34], a procedure that is required by pharmacopoeia for the identification of each API and its impurities [35]. Ultra-high-performance liquid chromatography tandem mass spectrometry (UHPLC–MS/MS) has also been applied for the simultaneous determination of the two APIs, not in the formulation but in different biological matrices (serum, urine, renal replacement therapy effluent) in regard to therapeutic drug monitoring (TDM) after administration [36,37]. Although this method requires microsample volumes and the run time does not exceed 5 min, the analytical procedure requires more expensive equipment and more specialized staff and cannot be applied directly to the as-received initial dry formulation before reconstitution. Furthermore, another vibrational technique, FT-IR/ATR spectroscopy, has been successfully applied for the identification and quantification of the two APIs in the commercial mixture. The methodology developed is more than promising for use in the clinical workflow; however, the whole procedure is partially invasive and acquires some more time to be completed [38]. UV/Vis spectroscopy has also been used, but its application proved to be complex since it requires selection of appropriate derivative order and smoothing factor [39,40]. One additional obstacle for the UV/Vis application in the current study is that extensive dilution of the drug solutions (far from the clinically relevant concentrations) is required, thus rendering such a process more prone to errors.

Taking all the above into consideration, the novelty of the current study lies in that there is no previous report on an analytical Raman methodology developed to be fully harmonized with the preparation and administration procedures of IV drugs in a clinical environment. Besides that, no known effort has been made for the simultaneous analysis of two APIs in the same formulation, which is in form of powder intended for reconstitution and not a solution, in both solid and liquid state. The results' subsections following involve:

- The non-invasive analysis of the solid formulation (powder) in its glass vial before reconstitution,
- The non-invasive analysis of the reconstituted liquid formulation in its glass vial, and
- The analysis of the final (further diluted in the IV infusion bag in order to lie in the therapeutic concentration range) formulation just before administration; either it is placed in the drip chamber of the IV infusion set used (on line analysis) or in a Raman optic cell. The latter would be filled with 1–2 mL of the liquid drug, removed from the IV bag using an aseptic syringe (at-line analysis).

2. Results

2.1. Analysis of the Non-Reconstituted Solid Formulation through the Commercial Glass Bottle

2.1.1. Identification of APIs

Raman spectra of pure APIs and the formulation (mixture of PIP and TAZ), in both solid and liquid state, were obtained using a high reflectiveness gold-coated slide (g.slide) as a sampleholder (Figures 1 and 2). Characteristic peaks of each API were detected and attributed to vibration modes of their molecules, according to the bibliography (Table 1) [41–50], and the formulation peaks were successfully attributed to the APIs without being shadowed by glass peaks.

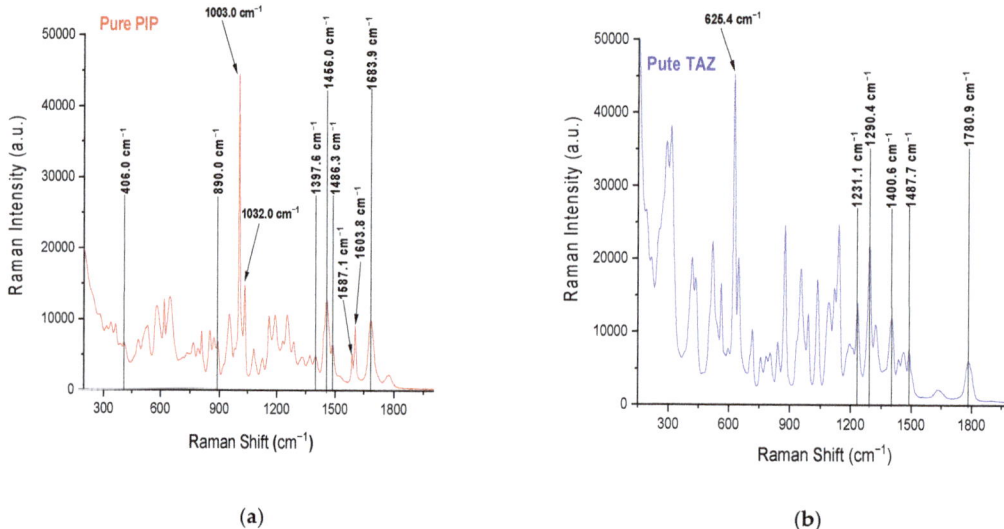

Figure 1. (**a**) Raman powder spectrum of pure piperacillin (PIP) and (**b**) Raman powder spectrum of pure tazobactam (TAZ). The peaks highlighted are the ones that are attributed to vibration modes in bibliography. The samples were placed on a g.slide.

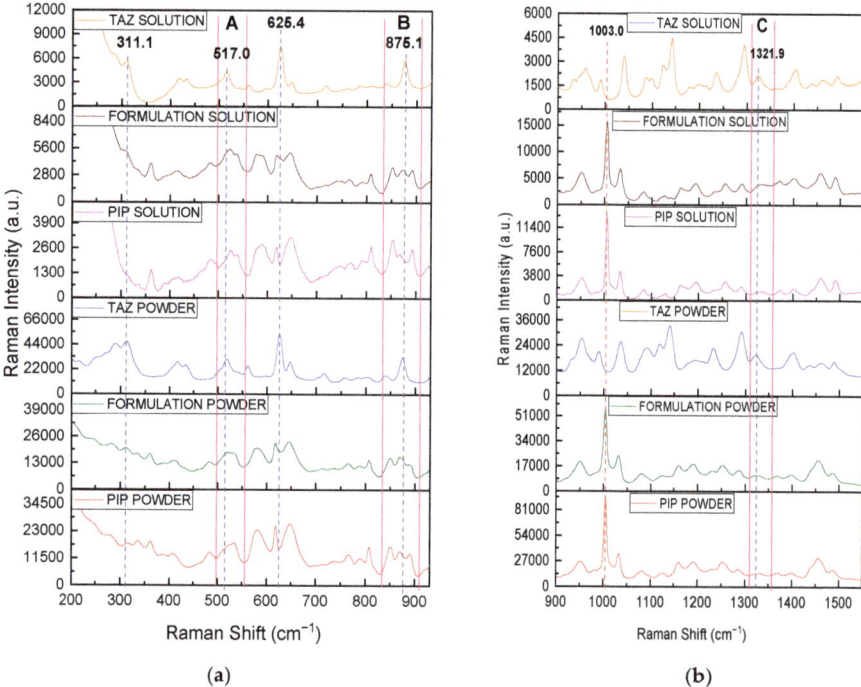

Figure 2. (**a**) Raman Spectra of PIP, TAZ, and the formulation, in both solid and liquid state: (**a**) from 200 to 930 cm^{-1} and (**b**) from 900 to 1550 cm^{-1}. The samples were placed on a g.slide. Differences between PIP and formulation spectra due to TAZ presence in the commercial product are shown with blue dotted lines and in the spectral regions A (494–551 cm^{-1}), B (826–906 cm^{-1}), and C (1307–1355 cm^{-1}). The characteristic peak of PIP at 1003 cm^{-1} is at the red dotted line.

Table 1. Characteristic peaks of pure APIs and their attribution to vibration modes according to bibliography.

API	Wavenumber (cm^{-1})	Vibration Mode
PIP	406.0	Carbonyl group (CO) bending
	890.0	C-H bending
	1003.0, 1032.0	Aromatic C-C-C bending
	1397.6	N-H stretching
	1456.0	C-C stretching
	1486.3	N=H bending
	1587.1, 1603.8, 1683.9	C=C stretching
TAZ	625.4	C-S stretching (Sulfone Ring)
	1231.1	N=N stretching (Triazole Ring)
	1290.4	Triazole Ring Bending
	1400.6	CO_2^- stretching
	1487.7	C=C stretching (Triazole Ring)
	1780.9	Carbonyl group (CO) stretching

A great similarity between the spectrum of the formulation and that of PIP is expected due to its high content in the mixture, while TAZ peaks are extensively overlapped by those of PIP, and its presence becomes mainly evident through small differences between formulation and PIP spectra, such as strengthening of PIP peaks at certain positions, broadening of certain peaks, and consequently, a slight shift of their center along the x axis or change in intensity ratio between neighboring PIP peaks. All these observations are summarized in Table 2.

Table 2. Spectral differences between PIP and formulation Raman spectra obtained from either solid or liquid samples of the substances on a g.slide. TAZ presence accounts for such differences.

Wavenumber (cm^{-1})	Observation
311.1	Intensity enhancement (solid state)/ detection of TAZ peak (liquid state, in this case PIP peak is not detected at the certain position)
517.0	Intensity enhancement and change in the morphology of the broad PIP peak in area A
625.4	Detection of TAZ shoulder
875.1	Intensity enhancement and change in intensity ratio between neighboring PIP peaks in area B
1321.9	Intensity enhancement and shift of the relatively broad, weak PIP peak in area C

Subsequently, the Raman spectrum of the solid formulation was obtained without removing it from the glass bottle (non-invasively) and it is presented along with pure solid APIs' spectra recorded through the glass container, as well as the spectrum of the empty bottle, which plays the role of the blank sample (Figure 3). All measurements were performed using the fiber optic probe of the Raman instrument, which was constantly in touch with the outer surface of the glass bottle, from six different marked regions of the latter. The commercial glass bottle had a diameter of 4.60 cm, while the mean thickness value coming from the six measuring positions was equal to 2.40 mm, with a standard deviation value of 0.19 mm.

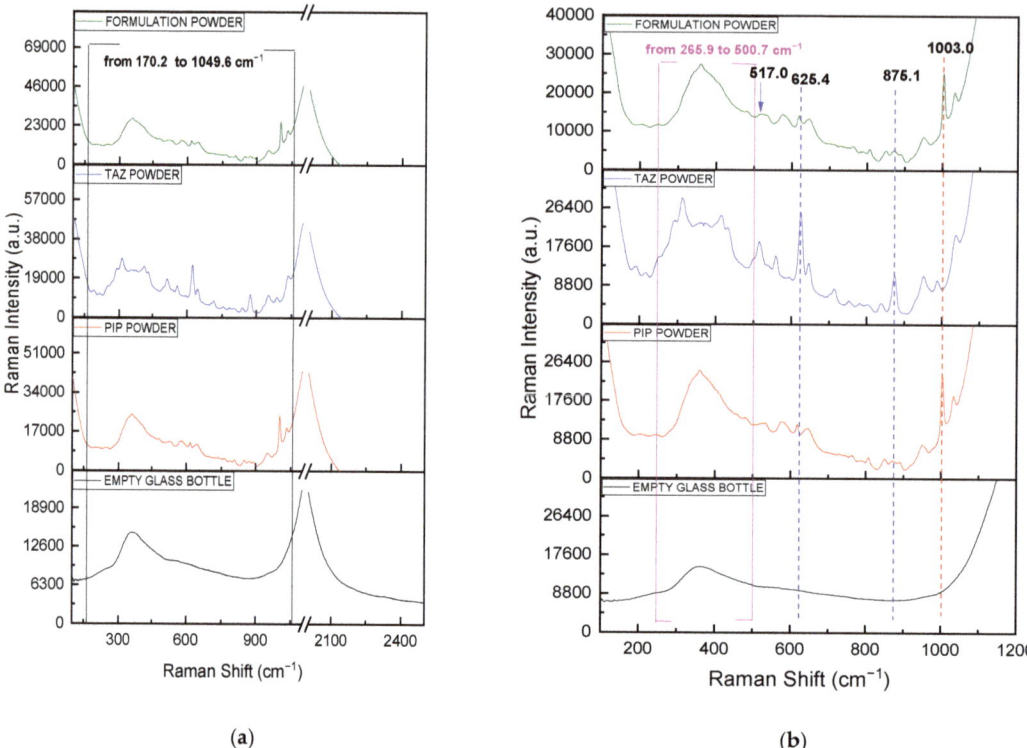

Figure 3. (a) Raman powder spectra of PIP, TAZ, and the formulation, obtained through the glass container of the formulation, as well as that of the empty glass bottle (blank sample): (a) from 100 to 2500 cm^{-1} and (b) from 100 to 1200 cm^{-1}. The framed spectral region from 170 to 1050 cm^{-1} (black frame on the left) corresponds to the wavenumber range where the sample peaks appear. The framed spectral region from 266 to 500 cm^{-1} (pink frame on the right) corresponds to the wavenumber range where the sample peaks can hardly be detected, mainly in case of PIP and formulation powder, as they are shadowed by a broad glass peak. Differences between PIP and formulation spectra due to TAZ presence in the commercial mixture are shown with blue dotted lines and the blue arrow at 517 cm^{-1}. The characteristic peak of PIP at 1003 cm^{-1} is at the red dotted line.

Analytes' peaks are detected in the spectral region ranging from 170 to 1050 cm^{-1} (Figure 3a, black framed region). The wavenumber range that has been cut out from the spectra is that between 1100 and 2000 cm^{-1}, where a very broad and intense glass peak appears, masking any of the sample peaks. Once more, the great similarity between the spectrum of the formulation powder and that of PIP powder is clearly observed. Contrary to the observations made when the spectra were obtained with the g.slide, the small differences between PIP and formulation spectra due to TAZ presence are in this case limited to the bands at 517, 625, and 875 cm^{-1} (Figure 3b), while the difference at 311 cm^{-1} cannot be detected, as the sample peaks between 266 and 500 cm^{-1} are shadowed by a relatively broad glass peak (Figure 3b, pink framed region). The characteristic PIP peak at 1003 cm^{-1} is effectively detected once more.

2.1.2. Quantification of APIs

Standard mixtures of the APIs, with a mass ratio ranging from 10% TAZ/90% PIP to 30% TAZ/70% PIP were prepared and measured after they had been placed in the glass bottle. The methodology included external usage of the fiber optic probe of the instrument, and six measurements were collected from different positions of the glass bottle. For the quantification of PIP, its characteristic peak at 1003 cm^{-1} was used. As for TAZ, no characteristic peak of it could be detected in the solid state. Nonetheless, after a variety of quantification efforts, the peak at 517 cm^{-1} proved to be the most appropriate for that purpose.

Worth mentioning is the fact that, as TAZ concentration in the standard solid mixtures increased, the intensity enhancement at 517 cm^{-1} became gradually more evident, and it ended up changing the morphology of the already existing PIP peak. The mean peak height ratio, $I(517)/I(1003)$ and relative standard deviation (RSD) values [51] resulting from the six replicate measurements were calculated for each standard mixture (Table 3).

Table 3. Mean peak height ratio, $I(517)/I(1003)$, and relative standard deviation (RSD) values corresponding to the six measurements performed, for each one of the standard mixtures placed in the commercial glass bottle.

% Mass Ratio TAZ/PIP	$I(517)/I(1003)$	RSD%
30:70	0.145	6.207
20:80	0.121	8.264
15:85	0.102	4.902
12.5:87.5	0.099	7.071
10:90	0.084	5.952

- Linearity

A minimum of five concentrations is required for the establishment of linearity, according to ICH guidelines for validation of analytical methods [52]. In this case, the linearity of the method was confirmed by a five point concentration calibration curve, achieved by preparing a series of separate weighings of synthetic mixtures of the drug product components.

The calibration graph plotted is reproduced in Figure 4 and was constructed through least square regression method, using the ratio $\frac{I(517)}{I(1003)}$ (y axis) against ratio $\frac{100}{C_{PIP}}$ (x axis) (see Appendix A).

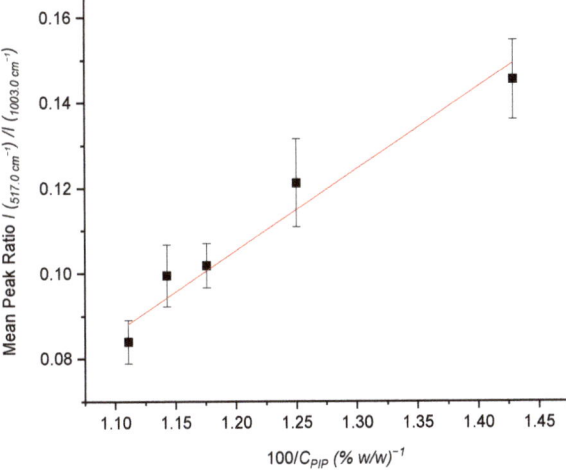

Figure 4. Calibration line for both PIP and TAZ for standard solid mixtures of them, with APIs' mass ratio ranging from 10% TAZ/90% PIP to 30% TAZ/70% PIP.

The equation describing the calibration line was:

$$\frac{I\,(517\ cm^{-1})}{I\,(1003\ cm^{-1})} = (0.192 \pm 0.025) \times \left(\frac{100}{C_{PIP}}\right) + (-0.125 \pm 0.029),\ \left(R^2 = 0.953\right) \quad (1)$$

- Intra- and Inter-day Precision

The intra-day (repeatability) and inter-day precision of the developed method were assessed using 6 determinations for each, at 100% of the test concentration, i.e., 11% TAZ/89% PIP (commercial drug product), characterized by a $\times (100/C_{PIP})$ value of 1.124. The repeatability study results gave a $100/C_{PIP}$ value of (1.108 ± 0.041) (% w/w)$^{-1}$, with a RSD value equal to 3.70%. Adidditionally, treating the precision test sample as unknown, PIP and TAZ concentration values, along with their total error [53], were found to be (90.17 ± 1.95) % w/w and (9.83 ± 1.95) % w/w, respectively. The relative error values (E_r) of the concentrations determined above were equal to 1.31% and 10.64%, respectively [54].

Accordingly, the inter-day precision study, tested over a period of 6 days, gave a x value of (1.110 ± 0.042) (% w/w)$^{-1}$, with a RSD value equal to 3.76%.

- Accuracy

The accuracy of the method was assessed using nine determinations over three concentrations (three determinations for each concentration), covering the specified range [52]. The results gave:

1. For 30% TAZ/70% PIP, x = (1.410 ± 0.070) (% w/w)$^{-1}$, RSD = 4.68%, and E_r = 1.33%.
2. For 15% TAZ/85% PIP, x = (1.189 ± 0.039) (% w/w)$^{-1}$, RSD = 3.32%, and E_r = 1.07%
3. For 10% TAZ/90% PIP, x = (1.080 ± 0.026) (% w/w)$^{-1}$, RSD = 2.38%, and E_r = 2.80%.

- LoD and LoQ

Since a minimum amount of tazobactam corresponds to a maximum amount of piperacillin, the limit of detection (LoD) of tazobactam was calculated after many blank (PIP powder) measurements [55]. Using Equation (1), LoD was found to be 3.01% w/w and LoQ (limit of quantification) equal to 9.03% w/w. The LoD value was confirmed by visual evaluation method [52].

- Calibration Range

The concentration range used for the construction of the calibration line corresponds to values from 10% TAZ/90% PIP to 30% TAZ/70% PIP, within which analytical procedure provides an acceptable degree of linearity, precision, and accuracy. The lower limit of the calibration range could not be less than 10% TAZ/90% PIP, as it is conformed to the LoQ value of TAZ API.

2.2. Analysis of the Reconstituted Formulation through the Commercial Glass Bottle

2.2.1. Identification of APIs

Adhering to the instructions of the usage leaflet of the product, 20 mL of NaCl 0.9% were used for the complete reconstitution of the solid formulation in the commercial glass bottle. According to calculations, the reconstituted sample had a nominal concentration of 200.00 mg PIP/mL and 25.00 mg TAZ/mL. Subsequently, its Raman spectrum was obtained without removing it out of the glass bottle (non-invasively) and was compared to spectra from pure APIs' solutions (80 mg PIP/mL and 30 mg TAZ/mL) that had been measured in the same way, i.e., using the fiber optic probe of the Raman instrument (Figure 5).

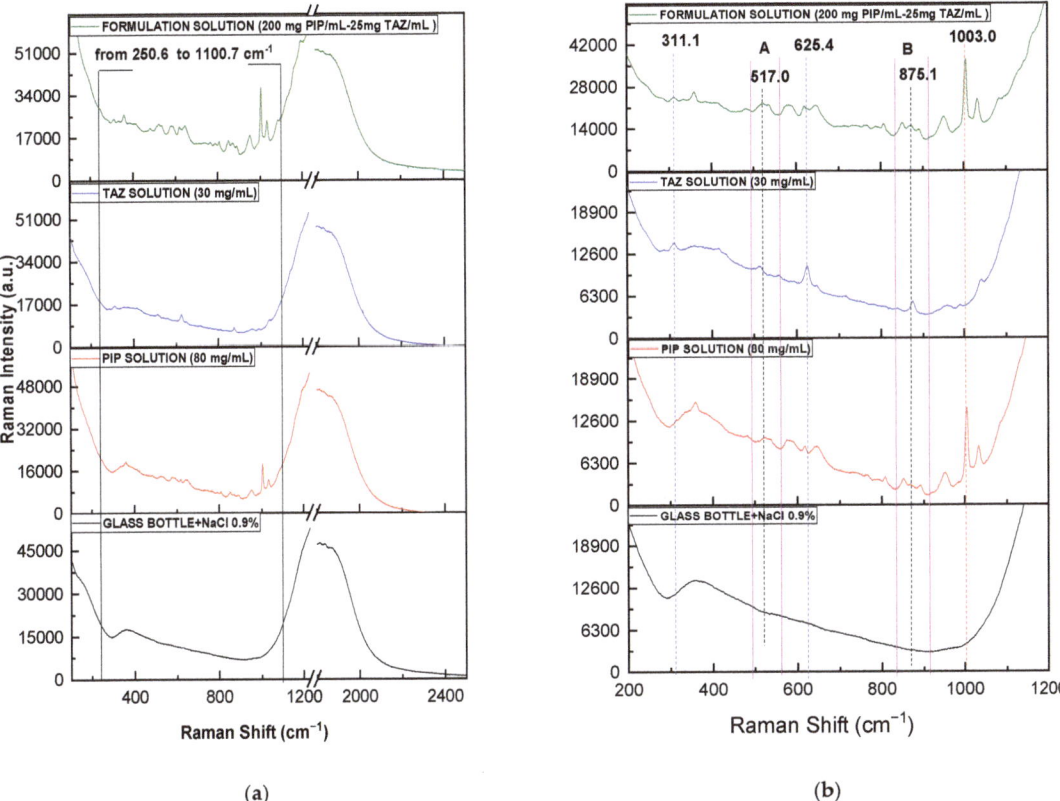

Figure 5. (**a**) Raman solution spectra of PIP (80 mg/mL), TAZ (30 mg/mL), and the formulation (200.00 mg PIP/mL and 25.00 mg TAZ/mL), obtained through the glass container of the formulation, as well as that of the glass bottle filled with NaCl 0.9% (blank sample): (**a**) from 100 to 2500 cm^{-1} and (**b**) from 200 to 1200 cm^{-1}. The framed spectral region from 250 to 1100 cm^{-1} (black frame on the left) corresponds to the wavenumber range where the sample peaks appear. Differences between PIP and formulation spectra due to TAZ presence are with blue dotted lines and in the spectral regions A (493–562 cm^{-1}) and B (834–913 cm^{-1}). The characteristic peak of PIP at 1003 cm^{-1} is with the red dotted line.

As it was expected, a great similarity is noticed between PIP and formulation spectrum, while above 1100 cm^{-1}, sample peaks are shadowed by the broad glass peak appearing. The characteristic PIP peak at 1003 cm^{-1} (vertical dashed red line) is highly detectable, while TAZ presence in the liquid formulation is confirmed through four different spectral positions (vertical dashed blue lines), at which the formulation spectrum is differentiated from that of pure PIP. Particularly, three of the four above-mentioned positions (517 (area A), 625, and 875 (area B) cm^{-1}) had been highlighted in Figure 3, which includes the powder spectra of the respective samples placed in the glass bottle. Furthermore, another peak at 311 cm^{-1} is clearly noticed, which, in this case, seems to be characteristic of TAZ API (see Figure 2 and Table 2).

2.2.2. Quantification of APIs

The reconstitution of the solid formulation in the commercial glass bottle was performed twice, once for preparing the concentrated, first standard solution of PIP with a nominal concentration of 266.67 mg/mL and a second time for preparing the respective solution for TAZ, with a nominal concentration of 60.00 mg/mL. Their Raman spectra were then obtained, in the same way as has already been described. By the method of successive dilutions for each of the concentrated solutions, another 10 standards were prepared, and their nominal concentration values ranged from 200.00 to 5.00 mg PIP/mL, as well as nine more standards with nominal concentration values ranging from 50.00 to 7.00 mg TAZ/mL. Each one of them was measured before it was further diluted.

Each API was quantified separately, using the peak at 1003 cm^{-1} for PIP and that at 311 cm^{-1} for TAZ. For every standard solution, the height/intensity value for both selected peaks was calculated for all of the six marked areas of the glass bottle (baseline correction method), after subtraction of the glass contribution to the final result for each area separately. The mean peak height values ($I1003(sample) - I1003(blank)$, $I311(sample) - I311(blank)$) along with the relative standard deviation values (RSD) were then calculated from the six replicate measurements performed (Table 4).

Table 4. Mean peak height values $I1003(sample) - I1003(blank)$ and $I311(sample) - I311(blank)$ and respective relative standard deviation (RSD) values corresponding to the six measurements taken from the six different marked areas on the commercial glass bottle.

API	C (mg/mL)	I(sample) − I(blank) (a.u.)	RSD%
PIP (1003.0 cm^{-1})	266.67	29028.38	8.24
	200.00	24576.15	9.16
	177.77	21479.97	6.90
	133.33	16696.92	6.62
	88.89	11867.34	8.07
	44.44	5848.70	8.78
	40.00	5654.34	8.77
	30.00	3981.94	4.11
	20.00	2634.56	3.81
	10.00	1350.39	4.60
	5.00	655.83	7.15
TAZ (311.1 cm^{-1})	60.00	3001.38	6.93
	50.00	2694.73	14.45
	40.00	2332.64	11.06
	33.33	2039.69	7.66
	25.00	1662.18	8.10
	22.22	1420.92	5.49
	16.67	1105.93	6.21
	11.11	840.31	10.09
	9.00	630.67	6.82
	7.00	494.35	9.46

Calibration curves, one for each API, were constructed using the intensity values (y axis) against concentration values (x axis). Particularly, for PIP quantification, the value $I1003(sample) - I1003(blank)$ was used against C_{PIP} and for TAZ quantification $I311(sample) - I1311(blank)$ against C_{TAZ}, respectively. The linear fitting of the standard points for the whole concentration range tested was not effective, proving that there is not a single straight line that can appropriately express the relationship between API concentration and Raman intensity. This was expected, taking into consideration the solubility values of PIP and TAZ, which are equal to 50 mg/mL for both APIs [56,57], i.e., part of each API was not dissolved using the reconstitution instructions of the manufacturer (see Section 2.2.2). Polynomial fitting of the standard points gave much better results (Figure 6).

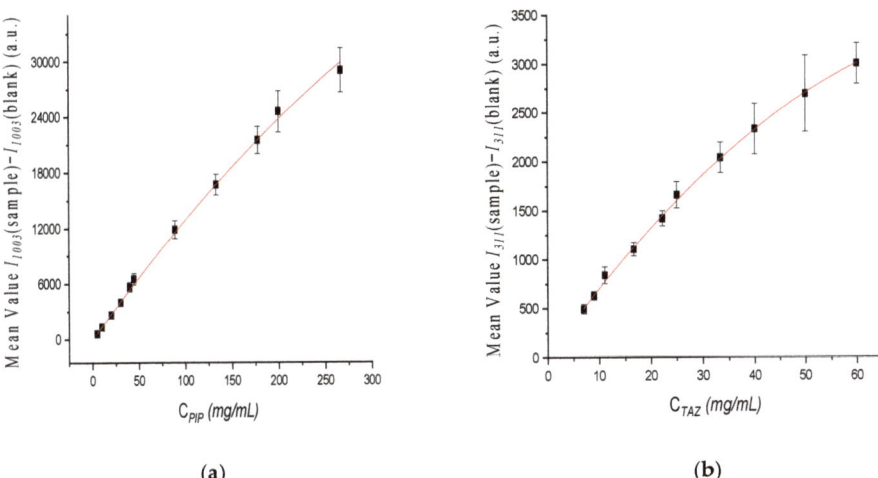

Figure 6. Calibration curve of: (**a**) PIP and (**b**) TAZ, after polynomial fitting of the points for the whole concentration range.

The respective equations describing the calibration curves resulting after polynomial fitting were:

$$I1003(sample) - I1003(blank) = (-0.10 \pm 0.02) \times (C_{PIP})^2 + (138.94 \pm 2.88) \times C_{PIP} + (-40.82 \pm 31.46), (R^2 = 0.999) \quad (2)$$

and

$$I311(sample) - I311(blank) = (-0.40 \pm 0.06) \times (C_{TAZ})^2 + (73.99 \pm 3.07) \times C_{TAZ} + (1.35 \pm 28.86), (R^2 = 0.997) \quad (3)$$

- Intra- and Inter-day Precision

The intra-day (repeatability) and inter-day precision, tested over a period of 6 days, of the developed method were assessed using six determinations for each at 100% of the test concentration directly after the reconstitution process, i.e., 200.00 mg PIP/mL and 25.00 mg TAZ/mL.

For PIP API, Equation (2) was used and the repeatability study results gave: C = (217.38 ± 17.92) mg/mL and RSD = 8.24%. The inter-day precision gave: C = (218.07 ± 18.02) mg/mL and RSD = 8.26%.

For TAZ API Equation (3) was used and the repeatability study results gave: C = (26.18 ± 2.52) mg/mL and RSD = 9.62%. The inter-day precision study resulted in: C = (27.23 ± 2.48) mg/mL and RSD = 9.12%.

- Accuracy

The accuracy of the method was assessed using 9 determinations over 3 concentration (3 determinations for each concentration), covering the specified range for each API. The accuracy test for PIP gave:

1. For 266.67 mg/mL, C = (264.12 ± 25.19) mg/mL, RSD = 9.54% and E_r = 0.96%.
2. For 44.44 mg/mL, C = (45.02 ± 1.23) mg/mL, RSD = 2.74% and E_r = 1.31% and
3. For 5.00 mg/mL, C = (5.02 ± 0.33) mg/mL, RSD = 6.66% and E_r = 0.40%.

For TAZ API, was respectively:

1. For 60.00 mg/mL, C = (63.70 ± 6.22) mg/mL, RSD = 9.76% and E_r = 6.17%.
2. For 33.33 mg/mL, C = (35.55 ± 3.02) mg/mL, RSD = 8.49% and E_r = 6.66% and
3. For 7.00 mg/mL, C = (7.31 ± 0.83) mg/mL, RSD = 11.37% and E_r = 4.43%.

- LoD and LoQ

Using Equation (2), and after many blank samples had been measured, PIP LoD was found to be 0.60 mg/mL, while LoQ was equal to 1.80 mg/mL. Respectively, using Equation (3), TAZ LoD was found to be 1.85 mg/mL, while LoQ was equal to 5.55 mg/mL. LoD values for both APIs were confirmed by visual evaluation method.

- Calibration Range

The concentration range used for the construction of the calibration curves corresponds to values from 5.00 to 266.67 mg/mL for PIP and from 7.00 to 60.00 mg/mL for TAZ, within which the analytical procedure provides an acceptable degree of precision and accuracy.

In order to test the performance of the polynomial calibration curves for each API once more, two different formulation solutions were prepared to be treated as unknown samples, and they were measured in the same way:

One solution containing 152.00 mg PIP/mL and 19.00 mg TAZ/mL. Using Equation (2), the PIP concentration was found to be (143.52 ± 10.52) mg/mL with a relative error of 5.58%. Using Equation (3), the TAZ concentration was found to be (20.42 ± 1.42) mg/mL with a relative error of 7.47%.

One solution containing 28.80 mg PIP/mL and 3.60 mg TAZ/mL. Using Equation (2), the PIP concentration was found to be (27.05 ± 2.02) mg/mL with a relative error of 6.08%. Using Equation (3), the TAZ concentration was found to be (3.30 ± 0.75) mg/mL with a relative error of 8.32%. Due to the fact that the TAZ concentration of the second unknown sample, i.e., 3.60 mg/mL, is below the LoQ value, and its quantification is more than satisfying, the robustness of the current methodology is confirmed, i.e., the validity of the analytical procedure is maintained regardless of small variations in the concentration range used.

Given the RSD values included in Tables 3 and 4 and the glass-induced thickness variability of the commercial bottle, six measurements for each sample would be needed in case of non-invasive analysis of either the solid drug before reconstitution or the liquid drug after reconstitution. Thus, the measurement process would take approximately 10 min.

2.2.3. Solubility of APIs after Reconstitution

In order to experimentally verify the incomplete dissolution of the solid formulation after reconstitution and quantify the percentage of drug that could not be diluted after reconstitution according to instructions, the solution being prepared was filtered. The filtrate was diluted five times in order for its concentration to be in the range applicable for Equations (2) and (3), while the filtering membrane was finally kept in a glass container with dehydrating material in order to remove the remaining moisture.

Subsequently, Raman spectra were obtained from ten different points/crystals of the remaining powder on the filter membrane, and the spectra found to be identical and compatible with the initial formulation powder, i.e., both APIs were identified in the spectra obtained from the undissolved material.

The filtrate that was diluted was then treated as an unknown sample and was measured in the same way as the standard solutions. Using Equations (2) and (3), it was found that the filtrate solution contained (33.79 ± 2.36) mg PIP/mL and (4.02 ± 0.66) mg TAZ/mL, respectively, i.e., approximately 15% of the initial solid formulation remained undissolved after the reconstitution process.

2.3. Analysis of the Liquid Formulation in the IV Bag

According to the instructions of use, 20 mL of the reconstituted formulation, including the undissolved APIs, is inserted aseptically in the IV infusion bag so as to be further diluted to a final volume of 100–150 mL. At that volume range, PIP concentration in the final solution will be between 26.67 and 40.00 mg/mL and that of TAZ between 3.33 and 5.00 mg/mL, i.e., both APIs will be completely dissolved. After the final preparation of the IV solution, the infusion bag content can be analyzed, either at-line or on-line. The at-line analysis requires the removal with an aseptic syringe of 1–2 mL of the solution and usage of an external optic Raman cell. The on-line analysis instead, which is preferable, requires the non-invasive analysis of the liquid in the IV drip chamber, a device that is used to allow air to rise out from a fluid so that it is not passed downstream, using Raman fiber optic.

2.3.1. Quantification of APIs through the Raman Optic Cell (At-line Analysis)

The Raman optic cell constitutes a special quartz cuvette with a mirror placed on the back part of it. The radiation path in the cell corresponds to a distance of 5 mm, while either the front quartz or the back mirror wall has a thickness of 1mm. In order to establish the ideal focus distance of the fiber optic probe on the Raman cell, a variety of focus positions of the incident laser beam were tested (Figure 7a). The cell was filled with a formulation solution of 40.00 mg PIP/mL and 5.00 mg TAZ/mL. The determining factor for the selection of the most appropriate focus position of the radiation was the intensity of the predominant, characteristic PIP peak at 1003 cm^{-1} (Figure 7b).

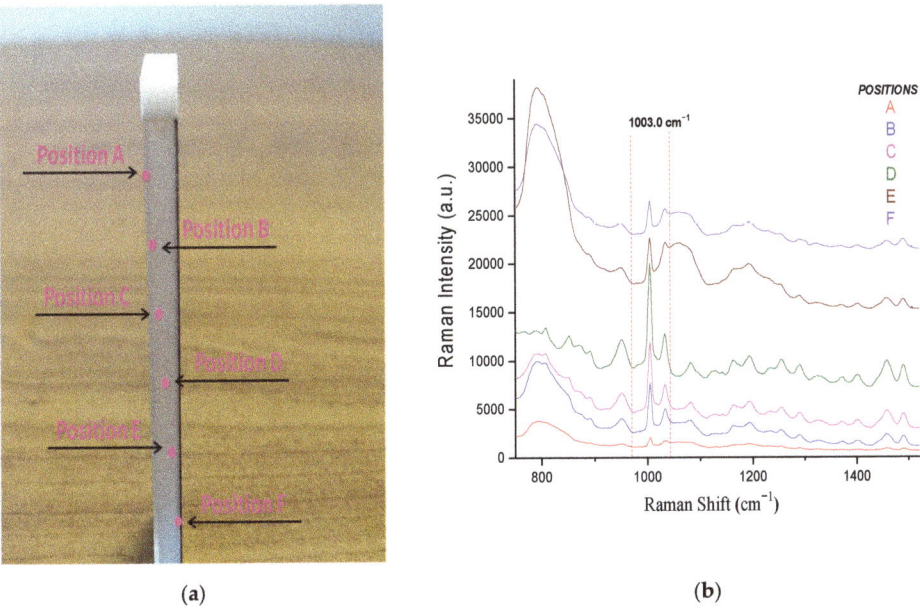

Figure 7. (**a**) Schematic illustration of the cross-section of the Raman cell depicting the depth of the six different focus positions of the laser beam, starting from the external quartz surface to the mirror surface for the spectra acquisition (for better clarity, each position of the focused beam is depicted along the diagonal of the cell), (**b**) Raman spectra of the liquid formulation (40.00 mg PIP/mL and 5.00 mg TAZ/mL) placed in the cell, each one corresponding to the radiation focus positions illustrated in part (**a**) of the figure.

In Table 5, comments can be found on the focus positions tested for the Raman optic cell, as well as on the quality of the respective spectrum received each time, with the latter being the determining factor for choosing the most appropriate position.

Table 5. Remarks on the position of the focused beam regarding the Raman optic cell, in correlation with the respective spectrum received. The peak mentioned in the remarks is the characteristic PIP peak at 1003 cm^{-1}.

Focus Position	REMARKS
A or F	front quartz surface or back mirror surface, the peak is detectable but quite weak especially in case of position A
B or E	at a depth of 1 or 4 mm (very close to the front quartz or the back mirror surface), the peak signal is stronger than before but is not as strong as it could be
C	at a depth of 2 to 2.5 mm (approximately in the middle of the radiation path), the peak signal is close to its maximum value
D	at a depth of 4 mm (closer to the back mirror than to the front quartz surface), the peak signal is the maximum that could be obtained

Based on Figure 7b, the laser beam should be focused on the D position, i.e., close to the mirror surface, which provides the strongest Raman signal, and thus all the following Raman spectra reported in the Raman cell section were obtained from the optimum position.

The spectrum of the liquid formulation (40.00 mg PIP/mL and 5.00 mg TAZ/mL), acquired through the Raman optic cell using the fiber optic probe, was compared to the spectra of pure APIs' solutions (50.00 mg PIP/mL and 30.00 mg TAZ/mL respectively) and that of the blank sample, which, in this case, is the optic cell filled with saline solution (NaCl 0.9%). Characteristic peaks of each API, at 1003 cm^{-1} for PIP and at 311 cm^{-1} for TAZ, were detected, as shown in Figure 8.

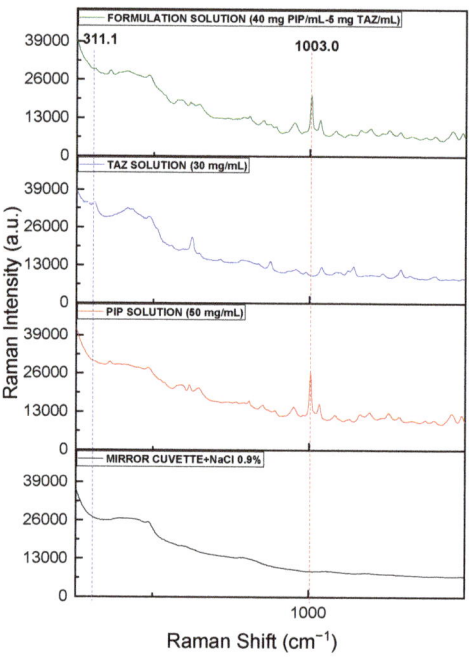

Figure 8. Raman spectra for solutions of pure PIP (50 mg/mL), pure TAZ (30 mg/mL), and the formulation (40.00 mg PIP/mL and 5.00 mg TAZ/mL), after the samples had been placed in the optic cell, as well as the blank sample spectrum (cell + NaCl 0.9%). PIP peak at 1003 cm^{-1} is highlighted by the red dashed line, whereas TAZ peak at 311 cm^{-1} is at the blue one.

It is apparent that the quartz material of which the cell is made, the thinner glass, and the vertical surfaces of the cell allow the detection of sample peaks in a wider spectral window compared to the common glass material of the commercial formulation bottle, which causes the extensive masking of them.

Standard solutions were prepared by the method of successive dilutions, and 1.7 mL of each one was transferred in the cell to be measured. The measurement was repeated three times for three different portions of the sample under test. The standards had concentration values that ranged from 44.44 to 5.00 mg/mL for PIP and 5.56 to 3.75 mg/mL for TAZ (therapeutic concentration range included).

Aiming at quantifying the APIs in the IV infusion bag used, the height/intensity values for both selected peaks (311 cm^{-1} for TAZ and 1003 cm^{-1} for PIP) were calculated for each standard solution, after baseline correction and subtraction of the quartz material contribution (resulting from a variety of blank measurements). The mean peak height values ($I1003(sample) - I1003(blank)$, $I311(sample) - I311(blank)$) along with the relative standard deviation values (RSD) were then calculated from the three replicate measurements performed (Table 6).

Table 6. Mean peak height values $I1003(sample) - I1003(blank)$ and $I311(sample) - I311(blank)$ and respective relative standard deviation (RSD) values corresponding to the three measurements obtained from the samples after they had been placed in the Raman optic cell.

API	C (mg/mL)	I(sample) − I(blank) (a.u.)	RSD%
PIP (1003.0 cm^{-1})	44.44	13800.36	1.77
	40.00	12062.34	1.60
	30.00	9463.25	0.80
	20.00	6354.26	0.75
	10.00	3084.76	1.51
	5.00	1561.74	1.66
TAZ (311.1 cm^{-1})	5.56	899.10	5.47
	5.00	761.14	9.22
	4.80	746.78	4.52
	4.20	626.54	4.08
	3.75	593.24	6.42

- Linearity

The linearity of the method was confirmed by a six point concentration calibration curve for PIP and a five point concentration calibration curve for TAZ, achieved by successive dilutions of an initial drug solution.

Calibration curves were constructed through least square regression method, using the intensity values (y axis) against concentration values (x axis) (Figure 9). Particularly, for PIP quantification the value $I1003(sample) - I1003(blank)$ was used against C_{PIP} and for TAZ quantification $I311(sample) - I1311(blank)$ against C_{TAZ} respectively.

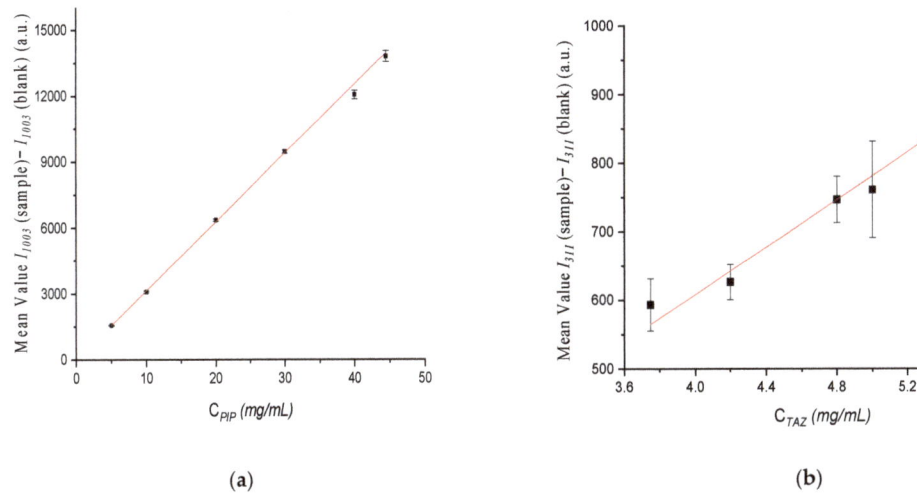

(a) (b)

Figure 9. Calibration curve of: (a) PIP and (b) TAZ, for standard solutions of liquid formulation placed in the Raman optic cell before administration, with concentration values ranging from 44.44 to 5.00 mg/mL for PIP and from 5.56 to 3.75 mg/mL for TAZ respectively.

The respective equations describing the calibration lines were:

$$I1003(sample) - I1003(blank) = (314.30 \pm 3.81) \times C_{PIP} + (-8.38 \pm 52.20), (R^2 = 0.999) \quad (4)$$

and

$$I311(sample) - I311(blank) = (173.05 \pm 19.13) \times C_{TAZ} + (-84.21 \pm 85.82), (R^2 = 0.965) \quad (5)$$

- Intra- and Inter-day Precision

The intra-day (repeatability) and inter-day precision, tested over a period of 6 days, of the developed method were assessed using six determinations for each at 100% of the test concentration, i.e., 40.00 mg PIP/mL and 5.00 mg TAZ/mL.

For PIP API, Equation (4) was used, and the repeatability study results gave: C = (40.34 ± 1.79) mg/mL and RSD = 4.44%. The inter-day precision gave: C = (39.34 ± 1.28) mg/mL and RSD = 3.27%.

For TAZ API, Equation (5) was used, and the repeatability study results gave: C = (5.29 ± 0.44) mg/mL and RSD = 8.26%. The inter-day precision gave: C = (4.79 ± 0.49) mg/mL and RSD = 10.33%.

- Accuracy

The accuracy of the method was assessed using nine determinations over three concentrations (three determinations for each concentration), covering the specified range for each API. The accuracy test for PIP gave:

1. For 44.44 mg/mL, C = (43.93 ± 0.78) mg/mL, RSD = 1.77% and E_r = 1.15%.
2. For 20.00 mg/mL, C = (20.24 ± 0.15) mg/mL, RSD = 0.75% and E_r = 1.20% and
3. For 5.00 mg/mL, C = (5.00 ± 0.08) mg/mL, RSD = 1.65% and E_r = 0.00%.

For TAZ API, was respectively:

1. For 5.56 mg/mL, C = (5.68 ± 0.28) mg/mL, RSD = 5.00% and E_r = 2.16%.
2. For 4.80 mg/mL, C = (4.80 ± 0.20) mg/mL, RSD = 4.06% and E_r = 0.00%.
3. For 3.75 mg/mL, C = (3.91 ± 0.22) mg/mL, RSD = 5.61% and E_r = 4.27%.

- LoD and LoQ

Using Equation (4), and after many blank samples had been measured, PIP LoD was found to be 0.24 mg/mL, while LoQ was equal to 0.72 mg/mL. Respectively, using Equation (5) TAZ LoD was found to be 0.94 mg/mL, while LoQ was equal to 2.81 mg/mL. LoD values for both APIs were confirmed by visual evaluation method.

- Calibration Range

The concentration range used for the construction of the calibration lines corresponds to values from 5.00 to 44.44 mg/mL for PIP and from 3.75 to 5.56 mg/mL for TAZ, within which analytical procedure provides an acceptable degree of linearity, precision, and accuracy. Additionally, this range includes the therapeutic concentration window and ensures the total dissolution of the APIs present in the formulation intended for administration.

In order to test the performance of the calibration lines for each API once more, a formulation solution of 23.00 mg PIP/mL and 2.87 mg TAZ/mL was prepared in order to be treated as unknown sample and it was measured in the same way. Using Equation (4), PIP concentration was found to be (22.46 ± 0.60) mg/mL with a relative error of 2.45%. Using Equation (5), TAZ concentration was found to be (3.05 ± 0.11) mg/mL with a relative error of 6.27%. Given the low RSD values included in Table 6 and the advantages the Raman cell offers, regarding its construction, it would be satisfying one measurement to be performed, thus the whole procedure (withdrawal of the substance and measurement) would not take more than 5 min.

2.3.2. Quantification of APIs through IV Drip Chamber (On-Line Analysis)

As in the case of the Raman optic cell, a variety of focus positions of the incident laser beam were tested (Figure 10) in order to establish the ideal focus position of the laser beam. The chamber was filled with a formulation solution of 40.00 mg PIP/mL and 5.00 mg TAZ/mL. The determining factor for the selection of the most appropriate focus position of the radiation was the intensity of the predominant, characteristic PIP peak at 1003 cm^{-1} (Figure 11).

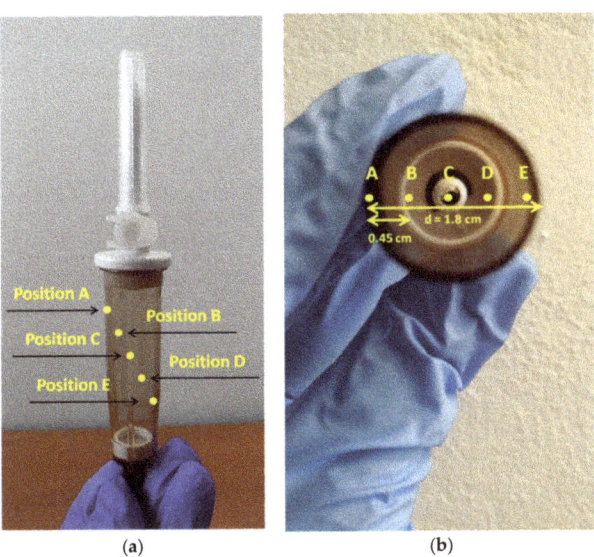

Figure 10. (a) Cross section schematic illustration and (b) top view of the drip chamber, along with the five different focus positions of the laser beam for the spectra acquisition. The diameter of the cylindrical drip chamber is equal to 1.8 cm, while the focus positions are equidistant, and the distance between them is 0.45 cm.

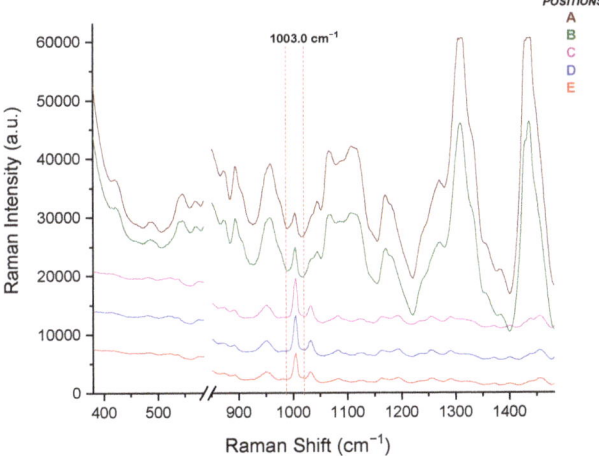

Figure 11. Raman spectra of the liquid formulation (40.00 mg PIP/mL and 5.00 mg TAZ/mL) placed in the drip chamber, each one corresponding to the radiation focus positions illustrated in Figure 10.

In Table 7, comments can be found on the focus positions tested for the IV drip chamber, as well as on the quality of the respective spectrum received each time, with the latter being the determining factor for choosing the most appropriate position.

Table 7. Remarks on the focus position of the laser beam on the drip chamber of the IV set, in correlation with the respective spectrum received. The peak mentioned in the remarks is the characteristic PIP peak at 1003 cm^{-1}.

Focus Position	REMARKS
A	outer front plastic surface, the peak is very weak
B	at a depth of 0.45 cm (close to the front plastic surface), the signal is stronger than before but not the maximum
C	at a depth of 0.9 to 1 cm (approximately in the middle of the radiation path), the peak signal is the maximum that could be obtained
D	at a depth of 1.35 cm (closer to the back than to the front plastic surface), the peak signal is close to its maximum value
E	at a depth larger than 1.60 cm (close to the back plastic surface), the signal is not as strong as it could be

Based on Figure 11, the laser beam should be focused on the C position, which has yielded the most intense Raman signal, and thus all the following Raman spectra reported in the drip chamber section were obtained from the optimum position.

The spectrum of the liquid formulation (40.00 mg PIP/mL and 5.00 mg TAZ/mL), acquired through the drip chamber using the Raman fiber optic probe, was compared to the spectra of pure APIs' solutions (50.00 mg PIP/mL and 30.00 mg TAZ/mL, respectively) and that of the blank sample, which, in this case, is the drip chamber filled with saline solution (NaCl 0.9%). Characteristic peaks of each API, at 1003 cm^{-1} for PIP and at 311 cm^{-1} for TAZ, were detected, as shown in Figure 12.

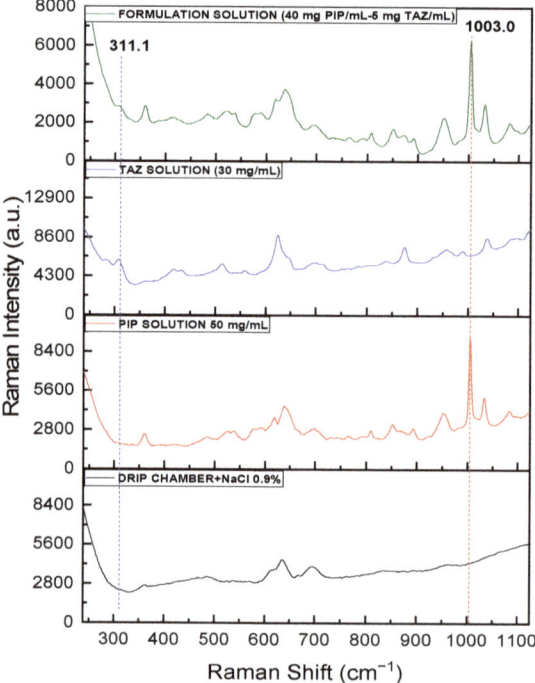

Figure 12. Raman spectra for solutions of pure PIP (50 mg/mL), pure TAZ (30 mg/mL), and the formulation (40.00 mg PIP/mL and 5.00 mg TAZ/mL), after the samples had been placed in the drip chamber, as well as the blank sample spectrum (drip chamber + NaCl 0.9%). PIP peak at 1003 cm^{-1} is highlighted with the red dashed line, whereas TAZ peak at 311 cm^{-1} with the blue one.

Although the plastic material, of which the drip chamber is made, contributes significantly to the obtained spectrum, the peaks attributed to the drip chamber are not overlapping significantly with API peaks.

Standard solutions were prepared by the method of successive dilutions and transferred in the drip chamber to be measured. The measurement was repeated three times for three different portions of the sample under test. The standards had concentration values that ranged from 44.44 to 5.00 mg/mL for PIP and 5.56 to 0.62 mg/mL for TAZ (therapeutic concentration range included).

Aiming at quantifying the APIs in liquid formulation using the current methodology, the height/intensity values for both selected peaks (311 cm^{-1} for TAZ and 1003 cm^{-1} for PIP) were calculated for each standard solution after baseline correction and subtraction of the plastic material contribution (resulting from a variety of blank measurements). The mean peak height values ($I1003(sample) - I1003(blank)$, $I311(sample) - I311(blank)$) along with the relative standard deviation values (RSD) were then calculated from the three replicate measurements performed (Table 8).

Table 8. Mean peak height values $I1003(sample) - I1003(blank)$ and $I311(sample) - I311(blank)$ and respective relative standard deviation (RSD) values corresponding to the three measurements obtained from the samples after they had been placed in the drip chamber of the IV set.

C (mg/mL)		I(sample) − I(blank) (a.u.)		RSD%	
PIP	TAZ	1003.0 cm^{-1}	311.1 cm^{-1}	1003.0 cm^{-1}	311.1 cm^{-1}
44.44	5.56	5708.64	349.73	2.00	5.77
40.00	5.00	5126.35	356.17	3.90	13.84
30.00	3.75	3905.66	185.09	2.69	26.59
20.00	2.50	2468.87	161.20	0.35	20.25
10.00	1.25	1298.05	not detected	1.38	not detected
5.00	0.62	583.98	not detected	4.99	not detected

Taking into account the unusually high RSD values corresponding to the replicate measurements of TAZ standards, as well as the inability of detecting the characteristic TAZ peak for the last two standard solutions, the investigation of TAZ LoD value, at that stage of the research, was triggered. The visual evaluation method gave the LoD value between 1.80 and 2.00 mg/mL, which meant that TAZ LoQ value was in no case less than 5.50 mg/mL. Therefore, the therapeutic TAZ concentration range (from 3.33 to 5.00 mg/mL) lies under LoQ value, and TAZ API cannot be reliably quantified through the IV drip chamber. Thus, only on-line analysis of PIP API is possible.

- Linearity

The linearity of the method was confirmed by a six point concentration calibration curve for PIP, achieved by successive dilutions of an initial drug solution. The calibration curve was constructed using the intensity values (y axis) against concentration values (x axis) (Figure 13). Particularly, the value $I1003(sample) - I1003(blank)$ was used against C_{PIP}.

The respective equation describing the calibration line was:

$$I1003(sample) - I1003(blank) = (123.07 \pm 3.09) \times C_{PIP} + (17.76 \pm 56.39), (R^2 = 0.997) \quad (6)$$

- Intra- and Inter-day Precision

The intra-day (repeatability) and inter-day precision, tested over a period of 6 days, of the developed method were assessed using six determinations for each, at 100% of the test concentration, i.e., 40.00 mg PIP/mL. Using Equation (6), the repeatability study results gave: C = (41.32 ± 2.10) mg/mL and RSD = 5.08%. The inter-day precision gave: C = (40.16 ± 1.13) mg/mL and RSD = 2.81%.

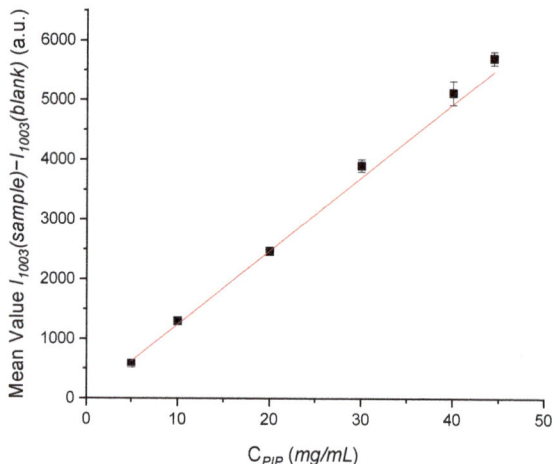

Figure 13. Calibration curve of PIP for standard solutions of liquid formulation placed in the drip chamber before administration, with concentration values ranging from 44.44 to 5.00 mg/mL.

- Accuracy

The accuracy of the method was assessed using nine determinations over three concentrations (three determinations for each concentration), covering the specified range. The accuracy test for PIP gave:

1. For 44.44 mg/mL, C = (46.24 ± 0.93) mg/mL, RSD = 2.00% and E_r = 4.05%.
2. For 20.00 mg/mL, C = (19.92 ± 0.07) mg/mL, RSD = 0.35% and E_r = 0.40% and
3. For 5.00 mg/mL, C = (4.60 ± 0.24) mg/mL, RSD = 5.14% and E_r = 8.00%.

- LoD and LoQ

Using Equation (6), and after many blank samples had been measured, PIP LoD was found to be 0.41 mg/mL, while LoQ was equal to 1.23 mg/mL. The LoD value was confirmed by visual evaluation method.

- Calibration Range

The concentration range used for the construction of the calibration line corresponds to values from 5.00 to 44.44 mg PIP/mL, within which analytical procedure provides an acceptable degree of linearity, precision, and accuracy. Additionally, this range includes the therapeutic concentration window, and ensures the total dissolution of the API present in the formulation intended for administration.

In order to test the performance of the calibration line once more, a formulation solution of 23.00 mg PIP/mL was prepared in order to be treated as unknown sample and it was measured in the same way. Using Equation (6), PIP concentration was found to be 22.19 ± 1.04 mg/mL with a relative error of 3.51%.

Given the RSD values for PIP standards included in Table 8, three measurements for each sample would be needed in this case, thus the measurement process, once the drug has been led into the drip chamber, would not take more than 5 to 6 min.

3. Discussion

The aim of the present study was to develop an appropriate methodology for the simultaneous identification and quantification of piperacillin and tazobactam, two APIs that coexist in an intravenously administered formulation, with a mass ratio of 8:1. This methodology was developed in such a way to comply with the procedures of preparation and administration of IV drugs, so as to be applicable in hospitals. Therefore, it was intended to be non-invasive, aseptic, non-destructive for the sample, simple to be applied, quick and effective, and preferably of low cost. Thus, the proposed methodology could be immediately applied in order to eliminate errors relating to the administration of the wrong drug or the incorrect concentration. Currently applied techniques in hospitals for such purposes include chromatography with flow injection analysis (FIA), HPLC linked to UV/DAD (diode array detector), and spectroscopic apparatus equipped with UV/FT-IR. However, most of them are time-consuming, and some of the methods require considerable consumables, e.g., mobile phase, chromatographic columns, or sample preparation, which leads to exposure of staff to hazardous materials, especially in case of cytotoxic drugs [58]. Additionally, UV spectroscopy is not efficient when the co-examined materials have the same UV signatures and infrared spectroscopy is not appropriate for aqueous solutions' analysis, as water considerably absorbs the infrared radiation [59]. So, the initial goal was approached through the utilization of Raman spectroscopy, while the formulation used was chosen due to the presence of two APIs with significant difference in their content in it, which itself rendered the desired analysis more complicated. Although the simultaneous identification and quantification of this pharmaceutical product has repeatedly been achieved, applying HPLC protocols, the whole procedure is not only time-consuming, destructive for the sample, and costly, but it also requires specialized staff. Additionally, it requires a large volume of solvents, thus excluding itself from the category of "green" techniques. UHPLC, on the other hand, takes only 5 min and few µL of sample to be applied, but the whole procedure still requires a mobile phase and a more specialized user. The above-mentioned drawbacks outweigh the high precision and the low limits of detection it offers.

The novel character of the approach presented here lies in that the developed methodology is fully compliant with the hospital routine procedures, two APIs in a single formulation are simultaneously analyzed, and both the solid and liquid formulation are measured. The proposed methodology involves testing procedures regarding not only the identity but also the composition of the formulation to be administered during three different preparation stages. More specifically, the drug analysis is possible either before or after the reconstitution process, as well as just before its administration to the patient, after it has been further diluted in the IV infusion bag.

The first part of the research was focused on the non-invasive (without removal of the substance out of the commercial glass bottle) qualitative and quantitative analysis of the initial dry powder of the formulation before reconstitution. For the identification of the solid drug at that stage, the spectrum of the solid formulation before reconstitution was compared with the spectra of pure APIs' powder. All spectra were obtained by the use of the fiber optic probe of the Raman instrumentation, which was constantly in contact with the outer surface of the glass bottle, after the samples had been placed in the latter. Although the masking of the sample peaks due to glass intervention was extensive, and the similarity between the spectra of the formulation and the PIP API was great, due to the high content of the latter in the drug, the two APIs were identified and TAZ API was effectively detected using spectral differences. Consequently, a number of standard solid mixtures of the APIs were prepared and measured in the same way, with the mass ratio of PIP and TAZ in them ranging from 10% TAZ/90% PIP to 30% TAZ/70% PIP (targeting the mass ratio of 11% TAZ/89% PIP in the commercial formulation). A calibration curve was then constructed, through which the% w/w content of each API in the final product could be determined. The analytical procedure was also validated according to characteristics including linearity, intra- and inter-day precision, accuracy, LoD, LoQ, and calibration range.

The second part of the research was focused on the non-invasive qualitative and quantitative analysis of the liquid formulation inside the original glass bottle after reconstitution. For the identification of the liquid drug at that stage, the spectrum of the reconstituted formulation was compared to the spectra of pure APIs' solutions. Once more, the measurements were performed by the aid of the fiber optic probe of the Raman instrumentation after the samples had been placed in the glass bottle. The extensive masking of sample peaks due to the nature of the methodology developed and the much higher content of PIP in the formulation did still make the detection of APIs, mainly in case of TAZ, more complex. Despite that, the identification was achieved once more. Standard formulation solutions were prepared, with nominal concentration values ranging from 266.67 to 5.00 mg/mL for PIP and from 60.00 to 7.00 mg/mL for TAZ. Calibration curves for each API separately were then constructed through polynomial regression method, while the relationship between the scattering signal and each API concentration was effectively expressed through quadratic equations. The whole analytical procedure was also validated according to characteristics including linearity, intra- and inter-day precision, accuracy, LoD, LoQ and calibration range. As it was further proved, on average, 15% of the initial dry powder of the formulation remained undissolved after the reconstitution process. The last-mentioned conclusion was the result of a study, which included the filtration of the reconstituted formulation, the quantitative analysis of the further diluted filtrate, as well as the characterization of the remaining powder on the filtering membrane.

The third part of the research was focused on the quantitative analysis of the liquid formulation, at-line and on-line, before administration to a patient. In this part, the sample was analyzed after it had been placed either in the drip chamber part of the IV administration set used to deliver the drug to the patient's vein (on-line test) or in a special Raman optic cell (at-line test). The latter could be filled with 1–2 mL of the liquid drug after this minimum amount had been withdrawn from the infusion bag by the nursing staff and the aid of a syringe. In all cases, the measurements were performed using the fiber optic probe of the Raman instrumentation. After the ideal focus position of the laser beam had been selected, solutions of pure APIs and the formulation, along with standard solutions of the latter, were placed in the sample holders in order to be measured. In case the Raman optic cell was used, the concentration values of the standards ranged from 44.44 to 5.00 mg PIP/mL and from 5.56 to 3.75 mg TAZ/mL respectively, values that belong to the therapeutic concentration range. One calibration curve for each API was constructed. In case the IV drip chamber was used, the quantification of PIP API only could be reliable, due to the fact that the therapeutic TAZ concentration range (from 3.33 to 5.00 mg/mL) lied under its LoQ value. The analytical procedure developed for the on-line determination of PIP API was also validated according to characteristics including linearity, intra- and inter-day precision, accuracy, LoD, LoQ and calibration range.

Finally, the spectral image, as well as the LoD values achieved for both APIs, obtained by each one of the three methodologies suggested (commercial glass bottle, drip chamber of the IV set, and the Raman optic cell, see in Table 9) are representative of the quality and efficacy of each one applied. Particularly, either the glass material of the commercial bottle or the polyethylene of which the drip chamber is made leads to extensive masking of analytes' peaks, and the LoD values achieved by these two methods are almost identical. On the other hand, the quartz material of which the Raman cell is made allows the detection of sample peaks in a wider spectral window; thus, the LoD value determined in this case is almost one-third that of the other sample holders. This is due to the fact that the quartz material of the outer cell surface is thinner compared to the glass and polyethylene surface of the remaining two sample holders used, and the mirror presence in the back part of it triggers the reflection of the radiation beam; thus, the enhancement of the signal intensity is achieved, and the sensitivity of the methodology was enhanced. The characteristics of the analytical alternatives presented in this study are summarized in the following comparative table.

Table 9. Stage and nature of analysis, the measurement recording time, the respective calibration curve/equation, LOD, and precision control results, regarding the analytical alternatives–methodologies suggested here, i.e, measurement through the commercial glass bottle, the drip chamber part of the IV set, or a Raman optic cell.

Method/ Container	Analysis Stage/ Nature/ Recording Time		Calibration Curve/Equation	LoD	Precision Control ($C_{unknown}$, E_r)	
Glass Bottle	Before Reconstitution/ Non-invasive/ $\cong$10 min		Linear Curve/ $\frac{I\,(517\,cm^{-1})}{I\,(1003\,cm^{-1})} = (0.192 \pm 0.025) \times \left(\frac{100}{C_{PIP}}\right) + (-0.125 \pm 0.029)$, ($R^2 = 0.953$)	TAZ 3.01% w/w	PIP (90.17 $\pm$ 1.95)% w/w, 1.31%	TAZ (90.17 $\pm$ 1.95)% w/w, 1.31%
	After Reconstitution/ Non-invasive/ $\cong$10 min	PIP	Polynomial Curve / $I1003(sample) - I1003(blank) = (-0.10 \pm 0.02)^*C_{PIP}{}^2 + (138.94 \pm 2.88)^*C_{PIP} + (-40.82 \pm 31.46)$, ($R^2 = 0.999$)	0.60 mg/mL	(143.52 $\pm$ 10.52) mg/mL, 5.58% (27.05 $\pm$ 2.02) mg/mL, 6.08%	
		TAZ	Polynomial Curve / $I1311(sample) - I1311(blank) = (-0.40 \pm 0.06)^*C_{TAZ}{}^2 + (73.99 \pm 3.07)^*C_{TAZ} + (1.35 \pm 28.86)$, ($R^2 = 0.997$)	1.85 mg/mL	(20.42 $\pm$ 1.42) mg/mL, 7.47% (3.30 $\pm$ 0.75) mg/mL, 8.32%	
Drip Chamber	Before Administration/ Non-invasive/ $\cong$5 min	PIP	$I1003(sample) - I1003(blank) = (123.07 \pm 3.09)^*C_{PIP} + (17.76 \pm 56.39)$, ($R^2 = 0.997$)	0.41 mg/mL	(22.19 $\pm$ 1.04) mg/mL, 3.51%	
Raman Cell	Before Administration/ Invasive or not/ $\cong$5 min (3 repeats) Or $\cong$1.5 min (1 repeat)	PIP	$I1003(sample) - I1003(blank) = (314.30 \pm 3.81)^*C_{PIP} + (-8.38 \pm 52.20)$, ($R^2 = 0.999$)	0.24 mg/mL	(22.46 $\pm$ 0.60) mg/mL, 2.45%	
		TAZ	$I1311(sample) - I1311(blank) = (173.05 \pm 19.13)^*C_{TAZ} + (-84.21 \pm 85.82)$, ($R^2 = 0.965$)	0.94 mg/mL	(3.05 $\pm$ 0.11) mg/mL, 6.27%	

4. Materials and Methods

In the present study, pure piperacillin and tazobactam APIs, were purchased in the form of sodium salts from Glentham Life Sciences (Corsham, Wiltshire, UK), which is a UK-based supplying company. The generic formulation used was Zobactam® (Vocate Pharmaceuticals S.A., Athens, Greece), a combination of piperacillin and tazobactam in a mass ratio of 8:1, respectively. The formulation was in the form of powder for injectable solution with no excipients and had been supplied by Aenorasis S.A., a Greece-based medical equipment company (Athens, Greece). In order for complete reconstitution of 4.5 g of the drug to be achieved, 20 mL of sodium chloride solution 0.9% w/v (Vioser S.A. Parenteral Solutions Industry, Trikala, Greece) were used. Further dilution of the reconstituted formulation resulted in a solution of a total volume between 100 and 150 mL. The clinically relevant concentration values of PIP and TAZ in the final liquid formulation ready for administration ranged from 26.67 to 40.00 mg PIP/mL and from 3.33 to 5.00 mg TAZ/mL, respectively, depending on the separate needs of each patient.

4.1. Preparation of Standard Mixtures

Pure APIs were used to prepare five dry mixtures. The mass ratio of TAZ:PIP ranged from 30:70 to 10:90. Using an analytical balance (Kern Inc., Grove City, OH, USA, ABJ 220-4NM) and depending on the desired ratio, an appropriate amount of each API was obtained in order a total mass of 100 mg of the standard mixture to be prepared (e.g., for the dry mixture 20:80, 20 mg TAZ and 80 mg PIP were obtained). Each mixture was placed in a special plastic container (NALGENE®, Rochester, NY, USA) with a magnetic rod and was homogenized by placing the latter on a magnetic stirrer (HANNA instruments, HI 190M (Woonsocket, RI, USA)) for 5 min.

4.2. Preparation of Standard Solutions

The reconstitution of the initial solid formulation took place twice by adding saline solution 0.9% through the rubbery cap of the commercial glass bottle using a syringe. Except from constant shaking of the bottle for 10 min (according to the instruction leaflet of the drug), Vortex (IKA®, Staufen, Germany, MS2, Minishaker) and an ultrasonic bath (Branson Ultrasonics, Danbury, CT, US, 2510E-MT) at a frequency of 60 HZ were also used for about 30 s and 15 min, respectively. According to calculations, the solution after the first reconstitution process had a nominal concentration of 266.67 mg PIP/mL, and that after the second reconstitution process had a nominal concentration value of 60.00 mg TAZ/mL. By the method of successive dilutions inside the commercial glass bottle, another 10 PIP standard solutions were prepared, and their concentration values ranged from 200.00 to 5.00 mg/mL, whereas the remaining 9 TAZ standard solutions prepared had concentration values ranging from 60.00 to 7.00 mg/mL. The solvent was added using automatic pipettes (Labnet International Inc., Edison, NJ, USA, Biopette™, Autoclavable Pipettes).

4.3. Experimental Methodology for the Dissolution Test after Reconstitution

The standard solution under test was filtered using a filtering device and a 0.22 μL nitrocellulose membrane (Merck Millipore Ltd., Burlington, MA, USA). Then, the filtrate was diluted five times, while the diluent was the solution of NaCl 0.9%. The filtering membrane was finally kept in a glass container (Bormioli Rocco, Fidenza, Italy) with dehydrating material for 24 h in order the remaining moisture to be removed.

4.4. Methodology of Raman Spectra Acquisition

Raman spectra were obtained using a portable Raman spectrometer combined with an optical microscope (i-Raman Plus®, B&W Tek Inc., Newark, DE, USA). The instruments' nominal power was equal to 350 mW, and the laser beam used was at 785 nm.

4.4.1. Method of High Reflectiveness Gold-Coated Slide

Pure APIs in their solid or liquid form, as well as the formulation, were measured by placing a small amount of the sample onto a high reflectiveness gold-coated slide (EMF Corporation, Ithaca, NY, USA). That slide was coated by a 50 Å titanium layer and, above this, a 1000 Å gold layer. The latter gives the slide high reflectiveness. The sample was placed thereon, and in case it was in its solid state, a low pressure was applied on top of it by the aid of a spatula, in order to make its surface flat and for a better signal-to-noise ratio to be achieved in the spectrum obtained. Each measurement was a result of 5 accumulations in the spectral region from 0 to 2814 cm^{-1}, while the accumulation time was 20 s and the laser power equal to 210 mW.

4.4.2. Method of Glass Bottle

Pure APIs in their solid or liquid form, their commercial mixture before reconstitution, as well as standard mixtures and standard solutions under test were analyzed without removing them from the glass bottle. Their spectra were obtained using the fiber optic probe of the instrument. Particularly, by properly adjusting the distance between the fiber optic probe and the outer surface of the glass bottle (the probe touched the outer surface of the glass), the laser beam was focused inside the bottle, thus enabling the capture of the spectral image of the sample. Each sample was measured 6 times from 6 differerent areas of the bottle, which had been marked and defined on it. In this way, the glass-induced variability, which could have had an effect on the results, would be eliminated. The commercial glass bottle had a diameter of 4.60 cm, while the mean thickness value coming from the six measuring positions was equal to 2.40 mm with a standard deviation value of 0.19 mm. Each measurement was a result of 5 accumulations in the spectral region from 0 to 2814 cm^{-1}, while the accumulation time was 20 s and the laser power equal to 280 mW.

4.4.3. Method of Drip Chamber

An intravenous gravity administration set (Primary PLUM^{TM} Set 14000, Hospira Inc., Lake Forest, IL, USA), with a length value of 272 cm and liquid capacity of 19 mL, which is used in a hospital workflow so the drug reaches a patient's vein, was filled with each of the standard solutions under test. Part of the IV set was the drip chamber, which, in this case, was the sample holder for the Raman spectra acquisition in the methodology developed. The main parts of the IV set, as well the drip chamber, were made of polyethylene. Each measurement was a result of 5 accumulations in the spectral region from 0 to 2814 cm^{-1}, while the accumulation time was 20 s and the laser power equal to 280 mW. As in the method of the glass bottle, the spectra were obtained using the fiber optic probe of the instrument, while the distance between the probe and the outer surface of the drip chamber played an important role in the quality of the received spectra.

4.4.4. Method of Raman Optic Cell

In this case, solutions of the liquid formulation under test were transferred in a Raman optic cell, which had a mirror placed on the back part of it (Hellma GmbH & Co. KG, Müllheim, Germany). The cell was made of synthetic quartz (Quartz Suprasil® 300), its thickness, which is the same as the optical path length, was equal to 5 mm, and it required 1.7 µL to be filled. Both the front quartz and the back mirror wall of the cell had a thickness value of 1 mm. Each measurement was a result of 5 accumulations in the spectral region from 0 to 2814 cm^{-1}, while the accumulation time was 20 s and the laser power equal to 280 mW. As in the method of the glass bottle, the spectra were obtained using the fiber optic probe of the instrument, while the distance between the probe and the outer surface of the cell played an important role in the quality of the received spectra.

4.5. Spectra Processing

Spectral data were processed using Origin (OriginPro8, OriginLabPro®, Northampton, MA, USA). The OriginPro8 integration method of measuring the absolute height (or intensity) of selected peaks was used. In the case of solid samples, different combinations of fingerprint APIs' peaks were tested so as to assess the potential to quantify the level of APIs present. After trials, the selected peaks were at 517 cm^{-1}, which was the result of both APIs' presence and the characteristic PIP peak at 1003 cm^{-1}. The baseline corresponding to the first peak ranged from 500 to 550 cm^{-1}, while the baseline for the second peak ranged from 987 to 1019 cm^{-1}. As for the liquid samples, either they were placed in the glass bottle or in the drip chamber and the mirror cuvette, and the peaks selected for the quantification were characteristic of each API. As before, the PIP peak used was at 1003 cm^{-1}, and the respective baseline ranged from 987 to 1019 cm^{-1}, while for TAZ, the peak used was found at 311 cm^{-1}, and its baseline ranged from 295 to 325 cm^{-1}.

5. Conclusions

The proposed methodology as a whole has the potential to be applied to intravenously administered formulations, for either on-line or at-line analysis, given that it can comply with the preparation and administration procedures followed. The important advantages Raman spectroscopy gathers, such as the fast analytical response, the absence of sample preparation necessity, the non-invasive analysis nature, the negligible maintenance costs, the reassurance of operator safety, the limited waste produced, the suppression of consumables, and the reduction in staff training cost, render it a valuable analytical tool. In terms of a robust transportable system design equipped with deported probes to handheld devices, large libraries of spectral data, and an inter-system calibration (quantitative algorithms), the gap between experimental setup and clinical implementation could be bridged.

Author Contributions: Conceptualization, I.C., M.O. and C.K.; methodology, I.C., M.O. and C.K.; investigation, I.C.; validation, I.C., M.O. and C.K.; data curation, I.C.; writing—original draft preparation, C.K.; review and editing, M.O. and C.K.; supervision, M.O. and C.K.; project administration, C.K. All authors have read and agreed to the published version of the manuscript.

Funding: This research received no external funding.

Informed Consent Statement: Not applicable.

Data Availability Statement: Data is contained within the article.

Acknowledgments: The generic formulation used (Zobactam®) was donated by Aenorasis S.A., a Greece based medical equipment company.

Conflicts of Interest: The authors declare no conflict of interest.

Sample Availability: Samples of pure APIs, PIP and TAZ, are available from the authors.

Appendix A

Appendix A is a section containing the mathematical proof of how the final equation of the calibration curve (Figure 4, Equation (1)) for the quantification of PIP and TAZ in the solid formulation before reconstitution was developed. Spectral observations arising from powder spectra of both pure active substances and the commercial formulation (Figures 2 and 3) confirmed that peak at 517 cm^{-1} is a result of not only PIP but also TAZ presence. Thus, signal intensity at that position is expected to be given from the equation:

$$I\left(517\ cm^{-1}\right) = k_1 \times C_{PIP} + k_2 \times C_{TAZ}, \quad (A1)$$

where k_1 and k_2 are standards correlating the value $I\left(517\ cm^{-1}\right)$ with C_{PIP} and $I\left(517\ cm^{-1}\right)$ with C_{TAZ}, respectively.

Due to the fact that the peak at 1003 cm^{-1} is characteristic of piperacillin, its intensity is expected to be given from the equation:

$$I\left(1003\ cm^{-1}\right) = k_3 \times C_{PIP}, \quad (A2)$$

where k_3 is a standard correlating the value $I\left(1003\ cm^{-1}\right)$ with C_{PIP}.

If (A1) is divided by (A2), then it turns out that:

$$\frac{I\left(517\ cm^{-1}\right)}{I\left(1003\ cm^{-1}\right)} = \frac{k_1}{k_3} + \frac{k_2}{k_3} \times \frac{C_{TAZ}}{C_{PIP}} \quad (A3)$$

Given that the formulation consists exclusively of the APIs, it is valid that:

$$C_{TAZ} = 100 - C_{PIP}, \quad (A4)$$

Finally, combining (A3) with (A4), a new equation is formed, where:

$$\frac{I\left(517\ cm^{-1}\right)}{I\left(1003\ cm^{-1}\right)} = \frac{k_1 - k_2}{k_3} + \frac{k_2}{k_3} \times \frac{100}{C_{PIP}} \quad (A5)$$

References

1. Doyle, G.R.; McCutcheon, J.A. *Clinical Procedures for Safer Patient Care*; BCcampus: Victoria, BC, Canada, 2015; pp. 445, 501.
2. Hicks, R.W.; Becker, S.C. An Overview of Intravenous-related Medication Administration Errors as Reported to MEDMARX®, a National Medication Error-reporting Program. *J. Infus. Nurs.* **2006**, *29*, 20–27. [CrossRef]
3. Hedlund, N.; Beer, I.; Hoppe-Tichy, T.; Trbovich, P. Systematic evidence review of rates and burden of harm of intravenous admixture drug preparation errors in healthcare settings. *BMJ Open* **2017**, *7*, e015912. [CrossRef]
4. Alemu, W.; Belachew, T.; Yimam, I. Medication administration errors and contributing factors: A cross sectional study in two public hospitals in Southern Ethiopia. *Int. J. Afr. Nurs. Sci.* **2017**, *7*, 68–74. [CrossRef]
5. Wirtz, V.; Barber, N.; Taxis, K. An observational study of intravenous medication errors in the United Kingdom and in Germany. *Pharm. World Sci.* **2003**, *25*, 104–111. [CrossRef] [PubMed]

6. Fahimi, F.; Ariapanah, P.; Faizi, M.; Shafaghi, B.; Namdar, R.; Ardakani, M.T. Errors in preparation and administration of intravenous medications in the intensive care unit of a teaching hospital: An observational study. *Aust. Crit. Care* **2008**, *21*, 110–116. [CrossRef] [PubMed]
7. Ding, Q.; Barker, K.N.; Flynn, E.A.; Westrick, S.; Chang, M.; Thomas, R.E.; Braxton-Lloyd, K.; Sesek, R. Incidence of Intravenous Medication Errors in a Chinese Hospital. *Value Health Reg. Issues* **2015**, *6*, 33–39. [CrossRef]
8. Barber, N.; Taxis, K. Incidence and severity of intravenous drug errors in a German hospital. *Eur. J. Clin. Pharmacol.* **2004**, *59*, 815–817. [CrossRef] [PubMed]
9. Anselmi, M.L.; Peduzzi, M.; Dos Santos, C.B. Errors in the administration of intravenous medication in Brazilian hospitals. *J. Clin. Nurs.* **2007**, *16*, 1839–1847. [CrossRef]
10. Westbrook, J.I.; Rob, M.I.; Woods, A.; Parry, D. Errors in the administration of intravenous medications in hospital and the role of correct procedures and nurse experience. *BMJ Qual. Saf.* **2011**, *20*, 1027–1034. [CrossRef] [PubMed]
11. Mendes, J.R.; Lopes, M.C.B.T.; Vancini-Campanharo, C.R.; Okuno, M.F.P.; Batista, R.E.A. Types and frequency of errors in the preparation and administration of drugs. *Einstein* **2018**, *16*. [CrossRef]
12. Wolf, Z.R. Medication Errors Involving the Intravenous Administration Route: Characteristics of Voluntarily Reported Medication Errors. *J. Infus. Nurs.* **2016**, *39*, 235–248. [CrossRef]
13. Bagheri-Nesami, M.; Esmaeili, R.; Tajari, M. Intravenous Medication Administration Errors and Their Causes in Cardiac Critical Care Units in Iran. *Mater. Socio-Medica* **2015**, *27*, 442–446. [CrossRef]
14. Gado, A.; Ebeid, B.; Axon, A. Accidental IV administration of epinephrine instead of midazolam at colonoscopy. *Alex. J. Med.* **2016**, *52*, 91–93. [CrossRef]
15. Aslan, N.; Yildizdas, D.; Arslan, D.; Horoz, O.O.; Yilmaz, H.L.; Bilen, S. Intravenous Paracetamol Overdose: A Pediatric Case Report. *Pediatr. Emerg. Care* **2019**, *35*, e42–e43. [CrossRef]
16. Nunes, A.; Magalhães, S. *Raman Spectroscopy Applied to Health Sciences, Raman Spectroscopy, Gustavo Morari do Nascimento*; IntechOpen: London, UK, 2018. [CrossRef]
17. Cui, S.; Zhang, S.; Yue, S. Raman Spectroscopy and Imaging for Cancer Diagnosis. *J. Healthc. Eng.* **2018**, *2018*, 8619342. [CrossRef]
18. Wang, W.; McGregor, H.C.; Short, M.A.; Zeng, H. Clinical utility of Raman spectroscopy: Current applications and ongoing developments. *Adv. Health Care Technol.* **2016**, *2*, 13–29. [CrossRef]
19. Pence, I.; Mahadevan-Jansen, A. Clinical instrumentation and applications of Raman spectroscopy. *Chem. Soc. Rev.* **2016**, *45*, 1958–1979. [CrossRef]
20. Bourget, P.; Amin, A.; Vidal, F.; Merlette, C.; Lagarce, F. Comparison of Raman spectroscopy vs. high performance liquid chromatography for quality control of complex therapeutic objects: Model of elastomeric portable pumps filled with a fluorouracil solution. *J. Pharm. Biomed. Anal.* **2014**, *91*, 176–184. [CrossRef] [PubMed]
21. Bourget, P.; Amin, A.; Moriceau, A.; Cassard, B.; Vidal, F.; Clément, R. La Spectroscopie Raman (SR): Un nouvel outil adapté au contrôle de qualité analytique des préparations injectables en milieu de soins. Comparaison de la SR aux techniques CLHP et UV/visible-IRTF appliquée à la classe des anthracyclines en cancérologie. *Pathol. Biol.* **2012**, *60*, 369–379. [CrossRef] [PubMed]
22. Lê, L.; Tfayli, A.; Zhou, J.; Prognon, P.; Baillet-Guffroy, A.; Caudron, E. Discrimination and quantification of two isomeric antineoplastic drugs by rapid and non-invasive analytical control using a handheld Raman spectrometer. *Talanta* **2016**, *161*, 320–324. [CrossRef]
23. Amin, A.; Bourget, P.; Vidal, F.; Ader, F. Routine application of Raman spectroscopy in the quality control of hospital compounded ganciclovir. *Int. J. Pharm.* **2014**, *474*, 193–201. [CrossRef]
24. Lê, L.M.M.; Berge, M.; Tfayli, A.; Zhou, J.; Prognon, P.; Baillet-Guffroy, A.; Caudron, E. Rapid discrimination and quantification analysis of five antineoplastic drugs in aqueous solutions using Raman spectroscopy. *Eur. J. Pharm. Sci.* **2018**, *111*, 158–166. [CrossRef]
25. Kiehntopf, M.; Mönch, B.; Salzer, R.; Kupfer, M.; Hartmann, M. Quality control of cytotoxic drug preparations by means of Raman spectroscopy. *Die Pharm.* **2012**, *67*, 95–96.
26. Lê, L.; Berge, M.; Tfayli, A.; Prognon, P.; Caudron, E. Discriminative and Quantitative Analysis of Antineoplastic Taxane Drugs Using a Handheld Raman Spectrometer. *BioMed Res. Int.* **2018**, *2018*, 8746729. [CrossRef]
27. Makki, A.A.; Massot, V.; Byrne, H.J.; Respaud, R.; Bertrand, D.; Mohammed, E.; Chourpa, I.; Bonnier, F. Vibrational spectroscopy for discrimination and quantification of clinical chemotherapeutic preparations. *Vib. Spectrosc.* **2021**, *113*, 103200. [CrossRef]
28. Makki, A.A.; Bonnier, F.; Respaud, R.; Chtara, F.; Tfayli, A.; Tauber, C.; Bertrand, D.; Byrne, H.J.; Mohammed, E.; Chourpa, I. Qualitative and quantitative analysis of therapeutic solutions using Raman and infrared spectroscopy. *Spectrochim. Acta Part A Mol. Biomol. Spectrosc.* **2019**, *218*, 97–108. [CrossRef]
29. Shorr, R.I.; Hoth, A.B.; Rawls, N. (Eds.) *Drugs for the Geriatric Patient*; W.B. Saunders: Philadelphia, PA, USA, 2007; pp. 930–1062.
30. Drawz, S.M.; Bonomo, R.A. Three Decades of β-Lactamase Inhibitors. *Clin. Microbiol. Rev.* **2010**, *23*, 160–201. [CrossRef]
31. Kuriyama, T.; Karasawa, T.; Williams, D.W. Chapter Thirteen—Antimicrobial Chemotherapy: Significance to Healthcare. In *Biofilms in Infection Prevention and Control*; Percival, S.L., Williams, D.W., Randle, J., Cooper, T., Eds.; Academic Press: Boston, MA, USA, 2014; pp. 209–244.
32. Veni, P.R.K.; Sharmila, N.; Narayana, K.; Babu, B.H.; Satyanarayana, P. Simultaneous determination of piperacillin and tazobactam in pharmaceutical formulations by RP-HPLC method. *J. Pharm. Res.* **2013**, *7*, 127–131. [CrossRef]

33. Atmakuri, L.R.; Krishna, K.; Kumar, C.H.; Raja, T.A. Simultaneous determination of piperacillin and tazobactum in bulk and pharmaceutical dosage forms by RP-HPLC. *Int. J. Pharm. Pharm. Sci.* **2011**, *3*, 134–136.
34. Pai, P.; Rao, G.; Murthy, M.; Prathibha, H. Simultaneous estimation of piperacillin and tazobactam in injection formulations. *Indian J. Pharm. Sci.* **2006**, *68*, 799–801. [CrossRef]
35. Council of Europe, European Directorate for the Quality of Medicines & Healthcare. *European Pharmacopoeia 10.4*; Council of Europe: Strasburg, France, 2020; p. 1168.
36. Naicker, S.; Guerra Valero, Y.C.; Ordenez Meija, J.L.; Lipman, J.; Roberts, J.A.; Wallis, S.C.; Parker, S.L. A UHPLC-MS/MS method for the simultaneous determination of piperacillin and tazobactam in plasma (total and unbound), urine and renal replacement therapy effluent. *J. Pharm. Biomed. Anal.* **2018**, *148*, 324–333. [CrossRef]
37. Zander, J.; Maier, B.; Suhr, A.; Zoller, M.; Frey, L.; Teupser, D.; Vogeser, M. Quantification of piperacillin, tazobactam, cefepime, meropenem, ciprofloxacin and linezolid in serum using an isotope dilution UHPLC-MS/MS method with semi-automated sample preparation. *Clin. Chem. Lab. Med.* **2015**, *53*, 781–791. [CrossRef] [PubMed]
38. Chrisikou, I.; Orkoula, M.; Kontoyannis, C. FT-IR/ATR Solid Film Formation: Qualitative and Quantitative Analysis of a Piperacillin-Tazobactam Formulation. *Molecules* **2020**, *25*, 6051. [CrossRef] [PubMed]
39. Karpova, S.P.; Blazheyevskiy, M.; Mozgova, O. Development and validation of UV spectrophotometric area under curve method quantitative estimation of piperacillin. *Int. J. Pharm. Sci. Res.* **2018**, *9*, 3556–3560.
40. Toral, M.I.; Nova-Ramírez, F.; Nacaratte, F. Simultaneous determination of piperacillin and tazobactam in the pharmaceutical formulation Tazonam® by derivative spectrophotometry. *J. Chil. Chem. Soc.* **2012**, *57*, 1189–1193. [CrossRef]
41. Domes, C.; Domes, R.; Popp, J.; Pletz, M.W.; Frosch, T. Ultrasensitive Detection of Antiseptic Antibiotics in Aqueous Media and Human Urine Using Deep UV Resonance Raman Spectroscopy. *Anal. Chem.* **2017**, *89*, 9997–10003. [CrossRef] [PubMed]
42. Bellamy, L.J.; Williams, R.L. 863. Infrared spectra and polar effects. Part VII. Dipolar effects in α-halogenated carbonyl compounds. *J. Chem. Soc. (Resumed)* **1957**, *0*, 4294–4304. [CrossRef]
43. Peesole, R.L.; Shields, L.D.; Cairus, T.; McWilliam, I.G. *Modem Method of Chemicailanalysis*; John Wiley & Sons: New York, NY, USA, 1976; p. 161.
44. Volpin, M.E.; Koreshkov, Y.D.; Dulova, V.G.; Kursanov, D.N. Three-membered heteroaromatic compounds—I. *Tetrahedron* **1962**, *18*, 107–122. [CrossRef]
45. Natarajan, A.; Savarianandam, A. IR and Laser Raman spectral analysis of dichlorodiphenylsulphone. *Asian J. Phys.* **1996**, *5*, 251–254.
46. Mohan, S.; Ilengovan, V. Fourier Transform Infrared (FTIR) and Raman studies on 3-aminopyridine molecule. *Proc. Natl. Acad. Sci. India* **1995**, *65*, III.
47. Boobyer, R.C. Intensities of C-H stretching modes for some small alicyclics. *Spectrochim. Acta* **1967**, *23*, 321–323. [CrossRef]
48. Higuchi, S.; Tsuyama, H.; Tanaka, S.; Kamoda, H. Some considerations on the out-of-plane vibration bands of Phnx type molecules in the 800-670 cm^{-1} region in relation to the estimation of the twist angle θ of benzene rings from their intensities. *Spectrochim. Acta* **1974**, *30*, 463–477. [CrossRef]
49. Kalp, M.; Totir, M.A.; Buynak, J.D.; Carey, P.R. Different intermediate populations formed by tazobactam, sulbactam, and clavulanate reacting with SHV-1 beta-lactamases: Raman crystallographic evidence. *J. Am. Chem. Soc.* **2009**, *131*, 2338–2347. [CrossRef]
50. Heidari Torkabadi, H.; Che, T.; Shou, J.; Shanmugam, S.; Crowder, M.W.; Bonomo, R.A.; Pusztai-Carey, M.; Carey, P.R. Raman spectra of interchanging β-lactamase inhibitor intermediates on the millisecond time scale. *J. Am. Chem. Soc.* **2013**, *135*, 2895–2898. [CrossRef] [PubMed]
51. Skoog, D.A.; West, D.M.; Holler, F.J. *Fundamentals of Analytical Chemistry*, 6th ed.; Saunders, H.J.B., Ed.; Elsevier (Saunders): Philadelphia, PA, USA, 1991; p. 14.
52. ICH Harmonized Tripartite Guideline Q2 (R1) Validation of Analytical Procedures. Available online: https://www.ema.europa.eu/en/ich-q2-r1-validation-analytical-procedures-text-methodology (accessed on 21 April 2021).
53. Miller, J.N.; Miller, J.C. Calibration methods: Regression and correlation. In *Statistics and Chemometrics for Analytical Chemistry*; Pearson Education Limited: London, UK, 2005; pp. 107–147.
54. Hadjiioannou, T.P.; Koupparis, M.A. *Instrumental Analysis*; University of Athens: Athens, Greece, 2005; Volume 17.
55. Anderson, D.J. Determination of the lower limit of detection. *Clin. Chem.* **1989**, *35*, 2152–2153. [CrossRef] [PubMed]
56. National Center for Biotechnology Information. PubChem Database. Piperacillin, CID=43672. Available online: https://pubchem.ncbi.nlm.nih.gov/compound/Piperacillin (accessed on 7 May 2021).
57. National Center for Biotechnology Information. PubChem Database. Tazobactam, CID=123630. Available online: https://pubchem.ncbi.nlm.nih.gov/compound/Tazobactam (accessed on 7 May 2021).
58. Bazin, C.; Cassard, B.; Caudron, E.; Prognon, P.; Havard, L. Comparative analysis of methods for real-time analytical control of chemotherapies preparations. *Int. J. Pharm.* **2015**, *494*, 329–336. [CrossRef] [PubMed]
59. Henriques, J.; Sousa, J.; Veiga, F.; Cardoso, C.; Vitorino, C. Process analytical technologies and injectable drug products: Is there a future? *Int. J. Pharm.* **2019**, *554*, 21–35. [CrossRef] [PubMed]

Article

Development of Ecofriendly Derivative Spectrophotometric Methods for the Simultaneous Quantitative Analysis of Remogliflozin and Vildagliptin from Formulation

Mahesh Attimarad [1,*], Katharigatta N. Venugopala [1,2], Bandar E. Al-Dhubiab [1], Rafea Elamin Elgack Elgorashe [3] and Sheeba Shafi [4]

1. Department of Pharmaceutical Sciences, College of Clinical Pharmacy, King Faisal University, Al Hofuf 31982, Saudi Arabia; kvenugopala@kfu.edu.sa (K.N.V.); baldhubiab@kfu.edu.sa (B.E.A.-D.)
2. Department of Biotechnology and Food Science, Faculty of Applied Sciences, Durban University of Technology, Durban 4000, South Africa
3. Department of Chemistry, College of Science, King Faisal University, Al Hofuf 31982, Saudi Arabia; relgorashe@kfu.edu.sa
4. Department of Nursing, College of Applied Medical Sciences, King Faisal University, Al Ahsa 31982, Saudi Arabia; sheeba@kfu.edu.sa
* Correspondence: mattimarad@kfu.edu.sa; Tel.: +966-55-3269799

Citation: Attimarad, M.; Venugopala, K.N.; Al-Dhubiab, B.E.; Elgorashe, R.E.E.; Shafi, S. Development of Ecofriendly Derivative Spectrophotometric Methods for the Simultaneous Quantitative Analysis of Remogliflozin and Vildagliptin from Formulation. *Molecules* **2021**, *26*, 6160. https://doi.org/10.3390/molecules26206160

Academic Editors: Franciszek Główka and Marta Karaźniewicz-Łada

Received: 11 September 2021
Accepted: 11 October 2021
Published: 12 October 2021

Publisher's Note: MDPI stays neutral with regard to jurisdictional claims in published maps and institutional affiliations.

Copyright: © 2021 by the authors. Licensee MDPI, Basel, Switzerland. This article is an open access article distributed under the terms and conditions of the Creative Commons Attribution (CC BY) license (https://creativecommons.org/licenses/by/4.0/).

Abstract: Three rapid, accurate, and ecofriendly processed spectrophotometric methods were validated for the concurrent quantification of remogliflozin (RGE) and vildagliptin (VGN) from formulations using water as dilution solvent. The three methods developed were based on the calculation of the peak height of the first derivative absorption spectra at zero-crossing points, the peak amplitude difference at selected wavelengths of the peak and valley of the ratio spectra, and the peak height of the ratio first derivative spectra. All three methods were validated adapting the ICH regulations. Both the analytes showed a worthy linearity in the concentration of 1 to 60 µg/mL and 2 to 90 µg/mL for VGN and RGE, respectively, with an exceptional regression coefficient ($r^2 \geq 0.999$). The developed methods demonstrated an excellent recovery (98.00% to 102%), a lower percent relative standard deviation, and a relative error (less than ±2%), confirming the specificity, precision, and accuracy of the proposed methods. In addition, validated spectrophotometric methods were commendably employed for the simultaneous determination of VGN and RGE from solutions prepared in the laboratory and the formulation. Hence, these methods can be utilized for the routine quality control study of the pharmaceutical preparations of VGN and RGE in pharmaceutical industries and laboratories. The ecofriendly nature of the anticipated spectrophotometric procedures was confirmed by the evaluation of the greenness profile by a semi-quantitative method and the quantitative and qualitative green analytical procedure index (GAPI) method.

Keywords: vildagliptin; remogliflozin; ratio derivative spectrophotometry; determination; formulation; ecofriendly

1. Introduction

Diabetes mellitus (DM) is a metabolic disease characterized by elevated blood glucose levels. It is estimated that one out of eleven people suffer from DM globally, and 90% of the population has type 2 DM (T2DM). Experts predict that the occurrence of cases of DM will increase to nearly 642 million by the end of 2040 from the current levels (422 million at present). The occurrence of T2DM is becoming common in the elderly population due to stress and physical inactivity. However, obesity, food habits, and hereditary factors predispose diabetes among the younger population [1,2]. The pathophysiological mechanisms and treatment of diabetes are complicated; hence, a multiple intervention approach such as the practice of healthy diets, physical activity, and various therapeutic strategies may help to minimize the complications of diabetes. Recently developed dipeptide peptidase-4 (DPP-4) and sodium-glucose transporter-2 (SGLT-2) inhibitors showed an enhanced HbA1c control when compared with conventional sulfonylureas and thiazolidinediones [3]. The Food and Drug Administration (FDA) has approved a fixed-dose remogliflozin and vildagliptin tablet for T2DM. Remogliflozin etabonate (Figure 1A, RGE) is an oral hypoglycemic drug [4], which acts by inhibiting the SGLT-2 enzyme and thereby decreasing the reabsorption of glucose from the glomerular filtrate back to the blood. SGLT-2 inhibitors reduce cardiovascular events, body weight, and also show a defensive effect on the renal system. These functional properties of SGLT-2 inhibitors considerably reduce the hospitalization of T2DM patients exclusively due to heart failure [5,6]. Vildagliptin (Figure 1B, VGN), a DPP-4 inhibitor, decreases the blood sugar level by protecting the incretins from degradation, which helps in the production of insulin after food and reduces glucagon formation in the liver. The protection of incretins also helps in reducing body weight by decreasing the appetite and prolonging the slow digestion of food [7,8].

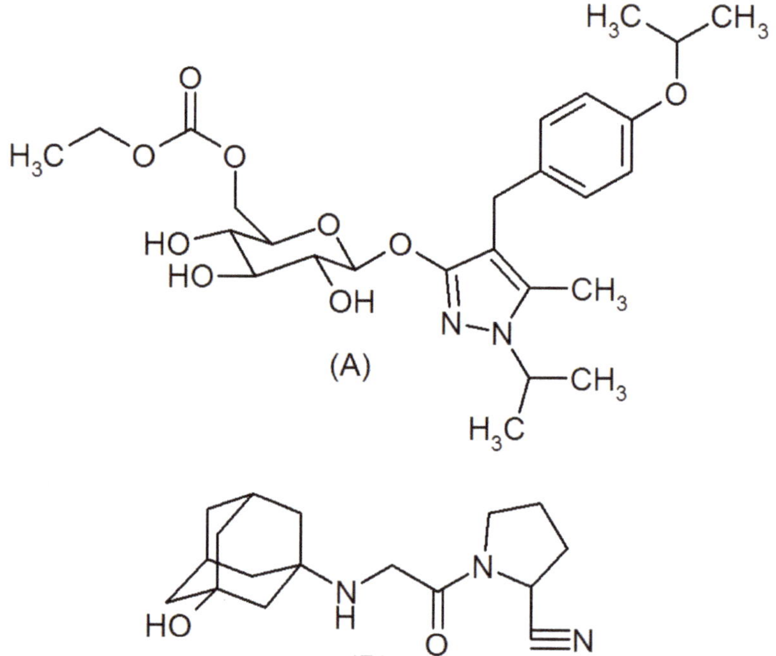

Figure 1. Chemical structure of remogliflozin etabonate (**A**) and vildagliptin (**B**).

Few quantitative analytical procedures are illustrated in the literature for the analysis of RGE and VGN alone and with metformin from medicines and biological fluids. A quantitative determination of RGE alone was explained using UV-Vis spectrophotometry [9,10], HPLC [11,12], and LCMS [13].Derivative UV spectrophotometry, RP-HPLC, and UPLC procedures were stated for the concurrent estimation of RGE with metformin [14–16].

Different analytical methods have been reported in the literature for the quantification of VGN from formulations and biological samples. VGN alone was determined using UV-Vis spectrophotometry [17–19], HPTLC [20], and UPLCMS [21]. Several spectrophotometric [22–24] and HPLC [24–27] methods were reported for the determination of VGN along with other drugs.

Several analytical methods have been reported for the determination of VGN and RGE alone and with other active ingredients. However, no quantitative analytical method has been described for the concurrent estimation of VGN and RGE from a formulation. Derivative UV spectrophotometric techniques are simple, accurate, fast, and may possibly be utilized for the quantification of multicomponent formulations showing overlapping spectra [28–32]. Hence, in the current work, three spectrophotometric methods were validated and applied to a concurrent determination of VGN and RGE from laboratory mixed solutions and formulations. Water was used as a dilution solvent for the samples, making the developed spectrophotometric methods ecofriendly.

2. Results and Discussion

UV-Vis spectroscopic methods are extensively used analytical techniques due to their simplicity, accuracy, and reproducibility. Salinas et al. reported a derivative spectroscopic method [29] for the analysis of multicomponent formulations with overlapping spectra (Figure 2A), which cannot be analyzed by a direct UV measurement. Many reports have demonstrated the use of derivative spectroscopic methods for the analysis of multicomponent formulations without a prior separation [28–32]. RGE showed UV absorption in the range of 200 to 300 nm due to the presence of individual five and six membered rings whereas VGN showed below 230 nm due to an absence of an aromatic ring and the presence of an aliphatic nitrile and carbonyl group. However, both analytes did not show any absorption above 300 nm; hence, in the present work, the analytes were scanned in the wavelength rage of 200 to 300 nm. In the present work, three processed UV spectroscopic procedures were validated for the concurrent quantification of VGN and RGE. The first procedure was established on the measurement of absorption at zero-crossings of one of the analytes where another analyte had a degree of absorption. For the determination of VGN and RGE, normal absorption spectra were processed into first derivative spectra utilizing 4 nm as Δλ and a scaling factor of 10. Different wavelengths of 2, 4, 8, and 10 nm were envisaged during the first derivative spectrum; however, 4 nm resulted in smooth spectra so 4 nm was selected. A scaling factor of 10 demonstrated a sufficient peak amplitude at a low concentration of VGN; hence, a scaling factor of 10 was selected. The first derivative spectra of VGN showed a good absorption at 213.7 and 225 nm where RGE had zero absorption (Figure 2B). Similarly, RGE showed good absorptions at 241.5, 265.7, and 287.7 nm where VGN had no absorption. However, at 213.7 and 287.7 nm, VGN and RGE showed a good recovery and reproducibility, respectively. The comparison of the first derivative spectra obtained from the pure and the combination of VGN and RGE showed the same amplitude. Hence, the calibration curves for the determination of VGN and RGE were computed by evaluating the peak height at 213.7 nm (Figure 2C) and 287.7 nm, respectively (Figure 2D).

2.1. Ratio Difference Absorption Method

The solutions of RGE and VGN were subjected to UV absorption in the wavelength range of 200 to 300 nm. The selection of the divisor analyte concentration is important in developing ratio spectra; hence, different concentrations of spectra were tried. However, no significant difference was observed in terms of the linearity range with different concentrations although a good intensity of ratio spectra and reproducibility were observed with 5 µg/mL spectra as the divisor. Hence, 5 µg/mL spectra of RGE and VGN were envisaged as the divisor spectra. For the estimation of REM, a mixture spectra consisting of 2 to 75 µg/mL of REM were divided with a 5 µg/mL spectrum of VGN to generate the ratio spectra for REM (Figure 3A). The peak amplitude differences were calculated by subtracting the peak amplitude at 230.3 and 251.4 nm, and linearity curves were generated against the corresponding concentration. Figure 3B shows the same peak height from the ratio spectra generated from the combination (REM and VGN). The pure REM indicated that REM could be determined in the presence of another analyte, VGN. Similarly, a mixture of REM and VGN consisting of 2 to 50 µg/mL of VGN was divided with a 5 µg/mL spectrum of RGE to generate the ratio spectra for VGN (Figure 3C). The two wavelengths selected were 207.2 and 230.6 nm, the peak amplitude difference was calculated, and the linearity curve was generated by drawing a graph against the respective concentration. Figure 3D shows the same peak amplitude in the ratio spectra generated from the mixture and the pure VGN spectra, demonstrating the application of the ratio difference absorption method for the quantification of VGN in the presence of RGE.

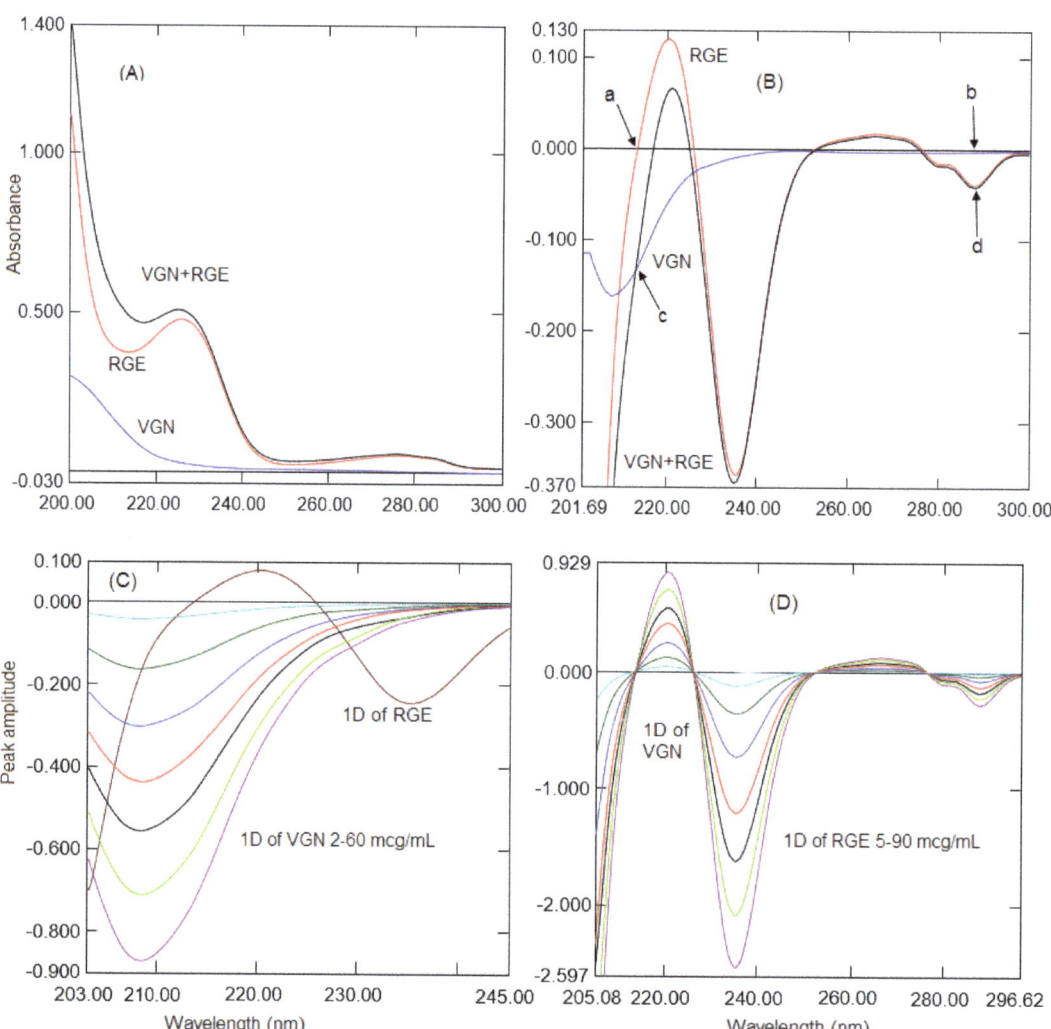

Figure 2. (**A**) Normal spectra of RGE, VGN, and a mixture. (**B**) First derivative spectra of RGE, VGN, and the mixture. Zero-crossings of RGE (a), zero-crossings of VGN (c), and the same amplitude points (b and d). (**C**) First derivative spectra of VGN 2 to 60 μg/mL with the RGE spectrum. (**D**) First derivative spectra of RGE 5 to 90 μg/mL with the VGN spectrum.

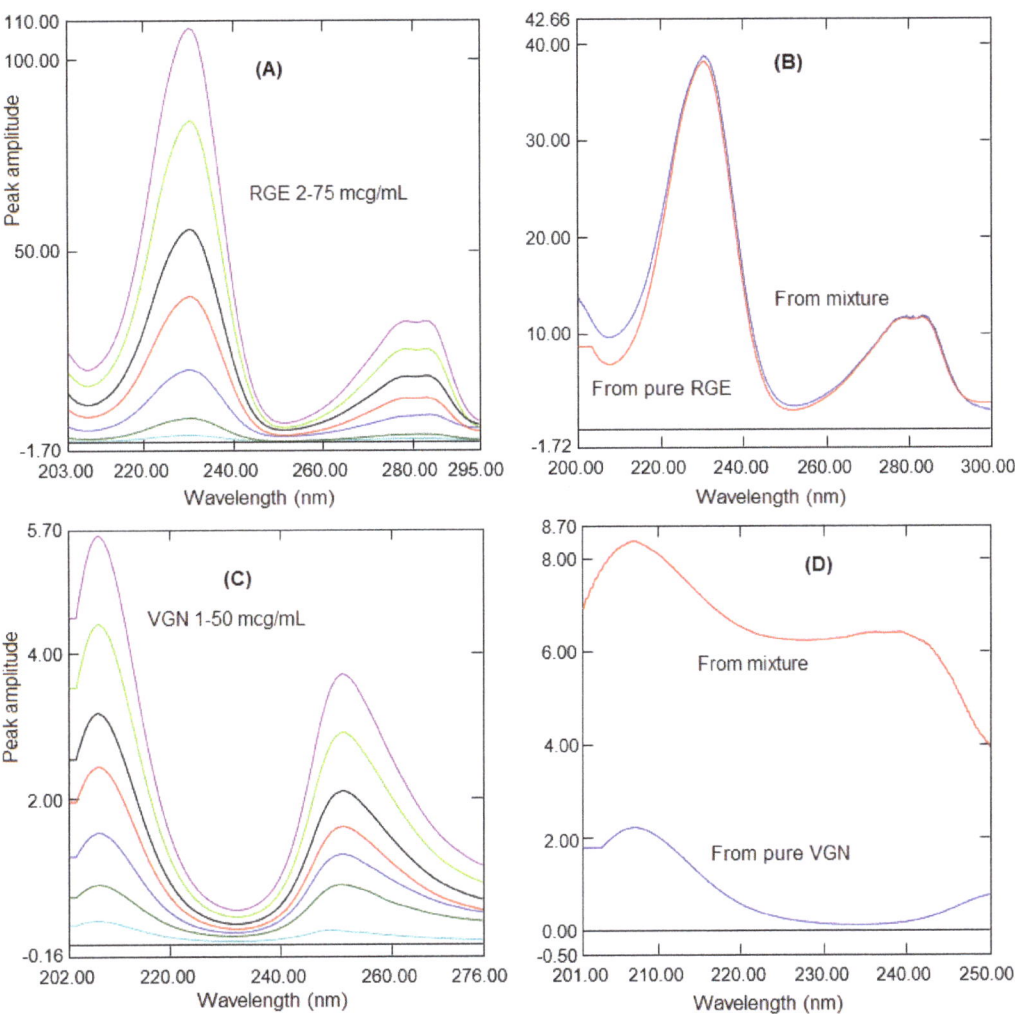

Figure 3. (**A**) Ratio absorption spectra of RGE (2 to 75 µg/mL) using a 5 µg/mL spectrum of VGN. (**B**) Ratio absorption spectra of pure RGE and a mixture. (**C**) Ratio absorption spectra of VGN (1 to 50 µg/mL) using a 5 µg/mL spectrum of RGE. (**D**) Ratio absorption spectra of pure VGN and a mixture.

2.2. Ratio First Derivative Method

From the equation $P_A = P_R + \delta$ (Materials and Methods), the constant (δ) could be excluded by changing the ratio spectrum into the derivative spectrum. The conversion of the derivative spectrum generated serval maxima and minima. The peak amplitude was proportional to the amount and the effects of another analyte and tablet excipients were excluded. This was a substitute technique to the ratio difference method to quantify VGN in the presence of RGE and vice versa. For the quantification of VGN, the ratio spectra of VGN were changed into first derivative spectra by applying 4 nm as $\Delta\lambda$ and a scaling factor of 10 (Figure 4A). Various wavelengths between 2 and 8 nm were tried as $\Delta\lambda$ although 4 nm exhibited an enhanced sensitivity. Three maxima at 204.4, 245.4, and 289.7 nm and two minima peaks at 213.7 and 257.7 nm were observed in the first derivative spectra of VGN. The peak amplitude was good at these wavelengths; however, the peak height was better at 213.7 nm. Hence, 213.7 nm was selected to generate the calibration curve. The same peak amplitudes were observed at 213.7 nm in the first derivative spectra generated from the mixture and the pure VGN spectra (Figure 4B). Similarly, for the quantification of RGE, the ratio absorption spectra of RGE were processed to develop first derivative spectra with 4 nm as Δλ and a scaling factor of 10 (Figure 4C). Two maxima at 221.8 and 272.6 nm and two minima peaks at 237.2 and 288.2 nm were observed, respectively, in the first derivative spectra of RGE. The reproducibility and sensitivity were better at 237.2 nm; hence, 237.2 nm was selected to construct the calibration curve. The same peak amplitudes were observed at 237.2 nm in the first derivative spectra generated from the mixture and the pure RGE spectra (Figure 4D).

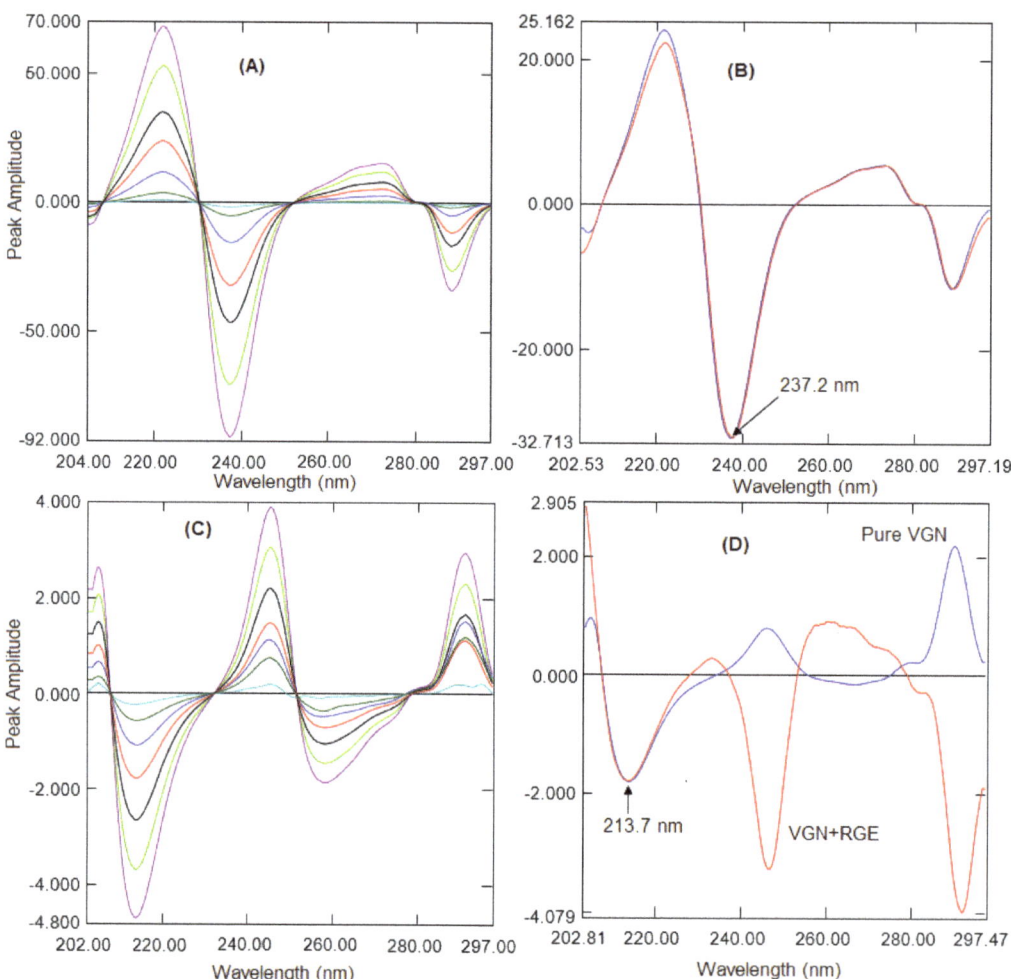

Figure 4. (**A**) Ratio first derivative spectra of RGE (2 to 75 μg/mL). (**B**) Ratio first derivative spectra of pure RGE and a mixture. (**C**) Ratio first derivative spectra of VGN (1 to 50 μg/mL). (**D**) Ratio first derivative spectra of pure VGN and a mixture.

2.3. Method Validation

Optimized spectroscopic methods were validated in terms of linearity, the limit of detection/quantification, accuracy, interday and intraday precision, selectivity, and stability of the standard solutions under experimental and storage conditions.

2.3.1. Linearity Range

The calibration curve was constructed using seven solutions of analytes using a series of solutions comprising 2 to 60 μg/mL for VGN and 5 to 90 μg/mL for RGE in triplicate by the first derivative spectroscopic method. The linearity was established in the range of 1 to 50 μg/mL for VGN and 2 to 75 μg/mL for RGE by the ratio difference and ratio first derivative methods. Both analytes did not show any absorption above 300 nm; hence, the solutions were scanned between the wavelengths of 200 to 300 nm. The molar absorptivity of the normal spectra of VGN and RGE was found to be 18,100.24 L M^{-1} cm^{-1} at 209.2 nm and 32,133.33 L M^{-1} cm^{-1} at 225.9 nm, respectively. The peak amplitudes were measured from the first derivative, ratio derivatives, and first derivative of the ratio spectra and calibration curves were constructed between the peak amplitudes and the corresponding concentration (Supplementary File S1). The linearity equations and coefficients are presented in Table 1. The intercept values for the RGE curves were negative, indicating that the model overestimated the average absorption; hence, there was a requirement to subtract the predicted concentrations. The intercept values for the VGE curves were positive, indicating that the model underestimated the average absorption; hence, there was a requirement for the predicted concentrations to be added.

Table 1. Validation parameter results of the proposed spectroscopic methods for the simultaneous determination of VGN and RGE.

Validation Parameters	Remogliflozin			Vildagliptin		
	FDS	RDS	RFD	FDS	RDS	RFD
Wavelength (nm)	287.7	230.3–251.4	237.2	213.7	207.2–230.6	213.7
Linearity range (μg/mL)	5–90	2–75	2–75	2–60	1–50	1–50
Slope	0.0031	1.328	1.226	0.012	0.0983	0.0920
Intercept	−0.0051	−1.221	−0.884	0.008	0.2489	0.0841
Regression coefficient (r^2)	0.9997	0.9998	0.9997	0.9996	0.9990	0.9994
LOD (μg/mL)	1.361	0.583	0.408	0.484	0.272	0.176
LOQ (μg/mL)	4.126	1.768	1.236	1.469	0.827	0.535
Accuracy (Mean % ± SD)	99.41 ± 1.170	99.13 ± 0.665	100.82 ± 0.910	99.52 ± 0.834	99.88 ± 1.655	101.07 ± 0.729
Precision (%RSD)						
Intraday	0.730	1.026	1.843	0.610	0.871	0.728
Interday	0.937	1.383	0.611	1.748	0.725	1.287

2.3.2. Accuracy

The accuracy of the optimized methods was performed by evaluating VGN and RGE at three different concentrations (low, medium, and high) covering the complete linearity concentration. The concentration of analytes was computed from the linearity equations generated from the three methods. The accuracy of the methods was presented as a percentage recovery and a percent relative error and is presented in Table 1. The mean percent recovery was found to be 99.13% to 100.82% for RGE and 99.52% to 101.07% for VGN, indicating the accuracy of the developed spectroscopic methods. The percent recovery was between 98.00% and 102.00%, indicating the accuracy of the anticipated procedures.

2.3.3. Precision

The reproducibility of the optimized procedures was assessed by performing an intraday analysis using three different concentrations comprising the complete linearity concentration. The solutions were analyzed on the same day in triplicate and the percent relative standard deviation (%RSD) was calculated. The intermediate precision of the procedures was performed by evaluating the above-prepared solution on three successive days. The interday precision was expressed as a percent relative standard deviation and is tabulated in Table 1. The intraday precision for VGN ranged from 0.610 to 0.871 and it was from 0.730 to 1.843 for RGE whereas the interday precision ranged from 0.725 to 1.748 for VGN and 0.611 to 1.383 for RGE. The %RSD was lower than 2 in all cases, representing the good precision of the UV derivative spectroscopic methods.

2.3.4. Limit of Detection and Quantification

One of the LOD and LOQ methods utilized the standard deviation of the intercept and slope of the calibration curve, which was 3.3 times the SD of the intercept/slope of the linearity curve for the LOD and 10 times the SD of the intercept/slope of the calibration curve for the LOQ. The LOD and LOQ determined for all three methods are tabulated in Table 1. Low LOQ values indicated the sensitivity of the anticipated procedures.

2.3.5. Specificity

Different ratios of VGN and RGE within the linearity range were prepared in the laboratory and investigated for testing the specificity of the proposed methods. The obtained outcomes by all three methods are tabulated in Table 2. The good analysis outcomes with a lower percent relative standard deviation confirmed the specificity of the optimized spectroscopic methods.

Table 2. Assay results of the laboratory mixed solutions of VGN and RGE.

Laboratory Prepared Mixture (µg/mL)		Remogliflozin (% Recovery)			Vildagliptin (% Recovery)		
RGE	VGN	FDS	RDS	RFD	FDS	RDS	RFD
10	50.00	99.20	101.60	98.50	98.30	98.22	101.04
40	50.00	97.93	100.70	99.43	98.58	99.06	100.78
40	5.00	100.30	96.48	100.20	98.60	99.40	99.80
70	5.00	99.24	98.06	99.03	101.20	100.40	98.60
70	25.00	100.67	98.73	98.21	99.24	99.04	101.36
Across Mean		99.47	99.11	99.07	99.18	99.27	100.32
SD		0.964	1.840	0.702	1.054	0.705	1.004

2.4. Application of the Optimized Methods for the Formulation

The proposed UV derivative spectroscopic methods were utilized for the simultaneous quantification of VGN and RGE in the formulation. The results, as presented in Table 2, confirmed the accuracy of the quantification of the analytes with a good agreement with the labeled quantity of the active ingredients of the medicine and confirmed the nonexistence of the influence of the formulation excipients on the assay of both analytes. The excipients used for preparation of the tablets were common and, in general, present in the formulation. The proposed methods eliminated the effects of the excipients; hence, they could be used for quality control.

The accuracy of the optimized procedures was further confirmed using the standard addition method. An assessed quantity of VGN and RGE was transformed to the previously evaluated formulation solution and the spiked solution was analyzed by all three proposed methods. The achieved results are tabulated in Table 3. The mean percent recovery (100 ± 2%) and the relative standard deviation (<2%) were well within the acceptable range, assuring the adequate accuracy of the proposed procedures.

Table 3. Assay results of the formulation and the standard addition method results.

Formulation Concentration		Remogliflozin (Mean %± SD)			Vildagliptin (Mean % ± SD)		
RGE	VGN	FDS	RDS	RFD	FDS	RDS	RFD
100 mg	50 mg	99.42 ± 0.985	99.62 ± 0.788	99.10 ± 0.76	99.77 ± 1.563	100.04 ± 1.384	100.19 ± 0.797
				Standard Addition Method			
Amount Added (µg/mL)		Remogliflozin (% Recovery)			Vildagliptin (% Recovery)		
10 µg/mL	5	98.70	100.70	98.30	99.20	100.40	99.80
20 µg/mL	10	100.95	98.85	99.05	101.90	98.20	101.30
30 µg/mL	15	96.47	99.30	100.23	98.20	101.53	99.47
Across Mean		98.71	99.62	99.19	99.77	100.04	100.19
%RSD		1.830	0.788	0.796	1.563	1.384	0.797

2.5. Greenness Evaluation of the Optimized Methods

The development of ecofriendly analytical methods is a requirement of the present day to save the environment. In the present work, two greenness evaluation methods were implemented to evaluate the ecofriendly nature of the proposed methods. Raynie et al. [33] developed a semi-quantitative method for a greenness evaluation (Supplementary material Table S1) that considered the nature of the chemicals and instruments used in the experiment as these are the main contributors to environmental pollution. In the current work, ethanol was used as a solvent to dissolve one of the analytes and water was used for further dilation, generating less than 5% alcohol in each solution. In addition, due to the use of dilute ethanol, the evaporation of ethanol was negligible. The measurement was made using a spectrophotometer and a computer; therefore, the energy consumed during the process was safe. Each solution prepared for the analysis contained less than 5% ethanol and a few micrograms of the drugs, generating less than 50 g of waste. Hence, the greenness profile (Figure 5A) of all three proposed spectroscopic methods was ecofriendly. The second method adopted for the greenness evaluation was based on the quantitative and qualitative method developed by Plotka-Wasylka, known as the green analytical procedure index (GAPI) [34]. The GAPI involves the grading of 15 parameters of analytical methods, starting from the sample preparation and the hazardous nature of the chemicals, solvents and instruments used along with the nature of the analysis to the waste treatment of the samples after the completion of experiments [35] (Supplementary material Table S2). The proposed spectroscopic methods involved a simple sample preparation during the analysis; hence, there was no need for the storage and transportation of the sample. A direct sample preparation technique was used during the sample preparation using green solvents without any extraction and pretreatment processes. Apart from water, another solvent used in the optimized methods was ethanol, which is considered to be a safe solvent with an NFPA irritant and the flammable score of one with no hazardous effect on animals and the environment. The total volume of each sample was 5 mL. A waste treatment procedure had not been established; hence, the parameter color codes for 14 and 15 were marked as yellow and red. However, the overall profile of the GAPI (shown in Table 4 and Figure 5B) confirmed the ecofriendly nature of the optimized spectroscopic methods.

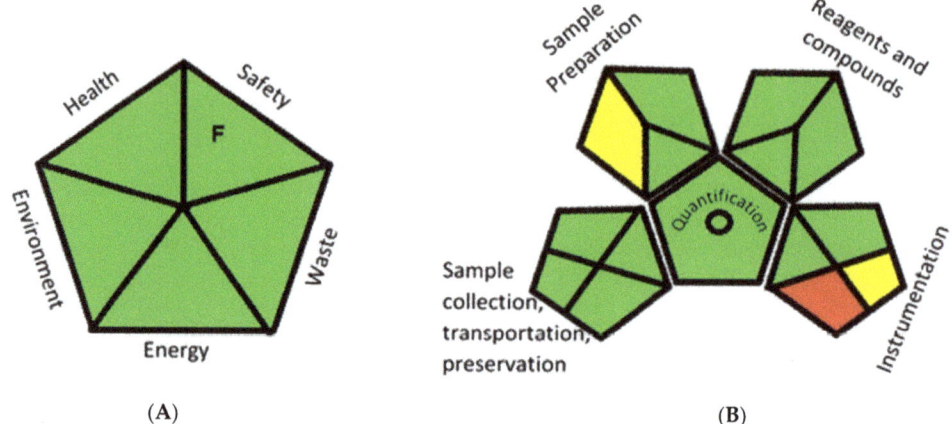

Figure 5. (**A**) Greenness evaluation results of the proposed method by Raynie et al.'s method [33]. (**B**) The GAPI method.

Table 4. The GAPI evaluation results for the spectrophotometric methods.

Category	UV Spectrophotometric Methods
Sample Preparation	
Collection (1)	In-line
Preservation (2)	Nil
Transport (3)	Nil
Storage (4)	Nil
Type of method: direct or indirect (5)	Direct (no sample preparation)
Scale of extraction (6)	Nil
Solvents/reagents used (7)	Green Solvents
Additional treatments (8)	Nil
Reagent and solvents	
Amount (9)	<10 mL
Health hazard (10)	Ethanol, slightly toxic and irritant NFPA score 1
Safety hazard (11)	Ethanol, instability score 0, flammability score 1, no special hazard
Instrumentation	
Energy (12)	≤0.1 kWh/sample
Occupational hazard (13)	Hermetic sealing of the analytical procedure
Waste (14)	1–10 mL
Waste treatment (15)	No treatment
Quantification	Yes

3. Materials and Methods

3.1. Instruments

The spectrophotometric determination was carried out using a UV-Vis spectrophotometer 1600 (Shimadzu, Japan) using 1 cm quartz cuvettes. The samples were scanned in the fast mode with a slit width of 0.1 nm. The processing of the scanned spectra was completed with UV-probe Ver 2.21 software (Shimadzu, Japan).

3.2. Chemicals

Standard drugs of VGN and RGE were purchased from Biokemics (Hyderabad, India). A fixed-dose combination of VGN (50 mg) and RGE (100 mg) was not available in the local pharmacies; hence, a geometrically mixed tablet was prepared by mixing a sufficient quantity of VGN and RGE with solid dosage form excipients of microcrystalline cellulose, hydroxypropyl cellulose, crospovidone, colloidal silicon dioxide, and stearic acid. Analytical grade ethanol was bought from Sigma Aldrich (St. Louis, MO, USA). Purified water prepared using a double-distil water purifier in our laboratory was utilized throughout the analysis.

3.3. Preparation of the Standard Solutions

The stock solutions of VGN and RGE were prepared by directly transferring 100 mg of VGN and RGE in 100 mL graduated flasks separately. VGN was dissolved using water and RGE was dissolved using ethanol to obtain a 1000 µg/mL solution. For preparing the calibration standards and the solutions for validation, the stock solutions were diluted with water by maintaining 200 µL of ethanol per sample in a total volume of 5 mL. A blank solution was prepared by adding 200 µL of ethanol in 5 mL of water.

3.4. Preparation of the Sample Solutions

The newly approved solid dosage forms of VGN (50 mg) and RGE (100 mg) were not accessible in the local pharmacies. Hence, a table mixture was prepared in the laboratory by mixing a sufficient quantity of VGN and RGE with tablet excipients (microcrystalline cellulose, hydroxypropyl cellulose, crospovidone, colloidal silicon dioxide, and stearic acid). A thoroughly mixed tablet powder consisting of 50 mg of VGN and 100 mg of RGE was transferred into a 100 mL graduated flask. A total of 25 mL of ethanol was transferred and sonicated for 15 min to completely extract both active ingredients. The solution was filtered into another flask, the residue was washed with fresh ethanol, and the final volume was attuned by adding a sufficient amount of ethanol. For the analysis of the solutions, the concentration of the sample solutions was brought in the range of the calibration curve using distilled water.

3.5. Theory of the Ratio Derivatization Technique

This was a simple and specific method adopted by Lotfy and Hegazy [36] for the simultaneous determination of a binary mixture showing complete overlap spectra without any separation. It was established by generating the ratio spectra by dividing the mixture spectra (VR) with one of the analyte spectra (R' or V') and determining the peak amplitude difference at two designated wavelengths, which was directly related to the concentration. According to the Beers law absorption of the mixture of components, A_{VR} is represented by Equation (1).

$$A_{VR} = \varepsilon_V \times C_V + \varepsilon_R \times C_R \quad (1)$$

where ε_V and ε_R are the molar extinction coefficients of VGN and RGE, respectively, at a particular wavelength whereas C_V and C_R are concentrations of VGN and RGE, respectively. The ratio spectra were generated by dividing the mixture spectra (VR) with an analyte spectrum ($A_{V'} = \varepsilon_{V'} \times C_{V'}$) at the concentration V' to determine another analyte (R), which is represented by Equation (2):

$$A_{VR}/A_{V'} = C_V/C_{V'} + A_R/A_{V'}. \quad (2)$$

The $C_V/C_{V'}$ is constant (δ); hence, the above equation can be simplified to:

$$P_A = P_R + \delta \quad (3)$$

where P_A is the $A_{VR}/A_{V'}$ absorption of the ratio spectra of a mixture of components to one of the components whereas P_R is the absorbance of the ratio spectra of one component to another component.

From Equation (3), the constant δ can be eliminated by determining the change in absorbance at two selected wavelengths from the ratio spectra as per Equation (4):

$$P_{A1} - P_{A2} = (P_{R1} + \delta) - (P_{R2} + \delta). \quad (4)$$

Equation (4) can then be simplified to Equation (5):

$$\Delta P = P_{R1} + P_{R2} \quad (5)$$

where P_{R1} and P_{R2} are peak amplitudes at wavelengths λ_1 and λ_2 of the ratio spectra. Equation (5) represents the absorption of only one analyte (RGE), eliminating the effects of the additional analyte (VGN). In the current effort, the mixture of RGE and VGN was prepared in the calibration curve range and the spectra were recorded.

3.6. Procedure

3.6.1. First Derivative Spectrophotometric Method (FDS)

A sufficient quantity of stock solutions of VGN and RGE was measured into 5 mL volumetric flasks separately to obtain the concentrations of 2, 10, 20, 30, 40, 50, and 60 µg/mL of VGN and 5, 15, 30, 45, 60, 75, and 90 µg/mL of RGE. The solutions were exposed to UV absorption using ethanol–water as a blank solution and the spectra were deposited into the computer. The RGE absorption spectra were transformed into first derivative spectra with 4 nm as $\Delta\lambda$ and a scaling factor of 10. The peak amplitude was measured at 287.7 nm and the linearity curve was created against the respective concentration. VGN spectra were also transformed into first derivative spectra with 4 nm as $\Delta\lambda$ and a scaling factor of 10. The peak height was measured at 213.7 nm and the linearity curve was developed using the respective concentrations. In addition, regression equations and regression coefficients were constructed from both curves.

3.6.2. Ratio Absorption Difference Method (RAD)

Standard solutions of VGN and RGE consisting of 5 µg/mL were prepared separately. A sufficient amount of standard solutions of VGN and RGE was transferred into seven 5 mL graduating flasks to obtain concentrations in the range of 1–50 µg/mL and 2–75 µg/mL. All the solutions were subjected to UV absorption at 200–300 nm and stored in the computer. The scanned spectra of the mixture were subjected to a division by the spectrum of RGE to generate the ratio spectra of VGN in the range of 1–50 µg/mL, which were smoothened using 4 nm and stored in the computer. The peak amplitude discrepancy was calculated by subtracting the peak amplitudes at 207.2 nm from 230.6 nm. Similarly, the mixture spectra were subjected to a division by the spectrum of VGN to generate the ratio spectra of RGE with concentrations of 2–75 µg/mL. The peak amplitude difference was recorded for each spectrum by evaluating the peak amplitudes at 230.3 and 251.4 nm. The calibration curves were constructed for both analytes by drawing a graph between the difference in the peak height and the respective concentration. The regression equations and coefficients were computed from the linearity curves.

3.6.3. Ratio First Derivative Absorption Method (RFA)

The above-recorded ratio spectra of VGN were modified into first derivative spectra utilizing 4 nm as $\Delta\lambda$ and a scaling factor of 10. The peak height was calculated at 213.7 nm and the linearity curve was created by drawing a graph between the peak height and the corresponding concentration. Similarly, the ratio spectra of RGE were processed into first derivative spectra utilizing 4 nm as $\Delta\lambda$ and a scaling factor of 10. The peak height of the spectra was measured at 237.2 nm and the linearity curve was created against the corresponding concentration.

3.6.4. Application of the Optimized Methods to the Formulation

The above-prepared formulation solution was diluted to obtain the amount of analytes in the range of the linearity range. The solutions were subjected to UV absorption at 200–300 nm and stored in the computer. For the FDS method, the stored spectra were transferred into first derivative spectra utilizing 4 nm as $\Delta\lambda$ and a scaling factor of 10. The concentration of RGE was computed by measuring the peak height of the first derivative spectra at 287.7 nm and using a regression equation. The peak height was calculated at

213.7 nm for VGN and the amount of analyte was estimated by means of a regression equation. For the ratio difference, the absorption method scanned spectrum of the formulation was divided by the VGN spectra to generate the RGE ratio spectra; similarly, the formulation spectrum was divided by the RGE spectrum to generate the VGN ratio spectra followed by smoothening with a wavelength of 4 nm. The peak amplitude variance was calculated by deducting the peak height at 207.2 nm from 230.6 nm for computing the concentration of VGN using the corresponding regression equation. Similarly, the peak amplitude difference was computed at 230.3 nm and 251.4 nm for determining the concertation of RGE. The above-generated ratio spectra of VGN and RGE were processed into first derivative spectra utilizing 4 nm as $\Delta\lambda$ and a scaling factor of 10. The peak height was measured at 213.7 nm for VGN and at 237.2 nm for RGE. The amounts of VGN and RGE were determined from the respective linearity equations.

4. Conclusions

Despite the complete overlapping of both analytes, derivative spectrophotometric methods were successfully developed for the concurrent estimation of VGN and RGE from the combined medicine. The proposed methods were simple, accurate, rapid, and validated according to the ICH guidelines. The greenness profile, evaluated by two different methods, confirmed the ecofriendly nature of the optimized spectroscopic methods. All three methods were utilized for a simultaneous determination without interference from the formulation excipients; hence, these procedures could be useful for the consistent quality control of medical dosage forms comprising VGN and RGE. In addition, the reported methods are considered to be economical as there was no need of expensive solvents and instruments in chromatographic methods. No additional software was purchased because the software provided with the UV spectrophotometer was used.

Supplementary Materials: The following are available online. Figure S1: UV derivative spectra of VGN and RGE of laboratory mixed solutions. Figure S2: UV derivative spectra of VGN and RGE for the formulation and standard addition method. File S1: Calibration curves of analytes by UV derivative methods. Table S1: Green assessment profile proposed by Raynie et al. Table S2: Green analytical procedure index parameters.

Author Contributions: Conceptualization, K.N.V. and M.A.; data curation, B.E.A.-D. and M.A.; formal analysis, M.A., K.N.V., R.E.E.E. and S.S.; investigation, B.E.A.-D.; methodology, M.A., K.N.V., R.E.E.E. and S.S.; project administration, M.A. and B.E.A.-D.; resources, K.N.V., B.E.A.-D. and R.E.E.E.; supervision, B.E.A.-D., R.E.E.E. and S.S.; validation, M.A. and S.S.; visualization, R.E.E.E. and S.S.; writing—original draft, M.A., K.N.V., R.E.E.E. and S.S.; writing—review and editing, M.A., K.N.V., B.E.A.-D., R.E.E.E. and S.S. All authors have read and agreed to the published version of the manuscript.

Funding: The authors are thankful to the Deanship of Scientific Research, King Faisal University, Al-Ahsa, Saudi Arabia for financial support under the Nasher track (Grant #206138). The APC was funded by the Deanship of Scientific Research, King Faisal University, Al-Ahsa.

Data Availability Statement: The data generated during this work were included in the manuscript and submitted as Supplementary files.

Acknowledgments: The authors are thankful to the Deanship of Scientific Research, King Faisal University, Al-Ahsa, Saudi Arabia for financial and moral support. The authors also thank Tameem Alyahian for his support.

Conflicts of Interest: The authors declare no conflict of interest.

References

1. Zheng, Y.; Ley, S.H.; Hu, F.B. Global aetiology and epidemiology of type 2 diabetes mellitus and its complications. *Nat. Rev. Endocrinol.* **2018**, *14*, 88–98. [CrossRef]
2. Klein, B.E.; Klein, R.; Moss, S.E.; Cruickshanks, K.J. Parental history of diabetes in a population-based study. *Diabetes Care* **1996**, *19*, 827–830. [CrossRef]

3. Ahsan, S. Effectiveness of Remogliflozin and Vildagliptin Combination in Type 2 Diabetes Mellitus Patients Uncontrolled on Triple Oral Drug Therapy. *Endocr. Pract.* **2021**, *27*, 6. [CrossRef]
4. Markham, A. Remogliflozin Etabonate: First global approval. *Drugs* **2019**, *79*, 1157–1161. [CrossRef]
5. Joshi, S.S.; Singh, T.; Newby, D.E.; Singh, J. Sodium-glucose co-transporter 2 inhibitor therapy: Mechanisms of action in heart failure. *Heart* **2021**, *107*, 1032–1038. [CrossRef]
6. Zelniker, T.A.; Wiviott, S.D.; Raz, I.; Im, K.; Goodrich, E.; Bonaca, M.P. SGLT2 inhibitors for primary and secondary prevention of cardiovascular and renal outcomes in type 2 diabetes: A systematic review and meta-analysis of cardiovascular outcome trials. *Lancet* **2019**, *393*, 31–39. [CrossRef]
7. Kawanami, D.; Takashi, Y.; Takahashi, H.; Motonaga, R.; Tanabe, M. Renoprotective Effects of DPP-4 Inhibitors. *Antioxidants* **2021**, *10*, 246. [CrossRef] [PubMed]
8. Schiapaccassa, A.; Maranhão, P.A.; de Souza, M.G.C. 30-days effects of vildagliptin on vascular function, plasma viscosity, inflammation, oxidative stress, and intestinal peptides on drug-naïve women with diabetes and obesity: A randomized head-to-head metformin-controlled study. *Diabetol. Metab. Syndr.* **2019**, *11*, 70. [CrossRef] [PubMed]
9. Dave, V.; Paresh, P. Method development and validation of UV spectrophotometric estimation of remogliflozin etabonate in bulk and its tablet dosage form. *Res. J. Pharm. Technol.* **2021**, *14*, 2042–2044. [CrossRef]
10. Tayade, A.B.; Patil, A.S.; Shirkhedkar, A.A. Development and Validation of Zero Order UV-Spectrophotometric Method by Area under Curve Technique and High Performance Thin Layer Chromatography for the Estimation of Remogliflozin Etabonate in Bulk and In-House Tablets. *Invent. Rapid Pharm. Anal. Qual. Assur.* **2019**, *3*, 1–5.
11. Bhatkar, T.; Badkhal, A.V.; Bhajipale, N.S. Stability Indicating RP-HPLC Method Development and Validation for the Estimation of Remogliflozin Etabonate in Bulk and Pharmaceutical Dosage Form. *Int. J. Pharm. Res.* **2020**, *12*, 4197–4207.
12. Shah, D.A.; Gondalia, I.I.; Patel, V.B.; Mahajan, A.; Chhalotiya, U.K. Stability indicating liquid chromatographic method for the estimation of remogliflozin etabonate. *J. Chem. Metrol.* **2020**, *14*, 125–132. [CrossRef]
13. Sigafoos, J.F.; Bowers, G.D.; Castellino, S.; Culp, A.G.; Wagner, D.S.; Reese, M.J.; Humphreys, J.E.; Hussey, E.K.; Semmes, R.L.C.; Kapur, A.; et al. Assessment of the drug interaction risk for remogliflozin etabonate, a sodium-dependent glucose cotransporter-2 inhibitor: Evidence from in vitro, human mass balance, and ketoconazole interaction studies. *Drug Metab. Dispos.* **2012**, *40*, 2090–2101. [CrossRef]
14. Attimarad, M.; Nair, A.B.; Sreeharsha, N.; Al-Dhubiab, B.E.; Venugopala, K.N.; Shinu, P. Development and Validation of Green UV Derivative Spectrophotometric Methods for Simultaneous Determination Metformin and Remogliflozin from Formulation: Evaluation of Greenness. *Int. J. Environ. Res. Public Health* **2021**, *18*, 448. [CrossRef] [PubMed]
15. Attimarad, M.; Elgorashe, R.E.E.; Subramaniam, R.; Islam, M.M.; Venugopala, K.N.; Nagaraja, S.; Balgoname, A.A. Development and Validation of Rapid RP-HPLC and Green Second-Derivative UV Spectroscopic Methods for Simultaneous Quantification of Metformin and Remogliflozin in Formulation using Experimental Design. *Separations* **2020**, *7*, 59. [CrossRef]
16. Tammisetty, M.; Challa, B.R.; Puttagunta, S.B. A novel analytical method for the simultaneous estimation of remogliflozin and metformin hydrochloride by UPLC/PDA in bulk and formulation. Application to the estimation of product traces. *Turk. J. Pharm. Sci.* **2020**, *39699*, 296–305. [CrossRef]
17. Dayoub, L.A.; Amali, F. Development of a new visible Spectrophotometric analytical method for determination of Vildagliptin in bulk and Pharmaceutical dosage forms. *Res. J. Pharm. Technol.* **2020**, *13*, 2807–2810. [CrossRef]
18. Kumari, B.; Khansili, A. Analytical Method Development and Validation of UV-visible Spectrophotometric Method for the Estimation of Vildagliptin in Gastric Medium. *Drug Res.* **2020**, *70*, 417–423. [CrossRef]
19. Waghulde, M.; Naik, J. Development and validation of analytical method for Vildagliptin encapsulated poly-ε-caprolactone microparticles. *Mater. Today Proc.* **2018**, *5*, 958–964. [CrossRef]
20. Patil, K.R.; Deshmukh, T.A.; Patil, V.R. A stability indicating HPTLC method development and validation for analysis of Vildagliptin as bulk drug and from its pharmaceutical dosage form. *Int. J. Pharm. Sci. Res.* **2020**, *11*, 2310–2316. [CrossRef]
21. Giordani, C.F.A.; Campanharo, S.; Wingert, N.R.; Bueno, L.M.; Manoel, J.W.; Garcia, C.V.; Volpato, N.M.; Steppe, M. UPLC-ESI/Q-TOF MS/MS Method for Determination of Vildagliptin and its Organic Impurities. *J. Chromatogr. Sci.* **2020**, *58*, 718–725. [CrossRef] [PubMed]
22. Abdel-Ghany, M.; Abdel-Aziz, O.; Ayad, M.F.; Tadros, M.M. Validation of Different Spectrophotometric Methods for Determination of Vildagliptin and Metformin in Binary Mixture. *Spectrochim. Acta Part A Mol. Biomol. Spectrosc.* **2014**, *125*, 175–182. [CrossRef]
23. Moneeb, M.S. Spectrophotometric and spectrofluorimetric methods for the determination of saxagliptin and vildagliptin in bulk and pharmaceutical preparations. *Bull. Fac. Pharm. Cairo Univ.* **2013**, *51*, 139–150. [CrossRef]
24. Shaikh, N.K.; Jat, R.; Bhangale, J.O. Analysis of vildagliptin and nateglinide for simultaneous estimation using spectro-chromatographic methods. *Eur. J. Mol. Clin. Med.* **2020**, *7*, 741–755.
25. Shakoor, A.; Ahmed, M.; Ikram, R.; Hussain, S.; Tahir, A.; Jan, B.M.; Adnan, A. Stability-indicating RP-HPLC method for simultaneous determination of metformin hydrochloride and vildagliptin in tablet and biological samples. *Acta Chromatogr.* **2020**, *32*, 39–43. [CrossRef]
26. Khalil, S.; Bushara, A.; Farag, H. New Analytical Methods for the Determination of Two Gliptin Drugs in Pharmaceutical Formulations and Urine Samples. *Am. J. Pharm. Technol. Res.* **2020**, *10*, 31–43.

27. Abdel Hady, K.K.; Abdel Salam, R.A.; Hadad, G.M.; Abdel Hameed, E.A. Simultaneous HPLC determination of vildagliptin, ampicillin, sulbactam and metronidazole in pharmaceutical dosage forms and human urine. *J. Iran. Chem. Soc.* **2021**, *18*, 729–738. [CrossRef]
28. Kamal, A.H.; El-Malla, S.F.; Hammad, S.F. A Review on UV spectrophotometric methods for simultaneous multicomponent analysis. *Eur. J. Pharm. Med. Res.* **2016**, *3*, 348–360.
29. Salinas, F.; Nevado, J.J.B.; Mansilla, A.E. A new spectrophotometric method for quantitative multicomponent analysis resolution of mixtures of salicylic and salicyluric acids. *Talanta* **1990**, *37*, 347–351. [CrossRef]
30. Chohan, M.S.; Elgorashe, R.E.E.; Balgoname, A.A.; Attimarad, M.; SreeHarsha, N.; Venugopala, K.N.; Nair, A.B.; Pottathil, S. Eco-friendly Derivative UV Spectrophotometric Methods for Simultaneous Determination of Diclofenac Sodium and Moxifloxacin in Laboratory Mixed Ophthalmic Preparation. *Indian J. Pharm. Edu. Res.* **2019**, *53*, 166–174. [CrossRef]
31. Bhatt, N.M.; Chavada, V.D.; Sanyal, M.; Shrivastav, P.S. Manipulating ratio spectra for the spectrophotometric analysis of diclofenac sodium and pantoprazole sodium in laboratory mixtures and tablet formulation. *Sci. World J.* **2014**, *2014*, 495739. [CrossRef] [PubMed]
32. Attimarad, M.; Chohan, M.S.; Balgoname, A.A. Simultaneous Determination of Moxifloxacin and Flavoxate by RP-HPLC and Ecofriendly Derivative Spectrophotometry Methods in Formulations. *Int. J. Environ. Res. Public Health* **2019**, *16*, 1196. [CrossRef] [PubMed]
33. Raynie, D.; Driver, J. Green Assessment of Chemical Methods. In Proceedings of the 13th Annual Green Chemistry and Engineering Conference, College Park, MD, USA; 2009.
34. Płotka-Wasylka, J. A new tool for the evaluation of the analytical procedure: Green Analytical Procedure Index. *Talanta* **2018**, *181*, 204–209. [CrossRef] [PubMed]
35. Abdalah, N.A.; Fathy, M.E.; Tolba, M.M.; El-Brashy, A.M.; Ibrahim, F.A. Green spectrofluorimetric assay of dantrolene sodium via reduction method: Application to content uniformity testing. *R. Soc. Open Sci.* **2021**, *14*, 210562. [CrossRef]
36. Lotfy, H.M.; Hegazy, M.A. Simultaneous determination of some cholesterol-lowering drugs in their binary mixture by novel spectrophotometric methods. *Spectrochim. Acta Part A Mol. Biomol. Spectrosc.* **2013**, *113*, 107–114. [CrossRef] [PubMed]

Article

Replicates Number for Drug Stability Testing during Bioanalytical Method Validation—An Experimental and Retrospective Approach

Elżbieta Gniazdowska [1,2], Wojciech Goch [3], Joanna Giebułtowicz [4] and Piotr J. Rudzki [5,*]

1. Łukasiewicz Research Network, Industrial Chemistry Institute, 8 Rydygiera, 01-793 Warsaw, Poland; elzbieta.gniazdowska@ichp.lukasiewicz.gov.pl or elzbieta.gniazdowska@wum.edu.pl
2. Department of Bioanalysis and Drugs Analysis, Doctoral School, Medical University of Warsaw, 61 Żwirki i Wigury, 02-091 Warsaw, Poland
3. Department of Physical Chemistry, Faculty of Pharmacy, Medical University of Warsaw, 1 Banacha, 02-097 Warsaw, Poland; wojciech.goch@wum.edu.pl
4. Department of Bioanalysis and Drugs Analysis, Faculty of Pharmacy, Medical University of Warsaw, 1 Banacha, 02-097 Warsaw, Poland; joanna.giebultowicz@wum.edu.pl
5. Celon Pharma S.A., Bioanalytical Laboratory, 15 Marymoncka, 05-152 Kazuń Nowy, Poland
* Correspondence: pj.rudzki@wp.pl

Abstract: Background: The stability of a drug or metabolites in biological matrices is an essential part of bioanalytical method validation, but the justification of its sample size (replicates number) is insufficient. The international guidelines differ in recommended sample size to study stability from no recommendation to at least three quality control samples. Testing of three samples may lead to results biased by a single outlier. We aimed to evaluate the optimal sample size for stability testing based on 90% confidence intervals. Methods: We conducted the experimental, retrospective (264 confidence intervals for the stability of nine drugs during regulatory bioanalytical method validation), and theoretical (mathematical) studies. We generated experimental stability data (40 confidence intervals) for two analytes—tramadol and its major metabolite (O-desmethyl-tramadol)—in two concentrations, two storage conditions, and in five sample sizes (n = 3, 4, 5, 6, or 8). Results: The 90% confidence intervals were wider for low than for high concentrations in 18 out of 20 cases. For n = 5 each stability test passed, and the width of the confidence intervals was below 20%. The results of the retrospective study and the theoretical analysis supported the experimental observations that five or six repetitions ensure that confidence intervals fall within 85–115% acceptance criteria. Conclusions: Five repetitions are optimal for the assessment of analyte stability. We hope to initiate discussion and stimulate further research on the sample size for stability testing.

Keywords: confidence interval; stability; retrospective analysis; sample size; regulatory bioanalysis; bioanalytical method validation

1. Introduction

Evaluation of drug or metabolite stability in biological samples in conditions reflecting sample handling and analysis during bioanalytical method validation is recommended by international regulatory guidelines [1,2] and ICH M10 draft guidelines [3]. This evaluation includes stability in the biological matrix (short-term, long-term, and freeze-thaw), in processed samples and solutions (stock and working solutions). Kaza et al. (2019) [4] discussed the differences and similarities in bioanalytical method validation guidelines [1,2], but the authors omitted to mention differences in the recommended sample size (number of samples) for stability testing. The European Medicines Agency (EMA) [1] does not recommend any specific sample size whereas the U.S. Food and Drug Administration (FDA) [2] and ICH [3] recommend a minimum of three quality control samples (QC) per level of concentration of low QC and high QC to assess the stability of an analyte in a biological matrix. A note from Health Canada does not recommend examining stability using only one repetition of a QC sample [5].

The analyte stability testing refers to other characteristics of the bioanalytical method. The calibration range helps to select studied concentrations (low- and high-quality control samples). However, method precision is important to compare reference samples (e.g., prepared ex tempore) and test samples (i.e., stored for a specified time in specified conditions). Before any regulatory bioanalytical method validation guideline was published, Timm et al. proposed a stability assessment incorporating the precision in the calculation of 95% confidence intervals [6]. However, its application was limited by the assumed equality of variances for the reference and test samples. Rudzki and Leś extended this method for datasets with unequal variances [7]. They also proposed the use of 90% confidence intervals instead of 95% [6] to make the probability equal to the bioequivalence recommendations [8]. Confidence intervals are a good tool for testing stability. Since their introduction by Jerzy Spława Neyman in 1936 [9] they became widely used, including clinical research—for example as bioequivalence criterium [8]. Briefly, the idea of confidence intervals is to define a range of values describing parameters of interest in the population, based on parameter estimates observed in the sample. This estimation has a defined probability—usually 90%, 95%, or 99%. For example, a 90% confidence interval of 85.1–105.2% for mean stability means that there is a 90% probability that the mean stability is between 85.1% and 105.2%. In the case of stability testing, the confidence interval combines central tendency (mean difference between stored and reference samples) and data dispersion (method precision) with a selected probability. This approach is not yet frequently used because it is more restrictive and labor intensive than the guidelines' recommendations. Nevertheless, the confirmation of analyte stability in a biological matrix using this method is associated with a low and pre-defined probability of true instability.

The stability assessment proposed in the draft of the ICH M10 bioanalytical method validation guideline [3] recommends analyzing stored and reference samples but does not include a description of any comparison between them. The lack thereof creates the risk of accepting the method regardless of the 29.8% instability of an analyte [4]. Moreover, there is an insufficient justification of sample size (number of samples) in the stability evaluation. Limiting testing to three samples in each dataset may lead to stability results biased by a single outlier. However, how much do additional analyses increase confidence in the stability results? Is this increase relevant? How to balance it with the cost of extra analyses? Although there may be no universal answer to these questions, further research on sample size for stability assessment is needed.

In this paper, we aim to evaluate the optimal sample size for drug stability testing in human plasma based on confidence intervals [6,7] by conducting an experimental study for tramadol and its major metabolite (O-desmethyl-tramadol), as well as a retrospective data analysis for nine drugs of different structure.

2. Materials and Methods
2.1. Materials

O-desmethyl-tramadol hydrochloride ($\leq$99%) was purchased from LoGiCal (Luckenwalde, Germany) and tramadol hydrochloride ($\leq$99%) was purchased from Saneca Pharmaceuticals (Hlohovec, Slovakia). O-desmethyl-tramadol-d6 ($\leq$98%) and tramadol-d6 hydrochloride ($\leq$99%) were purchased from TLC Pharmaceutical Standards (Newmarket, Ontario, Canada). All other reagents were of analytical grade. Methanol and formic acid were purchased from Merck KGaA (Darmstadt, Germany). Sodium hydroxide was obtained from Chempur (Piekary Śląskie, Poland). Human blank plasma with CPD (citrate, phosphate, dextrose) as an anticoagulant was obtained from the Regional Blood Donation and Blood Therapy Centre (Warsaw, Poland).

2.2. Mass Spectrometric and Chromatographic Conditions

The bioanalytical method was adapted from the previous study [10] with a different chromatographic column and the use of formic acid in the mobile phase instead of acetic acid. The adapted method was validated according to the EMA [1] guidelines, except for long-term stability which was confirmed previously. Instrumental analysis was performed on an Agilent 1260 Infinity (Agilent Technologies, Santa Clara, CA, USA), equipped with an autosampler, a degasser, and a binary pump coupled to a hybrid triple quadrupole/linear ion trap mass spectrometer QTRAP 4000 (ABSciex, Framingham, MA, USA). The Turbo Ion Spray source was operated in positive mode with voltage and source temperatures of 5500 V and 550 °C, respectively. The curtain gas, ion source gas 1, ion source gas 2, and collision gas (all high purity nitrogen) were set at 206.84 kPa, 275.79 kPa, 379 kPa, and "high" instrument units, respectively. The target compounds were analyzed in the Multiple Reaction Monitoring (MRM) mode (Table 1).

Table 1. Parameters of MS method.

	Retention Time (min)	MRM [m/z]	DP [V]	CE [V]	CXP [V]
tramadol	3.4	264.2 > 42.3	51	125	10
tramadol-d6	3.4	270.3 > 252.2	66	17	16
O-desmethyl-tramadol	2.6	250.2 > 232.2	71	17	18
O-desmethyl-tramadol-d6	2.6	256.0 > 238.3	61	17	14

MRM—multiple reaction monitoring; DP—declustering potential; CE—collision energy; CXP—cell exit potential.

Chromatographic separation was achieved with a Kinetex C18 column (100 mm × 4.6 mm, 2.6 µm, Phenomenex, Torrance, CA, USA) using isocratic elution with methanol and 0.1% formic acid in a ratio of 40:60 at a flow rate of 0.3 mL/min. The column and the autosampler temperature was 50 ± 1 °C and 20 ± 1 °C, respectively. The injection volume was 5 µL.

2.3. Stock Solution, Calibration Standards, and Quality Control Samples

The separate standard stock solutions of tramadol, O-desmethyl-tramadol, tramadol-d6, and O-desmethyl-tramadol-d6 were prepared in 50% methanol (v/v) and were stored at −20 °C. The standard working solution was prepared by mixing stock solutions with an appropriate volume of water. The internal standard working solution (250 ng/mL for tramadol-d6 and 75 ng/mL for O-desmethyl-tramadol-d6) was diluted with water and prepared by mixing both internal standards stock solutions.

All calibration standards and the quality control samples were prepared by spiking blank human plasma with a working solution containing both analytes. The calibration standards contained both tramadol and O-desmethyl-tramadol at eight concentrations ranging from 5.0 to 750 ng/mL and from 2.5 to 150 ng/mL. The quality control samples were prepared at concentrations of 15, 350, and 600 ng/mL for tramadol, and 7.5, 70, and 120 ng/mL for O-desmethyl-tramadol.

2.4. Sample Preparation

The liquid-liquid extraction with *tert*-butyl methyl ether and 1M sodium hydroxide was used for the sample preparation [10]. Internal standards were added in one solution. The ether phase was evaporated in nitrogen gas and the dry residue was reconstituted with 150 µL of the mobile phase.

2.5. Stability Evaluation and Statistical Methods

The short-term stability was evaluated with sets containing an equal number of test and reference-quality control samples (QC): 3, 4, 5, 6, and 8 for low QC (15/7.5 ng/mL tramadol and O-desmethyl-tramadol) and high QC (600/120 ng/mL tramadol and O-desmethyl-tramadol). The reference and test QC samples (plasma fortified with tramadol and O-desmethyl tramadol solution) were prepared. The test QC samples were stored at room temperature for 24 and 72 h before extraction and LC-MS analysis. Autosampler stability test during the validation method, confirmed that samples are stable for a minimum of 68 h at room temperature [10]. Reference samples were analyzed immediately after preparation, after 24 and 72 h storage in an autosampler at 20 ± 1 °C in the same sequence as test samples. Acceptance criteria were met when the whole confidence interval was within the acceptance range of 85–115%.

The statistical analysis of stability was based on the application of 90% confidence intervals [6,7]. The F-Snedecor test (significance level $\alpha = 0.01$) was applied to test the hypothesis on variance equality. The influence of the number of repetitions and analyte concentration on the position and width of the confidence interval was analyzed using an analysis of variance (ANOVA, $p = 0.05$) test with repeated measurements. Normal distribution of the stability was assumed in the estimation of the probability that the confidence interval width is below 30%. The probability $P(CI \subset [85; 115])$ was calculated using the equation:

$$P(CI \subset [85; 115]) = \chi_{n-1}\left(\frac{225\, n\,(n-1)}{k^2\, \sigma_S^2}\right)$$

where:

χ_{n-1}—cumulative distribution function of the chi-square distribution for degrees of freedom (df) = $n - 1$;
n—number of repetitions;
k—the value of the Student t-distribution quantile at a 0.1 significance level for $n - 1$ degrees of freedom (df);
σ_S—standard deviation in stability.

More details on mathematical calculations can be found in the Appendix A.

2.6. Retrospective Analysis

Stability results for nine drugs were recorded during method validations conducted under Good Laboratory Practice conditions at the former Pharmaceutical Research Institute in Warsaw, Poland ([11–15], and unpublished data). The following types of stability were studied: short-term stability, freeze and thaw stability, long-term stability at temperatures of −14 °C and −65 °C. Nine drugs with LC-MS and HPLC-UV methods of determination of varying precision were selected to create the data sets. For each drug and each stability test, $n = 6$ samples were recorded at each low and high QC concentration. To analyze the worst-case scenario, for each dataset a result lying nearest to the mean of $n = 6$ results was discarded to obtain $n = 5$ dataset. The same procedure was used to obtain datasets of $n = 4$ and $n = 3$. The final number of calculated confidence intervals was 264. Comparison of the width of the confidence intervals between low and high QC was made using a Wilcoxon signed-rank test (significance level $p < 0.05$). To analyze how differences in one variable (percentage of confidence intervals within acceptance criteria set at 85–115%) can be explained by a difference in a second variable (confidence width or the number of samples), the coefficient of determination was used.

3. Results

3.1. Experimental and Mathematical Studies

Thanks to the design of the experimental study (five sample sizes, two storage durations, two analytes in two concentrations each) we were able to calculate 40 confidence intervals (Figure 1). For 20 pairs of low and high QC concentrations, we recorded 18 cases (90%) where the 90% confidence interval was wider for low than for high concentration. Moreover, the variability of the confidence interval width—presented as relative standard deviation (RSD) in Table 2—was larger for low concentration. It shows the influence of method precision on stability evaluation, as lower concentrations were measured with worse precision.

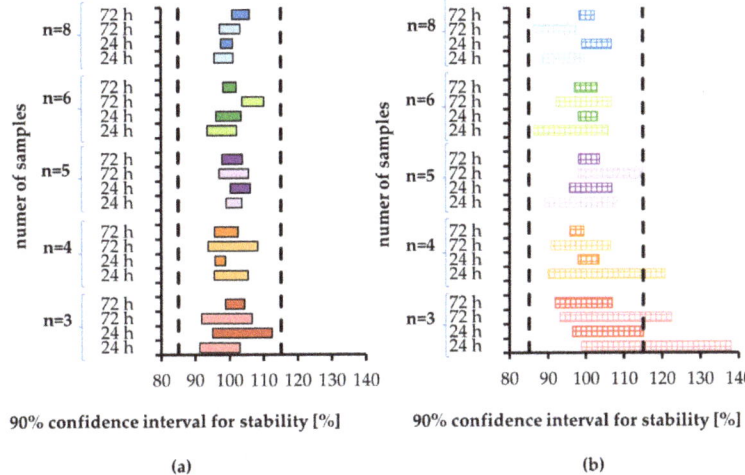

Figure 1. The 90% confidence intervals for the stability calculated according to [7] for (**a**) tramadol and (**b**) O-desmethyl-tramadol in human plasma stored at room temperature for 24 h and 72 h. Each sample size is associated with a different color, with light color indicating low concentration and dark color indicating high concentration. Vertical dashed lines indicate stability limits of 85–115%.

Table 2. Descriptive statistics for the width [%] of a 90% confidence interval. The number of pairs is the equal number of reference and study samples.

	Low QC					High QC				
Number of Pairs	3	4	5	6	8	3	4	5	6	8
Experimental Data for Tramadol and O-desmethyl-tramadol (n = 4 of results at each column)										
Mean	23.8	17.7	12.4	10.2	8.5	14.2	4.8	7.2	5.5	5.1
Geometric mean	21.1	16.2	10.7	9.7	8.1	12.9	4.6	6.9	5.4	4.9
Median	22.1	15.1	12.8	9.9	8.6	16.3	4.7	6.0	5.5	4.6
Min	11.7	9.9	4.7	6.4	5.5	5.7	3.1	5.5	3.9	3.6
Max	39.1	30.8	19.5	14.5	11.4	18.5	6.8	11.4	7.3	7.7
SD	12.9	9.1	6.9	3.5	3.1	5.8	1.7	2.8	1.4	1.8
RSD [%]	54	51	56	34	37	41	35	39	26	36
Retrospective Analysis (n = 33 of results at each column)										
Mean	21.5	14.9	11.4	9.1	-	11.9	8.4	6.4	5.1	-
Geometric mean	18.0	12.9	9.9	8.1	-	10.8	7.9	6.0	4.8	-
Median	18.5	12.8	10.1	7.7	-	10.9	7.6	5.8	4.6	-
Min	2.7	3.9	3.1	2.9	-	3.3	3.0	2.3	1.8	-
Max	54.3	37.6	28.2	23.2	-	28.8	19.5	14.6	12.9	-
SD	13.0	8.4	6.3	5.0	-	5.2	3.2	2.5	2.1	-
RSD [%]	57	54	52	53	-	43	38	38	40	-

Moreover, wider confidence intervals for low concentrations of O-desmethyltramadol than for low concentrations of O-tramadol indicate the importance of method precision. The precision of O-desmethyltramadol determination in quality control samples was 7.38% for low QC (7.5 ng/mL) and 2.90% for high QC (120 ng/mL). The precision of tramadol determination was 6.43% for low QC (15 ng/mL) and 3.07% for high QC (600 ng/mL). For each studied QC level, the mean extraction recovery was consistent for both analytes and their ISs—86.08–87.99% for tramadol, 85.55–86.99% for tramadol-d6, 74.45–78.75% for O-desmethyltramadol, and 74.61–79.07% for O-desmethyltramadol-d6. Thus, we do not expect that extraction recovery influenced stability results.

Visual assessment of low concentration data (Figure 2a) indicates that three and four repetitions are not appropriate due to the width of some confidence intervals over 30%. For five and six repetitions, width is below 20%, while for eight repetitions, width is below 12%. Visual assessment of high concentration data (Figure 2b) is a bit different. For three repetitions the confidence intervals width in 3/4 cases is over 15%, while for all other repetitions it is below 8%, with one exception of 11% ($n = 5$).

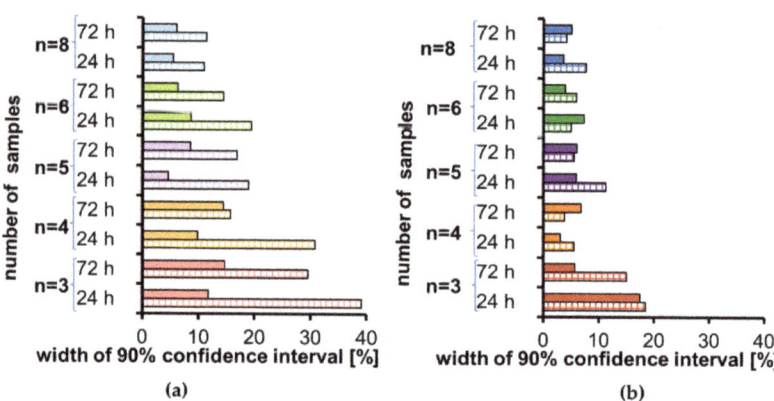

Figure 2. Width of a 90% confidence interval for stability calculated according to [7] for tramadol (full color) and O-desmethyl-tramadol (striped color) in human plasma stored at room temperature for 24 h and 72 h for each sample size: (**a**) low concentration, (**b**) high concentration.

ANOVA showed no dependence of the width of the confidence interval on the analyte concentration (Figure 3) ($p > 0.1187$). Results of the post-hoc least significant difference test (Fisher's LSD test) for sample size showed that the width of the confidence interval for $n = 3$ statistically significantly differs from more repetitions ($n = 4, 5, 6, 8$) (p from <0.0001 to 0.0249). The width of the confidence interval for $n = 4$ differs only from eight repetitions ($p < 0.05$).

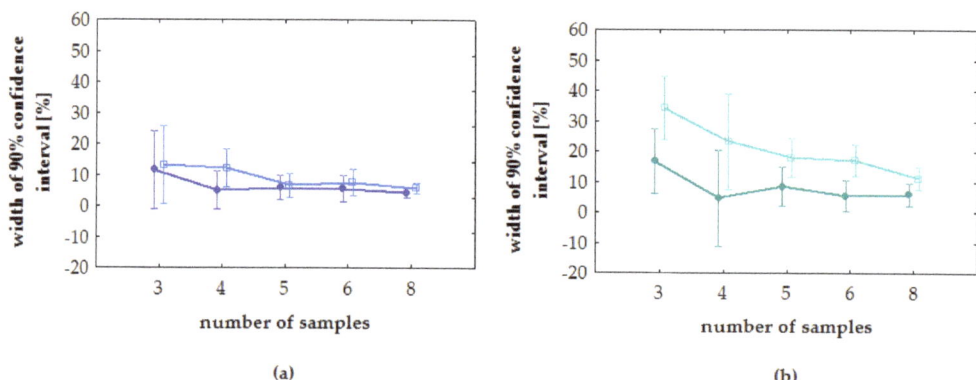

Figure 3. Post-hoc least significant difference test (Fisher's LSD test). Vertical bars means a 90% confidence interval: (**a**) tramadol; (**b**) O-desmethyl-tramadol. On each plot, light color indicates low concentration and dark color indicates high concentration.

Additionally, we have investigated the relation between precision, confidence interval, and the number of repetitions. The length of the confidence interval depends on the sample variance—the greater the n, the shorter the length of the interval (as it is inversely proportional to the square root of n), and the higher the chance the sample variance is assessed correctly. We calculated the probability that for a given precision, the confidence interval derived from n repetitions falls within a 30% range. As expected, the relation between precision and the number of repetitions is sharp (Figure 4). As an example, for 10% precision, the considered probability is 33% for $n = 3$, 51% for $n = 4$, 71% for $n = 5$, 86% for $n = 6$, and 98% for $n = 8$. In general, for a smaller number of repetitions, there is a significant probability that the measurements with even high precision may overestimate the sample variance and consequently the length of the confidence interval. The choice of five or six repetitions proves to be enough to ensure that the confidence intervals will fall within the 85–115% interval.

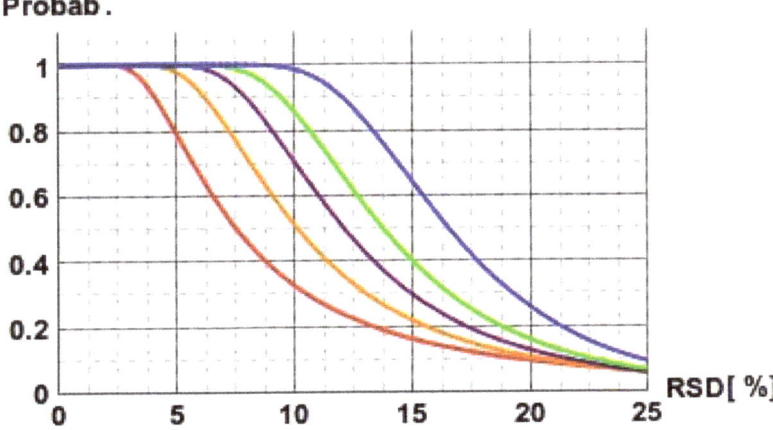

Figure 4. Dependence of the probability that the confidence interval width is below 30% on the precision in measurements. Equal precision for the reference and the studied measurements is assumed. Curves are defined for $n = 3$ (red), $n = 4$ (orange), $n = 5$ (purple), $n = 6$ (green), and $n = 8$ (blue).

We postulate that five repetitions of quality control samples at low and high concentration levels are optimal for stability tests during bioanalytical method validation. For each case with $n = 5$, the stability tests passed and the width of all confidence intervals was below 20%. For $n < 5$ some of the stability tests failed (part of the confidence interval outside of the acceptance criteria of 85–115%) due to the width of confidence intervals exceeding 30%. Moreover, for $n > 5$ all stability tests passed and the mean width of the confidence intervals decreased gradually (Table 2).

3.2. Retrospective Study

To verify observations from the experimental and the theoretical studies, we have analyzed human plasma stability data for nine validated bioanalytical methods (Figures 5 and A1). For all data, the percentage of confidence intervals lying within acceptance criteria was acceptable for $n = 5$ (88% for low and 93% for high concentration, respectively) and reached 100% for $n = 6$ (Figure 6a). For $n = 5$, only 5 of 66 results (including four for low QC) were outside of the acceptance limits. The greatest difference between the confidence interval limits and the acceptance criteria was 1.8%.

As expected, a strong positive correlation ($r^2 > 0.96$) was observed between the number of samples and the percentage of confidence intervals within the acceptance criteria (Figure 6a). Consequently, a strong negative correlation ($r^2 > 0.98$) was observed between the confidence interval width and the percentage of confidence intervals within the acceptance criteria (Figure 6b). Among confidence intervals for $n = 3$, 4, and 5, more than a 2-fold higher percentage of confidence intervals outside of acceptance criteria was observed for the low QC (Figure A2b) than for the high QC (Figure A2c) concentration ($p < 0.00001$). This observation is consistent with higher values of both width of the confidence interval and its variability expressed as RSD (Table 2, Figures 7 and A4).

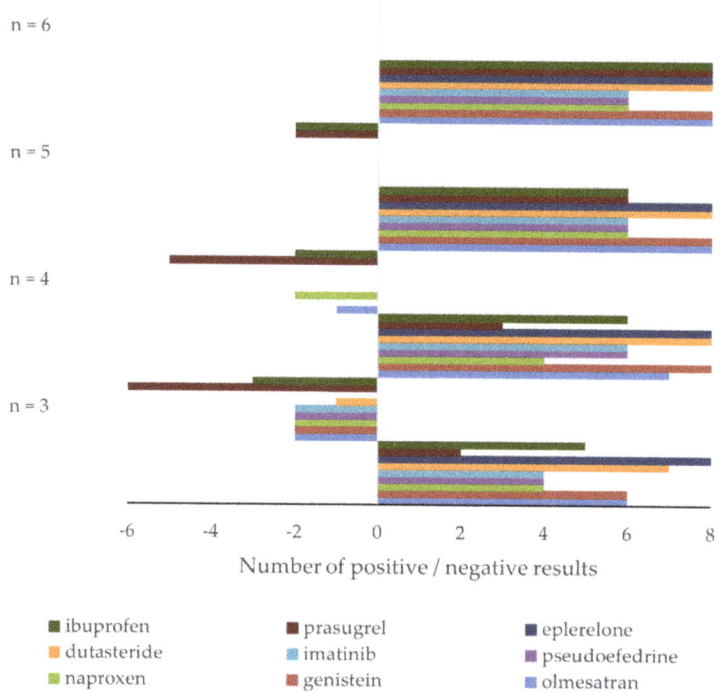

Figure 5. Retrospective study of nine drugs' stability in human plasma: number of confidence intervals within (positive results) and outside (negative results) acceptance criteria for nine drugs using n = 3, 4, 5, and 6 samples for stability testing. High and low concentration data are combined.

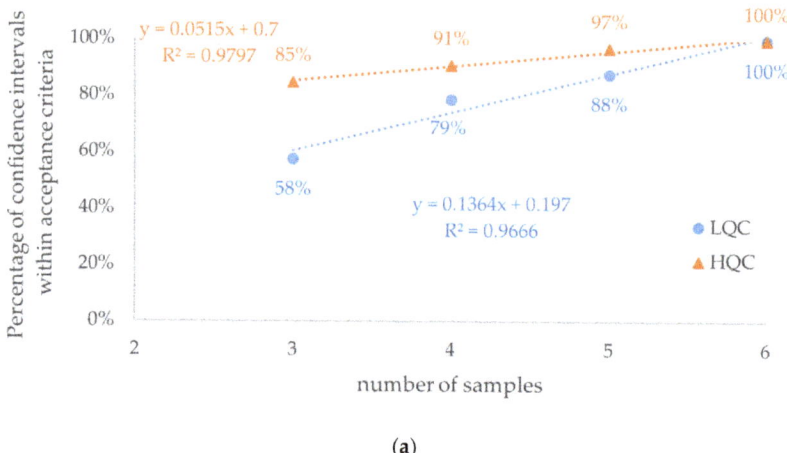

(a)

Figure 6. Cont.

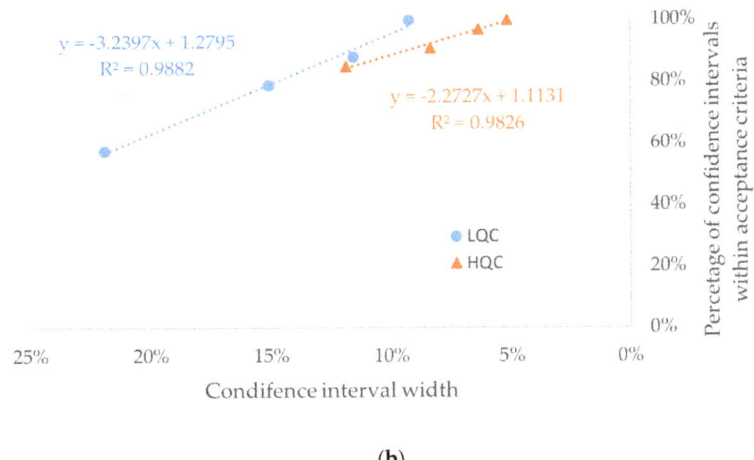

(b)

Figure 6. Retrospective study of nine drugs' stability in human plasma: percentage of confidence intervals within acceptance criteria in the function of (**a**) number of samples and (**b**) mean width of the confidence interval for each number of samples (see Table 2). The dataset consisted of 33 confidence intervals for each concentration level: LQC (circle)—low-quality control sample; HQC (triangle)—high-quality control sample.

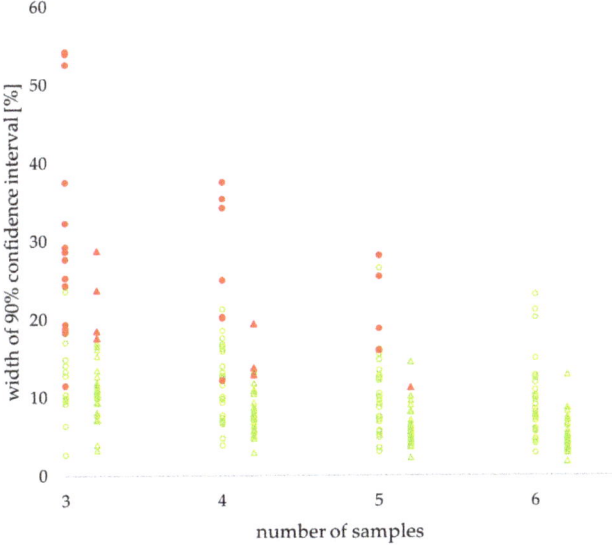

Figure 7. Retrospective study of nine drugs' stability in human plasma: individual values of confidence interval width. LQC (circle)—low-quality control sample, HQC (triangle)—high-quality control sample. Filled red figures indicate values outside acceptance criteria, unfilled green figures indicate values within acceptance criteria.

There were no relevant differences in confidence interval width between stability tests (Figure A3). The highest values for all sample numbers were recorded for the freeze and thaw test, but all other values for each sample number were only 1–2% lower.

4. Discussion

The results of the experimental, theoretical, and retrospective studies are in good agreement indicating that using 90% confidence intervals requires testing of at least five repetitions of quality controls as references and as stability samples. A retrospective study revealed that the percentage of the confidence intervals within acceptance criteria is strongly correlated with the number of samples used for stability testing (positively) and the mean of the width of confidence intervals (negatively). The statistically significant difference between low QC and high QC was observed between the percentage of confidence intervals within the acceptance criteria for a given sample number. The type of stability test did not influence confidence interval width. It seems that the excess work between $n = 5$ and $n = 8$ is not balanced with the benefit of a narrower width of the confidence interval. On the other hand, there would be 72 more analyses during full validation for one analyte, and this number does not include stability testing in solutions. The amount of excess work and resources for additional analyses may not be assessed in general, because it depends on particular method characteristics.

Our experimental study used a single bioanalytical method for the determination of two analytes in a single laboratory. To increase confidence in conclusions, we have reused previously generated stability data for nine drugs. Retrospective analyses are very popular in medicine [16,17], and slightly less popular in pharmacy [18,19]. On the contrary, in analytical chemistry retrospective analyses are used very rarely [20]. Over 20 years ago the concept of green analytical chemistry to protect the environment was established. Recently, its extension was proposed: white analytical chemistry in addition to green aspects also takes into account analytical and practical attributes [21]. Nevertheless, retrospective analysis has even greater ecological aspects since no chemical analysis is required and no waste is generated. Considering the high amount of analytical data produced each year in laboratories, it would be beneficial to explore them all deeply to draw some general conclusions, answer the emerging questions, and contribute to international guidelines development. The retrospective study enabled comparison of data generated using LC-MS and HPLC-UV methods (Table A3). It may be observed that narrower stability confidence intervals were recorded for HPLC-UV determined imatinib than for LC-MS/MS determined prasugrel (Figure A1). On the other hand, narrower stability confidence intervals were recorded for LC-MS determined eplerenone than for HPLC-UV determined ibuprofen (Figure A1). This indicates that the detector type and concentration range are not the appropriate indicator of confidence interval width, which is dependent on method precision.

We limited our study to plasma samples. For neat solutions, due to the lower probability of interferences and lack of variability introduced by sample preparation, the precision should be better and the optimal number of repetitions could be lower. We avoided the exclusion of outlying results. An alternative approach is to use a smaller number of replicates and remove outliers using statistical tests such as the Q-Dixon or Grubbs test. However, this approach—especially for a small number of replicates—may provoke questions from regulatory agencies. Additionally, it does not take into account the precision of the method. Therefore, we do not recommend this approach. The limitation of the retrospective study is that all confidence intervals for $n = 6$ were within the acceptance criteria as we used validated methods. The calculation of a 90% confidence interval may be considered as complicated compared to current bioanalytical method validation guidelines [1,2]. However, an extra effort in data analysis increases the reliability of stability evaluation.

We assumed a normal distribution of concentration data for stability and reference samples. However, stability is a ratio of stability samples over reference samples and the ratio of two normally distributed samples is never normally distributed itself. This statistical issue is taken into account for bioequivalence testing where the acceptance criteria of 80–125% does not center symmetrically around 100% but does so in log space. Thus, acceptance limits of 85–115% may not be appropriate for stability testing. An approach similar to bioequivalence suggests a criterion of 85.00–117.65%. We have opted to use 85–115% acceptance limits, which are well-established in regulatory guidelines [1,2], but their inconsistency with stability distribution needs further consideration.

Our results are important because the current recommendation of at least three samples for stability testing [2,3] is not sufficient. The proposed $n = 5$ is in line with reports from other laboratories [22–24] where five or six results were used to calculate the 90% confidence intervals for stability. Extending stability acceptance criteria from deviation from nominal concentration by adding a test-to-reference ratio may be considered as an increase of regulatory burden. On the other hand, the reliability of bioanalytical data is crucial for pharmacokinetic calculations and decisions on dosing schemes. The latter impacts drug efficiency and patient safety. Thus, the proper balance between too extensive testing and poor data quality requires further discussion. A possible answer may be a hybrid approach: hard criteria for deviation of the mean from nominal concentration combined with soft criteria for the 90% confidence interval for test-to-reference ratio.

Both experimental and retrospective studies suggest that an optimal number of repetitions is five, as also recommended by the European Bioanalysis Forum [25]. The proper assumption on the relationship between method precision and sample size may be a key factor for successful future simulations. We hope that this paper will initiate discussion and stimulate further research on optimal sample size for stability testing. We expect that further simulations and retrospective studies from other laboratories will support the need for bioanalytical guidelines update.

5. Conclusions

Five sample repetitions are optimal for the assessment of analyte stability during bioanalytical method validation. Experimental, theoretical, and retrospective study results led to similar conclusions. The number of three or four replicates, in spite of being acceptable in some guidelines, is insufficient (in some cases, the width of the confidence intervals for stability exceeded 30%, which precluded meeting the acceptance criteria). In contrast, the excess work between $n = 5$ and $n = 8$ was not balanced with any benefit of narrower confidence interval widths. We hope to initiate a discussion on sample size for stability studies. Such a discussion may result in updated bioanalytical method validation guidelines.

Author Contributions: Conceptualization, P.J.R. and J.G.; methodology, P.J.R. and W.G.; validation, E.G.; formal analysis, E.G. and W.G.; investigation, E.G. and W.G.; resources, E.G.; data curation, W.G., E.G. and P.J.R.; writing—original draft preparation, E.G., J.G., W.G., and P.J.R.; writing—review and editing, E.G., J.G. and P.J.R.; visualization, E.G., W.G. and P.J.R.; supervision, P.J.R.; project administration, J.G.; funding acquisition, E.G. and J.G. All authors have read and agreed to the published version of the manuscript.

Funding: This work was supported by the Polish Ministry of Science and Higher Education [DWD/3/6/2019, 21.11.2019]. Recording of data used in retrospective analyses was supported by the European Union (European Regional Development Fund) under the Polish Innovative Economy Operational Programme 2007-2013 [grants UDA-POIG.01.03.01-14-062/09, POIG.01.03.01-14-032/12]; and by the Polish National Centre for Research and Development [grants PBS1/B7/7/2012; INNOTECH-K1/IN1/14/159003/NCBR/12; INNOTECH-K2/IN2/65/182982/NCBR/13].

Institutional Review Board Statement: Not applicable.

Informed Consent Statement: Not applicable.

Data Availability Statement: Data presented in Appendix A.

Acknowledgments: The authors are grateful to Ryszard Marszałek for technical assistance in LC-MS/MS analyses, Edyta Pesta and Krzysztof Abramski for the critical reviewing of the manuscript. The authors gratefully acknowledge the contributions of research teams from the former Pharmaceutical Research Institute (led by Katarzyna Buś-Kwaśnik, Monika Filist, Michał Kaza, and Jacek Musijowski) to generate datasets used in the retrospective analysis. The authors acknowledge the anonymous reviewer for pointing out the inconsistency between acceptance limits and the distribution of variables. The views expressed in this paper are solely those of the authors and do not necessarily reflect their institutions' position on the subject.

Conflicts of Interest: The authors declare no conflict of interest.

Sample Availability: Not applicable.

Appendix A

Table A1. Concentrations of tramadol in human plasma during stability testing after storage for 24 h and 72 h at room temperature—low QC (nominal concentration of 15.0 ng/mL) and high QC (nominal concentration of 600 ng/mL).

Number of Samples (n)	Low QC (ng/mL)				High QC (ng/mL)			
	Reference for 24 h	Tested 24 h	Reference for 72 h	Tested 72 h	Reference for 24 h	Tested 24 h	Reference for 72 h	Tested 72 h
8	14.9	14.5	14.2	14.6	573	567	590	640
	15.2	14.7	14.6	14.6	582	569	594	627
	15.4	14.8	14.7	14.6	584	574	572	611
	15.5	15.4	14.9	14.9	586	575	589	578
	15.6	15.4	15.1	14.9	588	579	580	617
	15.8	15.5	15.3	14.9	590	587	609	609
	15.9	15.6	15.5	15.4	594	591	586	584
	16.0	16.3	15.7	16.2	597	615	582	589
6	14.7	14.0	13.9	14.4	568	559	574	582
	14.8	14.5	13.9	15.1	573	564	578	585
	15.1	14.9	14.4	15.3	577	567	583	587
	15.2	15.1	14.5	15.6	587	583	596	591
	15.7	15.3	14.5	15.7	598	591	603	594
	16.3	16.0	14.7	15.7	600	630	606	602
5	14.7	14.8	14.9	14.5	555	572	553	569
	14.9	14.9	15.0	14.8	559	579	585	581
	14.9	15.2	15.1	15.2	566	584	589	588
	15.0	15.4	15.2	15.8	578	601	590	596
	15.4	15.6	15.2	16.1	591	601	594	599
4	14.3	14.3	13.8	14.3	594	571	579	559
	14.9	14.3	14.9	14.7	594	582	579	567
	14.9	15.1	15.5	15.5	595	585	582	592
	15.0	15.7	15.7	15.8	606	588	597	597
3	15.0	14.9	14.6	14.7	570	565	575	584
	15.6	15.2	15.4	15.4	574	578	589	598
	16.4	15.5	16.2	15.6	580	642	592	603

Table A2. Concentrations of O-desmethyl tramadol in human plasma during stability testing after storage for 24 h and 72 h at room temperature—low QC (nominal concentration of 7.50 ng/mL) and high QC (nominal concentration of 120 ng/mL).

Number of Samples (n)	Low QC (ng/mL)				High QC (ng/mL)			
	Reference for 24 h	Tested 24 h	Reference for 72 h	Tested 72 h	Reference for 24 h	Tested 24 h	Reference for 72 h	Tested 72 h
8	7.59	6.87	6.74	6.40	112	132	117	118
	7.21	6.88	7.07	6.47	111	115	118	119
	7.65	6.90	7.55	6.66	112	117	118	121
	7.89	7.22	7.64	6.70	114	117	122	122
	7.93	7.41	7.79	6.95	117	117	122	122
	8.16	7.67	7.92	7.34	117	118	123	123
	8.36	8.02	7.94	7.68	119	119	123	124
	8.37	8.52	8.28	7.69	126	119	127	124
6	7.89	6.37	7.16	6.61	114	117	117	125
	7.35	6.44	7.21	6.81	116	118	118	116
	7.77	8.08	7.23	7.11	116	118	119	119
	8.11	8.19	7.46	7.45	120	120	124	122
	8.22	8.31	7.73	8.04	120	120	124	122
	8.25	8.36	7.82	8.32	124	121	125	123
5	7.02	7.85	6.64	6.87	119	124	117	119
	7.48	6.24	6.71	7.16	113	111	120	120
	7.52	7.70	6.72	7.33	113	114	121	121
	7.87	7.84	7.37	7.69	117	115	122	122
	8.20	7.84	7.68	8.22	122	127	124	127
4	6.98	7.66	6.94	6.40	117	116	122	118
	6.42	7.52	7.02	6.83	117	118	122	119
	8.39	7.55	7.09	6.97	119	119	124	120
	8.46	8.61	7.33	7.74	120	123	124	123
3	7.05	7.64	6.70	6.41	123	117	115	120
	6.39	8.54	6.84	7.54	116	125	127	122
	8.01	8.80	6.88	7.93	117	133	128	124

Table A3. Characteristics of the bioanalytical methods for the determination of the nine drugs used for retrospective analysis.

Drug	Method	Internal Standard	Low/High QC (ng/mL)	Type of Extraction	Source
Dutasteride	HPLC, ESI +	[$^{13}C_6$]-dutasteride	0.3/2.8	LLE	[24]
Eplerenon	HPLC-MS, ESI +	[$^2H_3$]-eplerenone	50/1500	LLE	[23]
Genistein	HPLC-MS, ESI −	[$^2H_4$]-genistein	50/2000	LLE	N/A
Ibuprofen	HPLC-UV, λ = 220 nm	naproxen	900/24,000	LLE	N/A
Imatinib	HPLC-UV, λ = 265 nm	propranolol hydrochloride	120/3200	LLE	[22]
Naproxen	HPLC-UV, λ = 265 nm	ibuprofen	1500/60,000	LLE	[20]
Olmesartan	HPLC-MS, ESI +	[$^2H_6$]-olmesarta	15/2000	LLE	[21]
Prasugrel	HPLC-MS/MS, ESI +	[$^{13}C_6$] R-138727	1.5/200	LLE	N/A
Pseudoephedrine	HPLC-MS/MS, ESI +	[$^2H_3$][$^{13}C_6$]-pseudoephedrine	4.5/240	LLE	N/A

LLE—liquid-liquid extraction; ESI—electrospray ionization; N/A—unpublished data.

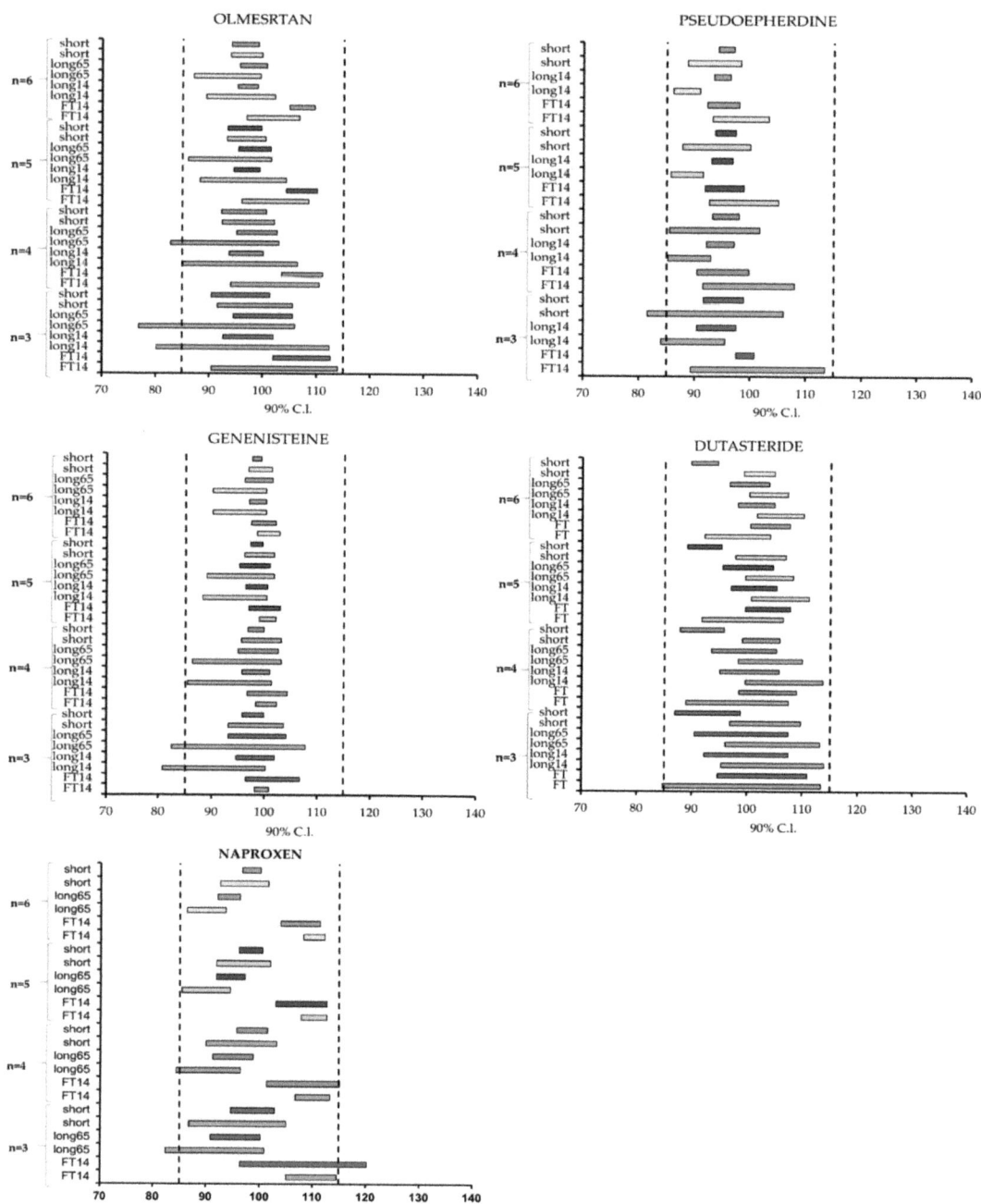

Figure A1. *Cont.*

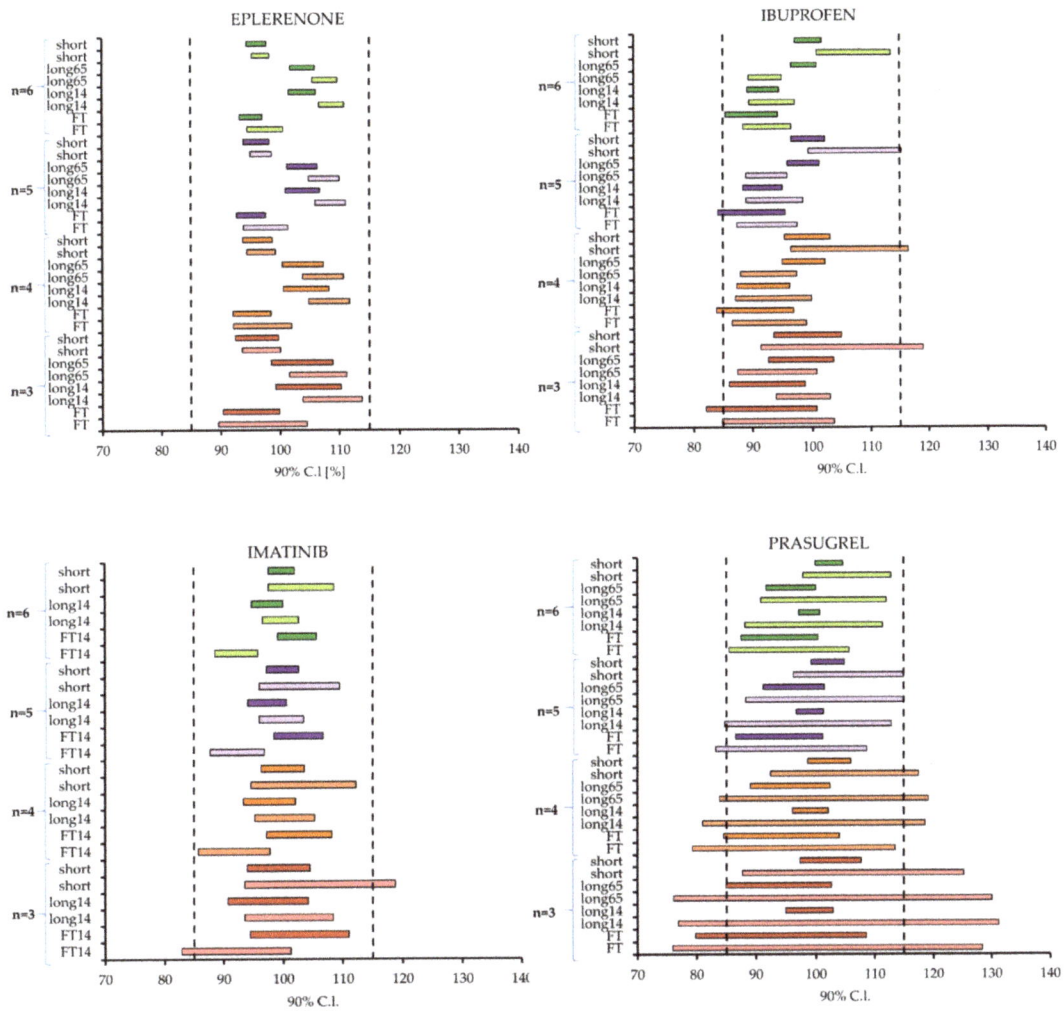

Figure A1. Retrospective study: 90% confidence intervals (90% C.I) for the stability of five drugs using n = 3, 4, 5, and 6. Abbreviations: Short—short-term stability, FT—freeze and thaw stability, long—long-term stability. Numbers 14 and 65 indicate the storage temperature of $-14\ °C$ and $-65\ °C$, respectively. Low QC—light color, high QC—intensive color.

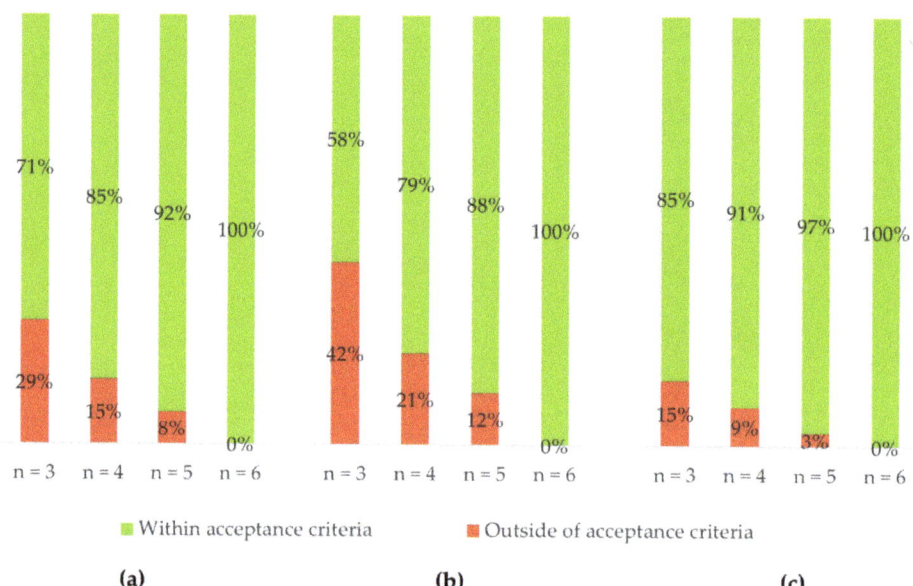

Figure A2. Retrospective study: Percentage of confidence intervals outside (red) and within (green) stability acceptance criteria for (**a**) all data, (**b**) low QC concertation, and (**c**) high QC concertation. Combined data from a retrospective study of nine drugs using n = 3, 4, 5, and 6.

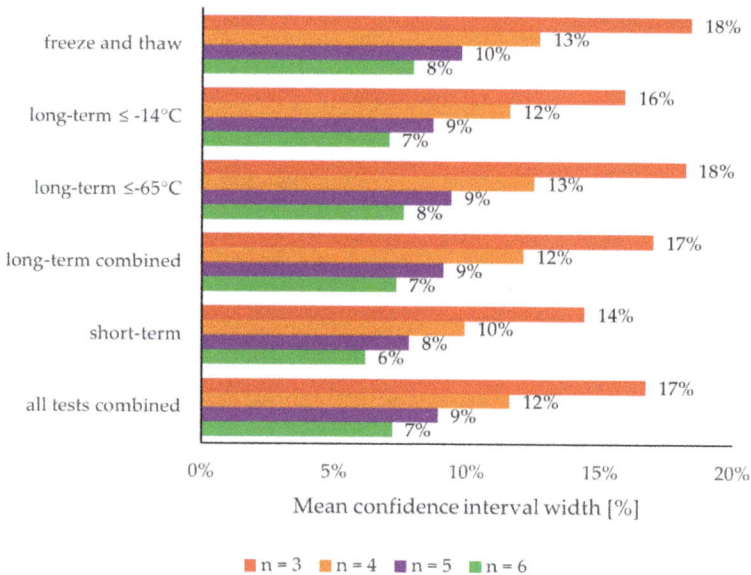

Figure A3. Retrospective study: Relation between mean confidence interval width and type of stability test.

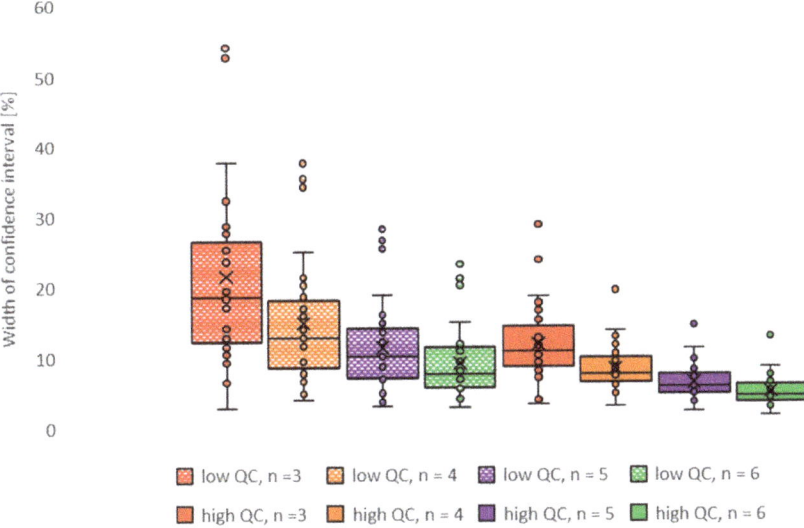

Figure A4. Retrospective study: box & whiskers plot; box indicates 2nd and 3rd quartile, whiskers indicate 1st and 4th quartile, points outside of whiskers indicate outliers.

Relation between precision in measurements and stability:

Stability S is determined as a ratio of two uncorrelated random variables X (tested samples) and Z (reference samples):

$$S = \frac{X}{Z}$$

Our goal is to derive the relation between the standard deviations in X and Z and the standard deviation in S. We start with linearization, which allows us to reformulate the Z variable as follows:

$$Z = \mu Z + \sigma_Z * \tilde{Z}$$

where $\tilde{Z}$ is the centralized Z variable (mean = 0, standard deviation = 1). Using linear approximation, we may obtain:

$$\frac{X}{\mu Z + \sigma_Z * \tilde{Z}} \approx \frac{X}{\mu Z} - \frac{1}{\mu Z^2}\sigma_Z * \tilde{Z} * X$$

Now:

$$\sigma_S^2 = E\left(\left(\frac{X}{Z}\right)^2\right) - \left(E\left(\frac{X}{Z}\right)\right)^2$$

$$\sigma_S^2 \approx E\left(\left(\frac{X}{\mu Z} - \frac{1}{\mu Z^2}\sigma_Z * \tilde{Z} * X\right)^2\right) - \left(E\left(\frac{X}{Z}\right)\right)^2$$

$$E\left(\left(\frac{X}{\mu Z} - \frac{1}{\mu Z^2}\sigma_Z * \tilde{Z} * X\right)^2\right) = \frac{1}{\mu Z^2}E(X)^2 - 2\frac{dZ}{\mu Z^2}E\left(X^2\tilde{Z}\right) + \frac{\sigma_Z^2}{\mu Z^4}E\left(X^2\tilde{Z}^2\right) = I + II + III$$

$$I. \quad E(X)^2 = \left(\sigma_X^2 + \mu X^2\right)$$

Variables are uncorrelated and expected value of $\widetilde{Z}$ is equal to 0:

$$II. \quad E\left(X^2\widetilde{Z}\right) = 0$$

Again, variables are uncorrelated and the standard deviation of $\widetilde{Z}$ is equal to 1:

$$III. \quad E\left(X^2\widetilde{Z}^2\right) \approx E\left(X^2\right)E\left(\widetilde{Z}^2\right) = \left(\sigma_X^2 + \mu X^2\right)$$

Using linearization, we can approximate:

$$E\left(\frac{X}{Z}\right) \approx E\left(\frac{X}{\mu Z} - \frac{1}{\mu Z^2}\sigma_Z * \widetilde{Z} * X\right) = \frac{\mu X}{\mu Z}$$

Finally:

$$\sigma_S^2 \approx \left(\sigma_X^2 + \mu X^2\right)\left(\frac{\sigma_Z^2}{\mu Z^4} + \frac{1}{\mu Z^2}\right) - \left(\frac{\mu X^2}{\mu Z^2}\right)$$

Probability for the confidence interval:

As demonstrated by Rudzki and Leś, measurements may follow a log-normal distribution [7]. In such a case, the confidence interval can be calculated using logarithmic transformation, which yields a normal distribution of the stability. In order to keep the model simple, from now on we will assume the normal distribution of the stability.

Let us denote the standard deviation in stability as s_S. Under the assumption o the normal distribution, the 90% confidence interval of stability has the following form:

$$CI = \mu_S \pm \frac{s_S\, k}{\sqrt{n}}$$

where:

μ_S is the mean value of stability and k is the value of the Student t-distribution quantile at a 0.1 significance level for $n-1$ degrees of freedom (df). In the presented work, we consider only stable analytes, i.e., $\mu_S = 100$. The probability that the confidence interval is in the 85–115% interval:

$$P(CI \subset [85;\ 115])$$

is equivalent to:

$$P\left(\frac{s_S\, k}{\sqrt{n}} < 15\right) = P\left(s_S^2 < \left(\frac{15\sqrt{n}}{k}\right)^2\right)$$

Assuming that the true standard deviation in stability is σ_S:

$$P\left(s_S^2 < \left(\frac{15\sqrt{n}}{k}\right)^2\right) = P\left(\frac{s_S^2(n-1)}{\sigma_S^2} < \frac{225\, n\, (n-1)}{\sigma_S^2 k^2}\right)$$

where:

$$\frac{s^2(n-1)}{\sigma_S^2} \sim chi^2(n-1)$$

As a result:

$$P(CI \subset [85;\ 115]) = \chi_{n-1}\left(\frac{225\, n\, (n-1)}{k^2\, \sigma_S^2}\right)$$

where χ_{n-1} is the cumulative distribution function of the chi-square distribution for $df = n - 1$.

References

1. *Guideline on Bioanalytical Method Validation*; EMEA/CHMP/EWP/192217/2009; Committee for Medicinal Products for Human Use (CHMP), European Medicines Agency: London, UK, 2011.
2. *Guidance for Industry: Bioanalytical Method Validation*; Food and Drug Administration; Center for Drug Evaluation and Research (CDER), Center for Veterinary Medicine (CVM): Rockville, MD, USA, 2018.
3. *Draft ICH Guideline M10 on Bioanalytical Method Validation*; EMA/CHMP/ICH/172948/2019; Committee for Human Medicinal Products, European Medicines Agency: London, UK, 2019.
4. Kaza, M.; Karaźniewicz-Łada, M.; Kosicka, K.; Siemiątkowska, A.; Rudzki, P.J. Bioanalytical method validation: New FDA guidance vs. EMA guideline. Better or worse? *J. Pharm. Biomed. Anal.* **2019**, *165*, 381–385. [CrossRef] [PubMed]
5. Health Canada's: Notice Clarification of Bioanalytical Method Validation Procedures. 2015. Available online: https://www.canada.ca/en/health-canada/services/drugs-health-products/drug-products/announcements/notice-clarification-bioanalytical-method-validation-procedures.html (accessed on 6 December 2021).
6. Timm, U.; Wall, M.; Dell, D. A New Approach for Dealing with the Stability of Drugs in Biological Fluids. *J. Pharm. Sci.* **1985**, *74*, 972–977. [CrossRef] [PubMed]
7. Rudzki, P.J.; Leś, A. Application of confidence intervals to bioanalytical method validation-drug stability in biological matrix testing. *Acta Pol. Pharm.* **2008**, *65*, 743–747. [PubMed]
8. *Guideline on the Investigation of Bioequivalence*; CPMP/EWP/QWP/1401/98/Rev. 1/Corr**; Committee for Human Medicinal Products, European Medicines Agency: London, UK, 2010.
9. Neyman, J. Outline of a theory of statistical estimation based on the classical theory of probability. *Philos. Trans. R. Soc. Lond. Ser. A Math. Phys. Sci.* **1937**, *236*, 333–380.
10. Rudzki, P.J.; Jarus-Dziedzic, K.; Filist, M.; Gilant, E.; Buś-Kwaśnik, K.; Leś, A.; Sasinowska-Motyl, M.; Nagraba, Ł.; Bujalska-Zadrożny, M. Evaluation of tramadol human pharmacokinetics and safety after co-administration of magnesium ions in randomized, single- and multiple-dose studies. *Pharmacol. Rep.* **2021**, *73*, 604–614. [CrossRef] [PubMed]
11. Filist, M.; Szlaska, I.; Kaza, M.; Pawiński, T. Validated HPLC-UV method for determination of naproxen in human plasma with proven selectivity against ibuprofen and paracetamol. *Biomed. Chromatogr.* **2016**, *30*, 953–961. [CrossRef] [PubMed]
12. Piórkowska, E.; Musijowski, J.; Buś-Kwaśnik, K.; Rudzki, P.J. Is a deuterated internal standard appropriate for the reliable determination of olmesartan in human plasma? *J. Chrom. B* **2017**, *1040*, 53–59. [CrossRef] [PubMed]
13. Kaza, M.; Piorkowska, E.; Filist, M.; Rudzki, P.J. HPLC-UV assay of imatinib in human plasma optimized for bioequivalence studies. *Acta Pol. Pharm.* **2016**, *73*, 1495–1503. [PubMed]
14. Buś-Kwaśnik, K.; Filist, M.; Rudzki, P.J. Environmentally friendly LC/MS determination of eplerenone in human plasma. *Acta Pol. Pharm.* **2016**, *73*, 1487–1493. [PubMed]
15. Gniazdowska, E.; Kaza, M.; Buś-Kwaśnik, K.; Giebułtowicz, J. LC-MS/MS determination of dutasteride and its major metabolites in human plasma. *J. Pharm. Biomed. Anal.* **2021**, *206*, 114362. [CrossRef] [PubMed]
16. Quaranta, L.; Micheletti, E.; Carassa, R.; Bruttini, C.; Fausto, R.; Katsanos, A.; Riva, I. Efficacy and Safety of PreserFlo® MicroShunt After a Failed Trabeculectomy in Eyes with Primary Open-Angle Glaucoma: A Retrospective Study. *Adv. Ther.* **2021**, *38*, 4403–4412. [CrossRef] [PubMed]
17. Wilde, H.; Dennis, J.M.; McGovern, A.P.; Vollmer, S.J.; Mateen, B.A. A national retrospective study of the association between serious operational problems and COVID-19 specific intensive care mortality risk. *PLoS ONE* **2021**, *16*, e0255377. [CrossRef] [PubMed]
18. Monakhova, Y.B.; Diehl, B.W.K. Retrospective multivariate analysis of pharmaceutical preparations using (1)H nuclear magnetic resonance (NMR) spectroscopy: Example of 990 heparin samples. *J. Pharm. Biomed. Anal.* **2019**, *173*, 18–23. [CrossRef] [PubMed]
19. Ko, Y.; Jeon, W.; Choi, Y.J.; Yang, H.; Lee, J. Impact of drug formulation on outcomes of pharmaceutical poisoning in children aged 7 years or younger: A retrospective observational study in South Korea. *Medicine* **2021**, *100*, e27485. [CrossRef] [PubMed]
20. Yoneyama, T.; Kudo, T.; Jinno, F.; Schmidt, E.R.; Kondo, T. Retrospective Data Analysis and Proposal of a Practical Acceptance Criterion for Inter-laboratory Cross-validation of Bioanalytical Methods Using Liquid Chromatography/Tandem Mass Spectrometry. *AAPS J.* **2014**, *16*, 1226–1236. [CrossRef] [PubMed]
21. Nowak, P.M.; Wietecha-Posłuszny, R.; Pawliszyn, J. White Analytical Chemistry: An approach to reconcile the principles of Green Analytical Chemistry and functionality. *TrAC Trends Anal. Chem.* **2021**, *138*, 116223. [CrossRef]
22. Watanabe, K.; Varesio, E.; Hopfgartner, G. Parallel ultra high pressure liquid chromatography–mass spectrometry for the quantification of HIV protease inhibitors using dried spot sample collection format. *J. Chrom. B* **2014**, *965*, 244–253. [CrossRef] [PubMed]
23. Pihl, S.; Huusom, A.K.T.; Rohde, M.; Poulsen, M.N.; Jørgensen, M.; Kall, M.A. Evaluation of an isochronic study design for long-term frozen stability investigation of drugs in biological matrices. *Bioanalysis* **2010**, *2*, 1041–1049. [CrossRef] [PubMed]
24. Bourgogne, E.; Mathy, F.X.; Boucaut, D.; Boekens, H.; Laprevote, O. Simultaneous quantitation of histamine and its major metabolite 1-methylhistamine in brain dialysates by using precolumn derivatization prior to HILIC-MS/MS analysis. *Anal. Bioanal. Chem.* **2012**, *402*, 449–459. [CrossRef] [PubMed]
25. Wilson, A.; Barker, S.; Freisleben, A.; Laakso, S.; Staelens, L.; White, S.; Timmerman, P. European Bioanalysis Forum recommendation on the best practices to demonstrate processed sample stability. *Bioanalysis* **2019**, *11*, 7–11. [CrossRef] [PubMed]

Article

Efficient Heparin Recovery from Porcine Intestinal Mucosa Using Zeolite Imidazolate Framework-8

Mahmood Karimi Abdolmaleki [1,*], Deepak Ganta [2], Ali Shafiee [3], Carlo Alberto Velazquez [1] and Devang P. Khambhati [1]

[1] Department of Biology and Chemistry, Texas A&M International University, Laredo, TX 78041, USA; carlovelazquez@dusty.tamiu.edu (C.A.V.); devang.khambhati@tamiu.edu (D.P.K.)
[2] School of Engineering, Texas A&M International University, Laredo, TX 78041, USA; deepak.ganta@tamiu.edu
[3] Department of Chemistry, Cape Breton University, Sydney, NS B1P 6L2, Canada; ali_shafiee@cbu.ca
* Correspondence: mahmood.abdolmaleki@tamiu.edu

Abstract: Heparin is one of the most valuable active pharmaceutical ingredients, and it is generally isolated from porcine intestinal mucosa. Traditionally, different types of commercial resins are employed as an adsorbent for heparin uptake; however, using new, less expensive adsorbents has attracted more interest in the past few years to enhance the heparin recovery. Zeolite imidazolate framework-8 (ZIF-8), as a metal–organic framework (MOF) with a high surface area, porosity, and good stability at high temperatures, was selected to examine the heparin recovery. In this research, we demonstrate that ZIF-8 can recover up to ~70% (37 mg g^{-1}) of heparin from porcine intestinal mucosa. A mechanistic study through kinetic and thermodynamic models on the adsorption revealed appropriate surface conditions for the adsorption of heparin molecules. The effect of different variables such as pH and temperature on heparin adsorption was also studied to optimize the recovery. This study is the first to investigate the usage of MOFs for heparin uptake.

Keywords: heparin; metal–organic framework; zeolite imidazolate framework-8; kinetic; thermodynamic

1. Introduction

Heparin (Scheme 1) is a linear, sulfur-rich polysaccharide in the glycosaminoglycan family with varying lengths and weights ranging from 2000–40,000 kDa; it is generally procured from the bovine and porcine intestinal mucosa. Heparin is predominantly used as an anticoagulant and antithrombotic agent to treat a variety of medical conditions, such as systemic embolism syndrome and deep vein thrombosis; however, it also has applications as an anti-inflammatory and antiviral drug. Furthermore, heparin has been demonstrated to inhibit cancer cell growth and delay the onset of the detrimental symptoms of Alzheimer's disease [1–4].

Heparin is isolated from porcine intestinal mucosa through a multi-step adsorption/desorption process after digestion has occurred; numerous parameters, namely, adsorbent surface properties, porosity, and thermodynamic and chemical stability, must be considered to securely deliver the highest amount of adsorbed heparin in the final step after elution [5–8]. In addition, the adsorbent needs to be easily packable (for continuous systems) or easy to separate (for batch systems). To date, several resin beads, such as Amberlite [5], Dowex [8], Lewatit [9], and DEAE [10], have been designed and commercially utilized to recover heparin from porcine intestinal mucosa. Although effective, the methods by which these materials are synthesized are expensive and, in some cases, not environmentally friendly. Recent research has shown the high efficiency and effectiveness of different nanomaterials in the adsorption/desorption of various pharmaceuticals from different matrices [11–15]. Among these nanomaterials, quaternized chitosan/polystyrene microbeads [16], magnetic clay nanotubes [17], cross-linked spherical polycationic [18], and quaternary ammonium-functionalized halloysite nanotubes [19] have especially been designed for the recovery of heparin.

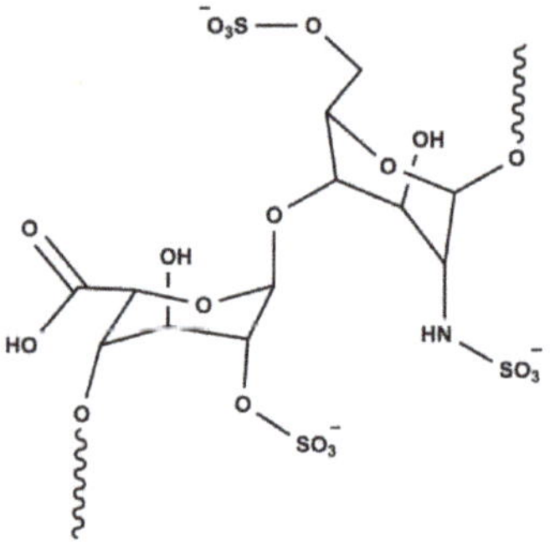

Scheme 1. Heparin structure.

Metal–organic frameworks (MOFs) are 3D nanoporous materials with an ultra-high surface area (exceeding 6000 m^2/g), porosity (up to 90% free volume), and thermodynamic/chemical stability. Moreover, their fabrication methods are simple, inexpensive, and environmentally friendly [20,21], making them suitable for commercial use [22–26]. Zeolite imidazole frameworks (ZIFs) are a new class of MOFs consisting of imidazolate linkers and metal ions. Due to their high surface area, pore volume, thermal and chemical stability, and adaptable functionalization, ZIFs have found applications in various fields as sensors, drug delivery agents, gas separation agents, and electronic device components [27–31].

In this research, we report the recovery of heparin from the porcine intestinal mucosa using two types of ZIFs (ZIF-8 and ZIF-67) with different surface areas. Our results demonstrated that the ZIF with a higher surface area (ZIF-8) showed a higher amount of heparin recovery. We also evaluated the effects of several variables, such as pH and temperature, to investigate the most optimal conditions for heparin recovery. Our results further illustrated that the heparin recovery with ZIF-8 stabilized after five cycles. We also utilized a sheep plasma test to measure and compare the anticoagulant potency of ZIF-8 with commercial Amberlite FPA98 Cl, the results of which confirmed the significant potential of MOFs in the heparin uptake industry.

2. Result and Discussion

2.1. ZIF-8 and ZIF-67 Characterization

Scanning electron microscope (SEM) images of ZIF-8 and ZIF-67 were taken at the ACS Materials Facility using a Hitachi S-4800 scanning electron microscope (SEM) at 10 kV. The SEM images are shown in Figure 1. The infrared spectrum of ZIF-8 is shown in Figure 2.

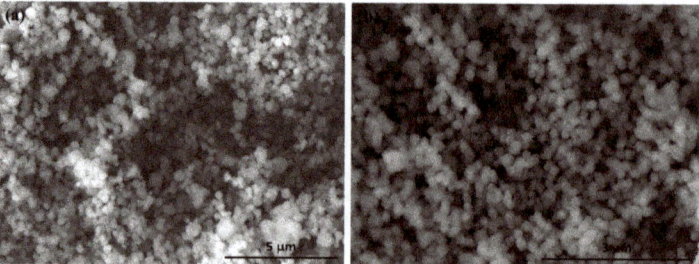

Figure 1. SEM images of (**a**) ZIF-67 (8000×) and (**b**) ZIF-8 (18,000×) nanomaterials.

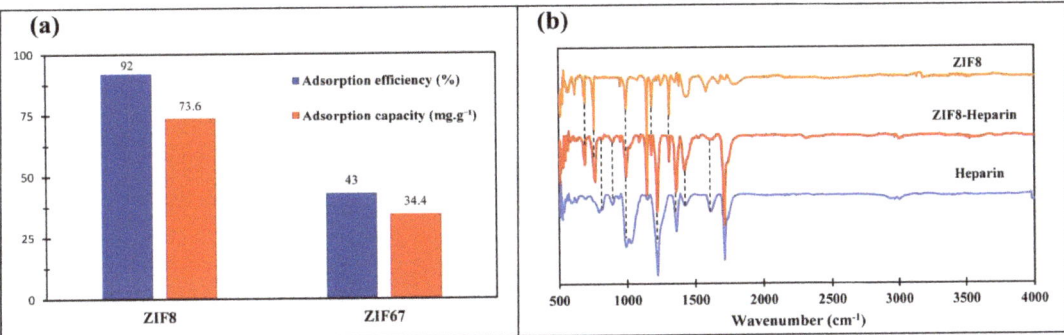

Figure 2. (**a**) Adsorption efficiency of heparin over ZIF-8 vs. ZIF-67 (conditions: 50 mL of 1000 mg L^{-1} heparin standard solution and 0.5 g of the adsorbent); (**b**) FTIR spectra of standard heparin, ZIF-8, and preabsorbed heparin on the ZIF-8.

2.2. Heparin Adsorption Studies

Prior to optimization, it was important to evaluate the affinity of heparin toward each absorbent in a pure heparin solution. The results (Figure 2a) demonstrated that ZIF-8 could adsorb more than 2-fold more heparin compared to ZIF-67 in a pure aqueous solution. The reported BET surface areas for ZIF-8 and ZIF-67 are >1300 and ~316 m^2/g, respectively [31,32], which are well correlated with their differences in adsorption efficiency. Due to the superior adsorption capabilities of ZIF-8 over ZIF-67, ZIF-8 was selected for further experiments to evaluate adsorption in real samples containing porcine intestinal mucosa. The FTIR spectrum of ZIF-8 shows a band at 1712 cm^{-1}, which confirms the presence of C=N bond stretching. The bands at 1100 cm^{-1} and 990 cm^{-1} are attributed to C-N stretching. Additionally, a preabsorbed sample of ZIF-8 was dried and analyzed via FTIR spectroscopy so as to ensure that heparin adsorption occurred (Figure 2b). The resulting spectrum was compared with the FTIR spectrum of pure sodium heparin salt and the FTIR spectrum of ZIF-8 powder. The appearance of a band at 1712 cm^{-1} in the ZIF-8–heparin mixture is attributed to the C=O group of heparin, and the appearance of a band at 1222 cm^{-1} is attributed to the S=O group of heparin, which confirms the adsorption of heparin on the surface of ZIF-8.

2.2.1. Effect of pH

One of the most important parameters to consider when studying any adsorption/ desorption system is pH, as pH fluctuations can have significant impacts on adsorption efficiency. For this reason, the effect of pH on the adsorption of heparin onto ZIF-8 was investigated by changing the pH of the real sample of the heparin solution from 3 to 11 in a series of experiments; adsorption efficiency and capacity were then evaluated. The results (Figure 3a) indicate that heparin adsorption increases when increasing the pH from 3 to 8. The highest adsorption was observed at a pH of 8, where an efficiency of 57.8% and a capacity of 30.4 mg g^{-1} were recorded; efficiency and capacity subsequently remained constant as more alkaline conditions were induced. Notably, increasing pH slightly deprotonates heparin, resulting in more negative charges that make it more receptive to the positive active sites of ZIF-8. We selected pH = 8 as the optimal pH since the stability of MOFs subsequently decreases at higher pH conditions.

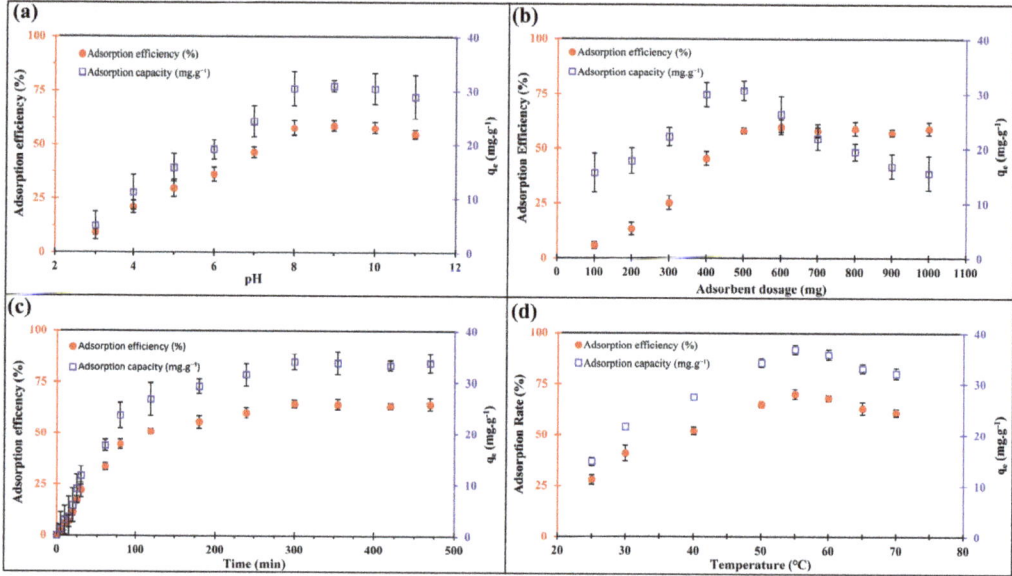

Figure 3. (**a**) Effect of pH (conditions: 50 mL of 1000 mg L^{-1} real sample, 0.5 g of ZIF-8, temperature 50 °C, and 240 min of contact time), (**b**) adsorbent dosage (conditions: 50 mL of 1000 mg L^{-1} real sample at pH = 8, temperature 50 °C, and 240 min of contact time), (**c**) contact time (conditions: 50 mL of 1000 mg L^{-1} real sample at pH = 8, 0.5 g of ZIF-8, and temperature 50 °C), and (**d**) temperature (conditions: 50 mL of 1000 mg L^{-1} real sample at pH = 8, 0.5 g of ZIF-8, and 300 min of contact time) of the adsorption efficiency of heparin over ZIF-8.

2.2.2. Effect of the Adsorption Dosage

ZIF-8 showed a remarkable capacity in the removal of contaminants. In this part of the study, we investigated the effect of the adsorbent dosage of ZIF-8 on heparin uptake by varying the amount of ZIF-8 from 100 to 1000 mg. The results (Figure 3b) indicated that increasing the adsorbent dosage to 500 mg caused the adsorption efficiency and capacity to increase to 58.2% and 30.6 mg g^{-1}, respectively. Increasing the dosage of the adsorbent further led to a slight decrease in the capacity while efficiency remained constant. Notably, a higher adsorbent dosage increased the adsorption due to a higher surface area and the presence of more active sites on ZIF-8 adsorbents; the adsorption capacity, however, decreased at dosages higher than 500 mg. This phenomenon is attributed to the interference of other components in the sample, namely, dermatan and chondroitin sulfate, which compete with heparin and prevent it from being adsorbed onto the surface of ZIF-8. Furthermore, the aggregation of ZIF-8 at a higher dosage may also serve to decrease the active sites of the adsorbents for heparin molecules. In addition, the observed decrease in the adsorption capacity from 35 to 15 mg g^{-1} is the result of increasing the mass (m) in Equation (4) [17].

2.2.3. Adsorption Time and Temperature Effects

The results from the contact time experiment (Figure 3c) indicated that when increasing the contact time, the adsorption efficiency and capacity increased to 64.5% and 33.9 mg g^{-1}, respectively, after 300 min of contact time. This trend was anticipated due to the length of time that it takes heparin molecules to diffuse from the solution and bond to ZIF-8 adsorbent surfaces. As observed in the experiment, adsorption approached stability after that, revealing that ZIF-8 had reached its full capacity at these experimental conditions. In addition to contact time, the solution temperature is another highly influential parameter in adsorption systems. Therefore, to study its effect on heparin uptake, a series of experiments were performed by varying the temperature from 25 °C to 75 °C while keeping all other parameters constant. The results (Figure 3d) demonstrated that by increasing the temperature to 55 °C, the adsorption efficiency and capacity increased to 70% and 36.8 mg g^{-1}, respectively. By increasing the temperature, the heparin molecules can move more quickly in the solution due to a decrease in the solution's viscosity, which facilitates their interaction with the adsorbent surface. The minimal decrease observed after 65 °C is likely due to the decomposition of heparin molecules from temperatures that were too high to support its structural stability.

2.2.4. Kinetic and Thermodynamic Studies

In order to better comprehend the observed adsorption phenomena of ZIF-8 and its associated mechanisms, the collected data were analyzed via the following kinetic models: pseudo-first-order, pseudo-second-order, intraparticle diffusion, and Elovich (Equations (5)–(8); Figure 4 and Table 1). The obtained data, which are presented in Figure 4 and Table 1, show that the adsorption of heparin onto ZIF-8 best fits the pseudo-second-order kinetic model since the R^2 value is much closer to 1.0 and $q_{e(cal)}$ is closest to $q_{e(exp)}$ out of all the examined models. Based upon this model, the rate-limiting step is the surface adsorption that involves chemisorption; therefore, the adsorption process is not limited by the concentration of heparin but rather the adsorption capacity of ZIF-8 [32]. This confirms the decrease in heparin uptake with increasing adsorption dosages (Figure 3b). Further, the Elovich model also exhibits a good fit, supporting the adsorption of heparin on the ZIF-8 heterogeneous surface since this model assumes that the rate of the adsorption of the solute decreases exponentially as the amount of adsorption increases.

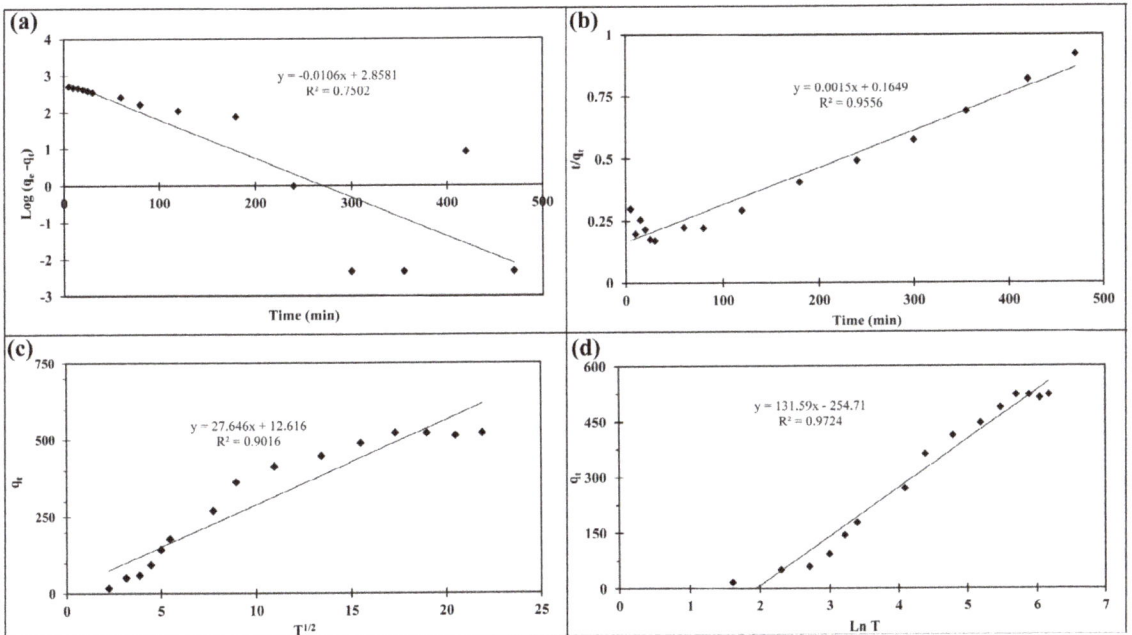

Figure 4. Kinetic study of heparin adsorption over ZIF-8 via (**a**) pseudo-first-order, (**b**) pseudo-second-order, (**c**) intraparticle diffusion, and (**d**) Elovich models.

Table 1. Calculated parameters from different kinetic study models of heparin adsorption over ZIF-8.

Model	Parameter	Value
Pseudo-first-order	$K_1\ (\min^{-1})$	0.024
	$q_{e\ (cal)}\ \left(\frac{mg}{g}\right)$	721.31
	R^2	0.8502
Pseudo-second-order	$K_2\ \left(\frac{g}{mg\ min}\right)$	1.34×10^{-5}
	$q_{e\ (cal)}\ \left(\frac{mg}{g}\right)$	672.22
	R^2	0.9556
Intraparticle Diffusion	$K_{diff}\ \left(\frac{mg}{g\ min^{0.5}}\right)$	27.64
	$C\ \left(\frac{mg}{g}\right)$	12.61
	R^2	0.901
Elovich	$\beta\ \left(\frac{g}{mg}\right)$	0.0076
	$\alpha\ \left(\frac{mg}{g\ min}\right)$	1.58
	R^2	0.9724
	$q_{e\ (exp)}$	522.11

The thermodynamic parameters of heparin adsorption on ZIF-8, including entropy change (ΔS), enthalpy change (ΔH), Gibbs free energy change (ΔG; calculated from ΔG = ΔH − TΔS), and activation energy (Ea), were calculated by using Equations (1) and (2) and plotting lnKc versus 1/T and ln(1 − θ) versus 1/T [33,34]. These data are presented in Figure 5 and Table 2.

$$\ln K_c = \frac{\Delta S}{R} - \frac{\Delta H}{RT} \quad (1)$$

$$\ln(1 - \theta) = \ln S^* + \frac{E_a}{RT} \quad (2)$$

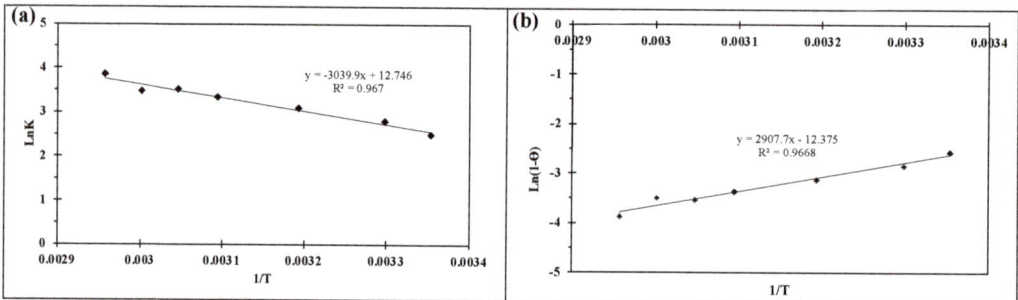

Figure 5. Thermodynamic study of heparin adsorption over ZIF-8 via (**a**) lnK vs. 1/T and (**b**) ln(1 − θ) versus 1/T.

Table 2. Calculated thermodynamic parameters of heparin adsorption over ZIF-8.

ΔH° (J/mol·K)	ΔS° (KJ/mol)	E_a (KJ/mol)	Temperature (K)						
			298	303	313	323	328	333	338
			ΔG° (KJ/mol)						
25.27	105.97	19.41	−31.57	−32.10	−33.16	−34.22	−34.74	−35.27	−35.80

The positive values of ΔH and ΔS indicate the endothermic nature of the adsorption process and the high affinity of heparin molecules toward ZIF-8. The positive value of E_a also further confirms the endothermic nature of the adsorption process, while the negative values of ΔG reflect the feasibility of the process and its spontaneous nature [35].

2.2.5. Sorbent Reusability

ZIF-8 showed impressive performance in heparin uptake in real biological samples, with a ~70% adsorption rate and a 37 mg g^{-1} capacity under optimized conditions. In order to evaluate the industrial potential of ZIF-8, we tested the stability and reusability of ZIF-8 through five adsorption/desorption cycles and harsh regeneration conditions. The results (Figure 6) indicate that the ZIF-8 adsorbent has high stability and reusability even after five adsorption/desorption cycles and exposure to harsh regeneration conditions.

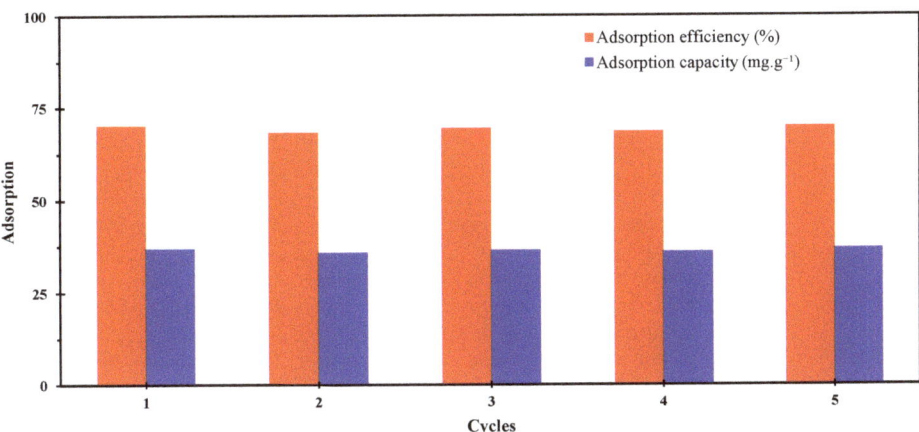

Figure 6. Stability test of heparin recovery over ZIF-8.

2.2.6. Sheep Plasma Clotting Assay

The porcine intestinal mucosa solution used for the sheep plasma clotting assay was prepared via a previously established and reported method [18]. The desired amount of adsorbent was mixed with the mucosa, followed by the stirring of the mixture at 55 °C for 3 h. Lastly, the adsorbent was filtered and washed with Milli-Q water. The solid heparin was then eluted from the adsorbent as described in a previous publication [18]. A similar method was employed for the commercial Amberlite resin. The anticoagulant potency of the recovered heparin from both ZIF-8 and the Amberlite resin was measured and compared [19]. The results indicated that the total activity of the eluted heparin utilizing the ZIF-8 adsorbent was 64 ± 1.8 U per gram of mucosa compared to 59 ± 2.6 U per gram of mucosa from the commercial Amberlite FPA98 Cl resin. These results highlight the promising potential of the ZIF-8 adsorbent in an industrial setting.

3. Experimental Procedures

3.1. Materials

Heparin sodium salt of analytical grade was purchased from VWR, USA. Commercially synthesized ZIF-8 and ZIF-67 were purchased from ACS Material, LLC (Pasadena, CA, USA). Milli-Q water was used for the preparation of all solutions and the adsorption experiments. Porcine intestinal mucosa (heparin content of ~1300 mg L^{-1}) was purchased from a local market. The subtilisin enzyme was purchased from STERM Company, Newburyport, MA, USA. Commercial Amberlite FPA98 Cl resin was purchased from Dow Chemical (Midland, MI, USA), USA. Sheep plasma, hydrochloric acid (37%), calcium chloride, analytical grade methanol, sodium hydroxide, and denatured ethanol were purchased from VWR and used as received. Heparin ELISA kits were purchased from MyBiosource, San Diego, CA, USA.

3.2. Instrumentation

A Spectrostar Nano microplate reader (BMG LABTECH, Cary, NC, USA) was used to measure the heparin concentration in both pure and real samples. MOFs, heparin sodium salt, and MOF–heparin were characterized by a Shimadzu IR Affinity FTIR instrument (Shimadzu Scientific Instruments, Columbia, MD, USA).

3.3. Solutions and Methods

A heparin stock solution (1000 ppm) was prepared by adding the required amount of heparin salt to Milli-Q water and used for standard experiments. This solution was used to investigate the adsorption capacity of each adsorbent in the standard condition. For this, 0.5 g of ZIF-8 and ZIF-67 were added to separate solutions, the two mixtures were stirred for 5 h at 55 °C in an incubator, and the adsorbed heparin was calculated for both mixtures by using the methylene blue method according to the previously published procedure [18,19]. Specifically, to examine the material's absorption capabilities, 250 mg of the MOFs was added to a 25 mL aqueous solution of 1000 ppm heparin sodium. In the incubator, the solution was stirred for 3 h at 55 °C. Five milliliters of the solution was removed and filtered using a syringe filter, and the concentration of heparin in the supernatant was measured using the methylene-blue-assisted spectrophotometric method. Methylene blue (MB) is a cationic metachromatic dye that can bind to heparin. First, 1 mL of the heparin solution was added to 1 mL of MB solution (10 mg L^{-1}). Then, the solution was mixed using a vortex mixer for 10 min, and the absorbance was measured by a plate reader based on the intensity of the band at λ_{max} = 663 nm, which is attributed to the free MB. When MB forms a dimer with heparin, the intensity of the band at 663 nm decreases due to the lower concentration of free MB. The concentration of heparin after adsorption was calculated using the pre-plotted standard calibration curve.

The adsorption efficiency of heparin on the ZIFs and the anticoagulant potency of the eluted heparin were evaluated in porcine intestinal mucosa by using sheep plasma and heparin ELISA kits. The detailed adsorption process can be found in our previous publication [18].

The adsorption efficiency and capacity were calculated via Equations (3) and (4):

$$\text{Adsorption efficiency (\%)} = \frac{(C_0 - C_e)}{C_0} \times 100 \quad (3)$$

$$\text{Adsorption capacity } (q_e) = \frac{(C_0 - C_e)}{m} \times V \quad (4)$$

where C_0 and C_e (mg L^{-1}) are the initial and equilibrium heparin concentrations (measured with the ELISA kit), V (L) is the volume of the mucosa solution that was used for heparin adsorption, and m is the mass of the adsorbent used (g). The experimental conditions of the adsorption were optimized through different influential parameters, such as pH, contact time, and temperature. Furthermore, in order to understand the mechanism of adsorption and also the commercial viability of the adsorbent, other parameters such as adsorbent dosage, kinetic and thermodynamic properties, and the reusability of the absorbent material were investigated. All experiments were carried out in triplicate, and the error bars in the figures indicate the standard deviation of each experiment.

Pseudo-first-order, pseudo-second-order, intraparticle diffusion, and Elovich models were calculated to have a better understanding of the adsorption phenomena. The linearized equation of each model is given in Equations (3)–(6), respectively. The pseudo-first-order model (Equation (5)) considers a proportional relation between the occupation of adsorption sites and the number of unoccupied sites.

$$\log(q_e - q_t) = \log q_e - \frac{K_1}{2.303} \times t \tag{5}$$

In Equation (5), the value of K_1 (min^{-1}; the rate constant of the second-order equation) and q_e can be obtained from the slope and intercept of the linear plot of $\log(q_e - q_t)$ versus t, respectively, where q_e (mg g^{-1}) and q_t (mg g^{-1}) are the adsorbed heparin at equilibrium and at any time t, respectively [36].

The pseudo-second-order kinetic model assumes that the adsorption rate is not dependent on the heparin adsorption but the adsorption capacity of ZIF-8.

$$\frac{t}{q_t} = \frac{1}{K_2 \times q_e^2} + \frac{t}{q_e} \tag{6}$$

In Equation (6), K_2 (g mg^{-1} min^{-1}) and q_e (mg g^{-1}) are the rate constant of the second-order equation and the maximum adsorption capacity, respectively, which can be calculated from the slope and intercept of the plot of t/q_t vs. t [37–39].

In the intraparticle diffusion model (Equation (7)), the limiting steps are considered to be mass transport.

$$q_t = K_{diff} \times t^{1/2} + C \tag{7}$$

In this model (Equation (8)), the values of K_{diff} (the rate constant of intraparticle diffusion (mg g^{-1} min^{-1})) and C (thickness of boundary) are calculated from the slope and intercept of the plot of q_t versus $t^{1/2}$ [40].

The Elovich kinetic model (Equation (4)) considers the sorbent to have heterogeneous active sites with varying active energies during the adsorption process.

$$q_t = \frac{1}{\beta \ln(\alpha\beta)} + \frac{1}{\beta \ln(t)} \tag{8}$$

In this model (Equation (8)), the values of α and β (Elovich constants) can be calculated from the slope and intercept of the plot of q_t versus $\ln(t)$ [41].

To evaluate the stability of the ZIF-8 adsorbent, after the first adsorption/desorption cycle, ZIF-8 adsorbent was regenerated by washing with a saturated NaCl solution for 3 h at 50 °C following the wash with Milli-Q water. The solution was tested with absorption spectroscopy to ensure that no heparin was carried to the next cycle. The same process was repeated across all five cycles [17].

4. Conclusions

Here, we report the use of ZIFs for heparin recovery, which is the first example of heparin recovery utilizing an MOF. To the best of our knowledge, these MOFs have been extensively used for various applications but not for heparin recovery. We employed two different MOFs with different surface areas, and the results met our expectations. ZIF-8, an MOF with a larger surface area than that of ZIF-67, proved to be more effective at adsorbing heparin, with approximately 70% adsorption capacity. Furthermore, our results revealed that the heparin recovered from ZIF-8 had higher total anticoagulant activity compared to the heparin recovered from the commercially available Amberlite FPA98 Cl resin (sheep plasma clotting assay). The mechanistic adsorption study demonstrated that the adsorption process adheres to the pseudo-second-order kinetic model and is both thermodynamically endothermic and feasible. We also observed the effects of different variables, such as pH, temperature, and adsorbent dosage, on heparin recovery so as to determine the most optimal conditions for heparin uptake.

Author Contributions: M.K.A.: supervision, conceptualization, methodology, validation, investigation, data curation, visualization, writing—original draft, writing—review and editing, and funding acquisition. D.G.: supervision, writing—original draft, writing—review and editing, visualization, validation, investigation, and data curation. A.S.: writing—original draft, writing—review and editing, visualization, methodology, validation, investigation, and data curation. C.A.V.: methodology, validation, investigation, data curation, and writing—review and editing. D.P.K.: methodology, validation, investigation, and data curation. All authors have read and agreed to the published version of the manuscript.

Funding: This research and the APC was funded by Welch Grant number BS-0051 and Texas A&M International University-University Research Grant.

Acknowledgments: The authors would like to thank Texas A & M International University, Department of Biology and Chemistry, for their generous support.

Conflicts of Interest: The authors declare no conflict of interest.

Sample Availability: Not applicable.

References

1. Rodriguez-Torres, M.D.P.; Acosta-Torres, L.S.; Díaz-Torres, L.A. Heparin-based nanoparticles: An overview of their applications. *J. Nanomater.* **2018**, *2018*, 9780489. [CrossRef]
2. Scott, L.J.; Perry, C.M. Tramadol: A review of its use in perioperative pain. *Drugs* **2000**, *60*, 139–176. [CrossRef] [PubMed]
3. Turpie, A.G.G.; Hirsh, J. Heparin. *Nova Scotia Med. Bull.* **1979**, *58*, 25–29. [CrossRef]
4. Urbinati, C.; Milanesi, M.; Lauro, N.; Bertelli, C.; David, G.; D'Ursi, P.; Rusnati, M.; Chiodelli, P. HIV-1 tat and heparan sulfate proteoglycans orchestrate the setup of in cis and in trans cell-surface interactions functional to lymphocyte trans-endothelial migration. *Molecules* **2021**, *26*, 7488. [CrossRef] [PubMed]
5. Vreeburg, J.-W.; Baauw, A. Method for Preparation of Heparin from Mucosa. Patent No. WO2010110654A12010, 30 September 2010.
6. Lee, M.S.; Kong, J. Heparin: Physiology, pharmacology, and clinical application. *Rev. Cardiovasc. Med.* **2015**, *16*, 189–199. [CrossRef]
7. Anderson, J.A.M.; Saenko, E.L. Heparin resistance. *Br. J. Anaesth.* **2002**, *88*, 467–469. [CrossRef]
8. Flengsrud, R.; Larsen, M.L.; Ødegaard, O.R. Purification, characterization and in vivo studies of salmon heparin. *Thromb. Res.* **2010**, *126*, e409–e417. [CrossRef]
9. Linhardt, R.J.; Ampofo, S.A.; Fareed, J.; Hoppensteadt, D.; Mulliken, J.B.; Folkman, J. Isolation and characterization of human heparin. *Biochemistry* **1992**, *31*, 12441–12445. [CrossRef]
10. Hoke, D.E.; Carson, D.D.; Höök, M. A heparin binding synthetic peptide from human HIP/RPL29 fails to specifically differentiate between anticoagulantly active and inactive species of heparin. *J. Negat. Results Biomed.* **2003**, *2*, 1. [CrossRef]
11. Shafiee, A.; Aibaghi, B.; Zhang, X. Reduced graphene oxide-cadmium sulfide quantum dots nanocomposite based dispersive solid phase microextraction for ultra-trace determination of carbamazepine and phenobarbital. *J. Braz. Chem. Soc.* **2021**, *32*, 833–841. [CrossRef]
12. Shafiee, A.; Aibaghi, B.; Zhang, X. Determination of ethambutol in biological samples using graphene oxide based dispersive solid-phase microextraction followed by ion mobility spectrometry. *Int. J. Ion Mobil. Spectrom.* **2020**, *23*, 19–27. [CrossRef]
13. Reddy, D.H.K.; Yun, Y.-S. Spinel ferrite magnetic adsorbents: Alternative future materials for water purification? *Coord. Chem. Rev.* **2016**, *315*, 90–111. [CrossRef]
14. Ghoraba, Z.; Aibaghi, B.; Soleymanpour, A. Application of cation-modified sulfur nanoparticles as an efficient sorbent for separation and preconcentration of carbamazepine in biological and pharmaceutical samples prior to its determination by high-performance liquid chromatography. *J. Chromatogr. B Anal. Technol. Biomed. Life Sci.* **2017**, *1063*, 245–252. [CrossRef]
15. de Fátima Alpendurada, M. Solid-phase microextraction: A promising technique for sample preparation in environmental analysis. *J. Chromatogr. A* **2000**, *889*, 3–14. [CrossRef]
16. Eskandarloo, H.; Godec, M.; Arshadi, M.; Padilla-Zakour, O.I.; Abbaspourrad, A. Multi-porous quaternized chitosan/polystyrene microbeads for scalable, efficient heparin recovery. *Chem. Eng. J.* **2018**, *348*, 399–408. [CrossRef]

17. Arshadi, M.; Eskandarloo, H.; Enayati, M.; Godec, M.; Abbaspourrad, A. Highly water-dispersible and antibacterial magnetic clay nanotubes functionalized with polyelectrolyte brushes: High adsorption capacity and selectivity toward heparin in batch and continuous system. *Green Chem.* **2018**, *20*, 5491–5508. [CrossRef]
18. Enayati, M.; Abdolmaleki, M.K.; Abbaspourrad, A. Synthesis of cross-linked spherical polycationic adsorbents for enhanced heparin recovery. *ACS Biomater. Sci. Eng.* **2020**, *6*, 2822–2831. [CrossRef]
19. Eskandarloo, H.; Arshadi, M.; Enayati, M.; Abbaspourrad, A. Highly efficient recovery of heparin using a green and low-cost quaternary ammonium functionalized halloysite nanotube. *ACS Sustain. Chem. Eng.* **2018**, *6*, 15349–15360. [CrossRef]
20. Zhou, H.-C.; Long, J.R.; Yaghi, O.M. Introduction to metal–organic frameworks. *Chem. Rev.* **2012**, *112*, 673–674. [CrossRef]
21. Aunan, E.; Affolter, C.W.; Olsbye, U.; Lillerud, K.P. Modulation of the thermochemical stability and adsorptive properties of MOF-808 by the selection of non-structural ligands. *Chem. Mater.* **2021**, *33*, 1471–1476. [CrossRef]
22. Luo, Y.; Estudillo-Wong, L.A.; Cavillo, L.; Granozzi, G.; Alonso-Vante, N. An easy and cheap chemical route using a MOF precursor to prepare Pd–Cu electrocatalyst for efficient energy conversion cathodes. *J. Catal.* **2016**, *338*, 135–142. [CrossRef]
23. Wang, C.; Kim, J.; Tang, J.; Na, J.; Kang, Y.; Kim, M.; Lim, H.; Bando, Y.; Li, J.; Yamauchi, Y. Large-scale synthesis of MOF-derived superporous carbon aerogels with extraordinary adsorption capacity for organic solvents. *Angew. Chem. Int. Ed.* **2020**, *59*, 2066–2070. [CrossRef] [PubMed]
24. Faust, T. MOFs move to market. *Nat. Chem.* **2016**, *8*, 990–991. [CrossRef] [PubMed]
25. Rubio-Martinez, M.; Avci-Camur, C.; Thornton, A.W.; Imaz, I.; Maspoch, D.; Hill, M.R. New synthetic routes towards MOF production at scale. *Chem. Soc. Rev.* **2017**, *46*, 3453–3480. [CrossRef] [PubMed]
26. Ryu, U.; Jee, S.; Rao, P.C.; Shin, J.; Ko, C.; Yoon, M.; Park, K.S.; Choi, K.M. Recent advances in process engineering and upcoming applications of metal–organic frameworks. *Coord. Chem. Rev.* **2021**, *426*, 213544. [CrossRef]
27. Chen, B.; Yang, Z.; Zhu, Y.; Xia, Y. Zeolitic imidazolate framework materials: Recent progress in synthesis and applications. *J. Mater. Chem. A* **2014**, *2*, 16811–16831. [CrossRef]
28. Jiao, C.; Li, M.; Ma, R.; Wang, C.; Wu, Q.; Wang, Z. Preparation of a Co-doped hierarchically porous carbon from Co/Zn-ZIF: An efficient adsorbent for the extraction of trizine herbicides from environment water and white gourd samples. *Talanta* **2016**, *152*, 321–328. [CrossRef]
29. Liang, C.; Zhang, X.; Feng, P.; Chai, H.; Huang, Y. ZIF-67 derived hollow cobalt sulfide as superior adsorbent for effective adsorption removal of ciprofloxacin antibiotics. *Chem. Eng. J.* **2018**, *344*, 95–104. [CrossRef]
30. Ganta, D.; Guzman, C.; Combrink, K.; Fuentes, M. Adsorption and Removal of Thymol from Water Using a Zeolite Imidazolate Framework-8 Nanomaterial. *Anal. Lett.* **2021**, *54*, 625–636. [CrossRef]
31. Bayati, B.; Ghorbani, A.; Ghasemzadeh, K.; Iulianelli, A.; Basile, A. Study on the separation of H_2 from CO_2 using a ZIF-8 membrane by molecular simulation and Maxwell-Stefan model. *Molecules* **2019**, *24*, 4350. [CrossRef]
32. Sahoo, T.R.; Prelot, B. Adsorption processes for the removal of contaminants from wastewater: The perspective role of nanomaterials and nanotechnology. In *Micro and Nano Technologies, Nanomaterials for the Detection and Removal of Wastewater Pollutant*; Bonelli, B., Freyria, F.S., Rossetti, I., Sethi, R., Eds.; Elsevier: Amsterdam, The Netherlands, 2020; pp. 161–222. [CrossRef]
33. Shokrollahi, A.; Alizadeh, A.; Malekhosseini, Z.; Ranjbar, M. Removal of bromocresol green from aqueous solution via adsorption on *Ziziphus nummularia* as a new, natural, and low-cost adsorbent: Kinetic and thermodynamic study of removal process. *J. Chem. Eng. Data* **2011**, *56*, 3738–3746. [CrossRef]
34. Ghadim, E.E.; Manouchehri, F.; Soleimani, G.; Hosseini, H.; Kimiagar, S.; Nafisi, S. Adsorption properties of tetracycline onto graphene oxide: Equilibrium, kinetic and thermodynamic studies. *PLoS ONE* **2013**, *8*, e79254. [CrossRef] [PubMed]
35. Ghaedi, M.; Shokrollahi, A.; Hossainian, H.; Kokhdan, S.N. Comparison of activated carbon and multiwalled carbon nanotubes for efficient removal of eriochrome cyanine R (ECR): Kinetic, isotherm, and thermodynamic study of the removal process. *J. Chem. Eng. Data* **2011**, *56*, 3227–3235. [CrossRef]
36. Ho, Y.S.; McKay, G. Pseudo-second order model for sorption processes. *Process Biochem.* **1999**, *34*, 451–465. [CrossRef]
37. Simonin, J.P. On the comparison of pseudo-first order and pseudo-second order rate laws in the modeling of adsorption kinetics. *Chem. Eng. J.* **2016**, *300*, 254–263. [CrossRef]
38. Robati, D. Pseudo-second-order kinetic equations for modeling adsorption systems for removal of lead ions using multi-walled carbon nanotube. *J. Nanostruct. Chem.* **2013**, *3*, 55. [CrossRef]
39. Azizi, S.; Shahri, M.M.; Mohamad, R. Green synthesis of Zinc oxide nanoparticles for enhanced adsorption of lead ions from aqueous solutions: Equilibrium, kinetic and thermodynamic studies. *Molecules* **2017**, *22*, 831. [CrossRef]
40. Wu, F.-C.; Tseng, R.-L.; Juang, R.-S. Initial behavior of intraparticle diffusion model used in the description of adsorption kinetics. *Chem. Eng. J.* **2009**, *153*, 1–8. [CrossRef]
41. Chien, S.H.; Clayton, W.R. Application of Elovich equation to the kinetics of phosphate release and sorption in soils. *Soil Sci. Soc. Am. J.* **1980**, *44*, 265–268. [CrossRef]

Article

Simultaneous Analysis of 19 Marker Components for Quality Control of Oncheong-Eum Using HPLC–DAD

Chang-Seob Seo * and Hyeun-Kyoo Shin

KM Science Research Division, Korea Institute of Oriental Medicine, Daejeon 34054, Korea; hkshin@kiom.re.kr
* Correspondence: csseo0914@kiom.re.kr; Tel.: +82-42-868-9361

Abstract: Oncheong-eum (OCE) is a traditional herbal prescription made by combining Samul-tang and Hwangryunhaedok-tang. It is primarily used to treat gynecological disorders such as metrorrhagia and metrostaxis. In the present study, we focused on developing and validating a simultaneous assay for the quality control of OCE using 19 marker components (gallic acid, 5-(hydroxymethyl)furfural, chlorogenic acid, geniposide, coptisine chloride, jatrorrhizine chloride, paeoniflorin, berberine chloride, palmatine chloride, ferulic acid, nodakenin, benzoic acid, baicalin, benzoylpaeoniflorin, wogonoside, baicalein, wogonin, decursin, and decursinol angelate). This analysis was performed using high-performance liquid chromatography coupled with a diode array detector, and chromatographic separation of the 19 markers was carried out using a SunFire™ C_{18} reversed-phase column and gradient elution conditions with two mobile phases (0.1% aqueous formic acid–0.1% formic acid in acetonitrile). The developed analytical method was validated through linearity, limits of detection and quantification, recovery, and precision. Under this assay, 19 markers in OCE samples were detected at not detected–9.62 mg/g. The analytical methods developed and validated in our research will have value as basic data for the quality control of related traditional herbal prescriptions as well as OCE.

Keywords: Oncheong-eum; traditional herbal prescription; method development; method validation; high-performance liquid chromatography

Citation: Seo, C.-S.; Shin, H.-K. Simultaneous Analysis of 19 Marker Components for Quality Control of Oncheong-Eum Using HPLC–DAD. *Molecules* **2022**, *27*, 2992. https://doi.org/10.3390/molecules27092992

Academic Editors: Franciszek Główka and Marta Karaźniewicz-Łada

Received: 25 April 2022
Accepted: 5 May 2022
Published: 6 May 2022

Publisher's Note: MDPI stays neutral with regard to jurisdictional claims in published maps and institutional affiliations.

Copyright: © 2022 by the authors. Licensee MDPI, Basel, Switzerland. This article is an open access article distributed under the terms and conditions of the Creative Commons Attribution (CC BY) license (https://creativecommons.org/licenses/by/4.0/).

1. Introduction

In general, traditional herbal prescriptions consist of two or more herbal medicines. They have been used for a long time in Asian countries, especially Korea, China, and Japan, because of the advantages of fewer side effects and multicomponent/multitarget compared with synthetic drugs or Western medicine [1–3]. However, despite their prolonged use, modern scientific validation of their biological activity and research on standardization to maintain the consistency of raw materials are still needed.

Oncheong-eum (OCE; "Wen-Qing-Yin" in Chinese and "Unsei-in" in Japanese), also called "Haedoksamultang", is a traditional herbal prescription that combines Samul-tang and Hwangryunhaedok-tang and consists of eight herbal medicines: *Angelica gigas* Nakai, *Cnidium officinale* Makino, *Paeonia lactiflora* Pall., *Rehmannia glutinosa* (Gaertn.) DC., *Coptis chinensis* Franch., *Phellodendron chinensis* C.K.Schneid., *Scutellaria baicalensis* Georgi, and *Gardenia jasminoides* Ellis, in the same weight ratio [4]. Since OCE was first recorded in the Manbyeonghoichun of Gong Tingxian, a physician in the Ming Dynasty, it has also been transmitted in Dongeuibogam of Heo Jun, a famous medical book of the Joseon Dynasty [4,5]. This prescription has been used to stop abnormal uterine bleeding, remove fever, and treat stomach pain. It is also used for skin diseases and metabolic diseases [5,6].

Various biological activities of OCE have been reported, such as the inhibition of cell proliferation in the human hepatocarcinoma cell line [7,8], inhibition of melanogenesis and tyrosinase activity in the murine melanoma cell line, skin regeneration and wrinkle improvement in the human normal fibroblast cell line [9], and pruritus [10,11].

OCE is composed of eight herbal medicines and contains numerous constituents. The major components are phenylpropanoids (e.g., chlorogenic acid) and coumarins (e.g., nodakenin, decursin, and decursinol angelate) from *A. gigas* [12], phenylpropanoids (e.g., ferulic acid) from *C. officinale* [13], monoterpenoids (e.g., albiflorin and paeoniflorin) from *P. lactiflora* [14], miscellaneous (e.g., 5-(hydroxymethyl)furfural) from *R. glutinosa* [15], alkaloids (e.g., berberine chloride) from *C. japonica* and *P. chinensis* [16,17], flavonoids (e.g., baicalin, and wogonoside) from *S. baicalensis* [18], and iridoid glycosides (e.g., geniposide) from *G. jasminoides* [19].

Several analytical methods are practiced for constant quality control of traditional herbal prescriptions using high-performance capillary electrophoresis [20], gas chromatography with mass spectrometry (GC–MS) [21], ultra-performance liquid chromatography (UPLC) [22], high-performance liquid chromatography combined with a diode array detector (HPLC–DAD) [23], HPLC with charged aerosol detector [24], and liquid chromatography–tandem mass spectrometry (LC–MS) [25]. Up until now, among various analytical techniques, the analytical method using HPLC–DAD equipment is the most widely used and recommended for the analysis of herbal medicines or herbal medicine prescriptions because of its wide selection of mobile phases, convenience of use, accuracy, and reproducibility of results [6]. A previous study reported a simultaneous analysis of six indicator compounds (berberine, baicalin, ferulic acid, geniposide, hydroxymethylfurfural, and paeoniflorin) in Wen-Qing-Yin using HPLC–DAD [6].

Thus, in this study, a simultaneous determination method for continuous quality control of OCE by HPLC–DAD using 19 markers—gallic acid (**1**), 5-(hydroxymethyl)furfural (**2**), chlorogenic acid (**3**), geniposide (**4**), coptisine chloride (**5**), jatrorrhizine chloride (**6**), paeoniflorin (**7**), berberine chloride (**8**), palmatine chloride (**9**), ferulic acid (**10**), nodakenin (**11**), benzoic acid (**12**), baicalin (**13**), benzoylpaeoniflorin (**14**), wogonoside (**15**), baicalein (**16**), wogonin (**17**), decursin (**18**), and decursinol angelate (**19**)—selected from each raw herbal medicine constituting OCE was developed, and the assay was validated.

2. Results and Discussion

2.1. Identification of the Major Components of Each Herbal Medicine Constituting OCE

To develop a simultaneous analysis method for the quality control of OCE using HPLC–DAD, the primary ingredients contained in eight raw herbal medicines were first explored. The ingredients selected for analysis in each raw herb are as follows: chlorogenic acid, nodakenin, decursin, and decursinol angelate from *A. gigas*; ferulic acid from *C. officinale*; albiflorin, paeoniflorin, benzoic acid, gallic acid, and benzoylpaeoniflorin from *P. lactiflora*; 5-(hydroxymethyl)furfural from *R. glutinosa*; berberine chloride, coptisine chloride, jatrorrhizine chloride, and palmatine chloride from *C. japonica* and *P. chinensis*; baicalin, baicalein, wogonin, and wogonoside from *S. baicalensis*; and geniposide and gardenoside from *G. jasminoides*. As a result of a comparative analysis of each constituent herbal medicine and their main components in HPLC using a reversed-phased C_{18} column and a distilled water (DW)–acetonitrile (ACN) mobile system (both containing 0.1% formic acid; FA), it was confirmed that the target components were present in each constituent herbal medicine (Figure S1).

2.2. Selection of Marker Analytes for Quality Control of OCE Using an HPLC Sytem

After performing as in Section 3.1 and performing the same in the OCE sample, 19 compounds were detected, whereas two components, gardenoside, and albiflorin, were not detected (Figure S2A,B). Of these two components, gardenoside showed a peak at the same retention time as the standard compound, but as a result of UV spectrum comparison, it was identified as a different component and thus excluded from analysis (Figure S2C). In addition, albiflorin was detected in the sample, but separation from other neighboring components was not complete, so it was not selected as a marker component. Finally, among the 21 components, 19 compounds completely isolated and identified from the OCE sample were selected as marker components suitable for the quality control of OCE. The chemical structures of compounds **1–19** selected as marker analytes for the quality control in OCE are shown in Figure S3.

2.3. Optimization of HPLC Analysis Conditions

We compared several conditions to search for efficient chromatographic separation conditions for the 19 marker analytes selected to achieve the development of an optimal HPLC simultaneous method for the quality control of OCE. First, the resolution of markers was compared by comparing reversed-phase C_{18} columns (4.6 mm × 250 mm, 5 µm) such as SunFire™ (Waters, Milford, MA, USA), Gemini (Phenomenex, Torrance, CA, USA), Capcellpak UG80 (Shiseido, Tokyo, Japan), and Hypersil GOLD (Thermo Fisher Scientific Inc., San Jose, CA, USA). Next, the types of acid added to the mobile phase (FA, trifluoroacetic acid; TFA, acetic acid; AA, and phosphoric acid; PA), and the temperatures of the column oven (30, 35, 40, and 45 °C) were also compared. Figure 1 shows the representative HPLC chromatogram measured under the optimal analysis conditions finally selected (SunFire™ C_{18} column, column temperature of 40 °C, and mobile phase system of 0.1% FA in DW–0.1% FA in ACN, based on the comparison and search conditions). All analytes were eluted within 65 min without the influence of neighboring peaks with a resolution of 16.9 or higher (Table S3).

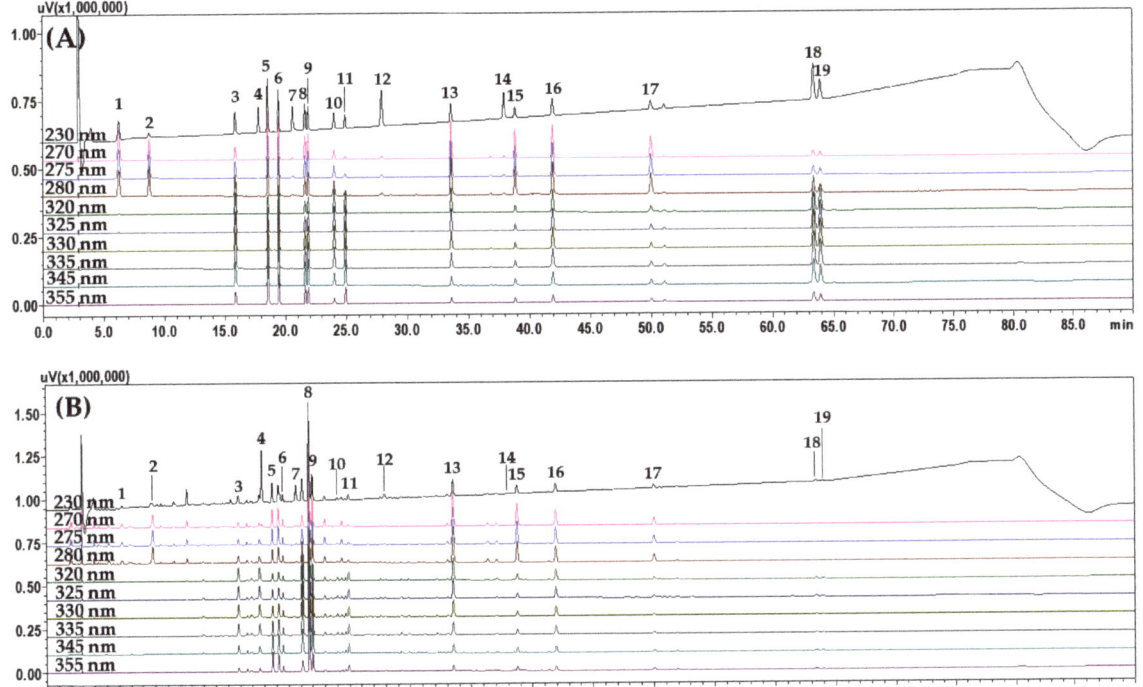

Figure 1. HPLC chromatograms of the standard solution (**A**) and OCE–1 sample (**B**). Gallic acid (**1**), 5-(hydroxymethyl)furfural (**2**), chlorogenic acid (**3**), genipiside (**4**), coptisine chloride (**5**), jatrorrhizine chloride (**6**), paeoniflorin (**7**), berberine chloride (**8**), palmatine chloride (**9**), ferulic acid (**10**), nodakenin (**11**), benzoic acid (**12**), baicalin (**13**), benzoylpaeoniflorin (**14**), wogonoside (**15**), baicalein (**16**), wogonin (**17**), decursin (**18**), and decursinol angelate (**19**). The concentration of each marker in the mixed standard solution is as follows: 10.00 µg/mL (compounds **2**, **8–10**, and **17**), 20.00 µg/mL (compounds **3**, **6**, **11**, **12**, **15**, and **16**), 30.00 µg/mL (compounds **1**, **4**, **5**, and **13**), 40.00 µg/mL (compounds **14**, **18**, and **19**), and 50.00 µg/mL (compound **7**).

2.4. Method Validation of the Developed HPLC Analytical Method

We developed an HPLC simultaneous analysis method using the 19 marker analytes for the efficient quality control of OCE. The developed assay was tested with several parameters, including linearity, the limit of detection (LOD), the limit of quantification (LOQ), and precision for validation. As shown in Table 1, the coefficient of determination (r^2), which evaluates linearity, showed excellent linearity from 0.9999 to 1.0000 in all markers based on the prepared calibration curve. The LODs and LOQs of the 19 investigated marker components were 0.005–0.094 µg/mL and 0.015–0.285 µg/mL, respectively (Table 1). Recoveries (%) of compounds **1–19** ranged from 95.27% to 105.44% and are summarized in Table 2. The validation of precision was evaluated by the relative standard deviation (RSD (%)). As a result, all RSD values in the repeatability and intraday and interday precisions of investigated markers were ≤2.40%, showing suitable precision validation results (Table S3 and Table 3). These validation data indicate that the developed HPLC assay is an appropriate and accurate method for determining the marker substances selected for the quality control of OCE.

Table 1. Detection wavelength, linear range, regression equation, r^2, LOD, LOQ, and retention time for simultaneous analysis of each marker component ($n = 3$).

Analyte	Detection Wavelength (nm)	Linear Range (µg/mL)	Regression Equation [a] $y = ax + b$	r^2	LOD (µg/mL)	LOQ (µg/mL)	Retention Time (min)
1	270	0.31–20.00	$y = 41{,}926.60x + 2559.71$	0.9999	0.034	0.102	6.24
2	280	0.47–30.00	$y = 91{,}675.22x + 4694.69$	0.9999	0.046	0.141	8.76
3	325	0.47–30.00	$y = 52{,}471.53x + 5007.02$	0.9999	0.075	0.228	15.84
4	230	0.47–30.00	$y = 21{,}695.37x + 3442.20$	0.9998	0.076	0.230	17.79
5	355	0.78–50.00	$y = 39{,}690.73x + 6737.30$	0.9999	0.083	0.252	18.61
6	345	0.31–20.00	$y = 53{,}377.60x + 2860.29$	1.0000	0.073	0.220	19.49
7	230	0.47–30.00	$y = 15{,}685.87x - 1318.46$	0.9999	0.006	0.018	20.61
8	345	0.47–30.00	$y = 51{,}279.59x + 5895.33$	0.9999	0.020	0.061	21.67
9	345	0.78–50.00	$y = 53{,}582.09x + 9136.34$	0.9999	0.094	0.285	21.95
10	320	0.31–20.00	$y = 84{,}394.31x + 7794.43$	0.9999	0.026	0.078	24.03
11	335	0.47–30.00	$y = 38{,}849.24x + 5106.67$	0.9999	0.045	0.138	24.95
12	230	0.31–20.00	$y = 70{,}000.98x + 4949.39$	0.9999	0.005	0.015	27.93
13	275	0.31–20.00	$y = 42{,}108.54x + 3779.66$	0.9999	0.078	0.236	33.55
14	230	0.31–20.00	$y = 21{,}287.06x + 626.36$	1.0000	0.052	0.156	37.94
15	275	0.78–50.00	$y = 56{,}245.50x + 12{,}765.82$	0.9999	0.079	0.239	38.81
16	275	0.78–50.00	$y = 64{,}528.60x + 6020.97$	0.9999	0.086	0.262	41.93
17	275	0.31–20.00	$y = 91{,}000.29x + 7853.01$	0.9999	0.029	0.089	50.01
18	330	0.31–20.00	$y = 41{,}162.22x + 2663.76$	0.9999	0.015	0.045	63.38
19	330	0.31–20.00	$y = 32{,}163.04x + 2483.09$	0.9999	0.034	0.102	63.94

[a] y: peak area of compounds; x: concentration (µg/mL) of compounds. Gallic acid (**1**), 5-(hydroxymethyl)furfural (**2**), chlorogenic acid (**3**), geniposide (**4**), coptisine chloride (**5**), jatrorrhizine chloride (**6**), paeoniflorin (**7**), berberine chloride (**8**), palmatine chloride (**9**), ferulic acid (**10**), nodakenin (**11**), benzoic acid (**12**), baicalin (**13**), benzoylpaeoniflorin (**14**), wogonoside (**15**), baicalein (**16**), wogonin (**17**), decursin (**18**), and decursinol angelate (**19**).

Table 2. Recovery (%) of the 19 marker components in the developed HPLC method ($n = 5$).

Analyte	Spiked Amount (µg/mL)	Found Amount (µg/mL)	Recovery (%)	RSD (%)
1	1.00	1.01	101.07	1.80
	2.00	2.01	100.62	1.29
	4.00	4.12	102.89	1.14
2	2.00	2.05	102.36	0.66
	5.00	5.15	103.08	0.33
	10.00	10.34	103.36	0.29

Table 2. Cont.

Analyte	Spiked Amount (μg/mL)	Found Amount (μg/mL)	Recovery (%)	RSD (%)
3	2.00	2.01	100.66	0.73
	5.00	5.11	102.28	0.66
	10.00	10.35	103.49	0.48
4	2.00	1.97	98.32	1.32
	5.00	4.95	98.99	1.26
	10.00	10.28	102.81	0.81
5	3.00	3.05	101.80	0.83
	7.50	7.15	95.27	0.15
	15.00	14.63	97.56	0.42
6	1.00	1.00	99.78	0.60
	2.00	1.97	98.66	1.31
	4.00	3.95	98.65	1.69
7	2.00	1.97	98.74	0.61
	4.00	3.96	98.98	0.65
	8.00	8.25	103.16	0.54
8	2.00	2.00	99.93	1.36
	4.00	3.96	99.09	0.71
	8.00	7.86	98.29	0.40
9	4.00	4.07	101.76	1.70
	10.00	9.94	99.35	1.30
	20.00	20.18	100.90	0.24
10	1.00	1.02	102.18	1.10
	2.00	2.04	101.84	0.30
	4.00	4.14	103.54	0.34
11	2.00	2.06	103.05	2.21
	5.00	5.05	100.98	0.47
	10.00	10.39	103.88	0.58
12	1.00	1.03	102.74	2.37
	2.00	2.01	100.65	0.82
	4.00	3.87	96.87	0.50
13	1.00	0.98	97.81	0.59
	2.00	1.95	97.49	0.88
	4.00	3.88	96.93	0.69
14	1.00	1.02	102.10	1.09
	2.00	2.01	100.65	1.06
	4.00	4.09	102.36	0.64
15	4.00	4.22	105.44	1.09
	10.00	9.67	96.72	1.80
	20.00	20.07	100.37	0.37
16	3.00	2.97	98.99	0.64
	7.50	7.46	99.50	0.51
	15.00	14.99	99.94	0.29
17	1.00	0.98	98.33	1.01
	2.00	1.92	95.98	0.82
	4.00	3.98	99.58	0.70
18	1.00	1.03	102.59	0.87
	2.00	1.95	97.32	0.37
	4.00	4.14	103.54	0.17
19	1.00	1.04	103.67	0.77
	2.00	1.96	98.19	0.53
	4.00	4.08	101.93	0.24

Gallic acid (1), 5-(hydroxymethyl)furfural (2), chlorogenic acid (3), geniposide (4), coptisine chloride (5), jatrorrhizine chloride (6), paeoniflorin (7), berberine chloride (8), palmatine chloride (9), ferulic acid (10), nodakenin (11), benzoic acid (12), baicalin (13), benzoylpaeoniflorin (14), wogonoside (15), baicalein (16), wogonin (17), decursin (18), and decursinol angelate (19).

Table 3. Precision test of marker compounds 1–19 in the developed HPLC method.

Analyte	Conc. (µg/mL)	Intraday (n = 5)			Interday (n = 5)		
		Observed Conc. (µg/mL) ± SD	Precision (RSD, %)	Accuracy (%)	Observed Conc. (µg/mL) ± SD	Precision (RSD, %)	Accuracy (%)
1	5.0	5.05 ± 0.04	0.72	101.04	5.03 ± 0.10	2.09	100.52
	10.0	10.12 ± 0.11	1.06	101.24	10.15 ± 0.20	2.00	101.49
	20.0	19.85 ± 0.1	0.74	99.26	19.91 ± 0.31	1.56	99.56
2	7.5	7.62 ± 0.02	0.32	101.61	7.56 ± 0.05	0.72	100.82
	15.0	14.99 ± 0.06	0.38	99.93	15.05 ± 0.17	1.14	100.34
	30.0	30.19 ± 0.08	0.28	100.62	30.00 ± 0.20	0.66	100.02
3	7.5	7.51 ± 0.04	0.56	100.18	7.45 ± 0.18	2.40	99.36
	15.0	15.13 ± 0.10	0.69	100.86	15.08 ± 0.32	2.11	100.53
	30.0	29.66 ± 0.23	0.79	98.87	29.57 ± 0.55	1.85	98.55
4	7.5	7.65 ± 0.03	0.41	101.97	7.68 ± 0.06	0.76	102.44
	15.0	15.25 ± 0.03	0.20	101.65	15.48 ± 0.18	1.18	103.18
	30.0	30.04 ± 0.18	0.61	100.13	30.31 ± 0.58	1.91	101.03
5	12.5	12.83 ± 0.12	0.94	102.68	12.92 ± 0.23	1.80	103.39
	25.0	25.68 ± 0.23	0.88	102.70	26.04 ± 0.51	1.96	104.17
	50.0	50.26 ± 0.47	0.93	100.53	50.68 ± 1.01	2.00	101.36
6	5.0	5.08 ± 0.04	0.75	101.68	5.14 ± 0.09	1.77	102.74
	10.0	10.19 ± 0.05	0.52	101.89	10.36 ± 0.20	1.91	103.55
	20.0	20.05 ± 0.18	0.92	100.25	20.21 ± 0.38	1.89	101.04
7	7.5	7.53 ± 0.06	0.77	100.39	7.56 ± 0.07	0.92	100.77
	15.0	15.13 ± 0.10	0.65	100.86	15.17 ± 0.26	1.70	101.14
	30.0	30.01 ± 0.16	0.53	100.03	30.31 ± 0.46	1.51	101.05
8	7.5	7.71 ± 0.06	0.84	102.82	7.77 ± 0.13	1.73	103.57
	15.0	15.41 ± 0.15	0.99	102.75	15.64 ± 0.31	1.96	104.27
	30.0	30.14 ± 0.26	0.86	100.46	30.40 ± 0.60	1.97	101.34
9	12.5	12.80 ± 0.11	0.82	102.41	12.91 ± 0.23	1.75	103.24
	25.0	25.61 ± 0.18	0.72	102.43	26.00 ± 0.50	1.91	104.00
	50.0	49.98 ± 0.44	0.88	99.96	50.55 ± 1.03	2.04	101.10
10	5.0	5.13 ± 0.02	0.45	102.67	5.17 ± 0.09	1.71	103.44
	10.0	10.28 ± 0.10	0.96	102.81	10.39 ± 0.21	2.01	103.93
	20.0	20.08 ± 0.19	0.94	100.42	20.21 ± 0.39	1.95	101.06
11	7.5	7.71 ± 0.04	0.49	102.74	7.78 ± 0.14	1.75	103.74
	15.0	15.43 ± 0.11	0.71	102.87	15.65 ± 0.29	1.88	104.36
	30.0	30.17 ± 0.25	0.84	100.57	30.41 ± 0.59	1.93	101.37
12	5.0	5.12 ± 0.02	0.48	102.45	5.11 ± 0.06	1.17	102.21
	10.0	10.18 ± 0.01	0.12	101.76	10.28 ± 0.10	0.98	102.79
	20.0	20.08 ± 0.13	0.65	100.41	20.14 ± 0.38	1.89	100.68
13	5.0	5.13 ± 0.04	0.82	102.69	5.17 ± 0.09	1.71	103.38
	10.0	10.31 ± 0.12	1.13	103.11	10.43 ± 0.19	1.84	104.26
	20.0	20.08 ± 0.22	1.11	100.41	20.21 ± 0.39	1.94	101.03
14	5.0	5.04 ± 0.03	0.54	100.74	5.08 ± 0.05	1.08	101.54
	10.0	9.91 ± 0.05	0.54	99.06	10.21 ± 0.23	2.28	102.06
	20.0	19.86 ± 0.10	0.49	99.28	20.11 ± 0.26	1.31	100.57
15	12.5	12.88 ± 0.11	0.88	103.05	12.97 ± 0.23	1.74	103.78
	25.0	25.81 ± 0.30	1.17	103.22	26.11 ± 0.48	1.85	104.45
	50.0	50.32 ± 0.54	1.08	100.63	50.64 ± 0.97	1.92	101.28
16	12.5	12.67 ± 0.11	0.86	101.36	12.76 ± 0.24	1.85	102.06
	25.0	25.42 ± 0.31	1.22	101.68	25.62 ± 0.48	1.86	102.47
	50.0	49.77 ± 0.45	0.90	99.54	49.87 ± 0.98	1.97	99.73
17	5.0	5.15 ± 0.04	0.84	102.99	5.18 ± 0.09	1.72	103.70
	10.0	10.32 ± 0.12	1.19	103.20	10.44 ± 0.19	1.85	104.38
	20.0	20.11 ± 0.21	1.05	100.53	20.24 ± 0.40	1.95	101.19
18	5.0	5.10 ± 0.05	0.89	102.08	5.14 ± 0.09	1.79	102.77
	10.0	10.28 ± 0.11	1.11	102.76	10.39 ± 0.19	1.85	103.93
	20.0	20.13 ± 0.19	0.96	100.63	20.24 ± 0.40	1.97	101.19
19	5.0	5.13 ± 0.04	0.84	102.58	5.17 ± 0.09	1.67	103.31
	10.0	10.31 ± 0.13	1.22	103.12	10.41 ± 0.19	1.83	104.15
	20.0	20.11 ± 0.20	0.98	100.55	20.22 ± 0.40	1.96	101.08

Gallic acid (**1**), 5-(hydroxymethyl)furfural (**2**), chlorogenic acid (**3**), geniposide (**4**), coptisine chloride (**5**), jatrorrhizine chloride (**6**), paeoniflorin (**7**), berberine chloride (**8**), palmatine chloride (**9**), ferulic acid (**10**), nodakenin (**11**), benzoic acid (**12**), baicalin (**13**), benzoylpaeoniflorin (**14**), wogonoside (**15**), baicalein (**16**), wogonin (**17**), decursin (**18**), and decursinol angelate (**19**).

2.5. System Suitability and Stability Tests

The system suitability values were 1.17–21.22 (capacity factor; k'), 1.01–1.75 (selectivity factor; α), 18,808.16–1,737,667.55 (number of theoretical plates; N), 1.69–28.98 (resolution; Rs), and 1.01–1.18 (tailing factor; Tf). furthermore, the stability of each marker analyte measured using a standard solution for three days (0, 24, and 48 h) showed an RSD value in the range of 0.68% to 2.36%. The detailed results for each marker component are summarized in Table S4.

2.6. Quantification of the 19 Marker Compounds in OCE Samples

The developed HPLC analytical assay was successfully applied to simultaneous quantitation of the 19 markers in the OCE samples. Nineteen marker compounds were simultaneously monitored at 230 nm (compounds **4**, **7**, **12**, and **14**), 270 nm (compound **4**), 275 nm (compounds **13** and **15–17**), 280 nm (compound **2**), 320 nm (compound **10**), 325 nm (compound **3**), 330 nm (compounds **18** and **19**), 335 nm (compound **11**), 345 nm (compounds **6**, **8**, and **9**), and 355 nm (compound **5**). The amounts of the 19 investigated components in one water extract (OCE–1) and four commercial samples (OCE–2 to OCE–5, Figure S4) are presented in Table 4.

Table 4. Amounts of the 19 marker components in OCE samples (n = 3).

Analyte	OCE–1 Mean (mg/g)	OCE–1 RSD (%)	OCE–2 Mean (mg/g)	OCE–2 RSD (%)	OCE–3 Mean (mg/g)	OCE–3 RSD (%)	OCE–4 Mean (mg/g)	OCE–4 RSD (%)	OCE–5 Mean (mg/g)	OCE–5 RSD (%)
1	0.67	0.20	0.20	2.20	0.75	0.15	0.44	0.41	0.24	0.46
2	1.10	0.21	0.08	0.24	0.04	1.38	<LOQ	–	<LOQ	–
3	1.20	0.17	0.11	0.85	0.03	0.73	<LOQ	–	0.32	0.70
4	9.62	0.32	1.61	0.20	1.87	0.33	1.76	1.73	2.73	0.12
5	1.72	0.65	<LOQ	–	0.04	2.05	0.26	0.43	0.79	0.38
6	0.56	0.25	<LOQ	–	<LOQ	–	0.07	0.99	0.18	0.12
7	6.14	0.63	2.76	0.23	2.42	0.29	2.85	0.14	4.14	0.37
8	7.21	0.67	2.01	0.18	1.68	0.04	1.59	0.06	3.90	0.03
9	1.89	0.24	0.07	0.97	0.09	1.27	0.32	0.36	0.60	0.33
10	0.23	0.32	0.07	0.21	<LOQ	–	0.02	0.51	0.14	0.59
11	1.21	0.23	0.18	0.65	0.07	0.24	0.03	0.61	0.02	2.03
12	0.41	0.47	0.13	0.86	0.24	1.04	0.21	1.09	0.17	0.49
13	4.43	1.28	6.12	0.25	8.06	0.04	8.59	0.19	7.51	0.08
14	0.16	0.75	0.09	1.10	0.05	1.53	0.05	2.06	0.14	2.05
15	2.20	0.66	1.41	0.17	0.24	0.24	1.52	0.95	1.45	0.98
16	1.46	2.08	0.37	1.21	0.21	0.19	0.09	0.76	0.01	1.66
17	0.57	0.51	0.10	0.18	0.06	0.22	0.05	0.60	0.04	0.39
18	0.24	0.69	0.09	0.44	0.05	0.83	0.01	2.16	ND	–
19	0.19	0.57	0.10	0.63	0.06	1.01	0.02	1.68	ND	–

Table 4 shows that compounds **4** (1.61–9.62 mg/g), **7** (2.42–6.14 mg/g), **8** (1.59–7.21 mg/g), and **13** (4.43–8.59 mg/g) have higher concentrations than other investigated analytes in all samples. These four components are the main components of *G. jasminoides*, *P. lactiflora*, *C. japonica*, *P. chinensis*, and *S. baicalensis*.

3. Materials and Methods

3.1. Plant Materials

The eight raw herbal ingredients shown in Table S1 were purchased from Kwangmyungdang Pharmaceutical (Ulsan, Korea), a specialized herbal medicine manufacturing company, in November 2017. Each herbal medicine was used after morphological identification according to the guideline "The Dispensatory on the Visual and Organoleptic Examination of Herbal Medicine" by Dr. Goya Choi, Korea Institute of Oriental Medicine (KIOM, Daejeon,

Korea) [26]. Each herbal medicine was deposited at the KM Science Research Division, KIOM (specimen no.: from 2018KE81–1 to 2018KE81–8).

3.2. Chemicals and Reagents

All reference standard compounds used for HPLC analysis were purchased from standard compound manufacturers: compounds **1** (CAS No. 149-91-7, 100.0%, Catalog No. G7384), **2** (CAS No. 67-47-0, ≥99.0%, Catalog No. W501808), **3** (CAS No. 327-97-9, 99.7%, Catalog No.PHL89175), and **12** (CAS No. 65-85-0, 99.9%, Catalog No. 242381) from KGaA (Darmstadt, Germany); compounds **4** (CAS No. 24512-63-8, >98.0%, Catalog No. 073-05891), **5** (CAS No. 6020-1804, ≥98.0%, Catalog No. 038-22001), **10** (CAS No. 1135-24-6, 98.0%, Catalog No. 086-04282), **13** (CAS No. 21967-41-9, 98.0%, Catalog No.024-15691), and **17** (CAS No. 632-85-9, ≥98.9%, Catalog No. 236-02321) from Fujifilm Wako Pure Chemical Co., Ltd. (Osaka, Japan); compounds **6** (CAS No. 6681-15-8, 98.4%, Catalog No. D91304201), **7** (CAS No. 23180-57-6, 99.4%, Catalog No. DR10579), **8** (CAS No. 633-65-8, 98.9%, Catalog No. DR10793), **14** (CAS No. 38642-49-8, ≥98.0%, Catalog No. DR10582), **15** (CAS No. 51059-44-0, 98.9%, Catalog No. DR10630), **16** (CAS No. 491-67-8, 99.4%, Catalog No. DR10625), and **18** (CAS No. 5928-25-6, 98.7%, Catalog No. DR11193) from Shanghai Sunny Biotech Co., Ltd. (Shanghai, China); compound **9** (CAS No. 10605-02-4, 99.3%, Catalog No. P2138) from Tokyo Chemical Industry Co., Ltd. (Tokyo, Japan); compound **11** (CAS No. 495-31-8, 99.5%, Catalog No. CFN90232) from ChemFaces Biochemical Co., Ltd. (Wuhan, China), and compound **19** (CAS No. 130848-06-5, 98.3%, Catalog No. BP1812) from Biopurify Phytochemicals (Chengdu, China). Methanol (MeOH), ACN, and DW (all HPLC grade) used for the preparation of test solutions, standard solutions, and chromatographic separation of marker analytes were purchased from JT Baker (Phillipsburg, NJ, USA). Acids, TFA (≥99.0%, HPLC grade) and AA (≥100.0%, ACS reagent grade) were purchased from Merck KGaA (Darmstadt, Germany), and FA (99.5%, HPLC grade) and PA (85.0%, ACS reagent grade) were purchased from Fujifilm Wako Pure Chemical Co., Ltd. (Osaka, Japan). All of these acids were used to add to the mobile phase. Dimethyl sulfoxide (DMSO, ≥99.9%, ACS reagent grade) was purchased from Merck KGaA (Darmstadt, Germany).

3.3. Preparation of OCE Water Extract

OCE water extract (OCE–1) was prepared at KIOM according to a previously reported manufacturing method [27,28]. That is, after mixing the same amount (each 625.0 g) of the eight herbal medicines shown in Table S1, 50 L of DW was added, and the mixture was extracted for 2 h at 100 °C using an electric extractor, COSMOS-660 (Kyungseo E&P, Incheon, Korea). According to the previously reported manufacturing method. Subsequently, the extracted water solution was filtered using a sieve (53 μm mesh). As a final step for producing a powder sample, the filtered extract was freeze-dried using an LP100R freeze dryer (IlShinBioBase, Yangju, Korea) (1232.6 g, yield 24.7%). Apart from the sample prepared by KIOM, the other four commercial samples (from OCE–2 to OCE–5) were purchased from different pharmaceutical companies, Kyungbang (Incheon, Korea), Jungwoo Medicines (Asan, Korea), Hankookshinyak (Nonsan, Korea), and Tsumura & Co. (Tokyo, Japan), respectively.

3.4. Preparation of Test Solutions and Standard Solutions for HPLC–DAD Analysis

For the test solutions for simultaneous determination of the 19 markers in OCE, about 100 mg was accurately taken of each prepared OCE water extract and commercially available products in a 10 mL volumetric flask, then filled with 70% MeOH (100 mg/10 mL). The continuously mixed samples were subjected to ultrasonic extraction at room temperature for 60 min. For the quantitative analysis of compounds **4**, **7**, **8**, and **13**, the prepared test solution was diluted 10-fold and used. All solutions were filtered before analysis using a 0.2 μm syringe filter (Pall Life Sciences, Ann Arbor, MI, USA) and then injected into the HPLC instrument.

Standard solutions of the 19 reference standard compounds were prepared at a concentration of 1.0 mg/mL using methanol or DMSO–MeOH solution (1:1), and then used while refrigerated.

3.5. HPLC Instrument and Analysis Conditions

The HPLC instrument used for simultaneous analysis of the 19 markers in OCE was a Prominence LC-20A modular system (Shimadzu Co., Tokyo, Japan) consisting of a quaternary pump (LC-20AT), DAD (SPD-M20A), autosampler (SIL-20A), and column oven (CTO-20A). The system is operated and controlled by LabSolution software (Version 5.53, SP3, Kyoto, Japan). Chromatographic separation for simultaneous analysis of marker analytes was performed using a reversed-phase column, SunFire™ C_{18} column (4.6 mm × 250 mm, 5 μm, Waters, Milford, MA, USA), and gradient elution with two mobile phases (0.1% FA in DW–0.1% FA in ACN). Other detailed analysis conditions, including gradient elution conditions, are shown in Table S2.

3.6. Validation of the Developed HPLC Analytical Method

The analytical HPLC method developed to be applied to the quality control of OCE was validated by measuring and confirming various parameters such as linearity, LOD, LOQ, recovery, and precision [29]. First, the linearity of each marker component was checked in the following concentration ranges (0.31–20.00 μg/mL for compounds **1**, **6**, **10**, **12–14**, and **17–19**; 0.47–30.00 μg/mL for compounds **2–4**, **7**, **8**, and **11**; 0.78–50.00 μg/mL for compounds **5**, **9**, **15**, and **16**) and evaluated through the r^2 of the prepared calibration curve.

Second, LOD and LOQ were calculated by the following equations, respectively. LOD = $3.3 \times \sigma/S$ and LOD = $10 \times \sigma/S$, where σ and S represent the standard deviation of the y-intercept and the slope of the calibration curve in the regression equation of each marker measured three times, respectively.

Third, the recovery was validated by the standard addition method. Briefly, after accurately taking a 100 mg OCE powder sample in a 10 mL volumetric flask, three concentration levels (low, medium, and high) of each known marker compound were added. After that, the pretreatment process was the same as the preparation of the sample solution in Section 2.4. Extraction recovery (%) was calculated using following equation: Recovery (%) = found amount/spiked amount × 100.

Finally, the precision was evaluated as the relative standard deviation (RSD) of intraday and interday precisions and repeatability. The intraday and interday precisions were evaluated by calculating the RSD after measuring five times each for one day and three consecutive days using a standard solution of three concentrations in which the 19 markers were mixed. Repeatability was evaluated by obtaining the RSD of each marker analyte's retention time and peak area after six repeated measurements.

3.7. System Suitability and Stability Tests

The system suitability test was validated by evaluating the k', α, Rs, N, and Tf to evaluate the normal operation of the analysis system [27]. Furthermore, the stability of each marker component was tested for three days (0, 24, and 48 h) at 21 ± 1 °C using a standard solution.

3.8. Statistical Analysis

Data were expressed as mean, SD, and RSD (%), using Microsoft Excel 2019 software (Microsoft Co., Redmond, WA, USA).

4. Conclusions

In this study, a simultaneous analysis method using convenient, accurate, and reproducible HPLC–DAD for the quality control of OCE was developed and validated. This analytical method has been successfully applied to qualitative and quantitative analysis for the quality control of OCE. These results can provide reference data for improving the quality standards of OCE and related traditional herbal prescriptions.

Supplementary Materials: The following supporting information can be downloaded at: https://www.mdpi.com/article/10.3390/molecules27092992/s1. Figure S1: HPLC chromatogram of constituent herbal medicines and their main components; Figure S2: HPLC chromatograms of the solution of the standard mixture (A) and 70% methanol solution of OCE–1 sample (B), and UV spectrum of gardenoside (C); Figure S3: Chemical structures of the selected 19 marker compounds for the quality control of OCE; Figure S4: HPLC chromatograms of OCE–2 to OCE–5 samples; Table S1: Composition of OCE; Table S2: HPLC chromatographic conditions for analyzing the 19 markers of OCE; Table S3. Repeatability of compounds **1–19** ($n = 6$); Table S4: System suitability and stability of compounds **1–19**.

Author Contributions: Conceptualization, C.-S.S. and H.-K.S.; performing experiments and analyzing data, C.-S.S.; writing—original draft preparation, C.-S.S.; funding acquisition, H.-K.S. All authors have read and agreed to the published version of the manuscript.

Funding: This research was supported by a grant from the Korea Institute of Oriental Medicine (No. KSN2022310).

Institutional Review Board Statement: Not applicable.

Informed Consent Statement: Not applicable.

Data Availability Statement: All data can be found in this paper.

Conflicts of Interest: The authors declare no conflict of interest.

Abbreviations

α	Selectivity factor
AA	Acetic acid
ACN	Acetonitrile
DMSO	Dimethyl sulfoxide
DW	Distilled water
FA	Formic acid
GC–MS	Gas chromatography with mass spectrometry
HPLC–DAD	High-performance liquid chromatography combined with a diode array detector
k'	Capacity factor
KIOM	Korea Institute of Oriental Medicine
LC-MS	Liquid chromatography–tandem mass spectrometry
LOD	Limit of detection
LOQ	Limit of quantification
MeOH	Methanol
N	Number of theoretical plates
OCE	Oncheong-eum
PA	Phosphoric acid
r^2	Coefficient of determination
Rs	Resolution
RSD	Relative standard deviation
Tf	Tailing factor
TFA	Trifluoroacetic acid
UPLC	Ultra-performance liquid chromatography

References

1. Chan, K. Chinese medicinal materials and their interface with Western medical concepts. *J. Ethnopharmacol.* **2005**, *96*, 1–18. [CrossRef]
2. Liu, S.; Yi, L.Z.; Liang, Y.Z. Traditional Chinese medicine and separation science. *J. Sep. Sci.* **2008**, *31*, 2113–2137. [CrossRef]
3. Mao, Q.; Xu, J.; Kong, M.; Shen, H.; Zhu, H.; Zhou, S.; Li, S. LC-MS-based metabolomics in traditional Chinese medicines research: Personal experiences. *Chin. Herb. Med.* **2017**, *9*, 14–21. [CrossRef]
4. Heo, J. *Donguibogam*; Namsandang: Seoul, Korea, 2007; pp. 159–161.

5. Han, J.M.; Lee, S.E.; Jung, H.J.; Choi, S.B.; Seo, H.S.; Jung, H.A.; Ko, W.S.; Yoon, H.J. Overseas clinical research trends of on Cheong Eum on skin disease. *J. Korean Med. Ophthalmol. Otolaryngol. Dermatol.* **2007**, *30*, 1–9.
6. Yeh, C.C.; Huang, S.S.; Liu, P.Y.; Wang, B.C.; Tsai, C.F.; Wang, D.Y.; Cheng, H.F. Simultaneous quantification of six indicator compounds in Wen-Qing-Yin by high-performance liquid chromatography-diode array detection. *J. Food Drug Anal.* **2019**, *27*, 749–757. [CrossRef]
7. Goo, I.M.; Shin, H.M. G1 arrest of the cell cycle by Onchungeum in human hepatocarcinoma cells. *Korean J. Orient. Physiol. Pathol.* **2008**, *22*, 821–828.
8. Goo, I.M.; Kim, G.W.; Shin, H.M. Change of ratio of Onchungeum composition induces different G1 arrest mechanisms in Hep3B cells. *Korean J. Orient. Physiol. Pathol.* **2008**, *22*, 1250–1255.
9. An, T.E.B.; Lim, D.C. *In vitro* cytotoxicity, skin regeneration, anti-wrinkle, whitening and *in vivo* skin moisturizing effects of Oncheongeum. *J. Korean Obstet. Gynecol.* **2016**, *29*, 14–34. [CrossRef]
10. Andoh, T.; Honma, Y.; Kawaharada, S.; Al-Akeel, A.; Nojima, H.; Kuraishi, Y. Inhibitory effect of the repeated treatment with Unsei-in on substance P-induced itch-associated responses through the downregulation of the expression of NK_1 tachykinin receptor in mice. *Biol. Pharm. Bull.* **2003**, *26*, 896–898. [CrossRef]
11. Andoh, T.; Al-Akeel, A.; Tsujii, K.; Nojima, H.; Kuraishi, Y. Repeated treatment with the traditional medicine Unsei-in inhibits substance P-induced itch-associated responses through downregulation of the expression of nitric oxide synthase 1 in mice. *J. Pharmacol. Sci.* **2004**, *94*, 207–210. [CrossRef]
12. Jeong, S.Y.; Kim, H.M.; Lee, K.H.; Kim, K.Y.; Huang, D.S.; Kim, J.H.; Seong, R.S. Quantitative analysis of marker compounds in *Angelica gigas*, *Angelica sinensis*, and *Angelica acutiloba* by HPLC/DAD. *Chem. Pharm. Bull.* **2015**, *63*, 504–511. [CrossRef]
13. Baek, M.E.; Seong, G.U.; Lee, Y.J.; Won, J.H. Quantitative analysis for the quality evaluation of active ingredients in Cnidium Rhizome. *Yakhak Hoeji* **2016**, *60*, 227–234. [CrossRef]
14. Bae, J.Y.; Kim, C.Y.; Kim, H.J.; Park, J.H.; Ahn, M.J. Differences in the chemical profiles and biological activities of *Paeonia lactiflora* and *Paeonia obovata*. *J. Med. Food* **2015**, *18*, 224–232. [CrossRef]
15. Lee, J.Y.; Lee, E.J.; Kim, J.S.; Lee, J.H.; Kang, S.S. Phytochemical studies on Rehmanniae Radix Preparata. *Korean J. Pharmacogn.* **2011**, *42*, 117–126.
16. Lv, X.; Li, Y.; Tang, C.; Zhang, Y.; Zhang, J.; Fan, G. Integration of HPLC-based fingerprint and quantitative analyses for differentiating botanical species and geographical growing origins of Rhizoma coptidis. *Pharm. Biol.* **2016**, *54*, 3264–3271. [CrossRef]
17. Ryuk, J.A.; Zheng, M.S.; Lee, M.Y.; Seo, C.S.; Li, Y.; Lee, S.H.; Moon, D.C.; Lee, H.W.; Lee, J.H.; Park, J.Y.; et al. Discrimination of *Phellodendron amurense* and *P. chinense* based on DNA analysis and the simultaneous analysis of alkaloids. *Arch. Pharm. Res.* **2012**, *35*, 1045–1054. [CrossRef]
18. Tong, L.; Wan, M.; Zhang, L.; Zhu, Y.; Sun, H.; Bi, K. Simultaneous determination of baicalin, wogonoside, baicalein, wogonin, oroxylin A and chrysin of Radix Scutellariae extract in rat plasma by liquid chromatography tandem mass spectrometry. *J. Pharm. Biomed. Anal.* **2012**, *70*, 6–12. [CrossRef]
19. Lee, E.J.; Hong, J.K.; Whang, W.K. Simultaneous determination of bioactive marker compounds from Gardeniae Fructus by high performance liquid chromatography. *Arch. Pharm. Res.* **2014**, *37*, 992–1000. [CrossRef]
20. Ku, Y.R.; Chang, Y.S.; Wen, K.C.; Ho, L.K. Analysis and confirmation of synthetic anorexics in adulterated traditional Chinese medicines by high-performance capillary electrophoresis. *J. Chromatogr. A* **1999**, *848*, 537–543. [CrossRef]
21. Su, S.; Hua, Y.; Duan, J.A.; Shang, E.; Tang, Y.; Bao, X.; Lu, Y.; Ding, A. Hypothesis of active components in volatile oil from a Chinese herb formulation, 'Shao–Fu–Zhu–Yu decoction', using GC-MS and chemometrics. *J. Sep. Sci.* **2008**, *31*, 1085–1091. [CrossRef]
22. Su, H.; Hui, H.; Xu, X.; Zhou, R.; Qin, L.; Shan, Q. Simultaneous determination of multiple components in formula and preparations of Xiaoyaosan. *Nat. Prod. Res.* **2021**, *35*, 1207–1211. [CrossRef]
23. Shu, X.; Tang, Y.; Jiang, C.; Shang, E.; Fan, X.; Ding, A. Comparative analysis of the main bioactive components of San-ao decoction and its series of formulations. *Molecules* **2012**, *17*, 12925–12937. [CrossRef]
24. Xie, M.; Yu, Y.; Zhu, Z.; Deng, L.; Ren, B.; Zhang, M. Simultaneous determination of six main components in Bushen Huoxue prescription by HPLC–CAD. *J. Pharm. Biomed. Anal.* **2021**, *201*, 114087. [CrossRef]
25. Seo, C.S.; Shin, H.K. Simultaneous analysis for quality control of traditional herbal medicine, Gungha-tang, using liquid chromatography–tandem mass spectrometry. *Molecules* **2022**, *27*, 1223. [CrossRef]
26. Lee, K.H. *The Dispensatory on the Visual and Organoleptic Examination of Herbal Medicine*; National Institute of Food and Drug Safety Evaluation: Seoul, Korea, 2013; pp. 143–728.
27. Seo, C.S.; Yoo, S.R.; Jeong, S.J.; Ha, H. Quantification of the constituents of the traditional Korea medicine, Samryeongbaekchul-san, and assessment of its antiadipogenic effect. *Saudi Pharm. J.* **2019**, *27*, 145–153. [CrossRef]
28. Seo, C.S.; Shin, H.K. Simultaneous determination of 12 marker components in Yeonkyopaedok-san using HPLC-PDA and LC-MS/MS. *Appl. Sci.* **2020**, *10*, 1713. [CrossRef]
29. International Conference on Harmonisation. *Guidance for Industry, Q2B, Validation of Analytical Procedures: Methodology*; Food and Drug Administration: Rockville, MD, USA, 1996.

Article

Wooden-Tip Electrospray Mass Spectrometry Characterization of Human Hemoglobin in Whole Blood Sample for Thalassemia Screening: A Pilot Study

Tingting Huang [1,2], Ting Huang [1,2], Yongyi Zou [1,2], Kang Xie [1,2], Yinqin Shen [1,2], Wen Zhang [1,2], Shuhui Huang [1,2,*], Yanqiu Liu [1,2,*] and Bicheng Yang [1,2,*]

- [1] Maternal and Child Health Affiliated Hospital of Nanchang University, Nanchang 330006, China; htt19871212@126.com (T.H.); huangting626@126.com (T.H.); zouyongyi@gmail.com (Y.Z.); ragexp@126.com (K.X.); syq2657454560@126.com (Y.S.); zw595581428@126.com (W.Z.)
- [2] Jiangxi Key Laboratory of Birth Defect Prevention and Control, Jiangxi Maternal and Child Health Hospital, Nanchang 330006, China
- * Correspondence: sinead321@163.com (S.H.); lyq0914@126.com (Y.L.); yangbc1985@126.com (B.Y.)

Abstract: Traditional analytical methods for thalassemia screening are needed to process complicated and time-consuming sample pretreatment. In recent decades, ambient mass spectrometry (MS) approaches have been proven to be an effective analytical strategy for direct sample analysis. In this work, we applied ambient MS with wooden-tip electrospray ionization (WT-ESI) for the direct analysis of raw human blood samples that were pre-identified by gene detection. A total of 319 whole blood samples were investigated in this work, including 100 α-thalassemia carriers, 67 β-thalassemia carriers, and 152 control healthy samples. Only one microliter of raw blood sample was directly loaded onto the surface of the wooden tip, and then five microliters of organic solvent and a high voltage of +3.0 kV were applied onto the wooden tip to generate spray ionization. Multiply charged ions of human hemoglobin (Hb) were directly observed by WT-ESI-MS from raw blood samples. The signal ratios of Hb chains were used to characterize two main types of thalassemia (α and β types) and healthy control blood samples. Our results suggested that the ratios of charged ions to Hb chains being at +13 would be an indicator for β-thalassemia screening.

Keywords: thalassemia; human hemoglobin; wooden-tip electrospray ionization; multiply charged ions; mass spectrometry; multiply charged ion

1. Introduction

Human thalassemia is an inherited blood disorder that causes patients to have less hemoglobin (Hb) than normal. The disease is commonly found worldwide, affecting part of the population in southern China and other Southeast Asian countries [1,2]. There are two main types of thalassemia, alpha thalassemia (α-thalassemia) and beta-thalassemia (β-thalassemia), in patients with thalassemia who are characterized by reduced α or β globin chain synthesis [3,4]. The absent production of α or β globin chains results in reducing the composition of Hb and the production of red blood cells and generating anemia symptoms, all of which are life-threatening conditions. Blood transfusions can usually improve patients' conditions [5]; however, there is currently no effective clinical method for thalassemia treatment, except stem cell transplantation and gene therapy [6]. Therefore, early detection and prevention play a key role in reducing thalassemia incidence through thalassemia screening.

Recently, there are many diagnosis methods for thalassemia screening [7,8], for example, routine blood analysis by mean corpuscular hemoglobin (MCH), mean corpuscular volume (MCV), erythrocyte osmotic fragility test (EOFT), serum iron test (SIT), Hb electrophoresis measurement (HEM), isoelectric focusing (IEF), liquid chromatography (LC), and next-generation sequencing (NGS). Among these methods, MCH and MCV are commonly used for thalassemia screening; however, these two methods are not specific enough. EOFT and SIT are typically used to identify microcytic hypochromic anemia with decreased MCH and MCV. HEM has been used for thalassemia screening, as this method can measure different electrophoretic behaviors of Hb tetramers. However, the principle of HEM is based on the structural integrity of intact Hb tetramers, and thus HEM could be affected by hemolysis and degradation of whole blood samples during storage. The combination of IEF and HPLC techniques could facilitate the identification of most known hemoglobinopathies. However, there is a limitation as it requires time-consuming and professional data interpretation, which is not amenable to thalassemia screening in wide clinical tests [9].

Mass spectrometry (MS) can measure the mass/charge ratio (m/z) and ion intensity of various clinical samples. To date, various MS methods have been developed as reference techniques for clinical analysis [10–12]. Moreover, MS-based methods have also been applied for diagnosing hemoglobinopathies [13,14]. Due to the complicated matrices in human blood samples, extensive sample preparation processes such as extraction and separation, which are labor intensive and time consuming, are usually required before LC-MS screening [7]. Therefore, reducing or removing the sample preparation and LC separation steps before MS analysis is highly demanded in clinical tests.

Ambient MS methods are of great interest for direct analysis of raw samples because analytes can be easily ionized with no or little sample pretreatment process [15–18]. Ambient MS is pioneered by desorption electrospray ionization (DESI), which allows direct determination of various analytes such as proteins from raw biological samples under ambient conditions [15]. Recently, various ambient MS methods have been developed for various applications [16]. Remarkable analytical properties were found in ambient ESI with different substrates [19]. Paper spray ionization is one of the powerful ambient ESI techniques using porous paper substrate to load and ionize raw samples. It is fact that paper spray is a great technological breakthrough in ambient ionization. Undoubtedly, paper spay can be used for the direct analysis of blood samples. Various excellent articles on blood analysis using paper spray have been published [20–22]. Similar to paper spray, electrospray ionization with a wooden tip (WT-ESI) is an ambient ESI method with the advantage that the ionization process occurs under ambient conditions in which the raw samples can be easily accessible during analytical processes [23]. WT-ESI has been successfully developed to analyze proteins, clinical samples, and other complex biological samples as a useful tool for direct analysis with high sensitivity and high specificity [24–28]. In the WT-ESI method, disposable and low-cost wooden tips are quite convenient for direct sampling and analysis without any hardware modifications. The wooden tip can be directly mounted on commercial nanoESI devices by replacing the nanoESI emitter [19,23], which is greatly beneficial for non-expert users, especially in clinical use. In general, raw samples can be directly loaded on the surface of a wooden tip under the strong electric field without the use of any gas. The compounds of interest are extracted from raw samples and then electrospray ions were directly generated from the wooden tip-end for MS analysis. In particular, WT-ESI-MS can also be used for the quantification of target analytes from complex samples with only little sample preparation and no chromatographic separation, and the analytical performances, including the linear range, accuracy, precision, and sensitivity, were well acceptable for the direct analysis of real samples [24,29,30]. It is reported that different protein structures can be recognized by ambient MS [31]. In previous studies [32,33], WT-ESI-MS has been successfully used for protein analysis. Furthermore, the mobility of proteins in wooden tips could be characterized under ESI conditions [34]. Therefore, WE-ESI-MS is expected to characterize the protein structures from raw biological samples.

To the best of our knowledge, the ambient MS analysis of blood samples for the application in thalassemia detection has rarely been investigated. Herein, we would like to make a new attempt at the direct analysis of blood samples for thalassemia detection with ambient MS with WT-ESI. In this work, a total of 319 whole blood samples were examined, and we aim to explore the detection of α and β chains from whole blood samples by WT-ESI-MS. We assume that α and β chains in thalassemia and healthy blood samples would have different ion responses during the spray ionization process and MS detection. Hence, the signal ratios of α and β chains between thalassemia and healthy blood samples would be significantly different, and thus the ratios can be used for detecting α-thalassemia and β-thalassemia

2. Results
2.1. Direct Analysis of Whole Blood Samples

Figure 1 shows a schematic diagram (Figure 1a) and picture (Figure 1b) of an experimental setup for WT-ESI-MS analysis of human whole blood samples. Figure 2a shows mass spectrum of human Hb in organic solution (i.e., methanol/water/formic acid, 50/50/0.1, $v/v/v$) obtained by conventional ESI-MS. Multiply charged ions of α-chains and β-chains with denatured Hb were observed. A wide range of charged-state distributions (CSD) of α-chains from +10 to +21 was found under a Gaussian distribution. The charged stats at +14~+16 were dominant in the mass spectrum, while there was a relatively narrow CSD ranging from +11 to +18 for β-chains where the main peaks are +13, +14, and +15. These peaks were also found under Gaussian distribution. Particularly, free heme at m/z 616.18 was detected; the heme detection confirmed that Hb is totally denatured under the organic solvent. Low-background-noise (Figure 1c) solvent is generated a from blank wooden material that is similar to a previous work using wooden-tip ESI [32–34].

Figure 2b shows a typical mass spectrum of a healthy whole blood sample using WT-ESI-MS. Various peaks of blood lipids at m/z 700–900 were observed. Lipids are common biometrics in blood samples and play an important role in life science [33]. Moreover, the heme and multiply charged ions of the α-chain and β-chain were clearly observed ranging from m/z 900 to m/z 1600, indicating that WT-ESI-MS has high ionization efficiency for direct ionization of Hb from raw biological samples. Unlike the pure Hb standard with a wide CSD in solution, only the main peaks of Hb were observed from raw blood samples. It is noted that the CSD of the α-chain was ranging from +10 to +16, while the main CSD of the β-chain was found from +11 to +13. It is also found that these CSDs of α-/β-chains from raw whole blood samples are narrower than those in Hb standard.

In particular, it is important to note that the peaks of protein ions from raw blood samples are wider than those obtained from Hb standard, probably because Hb chains combined with other small ligands, salt, and molecules from biological matrices, which is a common phenomenon in direct ionization of proteins from untreated biological samples [33]. By using WT-ESI-MS, direct analysis of a single blood sample can be completed within 1 min. Overall, these results show that Hb can be directly detected from raw blood samples without sample pretreatment.

Figure 2c shows a typical mass spectrum of whole blood from an α-thalassemia sample using WT-ESI-MS. Similar to healthy blood samples, there are abundant peaks observed in the mass spectrum. Interestingly, the heme at m/z 616.17 was dominated in the mass spectrum at the base peak, while the peaks of lipids and multiply charged ions of α-chains (CSD: +10~+18) and β-chains (CSD: +11~+16) were clearly also observed. Figure 2d shows a typical mass spectrum of the whole blood from a β-thalassemia sample using WT-ESI-MS. Similar to other raw blood samples, although the relative abundance is different, there are abundant peaks of heme, lipids, and Hb chains (α-chain: CSD: +10~+16; β-chain: CSD: +11~+14) observed in the mass spectrum.

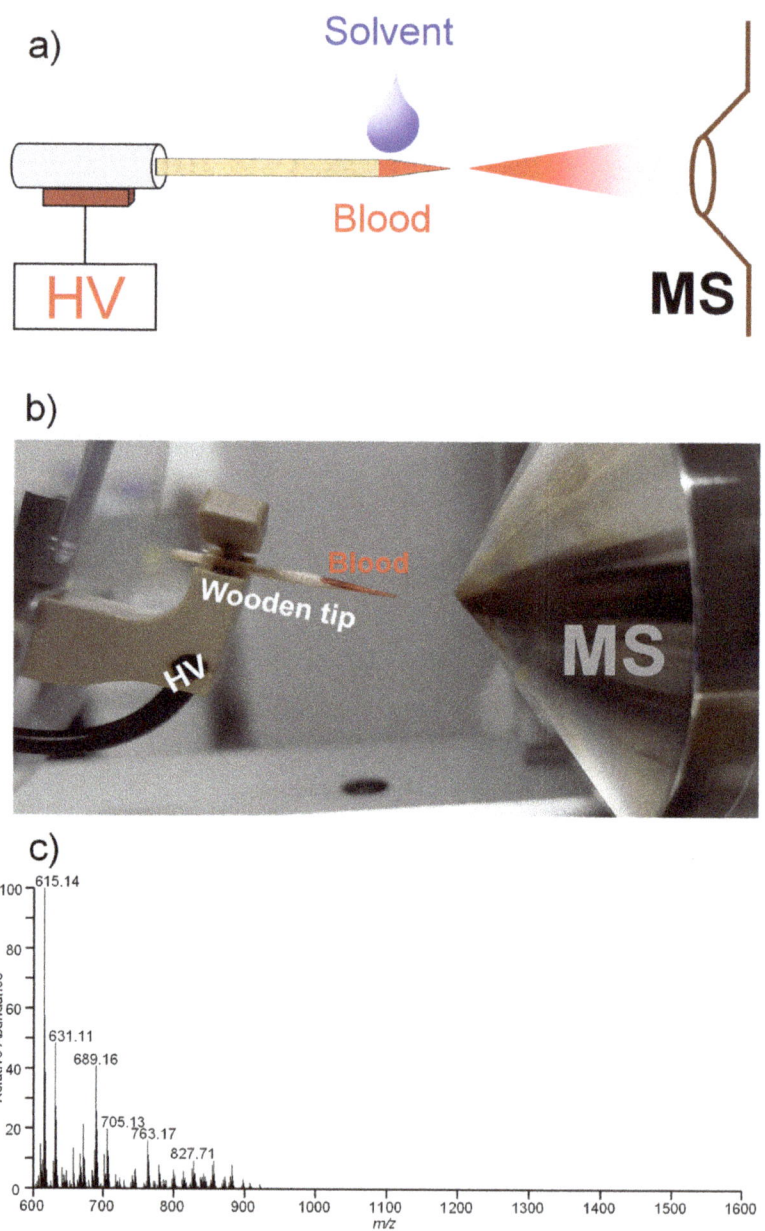

Figure 1. (**a**) Schematic diagram and (**b**) picture of WT-ESI-MS for direct analysis of blood samples; (**c**) background signal of a wooden tip with solvent only.

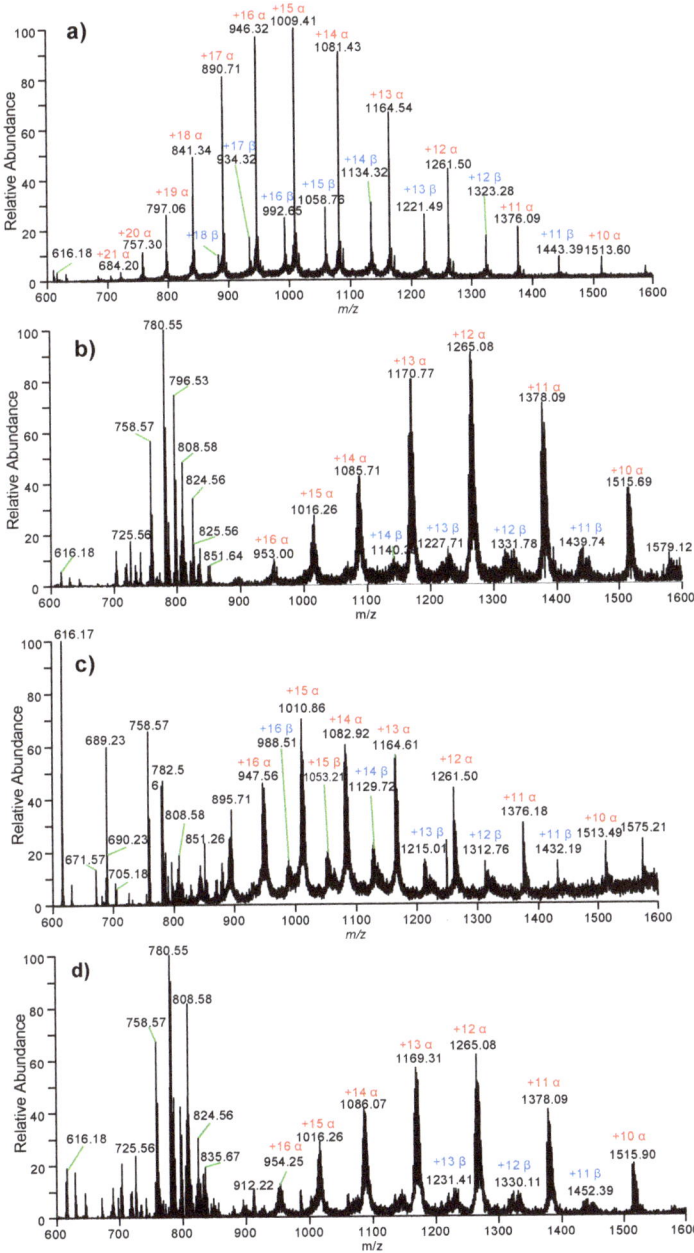

Figure 2. Mass spectra obtained by ESI-MS for the direct detection of human Hb from: (**a**) conventional ESI spectrum of pure Hb standard, (**b**) WT-ESI spectrum of healthy whole blood, (**c**) WT-ESI spectrum of α-thalassemia whole blood, and (**d**) WT-ESI spectrum of β-thalassemia whole blood.

2.2. Signal Ratios of Main Protein Ions

Figure 3a shows signal ratios of α-chain to β-chain at the charged state of +11 obtained from α-thalassemia and healthy control samples. It was found that the ratios are at about

4.0 for both samples. However, there was no significant difference between thalassemia and healthy control samples. For the β-thalassemia samples at the charged state of +11 (Figure 3b), it was also found that there was no significant difference between thalassemia and healthy control samples.

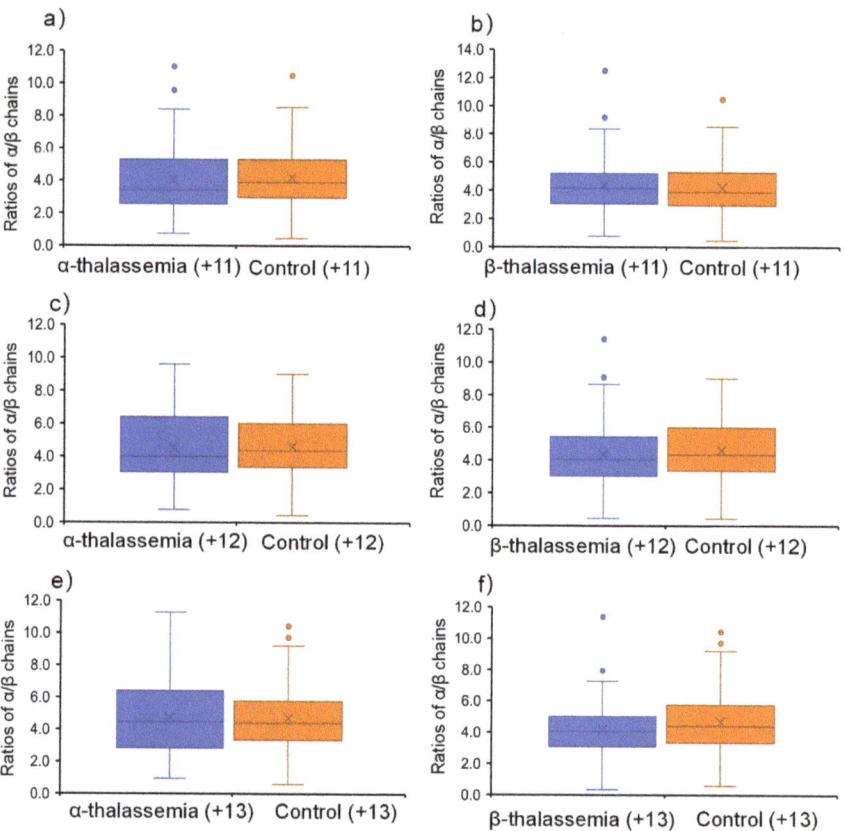

Figure 3. Ratios of Hb α-/β-chains from α-/β-thalassemia to healthly whole blood samples by wooden-tip ESI-MS under different charged states: (**a**) α-thalassemia (+11), (**b**) β-thalassemia (+11), (**c**) α-thalassemia (+12), (**d**) β-thalassemia (+12), (**e**) α-thalassemia (+13), and (**f**) β-thalassemia (+13).

When the α-chain and β-chain were at a charged state of +12, both α-thalassemia (Figure 3c) and β-thalassemia (Figure 3d) had no significant differences by comparing the healthy control samples. Interestingly, when the charged state was at +13, although there was no significant difference (Figure 3e) between α-thalassemia and healthy control samples, a significant difference was found between β-thalassemia and healthy control samples ($p < 0.05$), as shown in Figure 3f. Moreover, it was found that the ratios (median value: M; and average value: A) of healthy (A = 4.17, M = 4.06) and thalassemia (A = 4.69, M = 4.40) blood are different.

3. Discussion

The wooden tips were used for loading blood samples, and the organic solvent was loaded for extracting Hb chains to generate spray ionization. Multiply charged ions of Hb chains were obtained along with other biometrics. Considering multiply charged ions at different CSDs, each charged ion was compared in this work. As α chain and β chain are internal substances, the ratio would be stable for a normal blood sample, and thus the patients can be diagnosed. External and internal labels are not really needed. Our results suggested that the CDSs of α/β-chains at +13 ($p < 0.05$) and their ratios can be an indicator for β-thalassemia screening of a single blood sample. As 100 α-thalassemia, 67 β-thalassemia, and 152 healthy samples were analyzed in this work, some discrete data in Figure 3f are acceptable. In the electrophoresis process of chains of Hb, there are substantial significant differences obtained from β-thalassemia compared to α-thalassemia [35–38]. Therefore, by using a simple WT-ESI-MS method in this work, our results validated the results of the electrophoresis process of chains of Hb. Without any separation and sample pretreatment, discriminating β-thalassemia would be beneficial for developing a rapid diagnosis method based on ambient ionization methods in the future.

Hb is a tetramer with four polypeptide chains ($\alpha_2\beta_2$) that included two alpha (α) and two beta (β) chains. The α-chain of Hb is comprised of 141 amino acids (molecular weight: 15,126 Da) [39], while the β-chain is comprised of 147 amino acids and has a molecular weight of 15,867 Da. Compared to native MS analysis of human Hb [40], the observation of chains and heme ambiguously shows that Hb tetramer was totally denatured in organic solution, and thus the basic sites such as basic amino acids and N-terminal residues were protonated in the ESI process [40]. As there are different sequences and masses between α-chains and β-chains, the two chains at the same charged states have different ionization responses in the ESI process. Therefore, it is a reasonable hypothesis that if there is a lack of an α/β-chain in an Hb sample, the signal ratio of α-chains to β-chains would be changed accordingly. It is also noted that there are different responses of lipids in thalassemia in this work, suggesting that the blood lipids would also be an indicator of the diagnosis and research of thalassemia [41]. However, more investigations are highly needed to further validate blood lipids for thalassemia screening.

Furthermore, the tetramer is denatured and Hb chains are unfolded in organic solution, and amino acid residues of Hb chains can be exposed to solution. More protons can be attached onto exposed amino acid residues when the unfolding degree of the proteins is greater. Increased charged states indicated that more amino acids were ionized, and changed Hb chains have stronger detectability than native Hb. Therefore, it is reasonable that there is a significant difference between thalassemia and healthy samples at a charged state of +13. Therefore, it can be expected that higher charged states would give higher significant differences. However, there is a limitation to generating higher charged ions of Hb chains from raw samples. Supercharging agents were reported to further elevate peptide and protein charge states in ESI [42], which would be beneficial for increasing the charged states of Hb chains. Moreover, compared to the Hb standard, there is relatively lower signal-to-noise for detecting raw blood samples. Surface-modified wooden-tip and microextraction methods might give some hints to improving the signal-to-noise and enhancing the detection of target analytes by reducing the matrix effects [26,27,43,44].

4. Materials and Methods

4.1. Chemicals and Materials

Human Hb was bought from Sigma. HPLC-grade Methanol was bought from the Chinese Chemical Reagent Co., Ltd. (Shanghai, China). Water used in this work was Milli-Q water. Wooden tips (toothpicks) were purchased from the local supermarket (Nanchang, China). The length of the wooden tip was cut to ~2.0 cm in this study. All wooden tips were washed with methanol and water and were dried (100 °C) before use. A total of 319 whole blood samples from 319 women (person, their ages were from 21 to 41 and the average age was 28) were collected into EDTA-K2 tube (Shanghai Zhengbang Medical Treatment Technology Co., Ltd., Shanghai, China), including 100 α-thalassemia carriers, 67 β-thalassemia carriers, and 152 healthy samples (without any gene type of thalassemia carriers). All blood samples were collected from Jiangxi Maternal and Child Health Hospital (Nanchang, China). α/β-thalassemia blood samples were firstly confirmed by polymerase chain reaction (PCR) and flow-through hybridization technology analysis

(Hybribio Limited, Chaozhou, China) to assign the types of thalassemia carriers; the identification methods followed our previous studies [2,8].

4.2. WT-ESI-MS Analysis

Figure 1 shows the setup of wooden-tip ESI-MS. The wooden tips were cut to sharp points (tip size: 0.1 mm). The wooden tip was placed in the front of the mass spectrometer with distances of 0.8 cm horizontally from the wooden tip-end to the MS inlet via a nanoESI device (Thermo Fisher Scientific, Bremen, Germany). To load blood sample, whole blood samples (i.e., 1.0 µL) were directly loaded onto a wooden tip, and then organic solvent (5.0 µL of methanol/water/formic acid, 50/50/0.1, $v/v/v$), which is commonly used in spray ionization in ambient MS [45–47], was loaded onto the wooden tip in this work. With the application of a high voltage (3.0 kV) to wooden tip, spray ionization could be produced from wooden tip-end to acquire a mass spectrum. Mass spectra were acquired on an Orbitrap mass spectrometer (Thermo Fisher Scientific, Bremen, Germany). Data acquisition and instrumental control were conducted by using Xcalibur 3.0 software (Thermo Fisher Scientific, Bremen, Germany). The capillary temperature was set at 150 °C. The high voltage for spray ionization was supplied from the mass spectrometer to the wooden tip via a clip in a nanoESI device (Figure 1b). The acquisition speed was 4 scans/s. Typically, data from the first 1 min were averaged to generate the mass spectra. All the mass spectra were directly obtained. To determine the signal ratios of each ion, the absolute intensity of each ion was obtained to compare in this work. Conventional ESI-MS was also performed to analyze Hb standard (10 µM) in organic solvent (i.e., methanol/water/formic acid, 50/50/0.1, $v/v/v$). ESI-MS conditions were referred to ESI-MS analysis of protein solution [48]. Briefly, the ionization voltage wat at 3.0 kV under positive mode; the capillary temperature was set at 150 °C; flowrate of protein solution was at 3.0 µL/min; sheath gas, aux gas, and sweep gas were set at 20 bar, 10 bar and 2 L/min, respectively. Each sample was analyzed three times to measure the *p*-value in this work. *p*-value calculation with a two-tailed test is used in this work.

5. Conclusions

In conclusion, we applied an ambient MS method with a WT-ESI approach to explore the differentiation of thalassemia using Hb chains. Without any sample pretreatment and separation, WT-ESI-MS can be used for the direct analysis of raw blood samples. A single sample can be completed within 1 min. In this work, a significant difference is found between β-thalassemia and healthy control samples. Our pilot study suggests that WT-ESI-MS and other ambient MS methods would be potential new clinical tools for thalassemia screening in the future. However, more investigations on differentiating different types of thalassemia are needed to further validate this method, and more mechanical studies of thalassemia screening by ambient MS are also needed to improve the applicability of thalassemia screening.

Author Contributions: B.Y., Y.L. and S.H. conceived and designed this study. Y.Z., K.X., Y.S. and W.Z. collected the written informed consent and blood samples, carried out ESI-MS analysis, and analyzed the data. T.H. (Tingting Huang) and T.H. (Ting Huang) wrote and revised the manuscript. All authors have read and agreed to the published version of the manuscript.

Funding: This work was supported by the Superior scientific and technological innovation team of Jiangxi Province (No. 20181BCB24014).

Institutional Review Board Statement: This study was approved by the Ethics Committee of Jiangxi Provincial Maternal and Child Health Hospital and adhered to the Ethical Review Law of the Department of Law and Legislation of NHFPC of China and the tenets of The Declaration of Helsinki.

Informed Consent Statement: Written informed consent has been obtained from the volunteers and patients to publish this paper.

Data Availability Statement: Not applicable.

Conflicts of Interest: The authors declare no conflict of interest.

Sample Availability: Samples of the compounds are not available from the authors.

References

1. Srivorakun, H.; Fucharoen, G.; Changtrakul, Y.; Komwilaisak, P.; Fucharoen, S. Thalassemia and hemoglobinopathies in Southeast Asian newborns: Diagnostic assessment using capillary electrophoresis system. *Clin. Biochem.* **2011**, *44*, 406–411. [CrossRef] [PubMed]
2. Luo, H.; Zou, Y.; Liu, Y. A Novel β-Thalassemia Mutation [IVS-I-6 (T>G), HBB: C.92+6T>G] in a Chinese Family. *Hemoglobin* **2020**, *44*, 55–57. [CrossRef] [PubMed]
3. Viprakasit, V.; Ekwattanakit, S. Clinical Classification, Screening and Diagnosis for Thalassemia. *Hematol. Oncol. Clin. N. Am.* **2018**, *32*, 193–211. [CrossRef] [PubMed]
4. Muncie, H.L., Jr.; Campbell, J. Alpha and beta thalassemia. *Am. Fam. Physician.* **2009**, *80*, 339–344. [PubMed]
5. Shah, F.T.; Sayani, F.; Trompeter, S.; Drasar, E.; Piga, A. Challenges of blood transfusions in β-thalassemia. *Blood Rev.* **2019**, *37*, 100588. [CrossRef]
6. Karponi, G.; Zogas, N. Gene Therapy For Beta-Thalassemia: Updated Perspectives. *Appl. Clin. Genet.* **2019**, *12*, 167–180. [CrossRef]
7. Yu, C.; Huang, S.; Wang, M.; Zhang, J.; Liu, H.; Yuan, Z.; Wang, X.; He, X.; Wang, J.; Zou, L. A novel tandem mass spectrometry method for first-line screening of mainly beta-thalassemia from dried blood spots. *J. Proteom.* **2017**, *154*, 78–84. [CrossRef]
8. Liang, Q.; Gu, W.; Chen, P.; Li, Y.; Liu, Y.; Tian, M.; Zhou, Q.; Qi, H.; Zhang, Y.; He, J.; et al. A More Universal Approach to Comprehensive Analysis of Thalassemia Alleles (CATSA). *J. Mol. Diagn.* **2021**, *23*, 1195–1204. [CrossRef]
9. Ryan, K.; Bain, B.J.; Worthington, D.; James, J.; Plews, D.; Mason, A.; Roper, D.; Rees, D.C.; De La Salle, B.; Streetly, A.; et al. Significant haemoglobinopathies: Guidelines for screening and diagnosis. *Br. J. Haematol.* **2010**, *149*, 35–49. [CrossRef]
10. Yuan, Z.-C.; Hu, B. Mass Spectrometry-Based Human Breath Analysis: Towards COVID-19 Diagnosis and Research. *J. Anal. Test.* **2021**, *5*, 287–297. [CrossRef]
11. Banerjee, S. Empowering Clinical Diagnostics with Mass Spectrometry. *ACS Omega* **2020**, *5*, 2041–2048. [CrossRef] [PubMed]
12. Macklin, A.; Khan, S.; Kislinger, T. Recent advances in mass spectrometry based clinical proteomics: Applications to cancer research. *Clin. Proteom.* **2020**, *17*, 17. [CrossRef] [PubMed]
13. Boemer, F.; Ketelslegers, O.; Minon, J.M.; Bours, V.; Schoos, R. Newborn screening for sickle cell disease using tandem mass spectrometry. *Clin. Chem.* **2008**, *54*, 2036–2041. [CrossRef] [PubMed]
14. Traeger-Synodinos, J.; Harteveld, C.L. Advances in technologies for screening and diagnosis of hemoglobinopathies. *Biomark. Med.* **2014**, *8*, 119–131. [CrossRef]
15. Cooks, R.G.; Ouyang, Z.; Takats, Z.; Wiseman, J.M. Ambient Mass Spectrometry. *Science* **2006**, *311*, 1566–1570. [CrossRef]
16. Feider, C.L.; Krieger, A.; DeHoog, R.J.; Eberlin, L.S. Ambient Ionization Mass Spectrometry: Recent Developments and Applications. *Anal. Chem.* **2019**, *91*, 4266–4290. [CrossRef]
17. Pekov, S.I.; Zhvansky, E.S.; Eliferov, V.A.; Sorokin, A.A.; Ivanov, D.G.; Nikolaev, E.N.; Popov, I.A. Determination of Brain Tissue Samples Storage Conditions for Reproducible Intraoperative Lipid Profiling. *Molecules* **2022**, *27*, 2587. [CrossRef]
18. Shamraeva, M.A.; Bormotov, D.S.; Shamarina, E.V.; Bocharov, K.V.; Peregudova, O.V.; Pekov, S.I.; Nikolaev, E.N.; Popov, I.A. Spherical Sampler Probes Enhance the Robustness of Ambient Ionization Mass Spectrometry for Rapid Drugs Screening. *Molecules* **2022**, *27*, 945. [CrossRef]
19. Hu, B.; Yao, Z.-P. Electrospray ionization mass spectrometry with wooden tips: A review. *Anal. Chim. Acta* **2022**, *1209*, 339136. [CrossRef]
20. Shi, R.-Z.; El Gierari, E.T.M.; Faix, J.D.; Manicke, N.E. Rapid measurement of cyclosporine and sirolimus in whole blood by paper spray–tandem mass spectrometry. *Clin. Chem.* **2016**, *62*, 295–297. [CrossRef]
21. Carmany, D.O.; Mach, P.M.; Rizzo, G.M.; Dhummakupt, E.S.; McBride, E.M.; Sekowski, J.W.; Benton, B.; Demond, P.S.; Busch, M.W.; Glaros, T. On-substrate enzymatic reaction to determine acetylcholinesterase activity in whole blood by paper spray mass spectrometry. *J. Am. Soc. Mass Spectrom.* **2018**, *29*, 2436–2442. [CrossRef] [PubMed]
22. Frey, B.S.; Heiss, D.R.; Badu-Tawiah, A.K. Embossed Paper Platform for Whole Blood Collection, Room Temperature Storage, and Direct Analysis by Pinhole Paper Spray Mass Spectrometry. *Anal. Chem.* **2022**, *94*, 4417–4425. [CrossRef] [PubMed]
23. Hu, B.; So, P.-K.; Chen, H.; Yao, Z.-P. Electrospray ionization using wooden tips. *Anal. Chem.* **2011**, *83*, 8201–8207. [CrossRef] [PubMed]
24. Hu, B.; So, P.-K.; Yao, Z.-P. Analytical properties of solid-substrate electrospray ionization mass spectrometry. *J. Am. Soc. Mass Spectrom.* **2013**, *24*, 57–65. [CrossRef]
25. Yang, B.C.; Liu, F.Y.; Guo, J.B.; Wan, L.; Wu, J.; Wang, F.; Liu, H.; Huang, O.P. Rapid assay of neopterin and biopterin in urine by wooden-tip electrospray ionization mass spectrometry. *Anal. Methods* **2015**, *7*, 2913–2916. [CrossRef]
26. Hu, B.; So, P.-K.; Yang, Y.; Deng, J.; Choi, Y.-C.; Luan, T.; Yao, Z.-P. Surface-Modified Wooden-Tip Electrospray Ionization Mass Spectrometry for Enhanced Detection of Analytes in Complex Samples. *Anal. Chem.* **2018**, *90*, 1759–1766. [CrossRef]

27. Deng, J.; Yu, T.; Yao, Y.; Peng, Q.; Luo, L.; Chen, B.; Wang, X.; Yang, Y.; Luan, T. Surface-coated wooden-tip electrospray ionization mass spectrometry for determination of trace fluoroquinolone and macrolide antibiotics in water. *Anal. Chim. Acta.* **2017**, *954*, 52–59. [CrossRef]
28. Yang, Y.; Deng, J.; Yao, Z.P. Field-induced wooden-tip electrospray ionization mass spectrometry for high-throughput analysis of herbal medicines. *Anal. Chim. Acta* **2015**, *887*, 127–137. [CrossRef]
29. Yang, B.-C.; Wang, F.; Deng, W.; Zou, Y.; Liu, F.-Y.; Wan, X.-D.; Yang, X.; Liu, H.; Huang, O.-P. Wooden-tip electrospray ionization mass spectrometry for trace analysis of toxic and hazardous compounds in food samples. *Anal. Methods* **2015**, *7*, 5886–5890. [CrossRef]
30. So, P.-K.; Ng, T.-T.; Wang, H.; Hu, B.; Yao, Z.-P. Rapid detection and quantitation of ketamine and norketamine in urine and oral fluid by wooden-tip electrospray ionization mass spectrometry. *Analyst* **2013**, *138*, 2239–2243. [CrossRef]
31. Yao, Z.-P. Characterization of proteins by ambient mass spectrometry. *Mass Spectrom. Rev.* **2012**, *31*, 437–447. [CrossRef] [PubMed]
32. Wu, L.; Yao, Y.-N.; Yuan, Z.-C.; Di, D.; Li, L.; Hu, B. Direct detection of lysozyme in viscous raw hen egg white binding to sodium dodecyl sulfonate by reactive wooden-tip electrospray ionization mass spectrometry. *Anal. Sci.* **2020**, *36*, 341–346. [CrossRef] [PubMed]
33. Hu, B.; Yao, Z.-P. Detection of native proteins using solid-substrate electrospray ionization mass spectrometry with nonpolar solvents. *Anal. Chim. Acta* **2018**, *1004*, 51–57. [CrossRef] [PubMed]
34. Hu, B.; Yao, Z.-P. Mobility of proteins in porous substrates under electrospray ionization conditions. *Anal. Chem.* **2016**, *88*, 5585–5589. [CrossRef]
35. Giambona, A.; Passarello, C.; Renda, D.; Maggio, A. The significance of the hemoglobin A2 value in screening for hemoglobinopathies. *Clin. Biochem.* **2009**, *42*, 1786–1796. [CrossRef]
36. Mosca, A.; Paleari, R.; Ivaldi, G.; Galanello, R.; Giordano, P. The role of haemoglobin A2 testing in the diagnosis of thalassaemias and related haemoglobinopathies. *J. Clin. Pathol.* **2009**, *62*, 13–17. [CrossRef]
37. Huo, M.; Wu, W.-Y.; Liu, M.; Gan, Z.-B.; Mao, W.-Y.; Lin, R.-Y.; Liu, A.-Q.; He, G.-R. Analysis of Cut-off Value in Screening of Thalassemia by Capillary Hemoglobin Electrophoresis for Pregnant Women from Shenzhen region of China. *J. Exp. Hematol.* **2016**, *24*, 536–539.
38. Zou, J.; Huang, S.; Xi, H.; Huang, C.; Zou, L.; Qiu, L.; Nie, X.; Zhou, J.; Zhuang, Y.; Chen, Y.; et al. Application of an optimized interpretation model in capillary hemoglobin electrophoresis for newborn thalassemia screening. *Int. J. Lab. Hematol.* **2022**, *44*, 223–228. [CrossRef]
39. Mekecha, T.T.; Amunugama, R.; McLuckey, S.A. Ion trap collision-induced dissociation of human hemoglobin α-chain cations. *J. Am. Soc. Mass Spectrom.* **2006**, *17*, 923–931. [CrossRef]
40. Martin, N.J.; Griffiths, R.L.; Edwards, R.L.; Cooper, H.J. Native Liquid Extraction Surface Analysis Mass Spectrometry: Analysis of Noncovalent Protein Complexes Directly from Dried Substrates. *J. Am. Soc. Mass Spectrom.* **2015**, *26*, 1320–1327. [CrossRef]
41. Chrysohoou, C.; Panagiotakos, D.B.; Pitsavos, C.; Kosma, K.; Barbetseas, J.; Karagiorga, M.; Ladis, I.; Stefanadis, C. Distribution of serum lipids and lipoproteins in patients with beta thalassaemia major; an epidemiological study in young adults from Greece. *Lipids Health Dis.* **2004**, *3*, 3. [CrossRef] [PubMed]
42. Konermann, L.; Metwally, H.; Duez, Q.; Peters, I. Charging and supercharging of proteins for mass spectrometry: Recent insights into the mechanisms of electrospray ionization. *Analyst* **2019**, *144*, 6157–6171. [CrossRef] [PubMed]
43. Deng, J.; Yang, Y.; Fang, L.; Lin, L.; Zhou, H.; Luan, T. Coupling Solid-Phase Microextraction with Ambient Mass Spectrometry Using Surface Coated Wooden-Tip Probe for Rapid Analysis of Ultra Trace Perfluorinated Compounds in Complex Samples. *Anal. Chem.* **2014**, *86*, 11159–11166. [CrossRef]
44. So, P.-K.; Yang, B.-C.; Li, W.; Wu, L.; Hu, B. Simple Fabrication of Solid-Phase Microextraction with Surface-Coated Aluminum Foil for Enhanced Detection of Analytes in Biological and Clinical Samples by Mass Spectrometry. *Anal. Chem.* **2019**, *91*, 9430–9434. [CrossRef] [PubMed]
45. Yao, Y.-N.; Hu, B. Analyte-substrate interactions at functionalized tip electrospray ionization mass spectrometry: Molecular mechanisms and applications. *J. Mass Spectrom.* **2018**, *53*, 1222–1229. [CrossRef]
46. Wu, L.; Yuan, Z.-C.; Yang, B.-C.; Huang, Z.; Hu, B. In vivo solid-phase microextraction swab-mass spectrometry for multidimensional analysis of human saliva. *Anal. Chim. Acta* **2021**, *1164*, 338510. [CrossRef]
47. Yang, B.-c.; Wan, X.-d.; Yang, X.; Li, Y.-j.; Zhang, Z.-y.; Wan, X.-j.; Luo, Y.; Deng, W.; Wang, F.; Huang, O.-p. Rapid determination of carbendazim in complex matrices by electrospray ionization mass spectrometry with syringe filter needle. *J. Mass Spectrom.* **2018**, *53*, 234–239. [CrossRef]
48. Donnelly, D.P.; Rawlins, C.M.; DeHart, C.J.; Fornelli, L.; Schachner, L.F.; Lin, Z.; Lippens, J.L.; Aluri, K.C.; Sarin, R.; Chen, B. Best practices and benchmarks for intact protein analysis for top-down mass spectrometry. *Nat. Methods* **2019**, *16*, 587–594. [CrossRef]

Article

Analytical Method Development for 19 Alkyl Halides as Potential Genotoxic Impurities by Analytical Quality by Design

Kyoungmin Lee [1,2,†], Wokchul Yoo [1,2,†] and Jin Hyun Jeong [1,*]

1. College of Pharmacy, Yonsei Institute of Pharmaceutical Sciences, Yonsei University, 85 Songdogwahak-ro, Yeonsu-gu, Incheon 21983, Korea; lkm@hanmi.co.kr (K.L.); panda8838@hanmi.co.kr (W.Y.)
2. Analytical Research Department, Central Research Institute, Hanmi fine Chemical, 59 Gyeongje-ro, Siheung-si 15093, Korea
* Correspondence: organicjeong@yonsei.ac.kr
† These authors contributed equally to this work.

Abstract: Major issues in the pharmaceutical industry involve efficient risk management and control strategies of potential genotoxic impurities (PGIs). As a result, the development of an appropriate method to control these impurities is required. An optimally sensitive and simultaneous analytical method using gas chromatography with a mass spectrometry detector (GC–MS) was developed for 19 alkyl halides determined to be PGIs. These 19 alkyl halides were selected from 144 alkyl halides through an in silico study utilizing quantitative structure–activity relationship (Q-SAR) approaches via expert knowledge rule-based software and statistical-based software. The analytical quality by design (QbD) approach was adopted for the development of a sensitive and robust analytical method for PGIs. A limited number of literature studies have reviewed the analytical QbD approach in the PGI method development using GC–MS as the analytical instrument. A GC equipped with a single quadrupole mass spectrometry detector (MSD) and VF-624 ms capillary column was used. The developed method was validated in terms of specificity, the limit of detection, quantitation, linearity, accuracy, and precision, according to the ICH Q2 guideline.

Keywords: GC–MS; analytical QbD; genotoxic impurity; alkyl halide; (Q)SAR; analytical method development

1. Introduction

Alkyl halides, also known as haloalkanes or halogenoalkanes, are chemical compounds formed from alkanes containing one or more halogen atoms, such as chlorine, bromine, fluorine, or iodine. These compounds are mainly used in alkylation reactions via nucleophilic substitutions as starting materials or reagents in synthetic processes of active pharmaceutical ingredients (APIs) because of their good reactivity, ease of use, reasonable prices, and commercial availability. Alkyl halides are considered potential genotoxic impurities (PGIs) because they can possibly alkylate DNA bases (on N-7 of guanine and N-3 of adenine) [1–4]. Alkyl halides are potentially formed during API manufacturing processes during undesired chemical reactions or are carried over to APIs when used as starting materials or reagents. For these reasons, alkyl halides should be considered significant PGIs when developing manufacturing processes and designating quality control strategies for APIs to manage genotoxic impurities [5–8].

The concept of structural alerts for genotoxicity was established by Ashby and Tennant in the 1980s [9]. As a result of this establishment, mutagenic and genotoxic in vitro and in vivo studies for many substances have been conducted. However, these experiments had limitations (in that the results could not be obtained for all substances because of the costs and time requirements). In particular, there is a lack of genotoxic evaluation data for alkyl halides commonly used in the pharmaceutical industry. Additional studies and experiments are being conducted to mitigate these limitations. Regulatory authorities, such

as the FDA and EMEA, have also expanded their compilations of genotoxicity research data and are continuously publishing guidelines for the pharmaceutical industry as addendums. In silico studies using quantitative structure–activity relationship (Q-SAR) prediction programs could be effective alternative approaches in terms of cost- and time-saving. In many studies, Q-SAR prediction models showed excellent predictive performances and could serve as early warning systems for the prediction of genotoxicity in compounds.

Once the compounds are confirmed to have potential genotoxicity through the use of Q-SAR prediction programs, the manufacturing processes of APIs should be altered to remove or minimize these substances classified as PGIs. However, complete elimination is often not possible; in this case, the amount of PGIs present should be limited to specific levels because of their genetically-threatening behaviors. PGIs are controlled according to EMEA and FDA guidelines because of the different risks and behaviors they impose, unlike other general impurities that are controlled according to the ICH Q3A and Q3B guidelines. The limits of PGIs are established based on the experimental results of toxicities that may be applied according to the threshold of toxicological concern (TTC), which is set at 1.5 µg/day. Generally, the limits of PGIs are much lower than the typical limit of non-PGIs, as even a trace amount of PGIs can affect drug quality and human health. Therefore, it is essential to develop a highly sensitive analytical method capable of detecting limits of sub-ppm for potential genotoxic substances for manufacturing process research and product quality control. Several analytical methods have already been developed with high sensitivity for PGIs and alkyl halides to meet these requirements in other previous literature studies [10–14]. Most of these analysis methods were performed using a gas chromatography or liquid chromatography system, altered according to the characteristics of each impurity, alongside a mass spectrometry detector capable of detecting trace amounts of impurities. Other literature studies used detectors other than the mass spectrometry detector along with their respective chromatographic systems [15–21]. The analytical quality by design (QbD) approach can be a good tool for the development of an optimally robust analytical method for PGIs [22,23]. The analytical QbD is a systematic approach that applies the QbD concept to the analytical method development to facilitate regulatory flexibility and prevent the undesired risks of quality control, such as out of specification (OOS) and out of trend (OOT), by increasing the scientific understanding of analytical methods.

Although the genotoxic behaviors of alkyl halides are widely known, experimental mutagenicity and/or carcinogenicity results, as well as Q-SAR analysis results, are limited. control strategies based on risk assessment and classification according to the ICH M7 guidelines are required because alkyl halides are essential compounds in the pharmaceutical industry [24]. Applying the analytical QbD approach can be effective in developing an optimally robust analytical method, which is one of the most essential procedures in the control strategies for PGIs. Many researchers have adopted the analytical QbD principle in the development of analytical methods of APIs according to numerous literature studies [25–34]. However, the majority of these research studies focused on the purity and assay method development of non-PGIs. In particular, few literature studies have applied the analytical QbD approach to the development of analytical methods using gas chromatography as the source of instrumentation, which is an effective analytical method for volatile substances, such as alkyl halides.

In this study, an in silico experiment was performed using the Q-SAR approach for 144 alkyl halides with one to four carbons and one to two halogen atoms. As a result of the in silico experiment, the alkyl halides were classified from one to five according to the ICH M7 guideline. A total of 19 alkyl halides were selected as having the highest risks among the 144 alkyl halides evaluated through the risk assessment (Figure 1). The analytical method development using the analytical QbD approach and GC–MS focused on high sensitivity, specificity, and reproducibility for the selected 19 alkyl halides. In most literature studies about the analytical method development employing the analytical QbD approach, the resolution and analysis time were selected as critical method attributes

(CMAs). In this study, the peak area and resolution of each peak were selected as the CMAs due to the most important analytical target profiles (ATPs) in the analytical method for PGIs being detection sensitivity and specificity. In addition, the analytical method validation was conducted according to the ICH Q2 guidelines in terms of system suitability, specificity, linearity, range, accuracy, precision, and robustness [35].

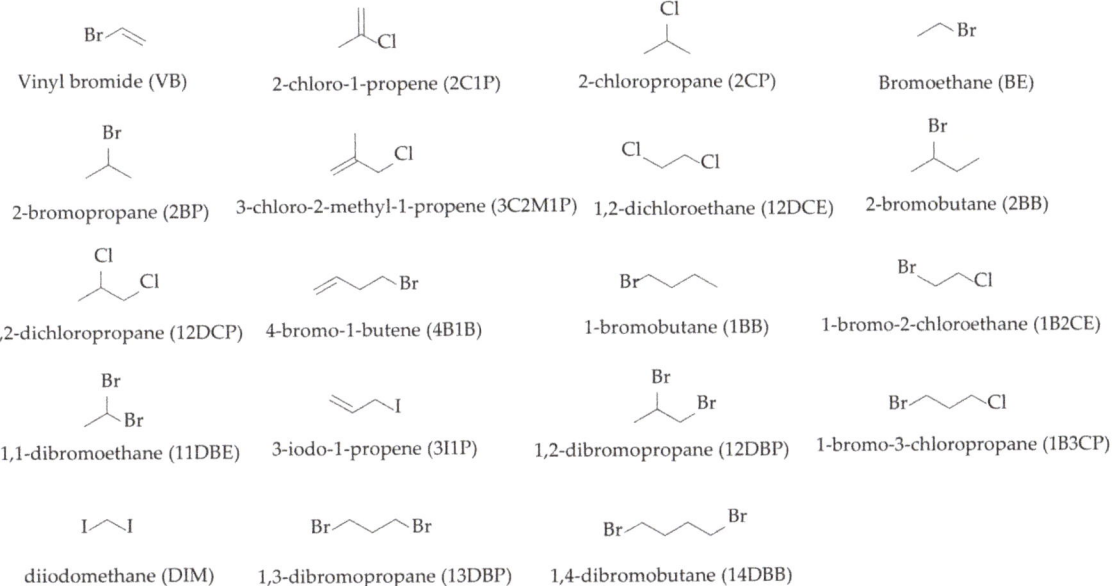

Figure 1. Chemical structures and abbreviations of the 19 selected alkyl halides.

2. Results and Discussion

2.1. In Silico Study for PGIs

Alkyl halides are chemicals widely known to be genotoxic. Alkyl halides consisting of one to four carbons were selected as PGIs for the in silico study because these alkyl halides have a high risk of genotoxicity due to their relatively small sizes. These alkyl halides also contain one or two halogens and/or oxygen atoms. After selecting a total of 144 alkyl halides, risk assessments using in silico programs were performed according to the ICH M7 guidelines. Derek Nexus, which is an expert, knowledge-based software, and Sarah Nexus, which is a statistical-based software, were applied complementary to one another to evaluate genotoxicity. Since the prediction results between the two programs may be the same or contradict one another, the results were carefully evaluated. Derek Nexus and Sarah Nexus were both developed by the same company called LHASA and run on the same interface. Since PGIs are automatically classified according to the ICH M7 guidelines (through the complementary combination of the two results from Derek and Sarah), there is an advantage in that the personal opinions of researchers on the classifications of the target analytes could be minimized. Both Derek Nexus and Sarah Nexus were properly validated according to OECD validation principles. VEGA Consensus, a free program, was used to verify the results of both programs.

From the results of the ICH M7 classification, it was found that among a total of 144 alkyl halides, 11 alkyl halides were known mutagenic carcinogens in class 1, 31 alkyl halides were known mutagens with unknown carcinogenic potential in class 2, 82 alkyl halides had alerting structures and no mutagenicity data in class 3, and 16 alkyl halides had no structural alerts or sufficient data to demonstrate a lack of mutagenicity or carcinogenicity in class 5. Only four alkyl halides were classified as inconclusive. Among the

144 alkyl halides, 124 alkyl halides were classified as classes 1, 2, and 3, which could be considered potential genotoxic impurities, and only 16 alkyl halides were classified as class 5, which could be considered non-genotoxic impurities. The Derek and Sarah prediction results showed minor differences from one another.

According to the literature review, many experimental studies, such as the Ames test, have been conducted on alkyl halides in which positive results were obtained in many cases [36–41].

Most alkyl halides can be considered hazardous and extra precautions must be taken when they are incorporated into any API manufacturing process. However, it is difficult to avoid the use of alkyl halides in API manufacturing processes. Therefore, it is essential to establish manufacturing processes that can reliably remove residual alkyl halides and develop a quality control strategy for these alkyl halides. The quantitative toxicity results for only seven alkyl halides were obtained from a literature review [42] and it was concluded that there were no quantitative toxicity results for the other 137 alkyl halides. Since there are no studies on the quantitative toxicities of most alkyl halides, alkyl halide quantification was controlled using conventional limits by the TTC concept to define an acceptable intake according to the ICH M7 guidelines.

For the selection of the target alkyl halides for this study, 144 compounds were classified from classes 1 to 5 through an in silico study. Among the compounds classified as classes 1 and 2, 19 target alkyl halides were selected based on the availability of the reference standards of these alkyl halides in South Korea. According to the ICH M7 guidelines, compounds in classes 1 and 2 must be heavily regulated as compounds in class 1 are known mutagenic carcinogens, and compounds in class 2 are known mutagens with unknown carcinogenic potential. The genotoxicity predictions for 19 selected alkyl halides are shown in Table 1.

Table 1. The genotoxicity predictions for 19 selected alkyl halides.

Abbreviation	Name	CAS No.	Derek Prediction	Sarah Prediction	VEGA Prediction	ICH M7 Class
1BB	1-Bromobutane	109-65-9	Plausible	Positive	Positive	Class 2
BE	Bromoethane	74-96-4	Plausible	Positive	Positive	Class 1
VB	Vinyl bromide	593-60-2	Probable	Positive	Positive	Class 1
2BP	2-Bromopropane	75-26-3	Plausible	Positive	Positive	Class 2
2BB	2-Bromobutane	78-76-2	Plausible	Positive	Positive	Class 2
4B1B	4-Bromo-1-butene	5162-44-7	Plausible	Positive	Positive	Class 2
2CP	2-Chloropropane	75-29-6	Plausible	Positive	Positive	Class 2
2C1P	2-Chloro-1-propene	557-98-2	Plausible	Positive	Positive	Class 2
3C2M1P	3-Chloro-2-methyl-1-propene	563-47-3	Plausible	Positive	Positive	Class 1
3I1P	3-Iodo-1-propene	513-48-4	Plausible	Positive	Positive	Class 2
1B2CE	1-Bromo-2-chloroethane	107-04-0	Plausible	Positive	Positive	Class 2
1B3CP	1-Bromo-3-chloropropane	109-70-6	Plausible	Positive	Positive	Class 2
12DCE	1,2-Dichloroethane	107-06-2	Plausible	Positive	Positive	Class 1
12DCP	1,2-Dichloropropane	78-87-5	Plausible	Positive	Positive	Class 1
13DBP	1,3-Dibromopropane	109-64-8	Plausible	Positive	Negative	Class 2
11DBE	1,1-Dibromoethane	557-91-5	Plausible	Positive	Negative	Class 2
12DBP	1,2-Dibromopropane	78-75-1	Plausible	Positive	Positive	Class 2
14DBB	1,4-Dibromobutane	110-52-1	Plausible	Positive	Positive	Class 2
DIM	Diiodomethane	75-11-6	Probable	Positive	Negative	Class 2

2.2. Analytical Method Development by Analytical QbD

An analytical method for PGIs should have good detection sensitivity to control PGIs with low acceptance criteria. Moreover, each peak should be well separated and not interfere with each other to analyze multiple PGIs simultaneously. For this reason, the ATP was set to achieve optimal detection sensitivity and separation between adjacent peaks. According to the established ATPs, CMAs were set as the peak areas (to increase sensitivity as much as possible) and the resolution (to secure sufficient separation between each peak).

The potential critical method parameters (CMPs) selected were the initial oven temperature, oven temperature hold time, oven temperature ramping rate, column flow rate, sample injection volume, final oven temperature, type of carrier gas, sample injector temperature, detector temperature, type of column, sample injection route, type of detector, and type of diluent. These parameters were selected from the fishbone diagram and the initial risk assessment based on a cause and effect (C&E) analysis. The experimental strategy for each parameter was classified under one-factor-at-a-time (OFAT), fixed, controlled, or design of experiment (DoE), based on the risk assessment performed using the failure mode and effect analysis (FMEA). The parameters classified under OFAT in the experimental strategy, such as column type, injection type, type of detector, and diluent were determined through the method scouting process. The effects on the results of the parameters, such as flow rate, initial temperature, ramping rate, and injector temperature, which are classified under DoE in the experimental strategy, were confirmed through the method screening process. The fishbone diagram is shown in Figure 2.

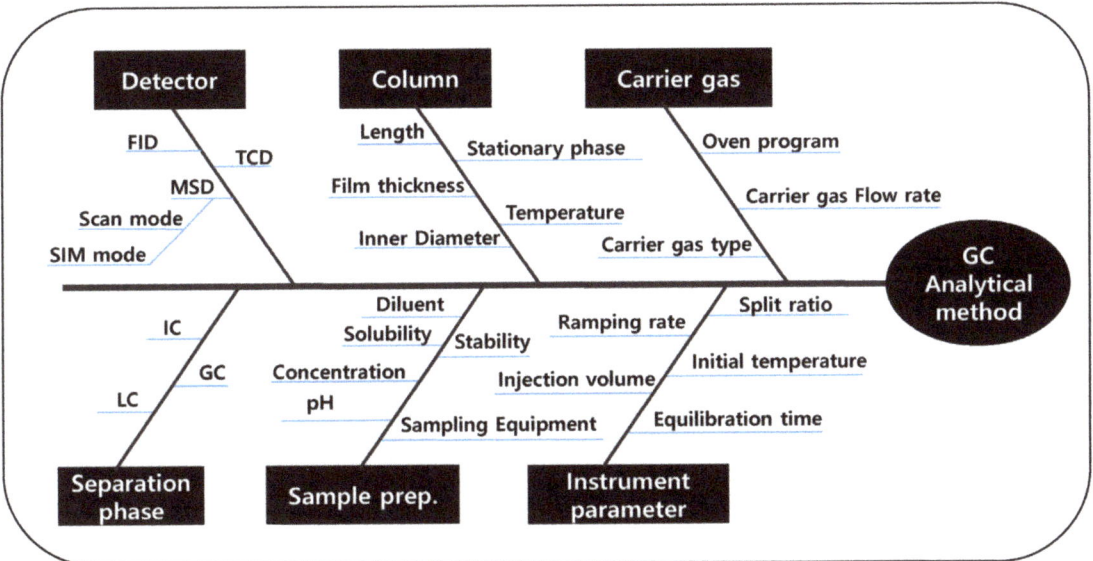

Figure 2. Fishbone diagram.

2.2.1. Method Scouting

The analytical method scouting was conducted by multiple OFAT experiments. The types of columns, injection equipment, detectors, and diluents were screened. Proper selection of the dilution solvent must be carried out to ensure complete solubility of all the target analytes and prevent other side reactions that can occur during analysis. When choosing the injection method, the boiling point of the target analytes must be considered, which is an important parameter that can particularly influence the detection sensitivity. The two common injection methods in gas chromatography are the auto-sampler method (in which the liquid analyte is directly injected) and the headspace method (in which the analyte is injected after vaporization). The type of column is a factor that has an overall effect on the results of gas chromatography. It mainly affects THE retention time, sensitivity, and resolution of THE analytes. The type of detector is also an important parameter; to find better reproducibility and sensitivity, a comparative test between the flame ionization detector (FID) and mass spectrometer detector (MSD) was conducted. VF-624 (6% cyanopropylphenyl/94% dimethylpolysiloxane) and DB-WAX (100% polyethylene glycol) capillary columns with different packing materials were tested by OFAT experiments;

VF-624 (with the best resolution) was selected. In the case of DB-wax, the column did not have sufficient retention, all impurities were detected before 10 min, and sufficient separation was not achieved.

The split ratio of 2:1 was chosen for the split mode because the splitless injection method caused overloading of peaks that disrupted the separation efficiency of the column, and sufficient sensitivity was not achieved at a split ratio greater than 2:1. To select an appropriate detector, the FID and MSD, which are generally used in gas chromatography, were compared. The FID is known to have better reproducibility and is more commonly used in pharmaceutical analyses. However, the FID did not produce the sufficient sensitivity needed to quantify trace amounts of PGIs, which are controlled at low limits using the concept of TTC. Therefore, the MSD was chosen as the detector using the selected ion monitoring (SIM) mode to maintain optimal sensitivity. The dilution solvent was selected based on its non-reactivity with the target compound and solubility. The chosen dilution solvent must also not interfere with the peak of the target impurities in the chromatograph. The selection process of the dilution solvent was carried out with dimethyl sulfoxide, dimethyl formamide, methanol, acetonitrile, purified water, dimethyl acetamide, and dichloromethane, which are commonly used in gas chromatography as dilution solvents. Because of their differences in solubility, all of the target alkyl halides were unable to dissolve in the same dilution solvents at high concentrations. Therefore, the analysis method was separated into methods A and B, with 13 alkyl halides dissolved in dimethyl sulfoxide, and 6 alkyl halides dissolved in acetonitrile, respectively.

2.2.2. Method Screening

Method screening was conducted to identify the significant parameters and interactions of the parameters critically affecting the pre-selected CMAs. The initial oven temperature, flow rate of carrier gas in column, oven temperature ramping rate, and sample injector temperature were selected as the CMPs through the method scouting process and risk assessment. Using Fusion QbD software, a DoE with a two-level-full factorial design with four center points was designed and tested with the CMPs that are expected to have critical effects on the selected response, resolution, and peak area. The DoE is tabulated in Table 2. A regression analysis of variance analysis (ANOVA) was evaluated to check the appropriate model fitting (Table 3). The statistical significance was confirmed using the F-ratio and p-value for each PGI analyzed. The calculated Pareto charts were reviewed to quantitatively identify the effects and interactions of each parameter on the established CMAs. The model terms ranking the Pareto charts are summarized in Figure 3. The Pareto charts show that the flow rate, initial temperature, or both parameters had the greatest effects on the resolution and peak area except for the curvature effect. A blue-colored bar corresponds to a positive effect, while a gray-colored bar corresponds to a negative effect. The flow rate was shown to have a positive effect on the peak area, while the initial temperature had a negative effect on the resolution. The selected responses were not affected or partially affected by the ramping rate of the oven and sample injector temperature. As a result, the flow rate of the carrier gas and the initial temperature of the oven were selected as the CMPs for the method optimization study.

Table 2. Design of experiment (DoE) for screening.

	DoE for Method A					DoE for Method B			
No. Run	Flow Rate	Initial Temp.	Ramping Rate	Injector Temp.	No. Run	Flow Rate	Initial Temp.	Ramping Rate	Injector Temp.
A-1	2.0	35	2.5	225	B-1	2.5	40	2.0	225
A-2	2.0	65	2.5	225	B-2	2.5	90	2.0	225
A-3	0.3	35	2.0	225	B-3	1.0	40	1.5	225
A-4	2.0	35	2.5	215	B-4	2.5	40	2.0	215
A-5	2.0	35	2.0	225	B-5	2.5	40	1.5	225
A-6	2.0	35	2.0	215	B-6	2.5	40	1.5	215
A-7	0.3	35	2.5	225	B-7	1.0	40	2.0	225
A-8	2.0	65	2.5	215	B-8	2.5	90	2.0	215
A-9	1.2	50	2.3	220	B-9	1.8	65	1.8	220
A-10	1.2	50	2.3	220	B-10	1.8	65	1.8	220
A-11	1.2	50	2.3	220	B-11	1.8	65	1.8	220
A-12	0.3	65	2.5	215	B-12	1.0	90	2.0	215
A-13	2.0	65	2.0	215	B-13	2.5	90	1.5	215
A-14	0.3	65	2.5	225	B-14	1.0	90	2.0	225
A-15	0.3	65	2.0	215	B-15	1.0	90	1.5	215
A-16	0.3	65	2.0	225	B-16	1.0	90	1.5	225
A-17	2.0	65	2.0	225	B-17	2.5	90	1.5	225
A-18	1.2	50	2.3	220	B-18	1.8	65	1.8	220
A-19	0.3	35	2.5	215	B-19	1.0	40	2.0	215
A-20	0.3	35	2.0	215	B-20	1.0	40	1.5	215

Table 3. Regression analysis of variance analysis (ANOVA) statistics results in the screening.

		Sum of Squares	DF*	Mean Square	F-Ratio	p-Value		Sum of Squares	DF	Mean Square	F-Ratio	p-Value
VB	A*	363,545	4	90,886	123.65	<0.01	3I1P	7,891,128	5	1,578,225	35.98	<0.01
	R*	2.31	9	0.25	115.00	<0.01	1BB	57.83	4	14.45	39.10	<0.01
2C1P	A	28,372,838	1	28,372,838	134.66	<0.01		703,483	1	703,483	84.49	<0.01
	R	2.50	6	0.41	723.67	<0.01	1B2CE	44.15	3	14.71	58.83	<0.01
2CP	A	1,617,766	2	808,883	55.08	<0.01		135,246	2	67,623	38.70	<0.01
	R	25.01	7	3.57	382.44	<0.01	11DBE	26.54	3	8.8497	62.24	<0.01
BE	A	5,578,546	1	5,578,546	171.11	<0.01		109,070	2	54,535	240.22	<0.01
	R	449.97	3	149.99	839.29	<0.01	12DBP	5,359	3	1,786	176.64	<0.01
2BP	A	174,573	2	87,286	58.73	<0.01		1,562,157	5	312,431	47.35	<0.01
	R	60.00	5	12.00	675.49	<0.01	1B3CP	31.74	10	3.17	122.00	<0.01
3C2M1P	A	313,435	1	313,435	133.11	<0.01		512,151	10	51,215	57.34	<0.01
	R	370.90	8	46.36	207.03	<0.01	DIM	308.94	6	51	133.18	<0.01
12DCE	A	336,637	6	56,106	185.06	<0.01		1,205,161	3	401,720	19.31	<0.01
	R	127.72	7	18.24	114.85	<0.01	13DBP	1.63	3	0.54	79.58	<0.01
2BB	A	1,122,779	2	561,389	56.16	<0.01		<0.0001	9	<0.0001	20.92	<0.01
	R	144.53	5	28.90	159.55	<0.01	14DBB	3,483	2	1,741	153.26	<0.01
12DCP	A	252,402	1	252,402	176.19	<0.01		12,128	7	1,732	8.42	<0.01
	R	10.91	3	3.63	66.17	<0.01						
4B1B	A	0.0003	4	<0.0001	36.66	<0.01						

A*: peak area, R*: resolution, DF*: degree of freedom.

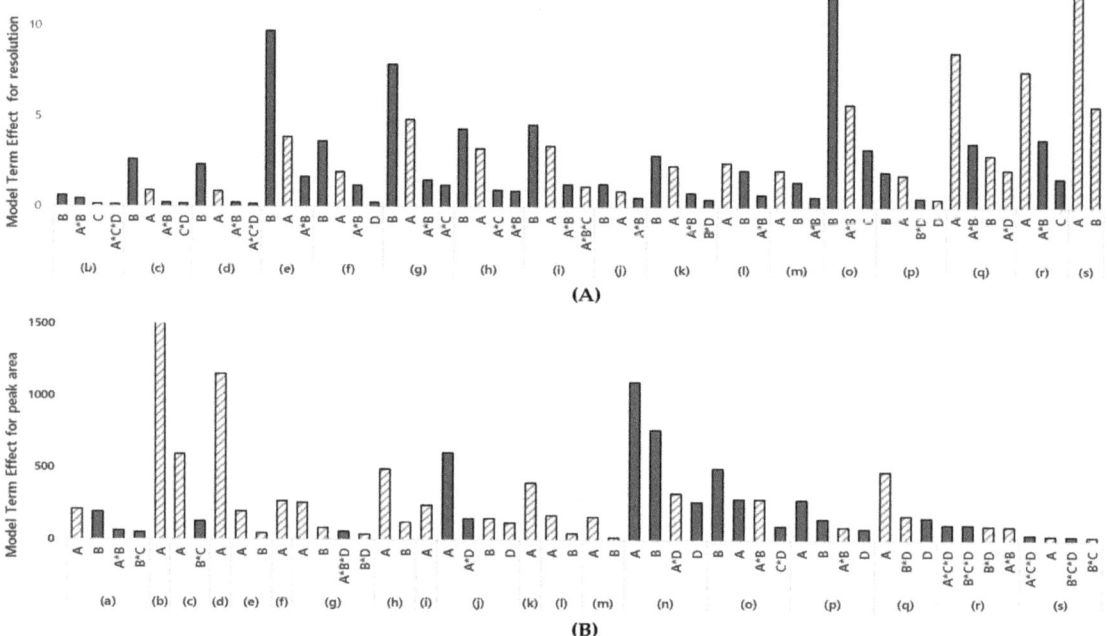

Figure 3. Model terms ranking the Pareto charts, depicting the influences of the initial oven temp., the flow rate of the carrier gas, the oven temp. ramping rate, and the sample injector temp. (**A**) Pareto chart for resolution; (**B**) Pareto chart for peak area; (**A**) initial oven temp., (**B**) the flow rate of the carrier, C oven temp. ramping rate D sample injector temp; (a) VB, (b) 2C1P, (c) 2CP, (d) BE, (e) 2BP, (f) 3C2M1P, (g) 12DCE, (h) 2BB, (i) 12DCP, (j) 4B1B, (k) 1BB, (l) 1B2CE, (m) 11DBE, (n) 3I1P, (o) 12DBP, (p) 1B3CP, (q) DIM, (r) 13DBP, (s) 14DBB. The height of the bar is the magnitude of the corresponding model term's effect on the response. A dashed bar corresponds to a positive effect, while a solid bar corresponds to a negative effect, and the interaction between parameters is marked with an asterisk (*).

2.2.3. Method Optimization by Analytical QbD

The response surface methodology (RSM), which is based on standard orthogonal arrays that contain three levels for each experiment variable, was employed for the method optimization study. To meet the goal of the analytical method optimization, the model can provide data from which linear and curvilinear variable behaviors can be quantified within the allowable ranges. The critical parameters should be properly selected for optimization because optimization studies require a relatively large number of experiments compared to screening studies. Based on the results obtained from the method screening, the flow rate of the carrier gas in the column and the initial temperature of the column were selected as the CMPs for method optimization studies. Although it was not possible in the method screening process, the RSM model could provide information on curvature effects, which help to understand the correlation between various parameters. The DoE is tabulated in Table 4. A regression ANOVA was evaluated to check the appropriate model fitting (Table 5). The regression ANOVA statistics result was shown to be significant with a p-value and F-ratio for all responses.

Table 4. DoE for optimization.

DoE for Method A			DoE for Method B		
No. Run	Flow Rate	Initial Temp.	No. Run	Flow Rate	Initial Temp.
A-1	1.7	45	B-1	1.8	90
A-2	1.7	40	B-2	1.8	65
A-3	2.0	40	B-3	2.5	65
A-4	2.0	45	B-4	2.5	90
A-5	1.3	35	B-5	1.0	40
A-6	1.3	40	B-6	1.0	65
A-7	1.7	40	B-7	1.8	65
A-8	2.0	35	B-8	2.5	40
A-9	1.7	40	B-9	1.8	65
A-10	1.7	40	B-10	1.8	65
A-11	1.7	35	B-11	1.8	40
A-12	1.3	45	B-12	1.0	90

Table 5. Regression ANOVA statistics results in optimization.

		Sum of Squares	DF*	Mean Square	F-Ratio	p-Value		Sum of Squares	DF	Mean Square	F-Ratio	p-Value
VB	A*	39,784	3	13,261.43	10.4579	<0.01	3I1P	5,270,439.51	4	1,317,609	56.0824	<0.01
	R*	0.3267	1	0.3267	26.6667	<0.01	1BB	2.1687	2	1.0844	7.0698	0.014
2C1P	A	5,078,587	3	1,692,862	10.3662	<0.01		171,839	3	57,279	52.615	<0.01
	R	3.1395	4	0.7849	16.018	<0.01	1B2CE	1.5	1	1.5	5.1546	0.047
2CP	A	658,727	3	219,575.	14.451	<0.01		23,564.55	1	23,564	48.6432	<0.01
	R	1.1267	1	1.1267	29.9778	<0.01	11DBE	1.9267	1	1.9267	6.0397	0.034
BE	A	125,017	3	41,672	13.1523	<0.01		4,647.75	3	1,549	4.2898	0.044
	R	27.9528	3	9.3176	29.3893	<0.01	12DBP	1,631.06	3	543.6883	52.3332	<0.01
2BP	A	<0.01	3	<0.01	41.9238	<0.01		6,255,223.87	4	1,563,805	12.6797	<0.01
	R	5.5787	3	1.8596	14.2526	<0.01	1B3CP	3.1758	2	1.5879	11.8351	<0.01
3C2M1P	A	49,069	3	16,356	32.7338	<0.01		496,361.17	2	248,180	12.5135	<0.01
	R	38.805	3	12.935	4.3709	0.042	DIM	53.0399	2	26.5199	10.9916	<0.01
12DCE	A	21,604	2	10,802	16.9309	<0.01		1,008,620.08	1	1,008,620	8.8133	0.014
	R	7.935	1	7.935	13.79	<0.01	13DBP	0.0158	2	0.0079	52.7139	<0.01
2BB	A	370,632	3	123,544	97.5273	<0.01		150,401.39	2	75,200	4.538	0.048
	R	17.2294	3	5.7431	10.4585	<0.01	14DBB	2185.18	3	728.3925	12.1323	<0.01
12DCP	A	42,511	3	14,170	23.6012	<0.01		172,456.65	2	86,228	8.2038	<0.01
	R	0.8388	2	0.4194	13.9608	<0.01						
4B1B	A	113,294	3	37,764	13.383	<0.01						

A*: peak area, R*: resolution, DF*: degree of freedom.

An RSM analysis was carried out employing 3D-response surface plots for identifying the underlying interaction among the selected parameters. A total of 36 3D response surface plots were obtained; among them, 8 PGIs with relatively low resolutions and peak areas were selected. Considering the parameter effects on PGIs with low resolutions and peak areas has a critical impact on the overall analytical method development. As shown in Figure 4, the effects of the flow rate and oven temperature on the resolution are different for each PGI. In the case of Figure 4A 2C1P, the resolution increases as the oven temperature decreases, whereas the increase in the flow rate of the carrier gas revealed negligible influence on the resolution. In the case of Figure 4A,C,D, a linear declining trend was observed for the resolution with an increase in the flow rate and a decrease in oven temperature. The response surface for Figure 4B 3C2M1P showed a curvilinear relationship between A and B, while the response surfaces for other PGIs showed linear relationships. The flow rate of the carrier gas and the oven temperature showed a more varied interaction with the change in the peak area. As shown in Figure 5, the peak area of Figure 5C 13DBP depends on the change of the flow rate and has a maximum point when the flow rate is 1.5 mL/min, while the effect of the column temperature on the peak area is negligible. Figure 5A VB, Figure 5B 2BP, and Figure 5D 14DBB portray considerably high levels of interaction between the flow rate of the carrier gas and the oven temperature for the peak area. Moreover, Figure 5A VB and Figure 5B 2BP produced a paraboloid that opened down and had a global maximum at its vertex within a given range, while Figure 5D 14DBB had a saddle point.

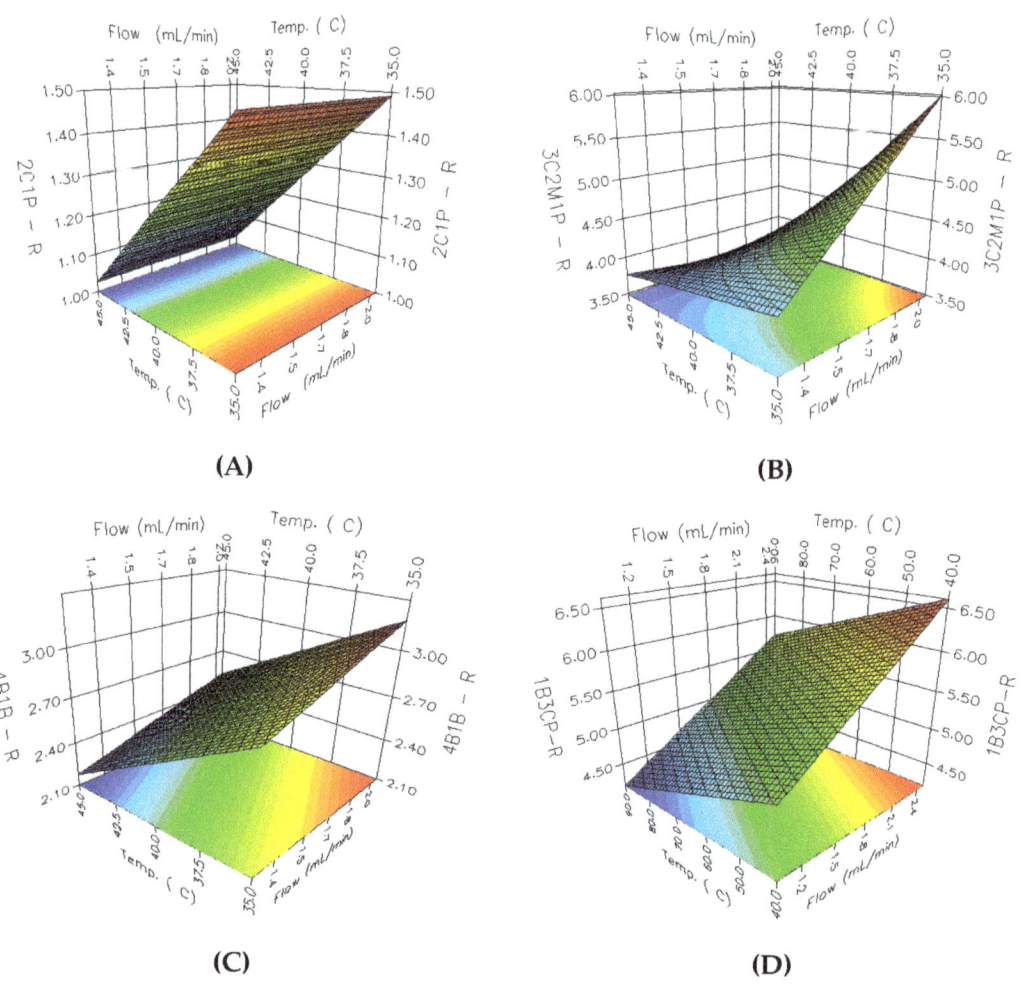

Figure 4. The 3D response surface plots for the resolution: (**A**) 2C1P; (**B**) 3C2M1P; (**C**) 4B1B; (**D**) 1B3CP.

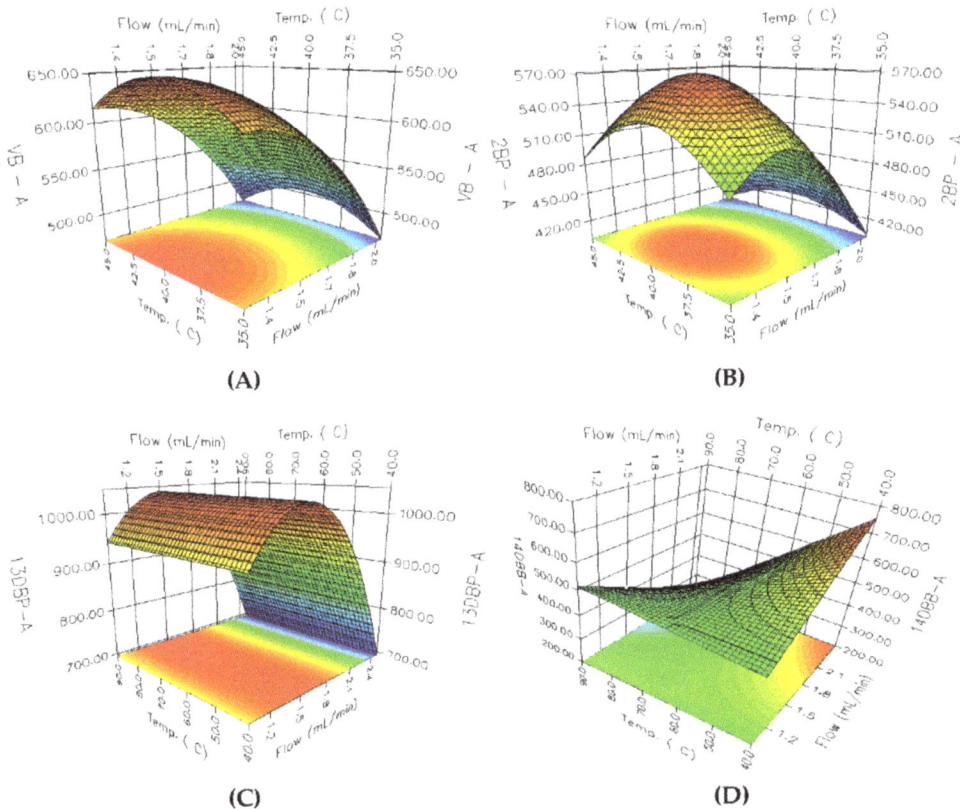

Figure 5. The 3D response surface plots for the peak area: (**A**) VB; (**B**) 2BP; (**C**) 13DBP; (**D**) 14DBB.

According to the ICH Q14 guideline draft, the method operable design region (MODR) represents a combination of analytical procedure parameter ranges in which the analytical procedure performance criteria are fulfilled and the quality of the measured results is assured [43]. The optimum region was established by overlaying contours of all responses (each having an acceptance criterion). The acceptance criteria of the responses were set to satisfy the purposes of the ATPs. The acceptance criteria of the peak area were set at 500 or more to ensure sufficient sensitivity to detect trace amounts of PGIs, and the acceptance criteria of the resolution were set to 10 or greater so that each peak was sufficiently separated and there were no interferences with each other. If the predefined acceptance criteria were not met because of the unique chemical properties of certain PGIs, the acceptance criteria were set at 90% of the expected values. For the set acceptance criteria (10 for resolution and 500 for the peak area), the shaded contour represents the unacceptable region, while the unshaded represents the acceptable region. In the final optimization study, the unshaded region may represent the design space for the variables being studied in terms of the graphed response. The MODR in Figures 6 and 7 show the regions for acceptable results for all responses, simultaneously. The proven acceptable range was defined inside of the predefined MODR, based on a linear combination of parameters, to the set center point of the parameter ranges. In the case of method A, the resolutions of 2CP, 3C2M1P, and 2C1P played important roles in determining the MODR. Since the sensitivity of 2BP is lower than that of other PGIs, the peak area of 2BP was confirmed as an important response. In the case of method B, the resolution of 1B3CP and the peak area of 14 DBB were important influences in determining the acceptable region. The typical GC−MS chromatograms using the scan mode and SIM mode are shown in Figures 8 and 9, respectively.

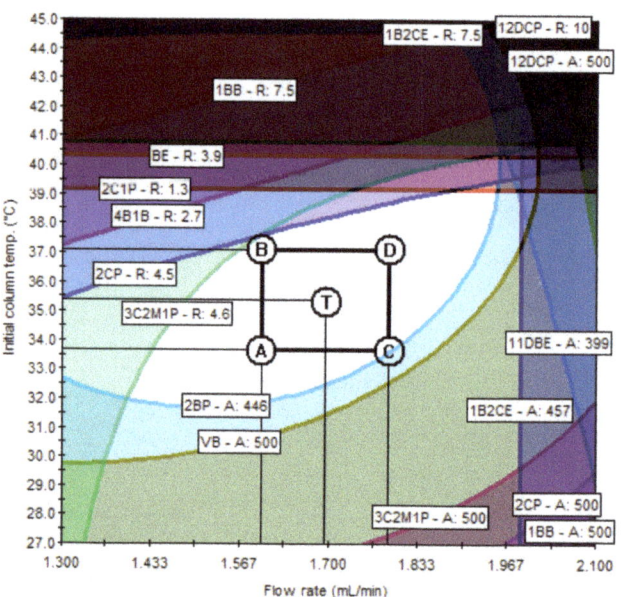

Figure 6. Method operable design region (MODR) and proven acceptable ranges (PARs) for method A.

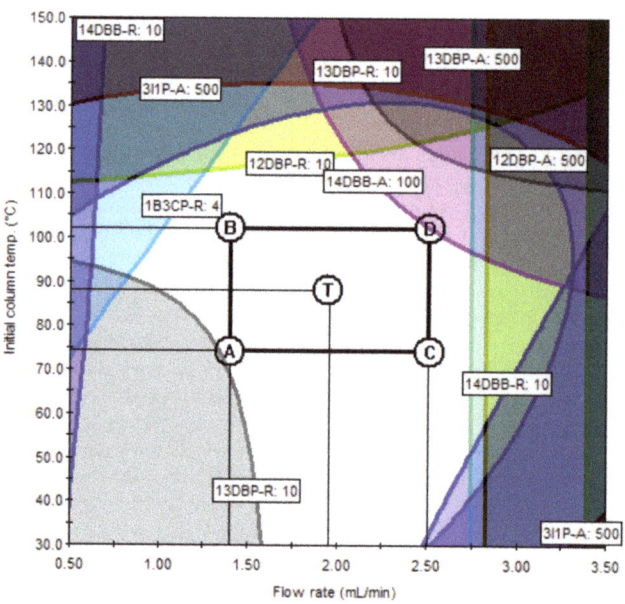

Figure 7. MODR and PARs for method B.

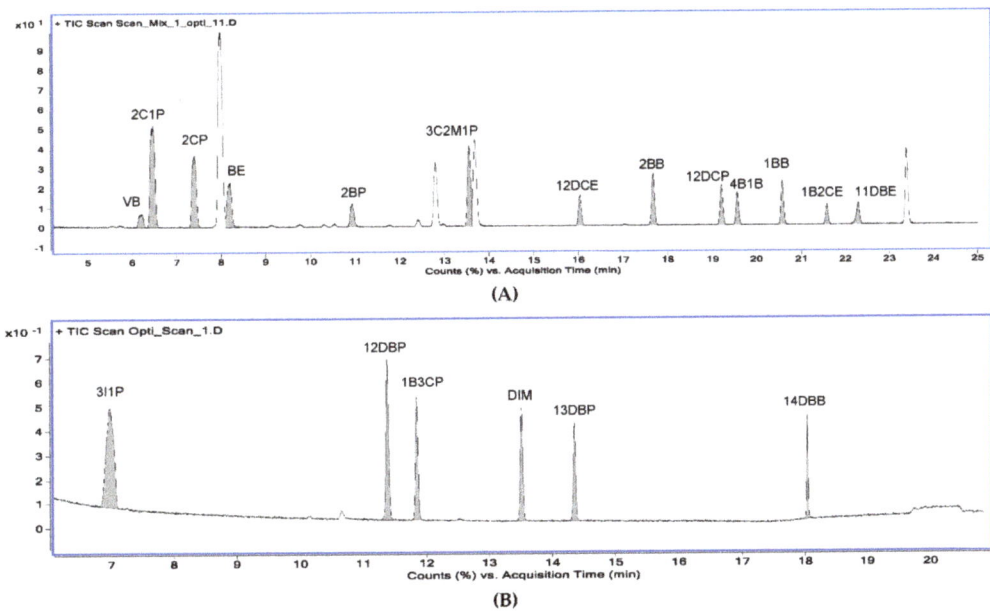

Figure 8. Typical GC−MS chromatograms of 19 alkyl halides using the scan mode. (**A**) Chromatogram for 13 alkyl halides obtained from method A; (**B**) chromatogram for 6 alkyl halides obtained from method B.

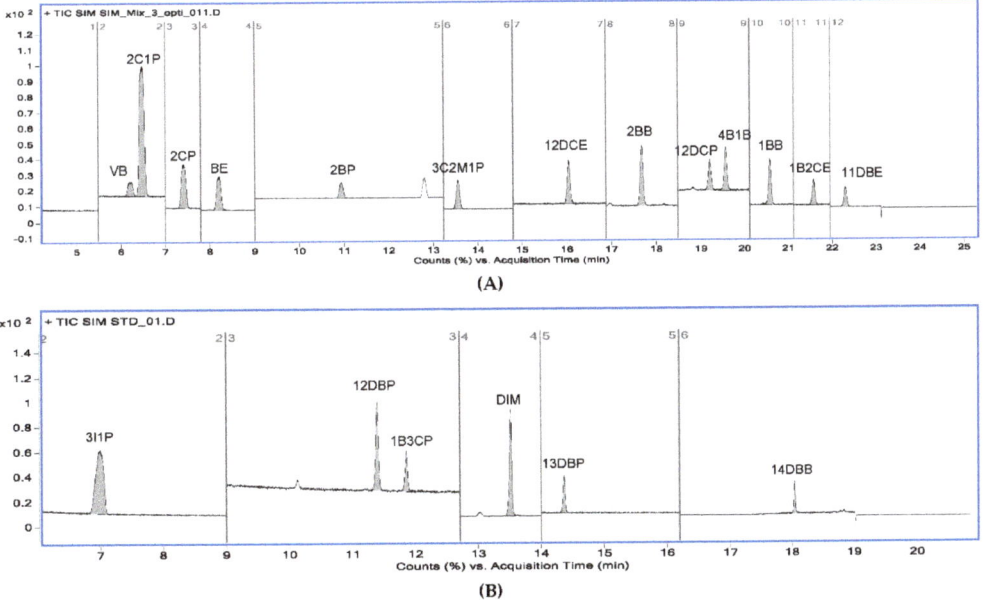

Figure 9. Typical GC−MS chromatograms of 19 alkyl halides using the selected ion monitoring (SIM) mode. (**A**) Chromatogram for 13 alkyl halides obtained from method A; (**B**) chromatogram for 6 alkyl halides obtained from method B.

2.3. Analytical Method Validation

The developed analytical method was validated according to the ICH Q2(R1) guidelines to demonstrate that the analytical procedure was suitable for its intended purpose. The analytical method was evaluated in the attributes of specificity, the limit of detection, the limit of quantitation, linearity, accuracy, and precision. A summary of the analytical method validation results is presented in Table 6.

Table 6. Summary of analytical method validation results.

Name	Specificity	Sensitivity (ppm)		Linearity			Accuracy (%)			Precision (%RSD)	
	Resolution	LOD*	LOQ*	R*	Slope	y-Intercept	Low	Mid	High	Repeat-Ability	LOQ Level
Acceptance Criteria	≥1.5	≤0.3 ppm	≤1.0 ppm	≥0.995	-	-	≥85.0 %	≥85.0 %	≥85.0 %	≤10 %RSD	≤10 %RSD
VB	-	0.09	0.29	0.9989	137.91	15.699	90.84	97.90	92.57	4.01	4.35
2C1P	1.5	0.01	0.03	0.9990	1215.68	109.155	87.97	97.64	92.75	5.45	3.35
2CP	5.4	0.03	0.10	0.9993	402.58	29.440	86.79	93.95	90.08	4.51	3.47
BE	4.6	0.01	0.04	0.9997	304.53	37.785	90.72	95.23	89.52	3.69	2.11
2BP	17.4	0.05	0.16	0.9997	170.84	4.359	90.73	96.36	91.21	2.99	1.80
3C2M1P	5.2	0.03	0.09	0.9994	298.47	−1.804	95.29	98.51	93.85	1.94	1.80
12DCE	18.4	0.04	0.13	0.9996	212.38	13.247	99.58	99.16	93.33	3.20	6.43
2BB	12.6	0.01	0.04	0.9996	529.32	8.433	100.64	103.81	98.26	2.07	1.61
12DCP	12.2	0.04	0.14	0.9975	192.29	24.495	98.07	97.78	90.49	3.33	3.31
4B1B	2.8	0.02	0.06	0.9996	347.26	58.163	97.93	97.02	91.62	2.91	1.34
1BB	8.2	0.02	0.05	0.9994	304.80	38.029	98.91	99.47	93.24	2.04	1.93
1B2CE	8.3	0.05	0.16	0.9998	153.73	6.450	100.21	96.61	92.26	4.48	4.31
11DBE	5.9	0.05	0.18	0.9993	158.29	−2.456	101.25	99.50	95.67	3.42	3.81
3I1P	-	0.07	0.25	0.9997	619.84	19.912	96.77	92.01	96.52	3.36	2.28
12DBP	27.8	0.07	0.25	0.9988	321.55	−15.625	94.64	93.67	93.67	6.60	3.97
1B3CP	5.3	0.10	0.33	0.9988	122.00	4.4667	95.70	97.30	95.51	5.41	1.82
DIM	20.6	0.07	0.24	0.9991	318.22	−5.9208	98.81	100.12	98.83	4.54	1.01
13DBP	11.7	0.11	0.38	0.9999	100.11	−0.7792	101.14	102.47	99.88	2.93	1.68
14DBB	63.4	0.07	0.29	0.9993	55.283	3.3083	96.09	97.97	93.34	3.38	2.95

LOD*: Limit of detection, LOQ*: limit of quantitation, R*: correlation coefficient.

2.3.1. Specificity

Specificity refers to the capability of the analytical method to selectively and accurately measure the target impurities in mixed states of impurities and/or degradants. Specificity was evaluated by the resolution between the adjacent peaks obtained from the standard solution. Specificity is a critical parameter in this method because multiple PGIs need to be analyzed simultaneously. The resolution between VB and 2C1P was determined to be the lowest with a resolution of 1.5. All of the resolution values between the adjacent peaks were above 1.5, which means that all peaks were well separated from each other for simultaneous analyses and the analytical procedure was proven to be specific for the target analytes.

2.3.2. Limit of Detection and Quantitation

The limit of detection (LOD) is the minimum detectable amount of each analyte present in the sample and the limit of quantitation (LOQ) is the minimum amount of each analyte to be analyzed among samples that could be expressed as quantitative values with appropriate precision. The test was conducted at a sufficiently low concentration to detect the signal-to-noise ratio of about 3 and 10 for the LOD and LOQ, respectively. The solutions with concentrations from about 0.5 to 0.8 ppm for each PGI were analyzed for the determinations of the LODs and LOQs. The calculated LODs for the PGIs were between 0.01 and 0.11 ppm and LOQs were between 0.03 and 0.38 ppm. The calculated LODs and LOQs were sufficiently low enough to control trace amounts of PGIs with low acceptance criteria.

2.3.3. Linearity

Linearity of the test method refers to the ability of the analytical method to obtain a linear measurement value within a specified range in proportion to the amount (or concentration) of each analyte. Linearity was tested over the concentration range from the LOQ level to 2.4 ppm. Linearity was evaluated as a correlation coefficient (R), with all results being no less than 0.995; the proposed analytical method was proven to be linear for the target analytes.

2.3.4. Accuracy

Accuracy refers to the degree to which a measured value is close to a known true or standard value. Accuracy was evaluated by the recovery study of a matrix API being spiked with a PGI standard. The spiked sample solutions were prepared in triplicates at three concentration levels with the addition of known amounts of each PGI standard to the chosen API matrix. Raloxifene (Cas No. 84449-90-1) was used as a matrix API because alkyl halides are used in the manufacturing process of Raloxifene. The analyte contents in the spiked sample solutions were determined using the proposed analytical procedure and the recovery was calculated for each solution. All results met the acceptance criteria, within $100 \pm 15\%$ of recoveries for all PGIs, and the proposed analytical method was proven to be accurate for all target analytes.

2.3.5. Precision

Precision refers to the proximity between each measured value by taking multiple samples from a homogeneous sample and analyzing them according to the specified conditions. Precision was evaluated by the %RSD of the peak areas from six replicate injections of the LOQ solution (LOQ level precision) with a concentration at the LOQ level, and the standard solution (repeatability) with a concentration of 2 ppm. %RSD at the LOQ level was determined to be from 1.01 to 6.43% and %RSD at 2 ppm was determined to be from 1.94 to 6.60%, and the proposed analytical method was proven to be precise for all target analytes.

2.4. Applicability of the Method to Real Sample

To ensure that the developed method applies to real APIs, raloxifene hydrochloride (Cas No. 82640-04-8), febuxostat (Cas No. 144060-53-7), sitagliptin phosphate (Cas No. 654671-78-0), amlodipine camsylate (Cas No. 652969-01-2), and ezetimibe (Cas No. 163222-33-1) were chosen and analyzed. All APIs were manufactured and supplied by Hanmi Fine Chemical in South Korea.

Since residual alkyl halides were not present in the chosen APIs, the matrix API samples were spiked with 19 of the alkyl halides at low concentrations.

Each test was repeated thrice, and the recovery, precision, specificity, LOD, and LOQ were analyzed each time using the matrix APIs. Recovery was determined to be between 86.0 and 102.7% for raloxifene hydrochloride, 77.9 and 110.2% for febuxostat, 77.0 and 106.4% for sitagliptin phosphate, 88.5 and 126.1% for amlodipine camsylate, and 81.8 and 117.6% for ezetimibe, respectively.

From the analyses of matrix APIs using the proposed analytical method, it was confirmed that the matrix APIs do not interfere with the analyses of alkyl halides and the applicability of the method to real samples was confirmed with accuracy and precision. The detailed test results are tabulated in the Supplementary Materials.

3. Materials and Methods

3.1. Reagents, Materials, and Standards

The reference standards for vinyl bromide (Cas No. 593-60-2, purity: 98%), 2-chloro-1-propene (Cas No. 557-98-2, purity: 98%), 2-chloropropane (Cas No. 75-29-6, purity: 98%), 1,2-dichloroethane (Cas No. 107-06-2, purity: 99.8%), 4-bromo-1-butene (Cas No. 5162-44-7, purity: 97%), 1,1-dibromoethane (Cas No. 557-91-5, purity: 98%), 3-iodo-1-propene (Cas No.

513-48-4, purity: 98%), and diiodomethane (Cas No. 75-11-6, purity: 99%) were purchased from Sigma-Aldrich (Burlington, NJ, USA).

The reference standards for bromoethane (Cas No. 74-96-4, purity: 99.0%), 2-bromopropane (Cas No. 75-26-3, purity: 99.0%), 3-chloro-2-methyl-1-propene (Cas No. 563-47-3, purity: 98.0%), 2-bromobutane (Cas No. 78-76-2, purity: 98.0%), 1,2-dichloropropane (Cas No. 78-87-5, purity: 98.0%), 1-bromobutane (Cas No. 109-65-9, purity: 98.0%), 1-bromo-2-chloroethane (Cas No. 107-04-0, purity: 98.0%), 1,2-dibromopropane (Cas No. 78-75-1, purity: 98.0%), 1-bromo-3-chloropropane (Cas No. 109-70-6, purity: 99.0%), 1,3-dibromopropane (Cas No. 109-64-8, purity: 98.0%), and 1,4-dibromobutane (Cas No. 110-52-1, purity: 98.0%) were purchased from TCI (Tokyo, Japan).

For the dilution solvents, N,N-dimethylacetamide (Cas No. 127-19-5, purity: 99.9%), N,N-dimethyl sulfoxide (Cas No. 67-68-5, purity: 99.0%), and dimethylformamide (Cas No. 75-12-7, purity: 99.5%) were purchased from Junsei Chemical (Tokyo, Japan). Dichloromethane (Cas No. 75-09-2, purity: 99.9%) and acetonitrile (Cas No. 75-05-8, purity: 99.9%) were purchased from Fisher Scientific (Pittsburgh, PA, USA). Methanol (Cas No. 67-56-1, purity: 99.8%) was purchased from Sigma-Aldrich (Burlington, USA). Purified water was provided from the Milli-Q water purification system (Burlington, MA, USA). The raloxifene hydrochloride (Cas No. 82640-04-8), febuxostat (Cas No. 144060-53-7), sitagliptin phosphate (Cas No. 654671-78-0), amlodipine camsylate (Cas No. 652969-01-2), and ezetimibe (Cas No. 163222-33-1) were obtained from Hanmi Fine Chemical (Gyeonggi-do, Korea).

For the method scouting process, a VF-624 ms capillary column (60 m × 0.25 mm i.d. × 1.4 μm film thickness, part no: CP9103) and DB-WAX capillary column (60 m × 0.53 mm i.d. × 1.0 μm film thickness, part no: 125-7062) were purchased from Agilent Technologies (Santa Clara, CA, USA).

3.2. In Silico Study

The program used for the in silico study was Derek Nexus (v 6.1.1, build 7 July 2021) from Lhasa Limited (Leeds, UK), which is an expert knowledge rule-based software according to the application of alerts and reasoning rules that cover structural alerts for various toxicological endpoints. For precise mutagenicity predictions, Sarah Nexus (v.3.1.1 built on 7 July 2021), also from Lhasa Limited, was used, which is a statistical-based software tool capable of calculating precise predictions of mutagenicity [44]. The ICH guidelines state that computational toxicity assessments can be conducted using two complementary methodologies for Q-SAR predictions that predict the bacterial mutation assay. One methodology can be expert rule-based and another can be statistical-based [45].

Vega (V 1.1.5-b48, built on 29 March 2021), free software in the VEGA HUB (www.vegahub.eu, accessed on 03 May 2022), was used as in silico prediction software for results regarding reliability [46].

3.3. Preparation of Solutions

Regarding the standard solution preparation for method A—0.1 g of vinyl bromide, 2-chloro-1-propene, 2-chloropropane, bromoethane, 2-bromopropane, 3-chloro-2-methyl-1-propene, 1,2-dichloroethane, 2-bromobutane, 1,2-dichloropropane, 4-bromo-1-butene, 1-bromobutane, 1-bromo-2-chloroethane and 1,1-dibromoethane—each was accurately weighed and transferred to a 100 mL volumetric flask, dissolved, and diluted with dimethylsulfoxide to the volume. Moreover, 0.5 mL of this solution was transferred to a 50 mL volumetric flask and diluted with dimethylsulfoxide to the volume. A total of 0.5 mL of this solution was transferred to a 50 mL volumetric flask and diluted with dimethylsulfoxide to volume. A total of 2.0 mL of this solution was transferred to a 10 mL volumetric flask and diluted with dimethylsulfoxide to volume.

A sample solution preparation for method A (0.1 g of the test specimen, accurately weighed) was transferred to a 10 mL volumetric flask, and then dissolved and diluted with dimethylsulfoxide to volume.

Regarding the standard solution preparation for method B—0.1 g of 3-iodo-1-propene, 1,2-dibromopropane, 1-bromo-3-chloropropane, diiodomethane, 1,3-dibromopropane and 1,4-dibromobutane—each was each accurately weighed, transferred to a 100 mL volumetric flask, dissolved, and diluted with acetonitrile to the volume. Moreover, 1.0 mL of this solution was transferred to a 100 mL volumetric flask and diluted with acetonitrile to the volume. A total of 0.5 mL of this solution was transferred to a 100 mL volumetric flask and diluted with acetonitrile to volume.

Regarding the sample solution preparation for method B—0.1 g of the test specimen was accurately weighed, transferred to a 10 mL volumetric flask, and then dissolved and diluted with acetonitrile to volume.

3.4. Analytical Condition and Equipment

GC condition. An Agilent 7890B gas chromatography instrument equipped with a 5977A Single Quadrupole MSD and FID (Agilent Technologies, Santa Clara, CA, USA) was used for this study. The sample injection was performed using an Agilent G7697A Headspace sampler and an 7693 autosampler (Agilent Technologies, Santa Clara, CA, USA). For the instrument control, data acquisition, and processing, Mass Hunter Qualitative Analysis ver. B.07.00 (Agilent Technologies) and OpenLAB CDS ChemStation edition Rev. C.01.07.SR4 (Agilent Technologies) were used.

The analyses of PGIs with MSD were performed using a VF-624 ms capillary column (60 m × 0.25 mm i.d. × 1.4 μm film thickness). The column oven temperature programming was as follows: Method A—the initial oven temperature was set at 35 °C, then ramped at 3 °C/min to 124 °C. Method B—the initial oven temperature was set at 90 °C, then ramped at 2 °C/min to 100 °C, then ramped at 5 °C/min to 130 °C, and then ramped at 10 °C/min to 160 °C. The flow rate of the helium carrier gas was set at 1.7~2.0 mL/min with a mode of constant flow. The split/splitless inlet (SSL) was used as the inlet type and the injection port was used in the split mode at a split ratio of 2:1. For the injection, a 4 mm ID straight inlet liner was used. The injection heater temperature was set at 280 °C.

The headspace sampler (HSS) and automatic liquid sampler (ALS), which are the most common GC sampling techniques, were reviewed. Since the target impurities are easily volatilized due to their relatively low boiling points, both the HSS and ALS were utilizable and showed excellent reproducibility. However, in the case of ALS, there was a problem with the sample being carried over in the gas chromatography system, so the HSS was selected as the sampling technique. The injection volume was set at 1000 μL as the gas phase using the headspace sampler.

MS condition. The mass spectrometer was operated in the SIM under high-efficiency EI source conditions. The m/z values of 55.1, 57.1, 61.9, 63.0, 76.0, 77.0, 77.9, 89.9, 105.9, 106.9, 108.0, 121.0. 122.0, 137.0 155.9. 168.0. 200.0, and 268.0 ions with 50 ms dwell times were selected for analysis. The source temperature was set at 230 °C, the quadrupole temperature was set at 150 °C, and the transfer line temperature was set at 280 °C. The gain factor was set at 1.0.

3.5. Method Development by Analytical QbD

The first step of the analytical QbD is to define a set of ATPs according to the purpose and scope of the analysis. The most important purpose of this analytical method is to analyze various and trace amounts of PGIs at once, so specificity and detection sensitivity were the ATPs assessed. The ATP corresponds to the quality target product profiles (QTPP) in a product QbD approach. To satisfy the defined ATPs, the CMAs need to be properly selected. Resolution and detection sensitivity were the chosen CMAs for the analytical method development for the PGIs assessed. The CMAs are equivalent to the critical quality attributes (CQAs) in a product QbD approach.

Risk assessment (RA). The risk assessment can be conducted in various ways. The potential risks related to the analytical method development were prioritized using the

FMEA method. Based on the risk assessment performed, the potential risks were grouped into CMPs and controlled parameters.

Method scouting. The method scouting process consisted of preliminary tests performed based on the risk assessment and knowledge gained through scientific experience using the OFAT method.

Method screening. Method screening using the DoE approach was performed for the parameters that were classified as having high risks through the risk assessment. The DoE was conducted using the Fusion QbD software in a two-level full factorial design with four center points using a quadratic design model. The significance of the DoE was verified using the ANOVA statistical method. An understanding of the main effects, interactions, and Pareto charts between CMPs can be obtained from the screening results.

Method optimization. Method optimization was conducted for the parameters, which are defined as CMPs based on the results from the method screening process. The DoE was conducted in a central composite full type with four center points using the RSM approach. The RSM approach is useful in understanding the correlation between various parameters and responses. The significance of the DoE was verified using the ANOVA statistical analysis. The 3D-response surface plots exhibiting the effects of the CMPs and MODR were obtained using Fusion QbD software.

3.6. Method Validation

The proposed analytical method was validated according to the ICH Q2 (R1) guideline. The validation parameters of specificity, the limit of detection, quantitation, accuracy, precision, and linearity of the analytical method were evaluated.

Specificity was evaluated based on the resolution between PGI peaks from the chromatograms. It was ensured that none of the blank peaks interfered with the peaks of interest for the standard and sample solutions (confirmed by visual inspection). The prepared standard solution was injected into the GC in six replicates. The resolution between peaks was assessed using the computerized data system after peak integration and analysis.

The LOD and LOQ were determined by calculating the signal-to-noise (S/N) ratio obtained from the comparison of the signal (height) of each of the analytes and noise in the given time range closest to the analyte. The S/N ratios of about 3 and 10 were considered acceptable for estimating the LOD and LOQ.

Accuracy was evaluated using the recovery study of spiked PGIs. By the addition of known amounts of PGIs to a single batch of the matrix sample, the accuracy solutions were prepared in triplicates at three levels of the nominal concentration, and the percent recoveries of the impurities were calculated, respectively.

Precision was evaluated to ensure that the proposed analytical procedure was able to achieve closeness in the data results between the series of measures from several injections of the standard solution over a short interval of time.

Linearity was evaluated by showing that test results were directly proportional to the concentrations of the analytes over the specific ranges from the reporting levels (=LOQ) of genotoxic impurities to 120% of the specifications. The results are reported as the slope, y-intercept of the linear regression line, and correlation coefficient obtained from the analysis of the linearity solutions at five concentration levels, including the specific range.

4. Conclusions

A sensitive and simultaneous analytical method using GC–MS was successfully developed employing the analytical QbD method for 19 alkyl halides as PGIs in APIs. This study emphasizes the usefulness and efficiency of the QbD approach implementation in the development of the analytical method using GC–MS to quantify PGIs, which require low detection limits. Moreover, 144 alkyl halides were classified according to the ICH M7 guidelines through an in silico study using an expert knowledge-based and statistical-based program. A total of 19 alkyl halides, which require a highly sensitive analytical method, were selected because of their high potential for genotoxicity. Based on these results, the optimized analytical conditions were developed using the design space for the range of parameters that could satisfy the predefined CMAs. The analytical method validation for the developed method was performed in terms of specificity, the limit of detection, the limit of quantitation, linearity, accuracy, and precision according to the ICH Q2 guidelines. The developed method was confirmed to be appropriate and validated for the analysis of trace amounts of the 19 target PGIs.

Supplementary Materials: The following supporting information can be downloaded at: https://www.mdpi.com/article/10.3390/molecules27144437/s1, Table S1: In silico prediction results for the 144 alkyl halides; Table S2: Screening data of the resolutions; Table S3: Screening data of the peak areas; Table S4: Optimization data of the resolutions; Table S5: Optimization data of the peak areas; Table S6: Applicability of the method on the API Raloxifene hydrochloride (Cas No. 82640-04-8); Table S7: Applicability of the method on the API Febuxostat (Cas No. 144060-53-7); Table S8: Applicability of the method on the Sitagliptin phosphate (Cas No. 654671-78-0); Table S9: Applicability of the method on the Amlodipine camsylate (Cas No. 652969-01-2); Table S10: Applicability of the method on the Ezetimibe (Cas No. 163222-33-1).

Author Contributions: Conceptualization, K.L. and W.Y.; methodology, K.L. and W.Y.; formal analysis, K.L. and W.Y.; data curation, K.L. and W.Y.; writing—original draft preparation, K.L.; writing—review and editing, W.Y.; visualization, W.Y.; supervision, J.H.J.; project administration, J.H.J. All authors have read and agreed to the published version of the manuscript.

Funding: This research received no external funding.

Institutional Review Board Statement: Not applicable.

Informed Consent Statement: Not applicable.

Data Availability Statement: Not applicable.

Conflicts of Interest: The authors declare no conflict of interest.

References

1. Eder, E.; Henschler, D.; Neudecker, T. Mutagenic properties of allylic and alpha, beta-unsaturated compounds: Consideration of alkylating mechanisms. *Xenobiotica* **1982**, *12*, 831–848. [CrossRef] [PubMed]
2. Sobol, Z.; Engel, M.E.; Rubitski, E.; Ku, W.W.; Aubrecht, J.; Schiestl, R.H. Genotoxicity Profiles of Common Alkyl Halides and Esters with Alkylating Activity. *Mutat. Res. Genet. Toxicol. Environ. Mutagenes.* **2007**, *633*, 80–94. [CrossRef] [PubMed]
3. Szekely, G.; de Sousa, M.C.A.; Gil, M.; Ferreira, F.C.; Heggie, W. Genotoxic Impurities in Pharmaceutical Manufacturing: Sources, Regulations, and Mitigation. *Chem. Rev.* **2015**, *115*, 8182–8229. [CrossRef] [PubMed]
4. Bolt, H.M.; Gansewendt, B. Mechanisms of Carcinogenicity of Methyl Halides. *Crit. Rev. Toxicol.* **1993**, *23*, 237–253. [CrossRef]
5. Müller, L.; Mauthe, R.J.; Riley, C.M.; Andino, M.M.; de Antonis, D.; Beels, C.; DeGeorge, J.; de Knaep, A.G.M.; Ellison, D.; Fagerland, J.A.; et al. A Rationale for Determining, Testing, and Controlling Specific Impurities in Pharmaceuticals That Possess Potential for Genotoxicity. *Regul. Toxicol. Pharmacol.* **2006**, *44*, 198–211. [CrossRef]
6. Bercu, J.P.; Dobo, K.L.; Gocke, E.; Mcgovern, T.J. Overview of Genotoxic Impurities in Pharmaceutical Development. *Int. J. Toxicol.* **2009**, *28*, 468–478. [CrossRef]
7. Raman, N.V.V.S.S.; Prasad, A.V.S.S.; Ratnakar Reddy, K. Strategies for the Identification, Control and Determination of Genotoxic Impurities in Drug Substances: A Pharmaceutical Industry Perspective. *J. Pharm. Biomed. Anal.* **2011**, *55*, 662–667. [CrossRef]
8. Looker, A.R.; Ryan, M.P.; Neubert-Langille, B.J.; Naji, R. Risk Assessment of Potentially Genotoxic Impurities within the Framework of Quality by Design. *Org. Process Res. Dev.* **2010**, *14*, 1032–1036. [CrossRef]
9. Ashby, J.; Tennant, R.W.; Zeiger, E.; Stasiewicz, S. Classification According to Chemical Structure, Mutagenicity to *Salmonella* and Level of Carcinogenicity of a Further 42 Chemicals Tested for Carcinogenicity by the U.S. National Toxicology Program. *Mutat. Res. Genet. Toxicol.* **1989**, *223*, 73–103. [CrossRef]
10. Borman, P.J.; Chatfield, M.J.; Crowley, E.L.; Eckers, C.; Elder, D.P.; Francey, S.W.; Laures, A.M.F.; Wolff, J.C. Development, Validation and Transfer into a Factory Environment of a Liquid Chromatography Tandem Mass Spectrometry Assay for the Highly Neurotoxic Impurity FMTP (4-(4-Fluorophenyl)-1-Methyl-1,2,3,6-Tetrahydropyridine) in Paroxetine Active Pharmaceutical. *J. Pharm. Biomed. Anal.* **2008**, *48*, 1082–1089. [CrossRef]
11. Li, S.; Dong, L.; Tang, K.; Lan, Z.; Liu, R.; Wang, Y.; Wang, R.; Lin, H. Simultaneous and Trace Level Quantification of Two Potential Genotoxic Impurities in Valsartan Drug Substance Using UPLC-MS/MS. *J. Pharm. Biomed. Anal.* **2022**, *212*, 114630. [CrossRef]
12. Wu, X.; Zhu, L.; Visky, D.; Xie, R.; Shao, S.; Liang, X. Derivatization of Genotoxic Nitroaromatic Impurities for Trace Analysis by LC-MS. *Anal. Methods* **2014**, *6*, 7277–7284. [CrossRef]

13. Venugopal, N.; Vijaya Bhaskar Reddy, A.; Madhavi, G.; Jaafar, J.; Madhavi, V.; Gangadhara Reddy, K. Trace Level Quantification of 1-(3-Chloropropyl)-4-(3-Chlorophenyl)Piperazine HCl Genotoxic Impurity in Trazodone Using LC–MS/MS. *Arab. J. Chem.* **2019**, *12*, 1615–1622. [CrossRef]
14. Liu, Z.; Fan, H.; Zhou, Y.; Qian, X.; Tu, J.; Chen, B.; Duan, G. Development and Validation of a Sensitive Method for Alkyl Sulfonate Genotoxic Impurities Determination in Drug Substances Using Gas Chromatography Coupled to Triple Quadrupole Mass Spectrometry. *J. Pharm. Biomed. Anal.* **2019**, *168*, 23–29. [CrossRef]
15. Khan, M.; Jayasree, K.; Reddy, K.V.S.R.K.; Dubey, P.K. A Validated CE Method for Determining Dimethylsulfate a Carcinogen and Chloroacetyl Chloride a Potential Genotoxin at Trace Levels in Drug Substances. *J. Pharm. Biomed. Anal.* **2012**, *58*, 27–33. [CrossRef]
16. Ji, S.; Gao, H.; Xia, X.; Zheng, F. A New HPLC-UV Derivatization Approach for the Determination of Potential Genotoxic Benzyl Halides in Drug Substances. *RSC Adv.* **2019**, *9*, 25797–25804. [CrossRef]
17. Wang, Y.; Feng, J.; Wu, S.; Shao, H.; Zhang, W.; Zhang, K.; Zhang, H.; Yang, Q. Determination of Methyl Methanesulfonate and Ethyl Methylsulfonate in New Drug for the Treatment of Fatty Liver Using Derivatization Followed by High-Performance Liquid Chromatography with Ultraviolet Detection. *Molecules* **2022**, *27*, 1950. [CrossRef]
18. Ho, T.D.; Yehl, P.M.; Chetwyn, N.P.; Wang, J.; Anderson, J.L.; Zhong, Q. Determination of Trace Level Genotoxic Impurities in Small Molecule Drug Substances Using Conventional Headspace Gas Chromatography with Contemporary Ionic Liquid Diluents and Electron Capture Detection. *J. Chromatogr. A* **2014**, *1361*, 217–228. [CrossRef]
19. Azzam, K.M.; Aboul-Enein, H.Y. Recent Advances in Analysis of Hazardous Genotoxic Impurities in Pharmaceuticals by HPLC, GC, and CE. *J. Liq. Chromatogr. Relat. Technol.* **2016**, *39*, 1–7. [CrossRef]
20. Liu, D.Q.; Sun, M.; Kord, A.S. Recent Advances in Trace Analysis of Pharmaceutical Genotoxic Impurities. *J. Pharm. Biomed. Anal.* **2010**, *51*, 999–1014. [CrossRef]
21. Elder, D.P.; Lipczynski, A.M.; Teasdale, A. Control and Analysis of Alkyl and Benzyl Halides and Other Related Reactive Organohalides as Potential Genotoxic Impurities in Active Pharmaceutical Ingredients (APIs). *J. Pharm. Biomed. Anal.* **2008**, *48*, 497–507. [CrossRef] [PubMed]
22. Dispas, A.; Avohou, H.T.; Lebrun, P.; Hubert, P.; Hubert, C. 'Quality by Design' Approach for the Analysis of Impurities in Pharmaceutical Drug Products and Drug Substances. *TrAC Trends Anal. Chem.* **2018**, *101*, 24–33. [CrossRef]
23. Tome, T.; Žigart, N.; Časar, Z.; Obreza, A. Development and Optimization of Liquid Chromatography Analytical Methods by Using AQbD Principles: Overview and Recent Advances. *Org. Process Res. Dev.* **2019**, *23*, 1784–1802. [CrossRef]
24. International Council for Harmonisation Guideline M7 (R1) on Assessment and Control of DNA Reactive (Mutagenic) Impurities in Pharmaceuticals to Limit Potential Carcinogenic Risk Step 4 Version. 2017. Available online: https://database.ich.org/sites/default/files/M7_R1_Guideline.pdf (accessed on 3 May 2022).
25. Kormány, R.; Rácz, N.; Fekete, S.; Horváth, K. Development of a Fast and Robust Uhplc Method for Apixaban In-Process Control Analysis. *Molecules* **2021**, *26*, 3505. [CrossRef]
26. Dinh, N.P.; Shamshir, A.; Hulaj, G.; Jonsson, T. Validated Modernized Assay for Foscarnet in Pharmaceutical Formulations Using Suppressed Ion Chromatography Developed through a Quality by Design Approach. *Separations* **2021**, *8*, 209. [CrossRef]
27. Kumar, K.Y.K.; Dama, V.R.; Suchitra, C.; Maringanti, T.C. A Simple, Sensitive, High-Resolution, Customized, Reverse Phase Ultra-High Performance Liquid Chromatographic Method for Related Substances of a Therapeutic Peptide (Bivalirudin Trifluoroacetate) Using the Quality by Design Approach. *Anal. Methods* **2020**, *12*, 304–316. [CrossRef]
28. Rub, R.A.; Beg, S.; Kazmi, I.; Afzal, O.; Almalki, W.H.; Alghamdi, S.; Akhter, S.; Ali, A.; Ahmed, F.J. Systematic Development of a Bioanalytical UPLC-MS/MS Method for Estimation of Risperidone and Its Active Metabolite in Long-Acting Microsphere Formulation in Rat Plasma. *J. Chromatogr. B Anal. Technol. Biomed. Life Sci.* **2020**, *1160*. [CrossRef]
29. Alkhateeb, F.L.; Wilson, I.; Maziarz, M.; Rainville, P. Ultra High-Performance Liquid Chromatography Method Development for Separation of Formoterol, Budesonide, and Related Substances Using an Analytical Quality by Design Approach. *J. Pharm. Biomed. Anal.* **2021**, *193*, 113729. [CrossRef]
30. Tome, T.; Obreza, A.; Časar, Z. Developing an Improved UHPLC Method for Efficient Determination of European Pharmacopeia Process-Related Impurities in Ropinirole Hydrochloride Using Analytical Quality by Design Principles. *Molecules* **2020**, *25*, 2691. [CrossRef]
31. Žigart, N.; Časar, Z. Development of a Stability-Indicating Analytical Method for Determination of Venetoclax Using AQbD Principles. *ACS Omega* **2020**, *5*, 17726–17742. [CrossRef]
32. Zhang, X.; Hu, C. Application of Quality by Design Concept to Develop a Dual Gradient Elution Stability-Indicating Method for Cloxacillin Forced Degradation Studies Using Combined Mixture-Process Variable Models. *J. Chromatogr. A* **2017**, *1514*, 44–53. [CrossRef]
33. Tome, T.; Casar, Z. Development of a Unified Reversed-Phase HPLC Method for E Ffi Cient Determination of EP and USP Process-Related Impurities in Celecoxib Using Analytical Quality by Design Principles. *Molecules* **2020**, *25*, 809. [CrossRef]
34. Kasagić-Vujanović, I.; Jančić-Stojanović, B. Quality by Design Oriented Development of Hydrophilic Interaction Liquid Chromatography Method for the Analysis of Amitriptyline and Its Impurities. *J. Pharm. Biomed. Anal.* **2019**, *173*, 86–95. [CrossRef]
35. International Council for Harmonisation Guideline on Q2(R1) Validation of Analytical Procedures: Text and Methodology Step 4 Version. 2005. Available online: https://database.ich.org/sites/default/files/Q2%28R1%29%20Guideline.pdf (accessed on 3 May 2022).

36. International Council for Harmonisation Guideline on M7 Assessment and Control of DNA Reactive (Mutagenic) Impurities in Pharmaceuticals to Limit Potential Carcinogenic Risk—Addendum Step 2b. 2021. Available online: https://www.ema.europa.eu/documents/scientific-guideline/draft-ich-guideline-m7-assessment-control-dna-reactive-mutagenic-impurities-pharmaceuticals-limit_en.pdf (accessed on 3 May 2022).
37. Lhasa Carcinogenicity Database. Available online: https://carcdb.lhasalimited.org (accessed on 3 May 2022).
38. National Institute of Health Sciences Website. Available online: http://www.nihs.go.jp/dgm/amesqsar.html (accessed on 3 May 2022).
39. Bercu, J.P.; Galloway, S.M.; Parris, P.; Teasdale, A.; Masuda-Herrera, M.; Dobo, K.; Heard, P.; Kenyon, M.; Nicolette, J.; Vock, E.; et al. Potential Impurities in Drug Substances: Compound-Specific Toxicology Limits for 20 Synthetic Reagents and by-Products, and a Class-Specific Toxicology Limit for Alkyl Bromides. *Regul. Toxicol. Pharmacol.* **2018**, *94*, 172–182. [CrossRef]
40. Barber, E.D.; Donish, W.H.; Mueller, K.R. A procedure for the quantitative measurement of the mutagenicity of volatile liquids in the Ames Salmonella/microsome assay. *Mutat. Res. Genet. Toxicol.* **1981**, *90*, 31–48. [CrossRef]
41. Seifried, H.E.; Seifried, R.M.; Clarke, J.J.; Junghans, T.B.; San, R.H.C. A Compilation of Two Decades of Mutagenicity Test Results with the Ames *Salmonella Typhimurium* and L5178Y Mouse Lymphoma Cell Mutation Assays. *Chem. Res. Toxicol.* **2006**, *19*, 627–644. [CrossRef]
42. International Council for Harmonisation Guideline on Q3C (R8) Residual Solvents Step 4 Version 2021. Available online: https://database.ich.org/sites/default/files/ICH_Q3C-R8_Guideline_Step4_2021_0422_1.pdf (accessed on 3 May 2022).
43. International Council for Harmonisation Guideline on Q14 Analytical Procedure Development Draft Version 2022. Available online: https://database.ich.org/sites/default/files/ICH_Q14_Document_Step2_Guideline_2022_0324.pdf (accessed on 3 May 2022).
44. Lhasa Website. Available online: https://www.lhasalimited.org (accessed on 3 May 2022).
45. Amberg, A.; Beilke, L.; Bercu, J.; Bower, D.; Brigo, A.; Cross, K.P.; Custer, L.; Dobo, K.; Dowdy, E.; Ford, K.A.; et al. Principles and Procedures for Implementation of ICH M7 Recommended (Q)SAR Analyses. *Regul. Toxicol. Pharmacol.* **2016**, *77*, 13–24. [CrossRef]
46. Vega Hub Website. Available online: https://www.vegahub.eu (accessed on 3 May 2022).

Article

A High-Throughput Clinical Laboratory Methodology for the Therapeutic Monitoring of Ibrutinib and Dihydrodiol Ibrutinib

Gellért Balázs Karvaly [1,*], István Vincze [1], Alexandra Balogh [2], Zoltán Köllő [1], Csaba Bödör [3,4] and Barna Vásárhelyi [1]

[1] Department of Laboratory Medicine, Semmelweis University, 4 Nagyvárad tér, 1089 Budapest, Hungary; vincze.istvan@pharma.semmelweis-univ.hu (I.V.); kollo.zoltan@med.semmelweis-univ.hu (Z.K.); vasarhelyi.barna@med.semmelweis-univ.hu (B.V.)

[2] Department of Internal Medicine and Hematology, Semmelweis University, 46 Szentkirályi Utca, 1088 Budapest, Hungary; balogh.alexandra@med.semmelweis-univ.hu

[3] Department of Pathology and Experimental Cancer Research, Semmelweis University, 26 Üllői út, 1085 Budapest, Hungary; bodor.csaba1@med.semmelweis-univ.hu

[4] HCEMM-SE Molecular Oncohematology Research Group, 26 Üllői út, 1085 Budapest, Hungary

* Correspondence: karvaly.gellert_balazs@med.semmelweis-univ.hu

Abstract: Ibrutinib (IBR) is an oral anticancer medication that inhibits Bruton tyrosine kinase irreversibly. Due to the high risk of adverse effects and its pharmacokinetic variability, the safe and effective use of IBR is expected to be facilitated by precision dosing. Delivering suitable clinical laboratory information on IBR is a prerequisite of constructing fit-for-purpose population and individual pharmacokinetic models. The validation of a dedicated high-throughput method using liquid chromatography–mass spectrometry is presented for the simultaneous analysis of IBR and its pharmacologically active metabolite dihydrodiol ibrutinib (DIB) in human plasma. The 6 h benchtop stability of IBR, DIB, and the active moiety (IBR + DIB) was assessed in whole blood and in plasma to identify any risk of degradation before samples reach the laboratory. In addition, four regression algorithms were tested to determine the optimal assay error equations of IBR, DIB, and the active moiety, which are essential for the correct estimation of the error of their future nonparametric pharmacokinetic models. The noncompartmental pharmacokinetic properties of IBR and the active moiety were evaluated in three patients diagnosed with chronic lymphocytic leukemia to provide a proof of concept. The presented methodology allows clinical laboratories to efficiently support pharmacokinetics-based precision pharmacotherapy with IBR.

Keywords: tyrosine kinase inhibitor; liquid chromatography–mass spectrometry; active metabolite; therapeutic drug monitoring; chronic lymphocytic leukemia; assay error equation; oral anticancer drug

1. Introduction

Ibrutinib (1-[(3R)-3-[4-Amino-3-(4-phenoxyphenyl)-1H-pyrazolo[3,4-d]pyrimidin-1-yl]-1-piperidinyl]-2-propen-1-one, chemical abstracts service number 936563-96-1, IBR) is a first-in-class, small-molecule, nonpeptide, nonnucleobase oral anticancer drug. First approved in 2013, its pharmacological indications include the treatment of mantle-cell lymphoma, chronic lymphocytic leukemia with or without 17p deletion, and Waldenström's macroglobulinemia. Its pharmacological action is exerted through the irreversible inhibition of Bruton tyrosine kinase, a signaling molecule of the B-cell antigen receptor and cytokine receptor pathways [1]. IBR is transformed extensively into a pharmacologically active metabolite dihydrodiol ibrutinib (1-[(3R)-3-[4-amino-3-(4-phenoxyphenyl)pyrazolo[3,4-d]pyrimidin-1-yl]piperidin-1-yl]-2,3-dihydroxypropan-1-one, chemical abstracts service number 1654820-87-7, DIB), with DIB/IBR concentration ratios of 1:1 to 3:1 being typically attained (Figure 1) [2]. IBR has been useful primarily in combination therapies

Citation: Karvaly, G.B.; Vincze, I.; Balogh, A.; Köllő, Z.; Bödör, C.; Vásárhelyi, B. A High-Throughput Clinical Laboratory Methodology for the Therapeutic Monitoring of Ibrutinib and Dihydrodiol Ibrutinib. *Molecules* **2022**, *27*, 4766. https://doi.org/10.3390/molecules27154766

Academic Editors: Franciszek Główka and Marta Karaźniewicz-Łada

Received: 27 May 2022
Accepted: 20 July 2022
Published: 25 July 2022

Publisher's Note: MDPI stays neutral with regard to jurisdictional claims in published maps and institutional affiliations.

Copyright: © 2022 by the authors. Licensee MDPI, Basel, Switzerland. This article is an open access article distributed under the terms and conditions of the Creative Commons Attribution (CC BY) license (https://creativecommons.org/licenses/by/4.0/).

(immunotherapy with obinutuzumab, ofatumumab, rituximab, or ublituximab; chemoimmunotherapy with fludarabine–cyclophosphamide–rituximab or bendamustine–rituximab; chimeric antigen receptor T-cell therapy; as well as concurrent treatment with the Bcl-2 protein inhibitor venetoclax, or with the phosphatidylinositol-3-kinase inhibitors duvelisib or idelalisib) [3].

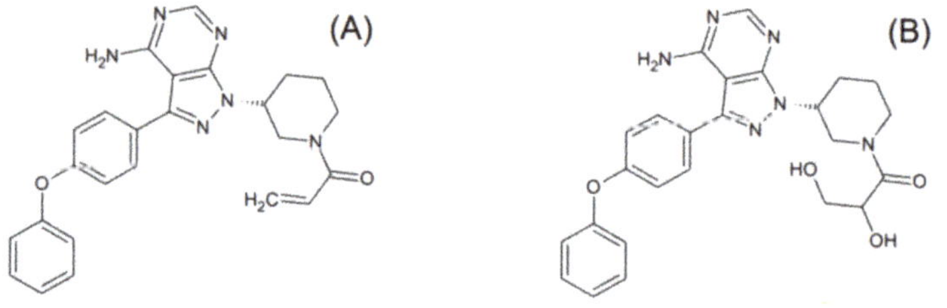

Figure 1. Structural formulae of (**A**) ibrutinib and (**B**) dihydrodiol ibrutinib.

Treatment with IBR requires careful guidance in dosing, primarily due to its severe adverse effects caused by off-target kinase inhibition and other mechanisms [4]. These include atrial fibrillation, the most common reason for the discontinuation of IBR therapy, major bleeding (occurring in 1–10% of patients), general debility, arthralgia, infection (especially pneumonitis), and secondary malignancy. Fatalities associated with these have been reported in up to 10%, while dose modification is prompted in 11–50% of cases [5]. A multicenter, retrospective chart study including adults treated with chronic lymphocytic leukemia revealed that—mainly due to the occurrence of adverse events—25% of patients experienced at least one dose reduction, while treatment discontinuation and dose holds impacted 20% and 34% of cases, respectively [6].

The poor solubility of IBR in water, its low permeability through membranes (Biopharmaceutics Classification System Class II), and extensive metabolism catalyzed by cytochrome P450 3A (CYP3A) results in considerable variability in its pharmacokinetic properties [7,8]. IBR has low oral bioavailability, high (>95%) affinity and special binding properties to plasma albumin, a large apparent volume of distribution, and changes in hepatic metabolism when coadministered with CYP3A inhibitors or inducers [8–10]. A population modeling study identified 67% interindividual and 47% intraindividual variability in the clearance of IBR, as well as 51% and 26% in the case of DIB, respectively. IBR exposure was higher in subjects with one copy of the CYP 3A4*22 variant. Nevertheless, when tested as candidate covariates, neither anthropometric or demographic properties of individuals, or the results of a wide range of laboratory tests, have proved to have a major impact on the pharmacokinetic behavior of IBR or DIB [10]. Recent discussions over IBR dose reduction and the clinical impact of related drug–drug and drug–food interactions have also highlighted the importance of individual therapy guidance [11–14].

The translation of therapeutic drug monitoring (TDM) results into clinically meaningful information using precision pharmacotherapy software is the most promising approach to addressing these issues and to optimizing the dosing regimens of IBR. Predicting IBR and DIB plasma concentrations using nonparametric pharmacokinetic modeling is a particularly attractive strategy, since the pharmacokinetic properties of IBR observed in each individual are retained instead of being melted into summary statistics [15–17]. The error of predictions, i.e., the differences between observed concentrations and those predicted by the model, is estimated as the combination of the measurable analytical error, derived from the standard deviation (SD) of each measurement result, and an unmeasurable "noise" of clinical and pharmaceutical origin [15,18]. Since the processing of each real-life TDM sample in several repeats is beyond clinical reality, the efficient estimation of the imprecision of measured IBR and DIB concentrations by applying empirical assay error equations is a key component of building their nonparametric pharmacokinetic models, and should be part of method validation [18,19]. Evidence shows that, concerning analysis relying on liquid chromatography–tandem mass spectrometry (LC-MS/MS) and the use of internal standards, the relationship between drug concentrations and SDs can be characterized with linear models [20–22]. Nonparametric population pharmacokinetic models incorporating linear assay error equations have been constructed for voriconazole, as well as for atorvastatin and its pharmacologically active metabolites [19,23].

An important prerequisite of reporting reliable assay results for efficient pharmacokinetic modeling is the evaluation of the stability of IBR and DIB in samples in the preanalytical and analytical phases. IBR has acceptable stability in heparinized plasma stored at 4 °C or lower, or when exposing the samples to multiple freeze–thaw cycles, but not at ambient temperature [24]. Rood et al. measured the concentrations of both IBR and DIB after keeping heparinized plasma samples at 0 °C for 2 h, or at −80 °C for 2 months, in addition to performing a freeze–thaw experiment and assessment of the autosampler stability of prepared samples [25]. However, no data have been published concerning IBR and DIB stability in whole blood or in plasma in the early preanalytical phase, i.e., before the samples reach the premises of the laboratory. In this early phase, patient samples are frequently kept on the bench at ambient temperature for an undefined length of time. Performing investigations for controlling for this phase is therefore pivotal.

Our aim is to present the results of experiments accomplished to attain comprehensive clinical laboratory information required for constructing nonparametric pharmacokinetic models that can be employed efficiently for individually optimized treatments with IBR. These experiments targeted (1) the development and validation of a high-throughput analytical method for the clinical analysis of IBR and DIB in human plasma, (2) the characterization of the stability of IBR, DIB, and the active moiety (represented by the sums of IBR and DIB concentrations) in the collected blood samples and in plasma separated in the early preanalytical phase, and (3) the construction of assay error equations of IBR, DIB, and the active moiety, which are incorporable into nonparametric pharmacokinetic models as the measurable error. A proof of concept of the developed methodology is provided by the evaluation of the pharmacokinetics of IBR, DIB, and the active moiety in three patients diagnosed with chronic lymphocytic leukemia.

2. Materials and Methods

2.1. Chemicals and Solutions

Ibrutinib (99%), dihydrodiol ibrutinib (99%), $^2H_5$-ibrutinib and $^2H_5$-dihydrodiol ibrutinib were purchased from Alsachim S.A.S. (Illkirch-Grafenstaden, France). LC-MS grade acetonitrile, formic acid, methanol and water were supplied by Reanal Labor (Budapest, Hungary).

Stock solutions (4 mg/mL) of the analytes, and 1 mg/mL stock solutions of the isotopically labeled internal standards (IS), respectively, were prepared in methanol. The solutions employed for spiking blank human plasma samples in the experiments conducted to establish the assay error equations contained IBR and DIB in the range of 0.011–26.1 µg/mL. The concentration of $^2H_5$-IBR and $^2H_5$-DIB in the IS working solution was 10 µg/mL. 1.4 µL IS working solution was added to each milliliter of acetonitrile employed for the deproteinization of plasma samples.

2.2. Sample Preparation

Deproteinizing solution (200 µL) was added to 50 µL plasma on a Phenomenex Impact 96-well protein precipitation plate (Gen-Lab, Budapest, Hungary). The plate was shaken at 1100 rpm for 10 min on an Allsheng TMS-200 thermoshaker incubator (Lab-Ex, Budapest, Hungary), and the supernatant was transferred to a collection plate (1 mL/slot). By applying nitrogen (purity rating 5.0) at gentle positive pressure using a Phenomenex Presston 100 positive-pressure manifold (Gen-Lab, Budapest, Hungary), the supernatant could be filtered successfully without carrying along solid particles. Further processing was therefore possible without centrifugation. Supernatant (150 µL) was mixed with 90 µL water, and the mixture was submitted for analysis.

2.3. Analysis

A modular CE-IVD certified liquid chromatograph–tandem mass spectrometer consisting of a Shimadzu DGU20 CL degasser, two LC30-AD CL pumps, a SIL-30-CL autosampler, a CTO-20AC column oven and an LCMS-8060 CL triple-quadrupole mass spectrometer (Simkon, Budapest, Hungary) was employed. Instrument control and data acquisition were performed using the LabSolutions CL (version 1.1) software. Chromatographic separation was accomplished using a Phenomenex Kinetex XB-C18, 50 × 2.1 mm (particle size 1.7 µm) stationary phase (Gen-Lab, Budapest, Hungary). The column temperature was set to 40 °C. The mobile phases were LC-MS grade water-formic acid 99.9:0.1 (v/v, mobile phase A) and methanol–formic acid 99.9:0.1 (v/v, mobile phase B). The following gradient program was applied (% mobile phase B): initial, 30%; 0.50 min, 30%; 3.00 min, 50%; 3.01 min, 90%; 5.50 min, 90%; and 5.51 min, 30%. The mobile phase flow rate was 0.25 mL/min, and the injected sample volume was 1.0 µL. The total run time was 7.00 min.

Mass spectrometry was performed using positive electrospray ionization and multiple reaction monitoring. Following the selection of the precursor [M+H]$^+$ and of the product ions, the detection of mass transitions was optimized by the instrument control software by adjusting quadrupole 1 bias, the collision energy, and quadrupole 3 bias. The positive-mode multiple-reaction monitoring-optimization reports of the analytes are provided in Supplementary File S1. The ion transitions of the internal standards providing optimally sensitive signal intensities were found by conducting chromatographic runs under the conditions described above where the precursor ions were defined as those of the analytes plus five (i.e., m/z = 445.9 for ^{2}H$_5$-IBR and m/z = 479.7 for ^{2}H$_5$-DIB), and the signal intensities of the product ions were monitored in the mass range starting with the masses of the target product ions of the analytes (m/z = 304.1), and ending with m/z = 309.1. The optimized mass spectrometry settings are summarized in Supplementary File S1.

2.4. Quantitation

Plasma samples spiked with known concentrations of the analytes were employed for calibration. Each calibration set contained 6–8 concentration levels of IBR and DIB. The target values were 0.2, 1.0, 2.5, 10, 40, 80, 100, and 150 ng/mL, with two additional calibrator samples (320 and 520 ng/mL) run on a single occasion for evaluating plasma samples spiked with IBR and DIB at concentrations higher than 150 ng/mL. These values corresponded to 0.454, 2.27, 5.68, 22.7, 90.8, 182, 227, 341, 726, and 1180 nmol/L for IBR and 0.422, 2.11, 5.27, 21.1, 84.3, 169, 211, 317, 674, and 1096 nmol/L for DIB. Calibration was performed at the beginning of each batch run by spiking pooled blank plasma in which the absence of the analytes had been verified earlier. Calibration models were established using 1/concentration^2-weighted linear regression.

The volumes of the analyte solutions spiked to calibrator and spiked plasma samples did not exceed 5% of that of plasma. Each calibrator, spiked plasma, and patient sample was measured in a single repeat.

2.5. Method Validation

Human plasma, separated from whole blood collected into 3-mL phlebotomy tubes containing tripotassium ethylene diamine tetraacetate (K$_3$-EDTA) as anticoagulant and left over from routine laboratory diagnostic tests, was provided by the Central Laboratory, Department of Laboratory Medicine, Semmelweis University following irreversible deidentification. A total of 110 deidentified plasma samples were used, 10 of which had been pooled for preparing the calibrators and for performing selectivity and sample carryover tests. No interaction was made with the donors of these samples. All deidentified samples underwent analysis before spiking to confirm the absence of IBR and DIB.

The developed method was validated by evaluating selectivity, sample carryover, the performance of calibration models, assay accuracy and imprecision (by establishing assay error equations), matrix effect, and the stability of IBR and DIB in whole blood and plasma [26].

Selectivity and sample carryover were evaluated by comparing the chromatographic peak areas obtained with the highest-level calibrators to those recorded in blank plasma, injected alternately in three cycles. The performance of calibration curves was assessed by back-calculating the accuracies of measured calibrator concentrations. No lower limits of quantitation (LLOQ) were defined, as one of the objectives of constructing assay error equations is to provide quantitative estimates of the SD all the way down to zero analyte concentration. This strategy allows the laboratory to report all TDM results in a pharmacokinetically informative manner, and without censoring sub-LLOQ assay results that may otherwise be important clinically [18,19].

Assay error equations were established by spiking a total of 100 independent plasma samples with the analytes in five experiments performed on separate days. Four spiking levels were prepared in each experiment, adding up to a total of 20 spiking levels in addition to the blanks. Twenty independent plasma samples were spiked at each concentration level. Equations were defined for IBR, DIB, and the active moiety.

Matrix factors corrected with the peak areas of the internal standards (IMF) were determined at two concentration levels (2.0 and 80 ng/mL for each analyte, i.e., 4.54 nmol/L and 182 nmol/L for IBR and 4.21 nmol/L and 169 nmol/L for DIB, respectively). To this end, 5.0 µL of the analyte solutions (12 ng/mL or 480 ng/mL) and of a 336 ng/mL IS solution, prepared in methanol, were added to 140 µL supernatant obtained following the deproteinization of 50 µL blank plasma using 200 µL acetonitrile as described in Section 2.2. Six independent plasma samples were processed. The reference solutions were 140 µL acetonitrile-water 4:1 (v/v) mixtures spiked as described above. The mixtures and the reference solutions were subsequently diluted with 90 µL water. Internal standard-corrected matrix factors were calculated as the peak area ratios of the analytes and the internal standards in prepared plasma versus those in a neat solution. In the stability studies, the recoveries of IBR and DIB were calculated as the ratio of the concentrations measured after incubation and those measured at the beginning of the study. The analytes were considered stable at time points where recoveries exceeded 85.0%.

The preanalytical stability of IBR and DIB was evaluated by adding 20 µL methanol solutions containing 500 ng/mL IBR and DIB to two 1.0 mL aliquots of blood freshly drawn into phlebotomy tubes containing K_3-EDTA. Three samples, 3 mL each, were taken from three healthy volunteers (manuscript authors G.B.K., I.V. and Z.K.). One of the fractions was centrifuged at 3000 rpm and 10 °C for 10 min immediately after spiking the analytes, and plasma was separated. Both spiked whole blood and the separated plasma were kept at ambient temperature for 6 h. At 0, 30, 60, 90, 180, and 360 min after sampling, whole blood was gently rotated five times and 150 µL whole blood pipetted into a 1.5 mL microcentrifuge tube that was subsequently centrifuged at 3000 rpm and 10 °C for 10 min. Fifty microliters was drawn from the supernatant of the whole blood sample as well as from the plasma, and was processed as described in Section 2.2.

2.6. Proof-of-Concept Experiments

In order to provide a proof of concept, IBR and DIB were assayed in the plasma samples of three patients treated with IBR. This evaluation was undertaken as part of a larger clinical study (ethical approval: 45371-2/2016/EKU, issued by the Scientific and Research Ethics Committee of the National Medical Research Council, Budapest, Hungary, Supplementary File S2). The criteria for inclusion were (1) age of ≥18 years, (2) treatment ongoing with IBR for more than 10 days, and (3) no concurrent administration of medications undergoing CYP3A4 metabolism. Detailed demographic and clinical information concerning the three participants is provided in Table 1. The subjects gave their written informed consent. Each participant took either two or three 140 mg Imbruvica capsules, as per the therapeutic provision, in the presence of the recruiting clinician. Blood was collected from the antecubital vein in a standard phlebotomy process by trained personnel into 3 mL tubes containing K_3-EDTA at 0.5 h, 1 h, 2 h, 4 h, 23 h and 24 h postdose.

Table 1. Demographic and clinical characteristics of the patients receiving IBR. CLL, chronic lymphocytic leukemia.

	Participant 1	Participant 2	Participant 3
Gender	female	male	female
Age	79	79	60
Diagnosis	CLL	CLL	CLL
Reported co-morbidities	melanoma malignum, hypertension	none	resected gall-bladder
Ibrutinib daily dose	420 mg	420 mg	280 mg
eGFR (mL/min/1.73 m^2)	50.9	57.6	>90
glutaryl oxaloacetate transaminase (U/L)	18	20	17
glutaryl pyruvate transaminase (U/L)	12	13	14
gamma-glutamyl transferase (U/L)	23	12	20
white blood cell count (G/L)	2.7	27.8	241
neutrophile (%)	59.2	0.0	0.0
eosinophile (%)	0.4	0.0	0.0
basophile (%)	1.5	0.0	0.0
monocyte (%)	9.6	0.0	0.0
lymphocyte (%)	29.3	0.0	0.0
immature granulocyte (%)	9.3	0.0	0.1

Blood samples were centrifuged at 3000 rpm for 10 min in a Hettich Universal 320R centrifuge (Auro-Science, Budapest, Hungary) at 10 °C. Plasma was separated and frozen at −70 °C until analysis was done within 2 weeks.

2.7. Data Evaluation

Data management and basic calculations were performed using Microsoft Excel 2016. Statistical evaluation was conducted in the R environment (version 4.0.5, 31 March 2021) using the following packages: "stats", "AICcmodavg", "NonCompart", and "ncar". Plots were created using the free academic version of ACD/ChemSketch (ACD Labs, Toronto, ON, Canada), Microsoft Excel, and the "ggplot2" package of R [27].

Assay accuracy was calculated as the ratio of the mean observed analyte concentration and the nominal concentration. Assay error equations were generated using four algorithms: Theil's regression with and without the Siegel estimator, as well as unweighted linear or second-degree polynomial least squares regression. A script written by one of the authors (G.B.K.) in the R environment was employed to perform the calculations (Supplementary File S3) [21]. The goodness of the fitted assay error equations was quantified as the normalized sums of the squared residuals (NSSR) using the following formula:

$$NSSR = \sum_{i=1}^{m} \frac{(SD_{observed,i} - SD_{predicted,i})^2}{SD_{predicted,i}^2} \quad (1)$$

where $SD_{observed,i}$ is the observed SD of the concentrations measured in spiked plasma samples containing the analytes at the i-th spiking level, $SD_{predicted,i}$ is the estimate of the SD of the analyte concentration at the i-th spiking level, as inferred from the fitted regression equation, and m is the number of spiking levels (m = 20) [21].

Noncompartmental pharmacokinetic calculations were performed for IBR and for the active moiety (IBR + DIB) based on the concentration series obtained in the three participants receiving IBR. The AUC() function of the "NonCompart" as well as the pdfNCA() function of the "ncar" package were used in the R environment with default settings. Since the subjects were in steady state concerning IBR and DIB concentration profiles, the 24 h concentrations were also employed for simulating 0 h predose levels. The employed R packages "NonCompart" and "ncar" are compatible with the Study Data Tabulation Model-formatted dataset of the Clinical Data Interchange Standards Consortium standard, and their performance had previously been demonstrated to yield results equivalent to those obtained using leading commercial pharmacokinetic modeling software [28,29].

3. Results

3.1. Bioanalytical Method Validation

IBR and DIB were eluted from the stationary phase as symmetrical peaks with retention times of 4.3 min and 4.2 min, respectively. The internal standards were eluted with the same retention times as their unlabeled analogues. The method was sensitive and selective, with no sample carryover observed (Supplementary File S4). The relationship between the concentrations of IBR and DIB and the analyte/internal standard peak area ratios was linear in the calibrated concentration range.

Method accuracy and precision are presented in Table 2. In the calibrated concentration range, the accuracy was 99.4–110% and 91.7–118%, while the relative standard deviation was 1.88–6.04% and 0.59–27.3% for IBR and DIB, respectively. Bioanalytical method-validation guideline criteria (accuracy 85–115%, or 80–120% at the lower limit of quantitation, relative standard deviation <15%, or <20% at the lower limit of quantitation) were met for IBR in the entire calibrated range, and for DIB at 11.3–1096 nmol/L [26]. The accuracy of the assay was 120–139% and 92.1–339%, with relative SDs of 8.3–16.3% and 11.9–54.2%, respectively, under the calibrated concentration range. The internal standard-corrected matrix factors were 92.3% (10.4%) and 103% (6.4%) for IBR, and 115% (12.4%) and 101% (6.0%) for DIB (Table 3).

Table 2. Performance of the assay method. The analytes were spiked to 20 independent human plasma samples at each concentration level. Experiments were conducted on five different days, indicated by different colors. N/D, not determined. RSD, relative standard deviation. SD, standard deviation. Experiments performed on different days are shown in different colors.

Ibrutinib				Dihydrodiol Ibrutinib			
Concentration (nmol/L)	Accuracy (%)	SD (nmol/L)	RSD	Concentration (nmol/L)	Accuracy (%)	SD (nmol/L)	RSD
0.488	139	0.068	10.0%	0.453	339	0.830	54.2%
0.976	134	0.213	16.3%	0.906	209	0.864	45.8%
1.99	120	0.200	8.31%	1.85	137	0.299	11.9%
2.30	101	0.091	3.91%	2.15	95.4	0.561	27.3%
5.96	102	0.325	5.33%	5.56	92.1	1.20	23.5%
12.2	104	0.334	2.63%	11.3	118	1.98	18.1%
23.0	99.3	0.904	3.96%	21.5	94.1	1.43	7.06%
57.2	99.2	3.43	6.04%	52.9	114	3.29	0.59%
92.0	104	5.59	5.85%	85.9	92.8	3.88	4.86%
146	106	0.386	1.88%	136	116	4.85	1.20%
184	107	10.3	5.19%	172	94.2	14.2	8.78%
230	108	10.6	4.24%	215	91.7	12.4	6.31%
320	108	13.3	3.84%	297	108	12.7	3.96%
343	107	17.9	4.88%	318	110	10.3	2.94%
388	110	19.4	4.55%	360	112	15.4	3.82%
411	106	19.7	4.52%	381	110	19.1	4.53%
434	107	20.3	4.39%	403	112	18.4	4.06%
481	102	23.6	4.80%	447	115	24.4	4.76%
731	99.4	32.3	4.45%	649	105	17.1	2.40%
1187	106	58.2	4.62%	1100	107	38.4	3.25%

Table 3. Internal standard-corrected matrix factors of ibrutinib and dihydrodiol ibrutinib. Six independent human serum matrices (A-F) and two spiking levels were used. SD, standard deviation. RSD, relative standard deviation.

	Ibrutinib		Dihydrodiol Ibrutinib	
Matrix Identifier	Low Level: 2.0 ng/mL (4.54 nmol/L)	High Level: 80 ng/mL (182 nmol/L)	Low Level: 2.0 ng/mL (4.21 nmol/L)	High Level: 80 ng/mL (169 nmol/L)
A	0.894	1.002	1.226	0.942
B	1.069	0.975	0.920	0.993
C	0.906	1.073	1.039	1.044
D	0.815	1.014	1.278	1.009
E	0.849	1.131	1.267	1.104
F	1.004	0.958	1.173	0.952
Mean ± SD	0.923 ± 0.096	1.03 ± 0.065	1.15 ± 0.143	1.01 ± 6.0
RSD (%)	10.4	6.4	12.4	6.4

3.2. Stability of IBR and DIB in the Early Preanalytical Phase

Fifteen analyses of IBR and DIB (3 observations at 5 time points) were conducted in whole blood and in plasma. In whole blood, recoveries lower than 85% were obtained in two cases and in one case concerning IBR and DIB, respectively, with only one of these occurring after 6 h, and with no identifiable trends of the recoveries seen. The recovery of the active moiety (IBR + DIB) exceeded 85% in all cases. In plasma, 85% recoveries or higher were attained in all analyses. The dispersion of the measured concentrations was larger in whole blood than in plasma, indicating that binding to cell components may have influenced the analytical results. The recoveries (t = 0 min: 100%) are shown in Figure 2.

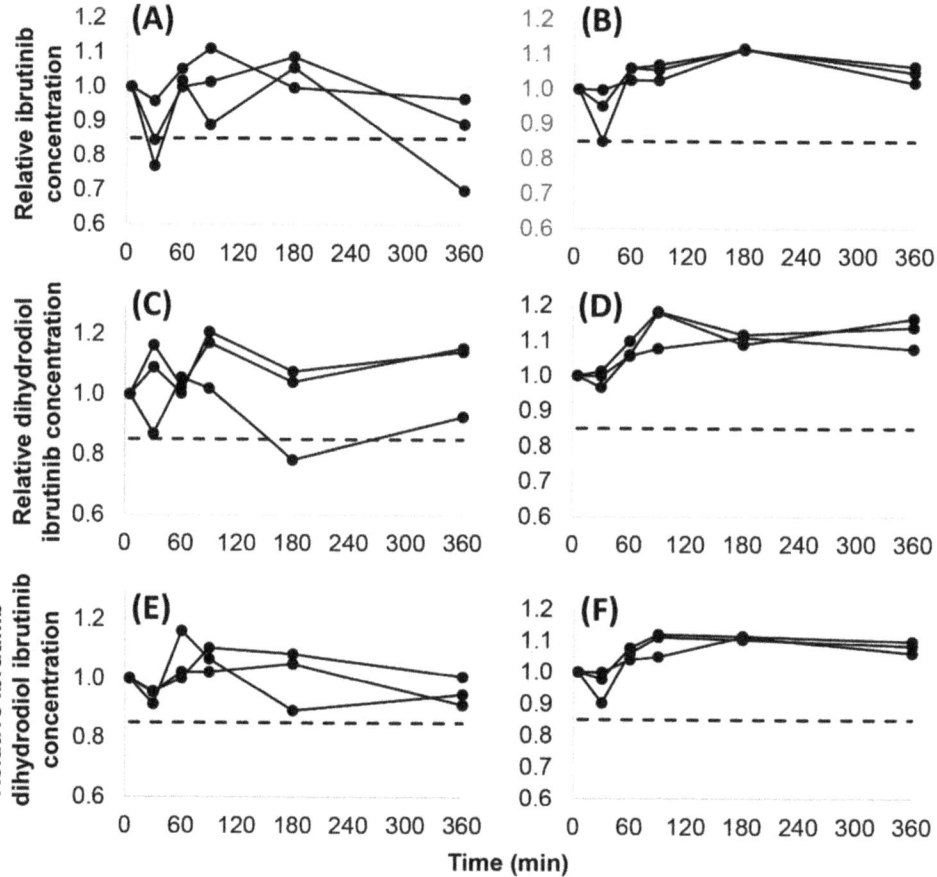

Figure 2. Stability of ibrutinib (**A,B**), dihydrodiol ibrutinib (**C,D**) and the active moiety (sum of ibrutinib and dihydrodiol ibrutinib concentrations) (**E,F**) in whole blood (**A,C,E**) and in plasma (**B,D,F**) at 25 °C over 6 h in 3 independent samples. The dashed line (- - -) displays the limit for judging analyte stability as acceptable (0.85).

3.3. Assay Error Equations of IBR, DIB, and the Active Moiety (IBR + DIB)

The results of various types of regression performed on the concentration–SD relationships are summarized in Table 4. The performance of unweighted linear least squares was unacceptable for IBR, as the predicted SDs were lower than 0 up to 5.96 nmol/L, with a negative intercept (−0.1285). The nonlinear coefficients of the unweighted second-degree least squares polynomials were <0.0001, confirming the linearity of the relationships. Theil's regression with the Siegel estimator delivered the best overall performance in view of the consistently low NSSR values, and of the low yet positive intercepts obtained. The assay error equations obtained using this algorithm were SD = 0.04721 × concentration + 0.05559 (IBR), SD = 0.04382 × concentration + 0.6814 (DIB) and SD = 0.03854 × sum of IBR + DIB concentrations + 0.3526 (active moiety, Figure 3). In the case of DIB and the active moiety, the differences in the performance of the four regression approaches were negligible.

Table 4. Performance of regression algorithms applied to the concentration–standard deviation relationships. IBR, ibrutinib. DIB, dihydrodiol ibrutinib. IBR + DIB, sum of IBR and DIB concentrations (active moiety). NSSR, normalized sum of squared residuals. OLS, unweighted linear least squares. 2nd LS, unweighted 2nd-degree least squares. Siegel, Theil's regression with the Siegel estimator. Theil, Theil's regression.

Algorithm	NSSR			Slope			Intercept		
	IBR	DIB	IBR + DIB	IBR	DIB	IBR + DIB	IBR	DIB	IBR + DIB
Theil	1.876	3.567	3.386	0.0479	0.0418	0.0387	0.06635	0.5308	0.4115
Siegel	2.352	2.516	4.682	0.0472	0.0438	0.0385	0.05559	0.6814	0.3526
OLS	106.9	4.428	1.986	0.0480	0.0342	0.0373	−0.1285	1.970	0.6084
2nd LS	1.667	2.615	1.934	0.0457 *	0.0447 *	0.0359 *	0.08408	1.071	0.8606

* Linear coefficients are shown.

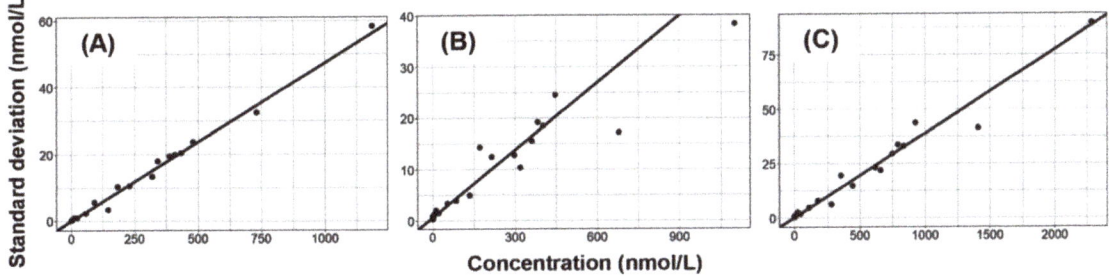

Figure 3. Linear regression applied to the concentration–standard deviation relationships using Theil's regression with the Siegel estimator. (A) Ibrutinib (standard deviation = 0.04721 × concentration + 0.05559. (B) Dihydrodiol ibrutinib (standard deviation = 0.04382 × concentration + 0.6814). (C) Active moiety (standard deviation = 0.03854 × sum of IBR + DIB concentrations + 0.3526).

3.4. 24 h Therapeutic Monitoring of IBR and DIB in the Plasma of Chronic Lymphocyte Leukemia Patients Receiving IBR

The concentration profiles of IBR and DIB obtained in adult chronic lymphocyte leukemia patients are displayed in Figure 4. Maximum concentrations of IBR, and also of DIB, were attained not later than 2 h after drug intake. The mean DIB/IBR concentration ratios were 0.96–1.19 (SD: 0.39–0.60) between 0.5–2 h, 2.36 ± 1.69 at 4 h, and 3.34–3.44 (SD: 1.47–1.56) at the trough (23–24 h). The primary determinant of the maximum concentrations and the areas under the concentration–time curves (AUC) was the dose. Noncompartmental pharmacokinetic characteristics calculated from these curves are shown in Table 5 (the reports of the evaluations are provided in Supplementary File S5).

Table 5. Calculated individual pharmacokinetic properties of IBR and the active moiety (IBR + DIB). AUC_{0-24}, area under the concentration–time curve from dose intake to 24 h postdose. $AUMC_{0-24}$, area under the first moment of the concentration–time curve from dose intake to 24 h postdose. CL/F, apparent clearance. c_{max}, peak concentration. K_e, terminal elimination rate constant. MRT_{0-24}, mean residence time from dose intake to 24 h postdose. $t_{1/2}$, systemic half-life. t_{max}, time to reach the peak concentration. V/F, apparent volume of distribution.

Parameter	Ibrutinib			Dihydrodiol Ibrutinib			Ibrutinib + Dihydrodiol Ibrutinib		
	Patient 1	Patient 2	Patient 3	Patient 1	Patient 2	Patient 3	Patient 1	Patient 2	Patient 3
AUC_{0-24} (nmol × L/h)	1786	1740	613	2347	2528	1800	4134	4268	2414
$AUMC_{0-24}$ (nmol × L)	7488	7434	2051	16,593	11,626	8172	24,082	19,071	10,230
c_{max} (nmol/L)	265.6	374.0	163.2	184.7	253.7	216.6	450.3	627.7	358.3
Dose-normalized c_{max} [nmol/(L × mmol)]	278.7	392.4	256.6	Cannot be calculated			472.5	658.6	563.3
t_{max} (h)	2.0	1.0	1.0	2.0	1.0	2.0	2.0	1.0	1.0
CL/F (L/h)	515	523	1008	Cannot be calculated			Cannot be calculated		
MRT_{0-24} (h)	4.19	4.28	3.34	7.07	4.60	4.54	5.82	4.47	4.24
k_e (1/h)	0.126	0.113	0.121	0.069	0.130	0.122	0.085	0.123	0.122
$t_{1/2}$ (h)	5.49	6.12	5.74	10.1	5.35	5.68	8.13	5.64	5.70
V/F (L)	4080	4620	8346	Cannot be calculated			Cannot be calculated		

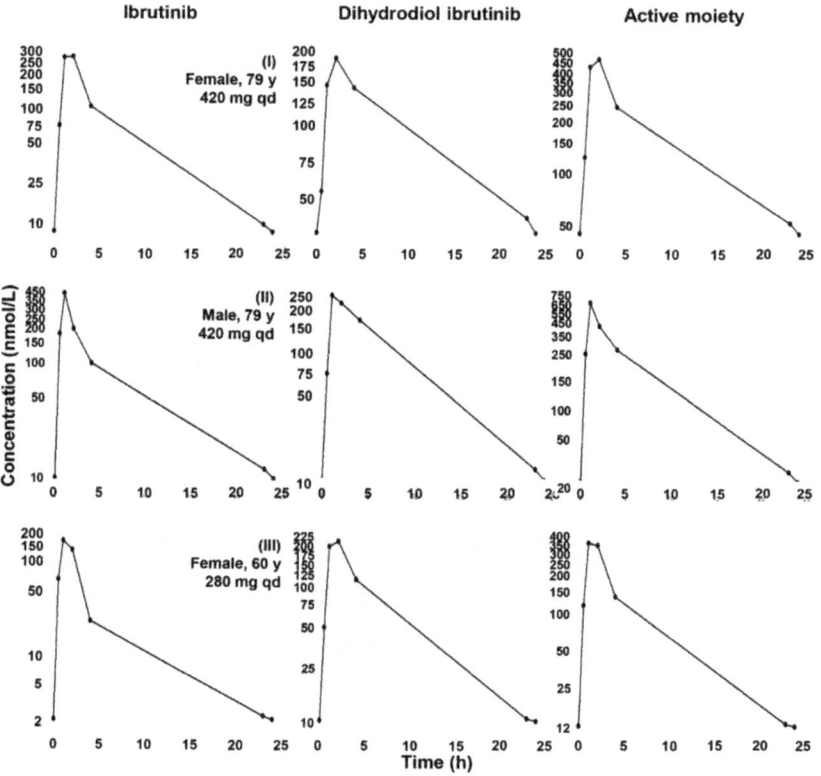

Figure 4. 24 h steady-state concentration profiles of IBR, DIB, and the active moiety (IBR + DIB) in three chronic lymphocyte leukemia patients taking 420 mg (Subjects **I** and **II**) or 280 mg (Subject **III**) IBR per day.

4. Discussion

Few publications have discussed the bioanalysis and the pharmacokinetics of IBR, especially together with its major active metabolite DIB, in humans. The available methodologies have been reviewed extensively [30]. So far, LC-MS/MS has been the only technology to be used for the simultaneous analysis of IBR and DIB, with positive electrospray ionization and the selection of the pseudomolecular ions as precursors. Taurocholic acid, which appears in the bloodstream at higher concentrations in patients with hepatic impairment, has been identified to interfere, requiring its separation from DIB either chromatographically or by high-resolution mass spectrometry [31]. Deproteinization with acetonitrile and solvent exchange have been selected most often for preparing samples [8,25,32–34]. Others have employed simplified liquid extraction and solvent exchange [31,35]. In a single case, liquid–liquid extraction was performed [36]. An approach to the simple and rapid analysis of IBR and DIB in cerebrospinal fluid has also been published [33]. The starting sample volume was 20–200 µL in all of these works.

The presented bioanalytical method, which has been implemented successfully for routine TDM in our laboratory, has been designed specifically to support clinical pharmacokinetic modeling and precision pharmacotherapy. The method allows the rapid assessment of IBR and DIB concentrations as components of a broader panel of tyrosine kinase inhibitors. The employed high-throughput approach relies on sample preparation consisting of two rapid, cost-efficient steps: deproteinization with acetonitrile and dilution. The preparation of a full 96-well plate for analysis requires less than 1 h. With an analysis time of 7 min, up to 170 test results can be reported within 24 h. Provided the 3-month stability of IBR and DIB in plasma at −70 °C and the lack of availability of CE-IVD lyophilized plasma controls, blank plasma spiked at various concentrations or patient samples collected before drug intake and at 1.5 h postdose can serve as control samples for the analysis. In our routine assays, we used plasma spiked at 10 ng/mL and 100 ng/mL to this end. As the range of drugs with a clinical demand to monitor their concentrations in patients belonging to high-risk populations is growing rapidly, the use of in-house calibrators and internal controls is becoming more common and accepted in the absence of commercially available preparations [37].

Recently, the differential absorption of IBR from its isotopically labeled analogues to polymeric surfaces, including the walls of containers used during sample preparation and the polyether ether ketone components of LC-MS/MS systems, has been reported by Mzik et al. [38]. This differential absorption of the analyte led to large SDs and remarkable carryover at low concentrations (0.25 ng/mL, corresponding to 0.567 nmol/L). The SDs of IBR obtained in our study did not confirm this finding, with the relative SDs never exceeding 16.3% from as low as 0.488 nmol/L. In addition, we did not observe appreciable carryover of IBR or DIB in our experiments. Nevertheless, Mzik et al. demonstrated that the components of the employed liquid chromatograph may have a profound impact on this phenomenon. Therefore, a potential reason for this discrepancy is that inside the liquid chromatograph used in our research, the analytes could only get in contact with plastic material after being eluted from the chromatographic column by an eluent composition containing a relatively high fraction of organic component. In addition, we used methanol as the organic solvent, while Mzik et al. used acetonitrile, a less potent solvent for IBR.

It should be noted that the RSDs we recorded for DIB exceeded 20% at all spiking levels, except one in the range of 0.453–5.56 nmol/L (0.217–2.67 ng/mL). In addition, RSDs changed stepwise from 45.8–54.2% (0.453 and 0.906 nmol/L) to 11.9–23.5% (1.89–11.3 nmol/L) and then to 0.59–8.78% (all spiking levels higher than 11.3 nmol/L). While the octanol–water partitioning of DIB is similar to that of IBR, the presence of a primary and a secondary hydroxyl group in the structure may increase the affinity of DIB to slightly polar polymeric surfaces. In conclusion, the sharp differences between the RSDs obtained for IBR and DIB and the fact that large RSDs were obtained for DIB only at low concentrations support the assumption of underlying causes similar to those described by Mzik et al.

The stability of IBR and DIB in whole blood and plasma kept at ambient temperature for 6 h was acceptable for clinical use. The recoveries of DIB were higher than those of IBR at late time points, probably as a result of IBR's covalent binding to endogenous thiols, such as glutathione [39]. Huynh et al. found that the degradation of IBR was considerable (with recoveries of 46.7–72.9%) after plasma was kept at ambient temperature for 24 h [24]. Recoveries of 87–100% were reported after keeping plasma at 0 °C for 2 h or at −80 °C for 2 months, as well as following multiple freeze–thaw cycles. Thermostatting the autosampler tray at 4 °C for 48 h resulted in all reanalyses yielding results within the 85–115% relative concentration range (80–120% at LLOQ), the recommended range of acceptability according to international bioanalytical method-validation guidelines [26]. It can be concluded that blood samples collected for the analysis of IBR and DIB should be centrifuged as soon as possible, preferably within 6 h, and the supernatant should be separated and kept frozen until the analysis.

The assay error equation is the experimental basis for determining the optimal weight (1/variance, also called the Fisher information) of each observation employed for nonlinear curve fitting during the construction of the pharmacokinetic model. Based on mathematical theory and the obtained unbiased NSSR indicators, Theil's regression with the Siegel estimator, a nonparametric linear regression method, which is 100% resistant to outliers regarding the identification of a linear trend, was the most consistently accurate for describing the quantitative relationship between IBR and DIB concentrations and assay SD. A disadvantage of the unweighted (ordinary) least squares method, also demonstrated by our results, is that negative SDs, which are nonsense, are frequently predicted for concentrations below the lower limit of the calibrated concentration range. Theil's regression with the Siegel estimator, on the other hand, is entirely resistant to the outliers of any linear trend and does not yield negative intercepts when evaluating concentration–SD relationships obtained by applying methods based on LC-MS/MS and the use of internal standards.

While determining the assay error equation experimentally is closely linked to the construction of nonparametric pharmacokinetic models, it also brings other important benefits. Assay accuracy and precision is evaluated by running 400 samples (in addition to the blanks) from zero concentration to the high end of the calibrated concentration range, in contrast to assaying 24 samples typically in a conventional within-run (intra-day) study and the reanalysis of a fraction of these in the between-run (interday) experiments. Concentration points can be retested and further concentration points can be added flexibly, providing a suitable context for partial method revalidation with an experimentally established SD acceptability range. In addition, knowledge of the SD and accepting that it can be relatively high at low concentrations allows the laboratory to avoid the use of a lower numerical limit for reporting drug and metabolite levels.

The pharmacokinetic values obtained in the three CLL patients on IBR were overall comparable to earlier findings [10,31,32]. It is also apparent that, due to the high concentrations it attains, the inclusion of DIB in the models, either as a metabolite or a component of the active moiety, is crucial.

Individualized pharmacotherapy relying on model-informed precision dosing is a multidisciplinary approach. In order to provide reliable and justified reports for supporting individual decisions, TDM laboratories need to exert dedicated knowledge and activities (an exception is when precision dosing is based on the use of physiologically based pharmacokinetic–pharmacodynamic models that do not require continuous TDM [40]). Failure to establish and periodically revise an experimentally determined error model or to be informed on the stability of the analytes is likely to contribute to the enormous differences in the pharmacokinetic estimates made by various research groups. Nevertheless, the maintenance of such dedicated TDM laboratories is affordable mainly to academic facilities, presenting a very large barrier to the broader application of model-informed precision dosing. It must also be emphasized that model-informed precision dosing is not equivalent to TDM or to population pharmacokinetic modeling.

A limitation of this research is that the direct application of the developed methodology to constructing nonparametric pharmacokinetic models of IBR, DIB, and the active moiety (IBR + DIB) or the clinical validation of these models could not be accomplished, due to the small number of available patients. In addition, evaluating the method using other types of blood samples should be useful for the optimization of method performance.

5. Conclusions

The presented high-throughput methodology allows TDM laboratories to assist precision pharmacotherapy with IBR efficiently. Blood samples are recommended to be centrifuged, with the supernatant separated and cooled no later than 6 h following sample collection. DIB concentrations should be monitored along with the parent drug. The presented assay error equations can be employed for estimating the imprecision of nonparametric pharmacokinetic models of IBR, DIB, and the active moiety (IBR + DIB). The correct timing of sample collection related to dosing is essential to capture information relevant for constructing efficient pharmacokinetic models of IBR, DIB, and IBR + DIB.

Supplementary Materials: The following supporting information can be downloaded at: https://www.mdpi.com/article/10.3390/molecules27154766/s1, Supplementary File S1: Multiple reaction monitoring optimization reports and the final detection settings employed for the high-throughput clinical analysis of ibrutinib and dihydrodiol ibrutinib. Note that the plus ('+') signs in the "Optimality condition" sections refer to employing positive polarity. Under this setting, the instrument control software displays negative values for Q1 Pre-bias, CE and Q3 Pre-bias. Supplementary File S2: Electronic copy of the ethical approval 45371-2/2016/EKU. Supplementary File S3: Scripts run in the R environment. Supplementary File S4: Representative ion chromatograms of ibrutinib (IBR), dihydrodiol ibrutinib (DIB), and their respective internal standards $^2H_5$-ibrutinib ($^2H_5$-IBR) and $^2H_5$-dihydrodiol ibrutinib ($^2H_5$-DIB). (A), IBR in the level 1 calibration sample (2.27 nmol/L), (B) $^2H_5$-IBR in the level 1 calibration sample, (C) IBR in a blank sample, (D) $^2H_5$-IBR in a blank sample, (E) DIB in the level 1 calibration sample (2.11 nmol/L), (F) $^2H_5$-DIB in the level 1 calibration sample, (G) DIB in a blank sample, (H) $^2H_5$-DIB in a blank sample. Supplementary File S5: Individual noncompartmental analysis results.

Author Contributions: Conceptualization, G.B.K. and I.V.; methodology, G.B.K. and I.V.; software, G.B.K.; validation, I.V., G.B.K. and A.B.; formal analysis, G.B.K.; investigation, A.B. and C.B.; resources, B.V.; data curation, A.B. and C.B.; writing—original draft preparation, G.B.K.; writing—review and editing, B.V. and Z.K.; visualization, G.B.K.; supervision, B.V.; project administration, Z.K.; funding acquisition, B.V. All authors have read and agreed to the published version of the manuscript.

Funding: This research was funded by the National Research, Development, and Innovation Office (National Bionics Program, Government Decree 1336/2017) of Hungary.

Institutional Review Board Statement: The study was conducted in accordance with the Declaration of Helsinki and was approved by the Scientific and Research Ethics Committee of the National Medical Research Council, Budapest, Hungary (45371-2/2016/EKU).

Informed Consent Statement: Informed consent was obtained from all subjects involved in the study.

Data Availability Statement: Additional data are available from the corresponding author on reasonable request.

Conflicts of Interest: The authors declare no conflict of interest.

Sample Availability: IBR and DIBR were purchased from a commercial supplier. Samples of the compounds are not available from the authors.

References

1. IMBRUVICA–Full Prescribing Information. Revised 01/2015. Available online: https://www.accessdata.fda.gov/drugsatfda_docs/label/2015/205552s002lbl.pdf (accessed on 12 February 2022).
2. Scheers, E.; Leclercq, L.; de Jong, J.; Bode, N.; Bockx, M.; Laenen, A.; Cuyckens, F.; Skee, D.; Murphy, J.; Sukbuntherng, J.; et al. Absorption, metabolism, and excretion of oral 14 C radiolabeled ibrutinib: An open-label, phase I, single-dose study in healthy men. *Drug Metab. Dispos.* **2015**, *43*, 289–297. [CrossRef] [PubMed]

3. Timofeeva, N.; Gandhi, V. Ibrutinib combinations in CLL therapy: Scientific rationale and clinical results. *Blood Cancer J.* **2021**, *11*, 79. [CrossRef] [PubMed]
4. Zimmermann, S.M.; Peer, C.J.; Figg, W.D. Ibrutinib's off-target mechanism: Cause for dose optimization. *Cancer Biol. Ther.* **2021**, *22*, 529–531. [CrossRef] [PubMed]
5. Paydas, S. Management of adverse effects/toxicity of ibrutinib. *Crit. Rev. Oncol. Hematol.* **2019**, *136*, 56–63. [CrossRef] [PubMed]
6. Hou, J.-Z.; Ryan, K.; Du, S.; Fang, B.; Marks, S.; Page, R.; Peng, E.; Szymanski, K.; Winters, S.; Le, H. Real-world ibrutinib dose reductions, holds and discontinuations in chronic lymphocytic leukemia. *Future Oncol.* **2021**, *17*, 4959–4969. [CrossRef]
7. Bose, P.; Gandhi, V.V.; Keating, M.J. Pharmacokinetic and pharmacodynamic evaluation of ibrutinib for the treatment of chronic lymphocytic leukemia: Rationale for lower doses. *Expert Opin. Drug Metab. Toxicol.* **2016**, *12*, 1381–1392. [CrossRef]
8. Shakeel, F.; Iqbal, M. Bioavailability enhancement and pharmacokinetic profile of an anticancer drug ibrutinib by self-nanoemulsifying drug delivery system. *J. Pharm. Pharmacol.* **2016**, *68*, 772–780. [CrossRef]
9. Tang, B.; Tang, P.; He, J.; Yang, H.; Li, H. Characterization of the binding of a novel antitumor drug ibrutinib with human serum albumin: Insights from spectroscopic, calorimetric and docking studies. *J. Photochem. Photobiol. B* **2018**, *184*, 18–26. [CrossRef]
10. Gallais, F.; Ysebaert, L.; Despas, F.; De Barros, S.; Dupré, L.; Quillet-Mary, A.; Protin, C.; Thomas, F.; Obéric, L.; Alla, B.; et al. Population pharmacokinetics of ibrutinib and its dihydrodiol metabolite in patients with lymphoid malignancies. *Clin. Pharmacokinet.* **2020**, *59*, 1171–1183. [CrossRef]
11. Chen, L.S.; Bose, P.; Cruz, N.D.; Jiang, Y.; Wu, Q.; Thompson, P.A.; Feng, S.; Kroll, M.H.; Qiao, W.; Huang, X.; et al. A pilot study of lower doses of ibrutinib in patients with chronic lymphocytic leukemia. *Blood* **2018**, *132*, 2249–2259. [CrossRef]
12. Yasu, T.; Momo, K.; Kuroda, S.; Kawamata, T. Fluconazole increases ibrutinib concentration. *Am. J. Ther.* **2020**, *27*, e620–e621. [CrossRef] [PubMed]
13. Eisenmann, E.D.; Fu, Q.; Muhowski, E.M.; Jin, Y.; Uddin, M.E.; Garrison, D.A.; Weber, R.H.; Woyach, J.A.; Byrd, J.C.; Sparreboom, A.; et al. Intentional modulation of ibrutinib pharmacokinetics through CYP3A inhibition. *Cancer Res. Commun.* **2021**, *1*, 79–89. [CrossRef] [PubMed]
14. Gribben, J.G.; Bosch, F.; Cymbalista, F.; Geiser, C.H.; Ghia, P.; Hillmen, P.; Moreno, C.; Stilgenbauer, S. Optimising outcomes for patients with chronic lymphocytic leukaemia on ibrutinib therapy: European recommendations for clinical practice. *Br. J. Haematol.* **2018**, *180*, 666–679. [CrossRef] [PubMed]
15. Goutelle, S.; Woillard, J.B.; Neely, M.; Yamada, W.; Bourguignon, L. Nonparametric methods in population pharmacokinetics. *J. Clin. Pharmacol.* **2022**, *62*, 142–157. [CrossRef] [PubMed]
16. Neely, M.N.; van Guilder, M.G.; Yamada, W.M.; Schumitzky, A.; Jelliffe, R.W. Accurate detection of outliers and subpopulations with Pmetrics, a nonparametric and parametric pharmacometric modeling and simulation package for R. *Ther. Drug Monit.* **2012**, *34*, 467–476. [CrossRef]
17. Jelliffe, R. Using the BestDose clinical software—Examples with aminoglycosides. In *Individualized Drug Therapy for Patients: Basic Foundations, Relevant Software, and Clinical Applications*, 1st ed.; Jelliffe, R.W., Neely, M., Eds.; Elsevier: Amsterdam, The Netherlands, 2017; pp. 59–75.
18. Jelliffe, R.W.; Tahani, B. Pharmacoinformatics: Equations for serum drug assay error patterns; implications for therapeutic drug monitoring and dosage. In Proceedings of the Annual Symposium on Computer Application in Medical Care, Washington, DC, USA, 30 October–3 November 1993; pp. 517–521.
19. Jelliffe, R.W.; Schumitzky, A.; Bayard, D.; Fu, X.; Neely, M. Describing assay precision—Reciprocal of variance is correct, not CV percent: Its use should significantly improve laboratory performance. *Ther. Drug Monit.* **2015**, *37*, 389–394. [CrossRef]
20. Gu, H.; Liu, G.; Wang, J.; Aubry, A.-F.; Arnold, M.E. Selecting the correct weighting factors for linear and quadratic calibration curves with least-squares regression algorithm in bioanalytical LC-MS/MS assays and impacts of using incorrect weighting factors on curve stability, data quality, and assay performance. *Anal. Chem.* **2014**, *86*, 8959–8966.
21. Karvaly, G.B.; Neely, M.N.; Kovács, K.; Vincze, I.; Vásárhelyi, B.; Jelliffe, R.W. Development of a methodology to make individual estimates of the precision of liquid chromatography-tandem mass spectrometry drug assay results for use in population pharmacokinetic modeling and the optimization of dosage regimens. *PLoS ONE* **2020**, *15*, e0229873. [CrossRef]
22. Karvaly, G.B.; Vincze, I.; Karádi, I.; Vásárhelyi, B.; Zsáry, A. Sensitive, high-throughput liquid chromatography-tandem mass spectrometry analysis of atorvastatin and its pharmacologically active metabolites in serum for supporting precision pharmacotherapy. *Molecules* **2021**, *26*, 1324. [CrossRef]
23. Karvaly, G.B.; Karádi, I.; Vincze, I.; Neely, M.N.; Trojnár, E.; Prohászka, Z.; Imreh, É.; Vásárhelyi, B.; Zsáry, A. A pharmacokinetics-based approach to the monitoring of patient adherence to atorvastatin therapy. *Pharmacol. Res. Perspect.* **2021**, *9*, e00856. [CrossRef] [PubMed]
24. Huynh, H.H.; Pressiat, C.; Sauvageon, H.; Madelaine, I.; Maslanka, P.; Lebbé, V.; Thieblemont, C.; Goldwirt, L.; Mourah, S. Development and validation of a simultaneous quantification method of 14 tyrosine kinase inhibitors in human plasma using LC-MS/MS. *Ther. Drug Monit.* **2017**, *39*, 43–54. [CrossRef] [PubMed]
25. Rood, J.J.M.; Dormans, P.J.A.; van Haren, M.J.; Schellens, J.H.M.; Beijnen, J.H.; Sparidans, R.W. Bioanalysis of ibrutinib, and its dihydrodiol- and glutathione cycle metabolites by liquid chromatography-tandem mass spectrometry. *J. Chromatogr. B* **2018**, *1090*, 14–21. [CrossRef] [PubMed]

26. European Medicines Agency. Guideline on Bioanalytical Method Validation. EMEA/CHMP/192217/2009 Rev. 1 Corr. 2**. 21 July 2011. Available online: https://www.ema.europa.eu/en/documents/scientific-guideline/guideline-bioanalytical-method-validation_en.pdf (accessed on 7 March 2022).
27. R Core Team. *R: A Language and Environment for Statistical Computing*; R Foundation for Statistical Computing: Vienna, Austria, 2022. Available online: https://www.R-project.org/ (accessed on 7 March 2022).
28. Kim, H.; Han, S.; Cho, Y.-S.; Yoon, S.-K.; Bae, K.-S. Development of R packages: 'NonCompart' and 'ncar' for noncompartmental analysis (NCA). *Transl. Clin. Pharmacol.* **2018**, *26*, 10–15. [CrossRef]
29. Clinical Data Interchange Standards Consortium (CDISC). Study Data Tabulation Model Implementation Guide: Human Clinical Trials Version 3.2. Available online: https://www.cdisc.org/standards/foundational/sdtmig/sdtmig-v3-2 (accessed on 31 March 2022).
30. Retmana, I.A.; Beijnen, J.H.; Sparidans, R.W. Chromatographic bioanalytical assays for targeted covalent kinase inhibitors and their metabolites. *J. Chromatogr. B* **2021**, *1162*, 122466. [CrossRef]
31. De Vries, R.; Smit, J.W.; Hellemans, P.; Jiao, J.; Murphy, J.; Skee, D.; Snoeys, J.; Sukbuntherng, J.; Vliegen, M.; de Zwart, L.; et al. Stable isotope-labelled intravenous microdose for absolute bioavailability and effect on grapefruit juice on ibrutinib in healthy adults. *Br. J. Clin. Pharmacol.* **2015**, *81*, 235–245. [CrossRef]
32. Marastaci, E.; Sukbuntherng, J.; Loury, D.; de Jong, J.; de Trixhe, X.W.; Vermeulen, A.; De Nicolao, G.; O'Brien, S.; Byrd, J.C.; Advani, R.; et al. Population pharmacokinetic model of ibrutinib, a Bruton tyrosine kinase inibitor, in patients with B cell malignancies. *Cancer Chemother. Pharmacol.* **2015**, *75*, 111–121.
33. Beauvais, D.; Goossens, J.-F.; Boyle, E.; Allal, B.; Lafont, T.; Chatelut, E.; Herbaux, C.; Morschhauser, F.; Genay, S.; Odou, P.; et al. Development and validation of an UHPLC-MS/MS method for simultaneous quantification of ibrutinib and its dihydrodiol-metabolite in human cerebrospinal fluid. *J. Chromatogr. B* **2018**, *1093–1094*, 158–166. [CrossRef]
34. Jiang, Z.; Shi, L.; Zhang, Y.; Lin, G.; Wang, Y. Simultaneous measurement of acalabrutinib, ibrutinib, and their metabolites in beagle dog plasma by UPLC-MS/MS and its application to a pharmacokinetic study. *J. Pharm. Biomed. Anal.* **2020**, *191*, 113613. [CrossRef]
35. Mukai, Y.; Yoshida, T.; Kondo, T.; Inotsume, N.; Toda, T. Novel high-performance liquid chromatography-tandem mass spectrometry method for simultaneous quantification of BCR-ABL and Bruton's tyrosine kinase inhibitors and their three active metabolites in human plasma. *J. Chromatogr. B* **2020**, *1137*, 121928. [CrossRef]
36. Veeraraghavan, S.; Viswanadha, S.; Thappali, S.; Govindarajulu, B.; Vakkalanka, S.; Rangasamy, M. Simultaneous quantification of lenalidomide, ibrutinib and its active metabolite PCI-45227 in rat plasma by LC-MS/MS: Application to a pharmacokinetic study. *J. Pharm. Biomed. Anal.* **2015**, *107*, 151–158. [CrossRef] [PubMed]
37. Feliu, C.; Konecki, C.; Candau, T.; Vautier, D.; Haudecoeur, C.; Gozalo, C.; Cazaubon, Y.; Djerada, Z. Quantification of 15 antibiotics widely used in the critical care unit with a LC-MS/MS system: An easy method to perform a daily therapeutic drug monitoring. *Pharmaceuticals* **2021**, *14*, 1214. [CrossRef] [PubMed]
38. Mzik, M.; Vánová, N.; Kriegelstein, M.; Hroch, M. Differential adsorption of an analyte and its D4, D5 and 13C6 labeled analogues combined with instrument-specific carry-over issues: The Achilles' heel of ibrutinib TDM. *J. Pharm. Biomed. Anal.* **2021**, *206*, 114366. [CrossRef]
39. Rood, J.J.M.; Jamalpoor, A.; van Hoppe, S.; van Haren, M.J.; Wasmann, R.E.; Janssen, M.J.; Schinkel, A.H.; Masereeuw, R.; Beijnen, J.H.; Sparidans, R.W. Extrahepatic metabolism of ibrutinib. *Investig. New Drugs* **2021**, *39*, 1–14. [CrossRef] [PubMed]
40. Yeo, K.R.; Ventakrishnan, K. Physiologically-based pharmacokinetic models as enablers of precision dosing in drug development: Pivotal role of the human mass balance study. *Clin. Pharm. Ther.* **2021**, *109*, 51–54.

Article

Development and Validation of a Sensitive and Specific LC-MS/MS Method for IWR-1-Endo, a Wnt Signaling Inhibitor: Application to a Cerebral Microdialysis Study

Sreenath Nair, Abigail Davis, Olivia Campagne, John D. Schuetz and Clinton F. Stewart *

Department of Pharmacy and Pharmaceutical Science, St. Jude Children's Research Hospital, Memphis, TN 38105, USA
* Correspondence: clinton.stewart@stjude.org; Tel.: +1-(901)-595-3665; Fax: +1-(901)-595-3125

Citation: Nair, S.; Davis, A.; Campagne, O.; Schuetz, J.D.; Stewart, C.F. Development and Validation of a Sensitive and Specific LC-MS/MS Method for IWR-1-Endo, a Wnt Signaling Inhibitor: Application to a Cerebral Microdialysis Study. *Molecules* **2022**, *27*, 5448. https://doi.org/10.3390/molecules27175448

Academic Editors: Franciszek Główka and Marta Karaźniewicz-Łada

Received: 20 July 2022
Accepted: 22 August 2022
Published: 25 August 2022

Publisher's Note: MDPI stays neutral with regard to jurisdictional claims in published maps and institutional affiliations.

Copyright: © 2022 by the authors. Licensee MDPI, Basel, Switzerland. This article is an open access article distributed under the terms and conditions of the Creative Commons Attribution (CC BY) license (https://creativecommons.org/licenses/by/4.0/).

Abstract: IWR-1-endo, a small molecule that potently inhibits the Wnt/β-catenin signaling pathway by stabilizing the AXIN2 destruction complex, can inhibit drug efflux at the blood–brain barrier. To conduct murine cerebral microdialysis research, validated, sensitive, and reliable liquid chromatography–tandem mass spectrometry (LC-MS/MS) methods were used to determine IWR-1-endo concentration in the murine plasma and brain microdialysate. IWR-1-endo and the internal standard (ISTD) dabrafenib were extracted from murine plasma and microdialysate samples by a simple solid-phase extraction protocol performed on an Oasis HLB μElution plate. Chromatographic separation was executed on a Kinetex C_{18} (100A, 50 × 2.1 mm, 4 μm particle size) column with a binary gradient of water and acetonitrile, each having 0.1% formic acid, pumped at a flow rate of 0.6 mL/min. Detection by mass spectrometry was conducted in the positive selected reaction monitoring ion mode by monitoring mass transitions 410.40 > 344.10 (IWR-1-endo) and 520.40 > 307.20 (ISTD). The validated curve range of IWR-1-endo was 5–1000 ng/mL for the murine plasma method ($r^2 \geq 0.99$) and 0.5–500 ng/mL for the microdialysate method ($r^2 \geq 0.99$). The lower limit of quantification (LLOQ) was 5 ng/mL and 0.5 ng/mL for the murine plasma and microdialysate sample analysis method, respectively. Negligible matrix effects were observed in murine plasma and microdialysate samples. IWR-1-endo was extremely unstable in murine plasma. To improve the stability of IWR-1-endo, pH adjustments of 1.5 were introduced to murine plasma and microdialysate samples before sample storage and processing. With pH adjustment of 1.5 to the murine plasma and microdialysate samples, IWR-1-endo was stable across several tested conditions such as benchtop, autosampler, freeze–thaw, and long term at −80 °C. The LC-MS/MS methods were successfully applied to a murine pharmacokinetic and cerebral microdialysis study to characterize the unbound IWR-1-endo exposure in brain extracellular fluid and plasma.

Keywords: IWR-1-endo; Wnt signaling inhibitor; LC-MS/MS; solid-phase extraction; pharmacokinetics; cerebral microdialysis; bioanalysis

1. Introduction

Group 3 medulloblastoma (MB) accounts for 25% of all MB and is considered the most aggressive form of this common malignant childhood brain tumor [1,2]. Patients with Group 3 MB have a poor prognosis, with a 5-year overall survival of less than 60% [1,2]. Thus, the development of novel therapeutic approaches is crucial for this MB subgroup.

Effective therapy for MB should overcome the blood–brain barrier (BBB) and blood–tumor barrier (BTB), as well as the efflux pumps associated with these barriers [3]. The ABC transporters are the main efflux transporters expressed at the BTB. Further, we previously showed that the ABCG2 transporter was highly expressed in Group 3 MB and inhibition of ABCG2 intensified antitumor activity [4]. Several studies have investigated the coadministration of efflux transporter inhibitors as a therapeutic strategy to improve penetration across both the BBB and BTB [3].

Recent studies show that inhibitors of the canonical Wnt/β-catenin signaling pathway can regulate the expression of ABC transporter expression in various cancers [5–7]. One of these therapeutic compounds, IWR-1-endo, is a small molecule that potently inhibits the Wnt/β-catenin signaling pathway by stabilizing the AXIN2 destruction complex [7,8]. IWR-1-endo also blocked doxorubicin efflux in a drug-resistant model of osteosarcoma [7]. An ABCG2 structural-based pharmacophore model provided the evidence to suggest that IWR-1-endo fits the ABCG2 pharmacophore and has the potential to block the efflux of anticancer drugs at the BBB [9]. Thus, to understand the in vivo central nervous system (CNS) penetration of IWR-1-endo, we performed cerebral microdialysis studies in murine models.

Performing these studies required that we have a sensitive and specific LC-MS/MS method for IWR-1-endo in murine plasma and microdialysates; however, to our knowledge, no bioanalytical methods for IWR-1-endo have been published. In this study, we reported LC-MS/MS methods for murine plasma and microdialysates that were developed for use in our murine cerebral microdialysis experiments.

2. Results and Discussion

2.1. Optimization of Mass Spectrometric and Chromatographic Conditions

The compound IWR-1-endo (Figure 1A) is in the preclinical development stage, and no data are published on LC-MS/MS determination of IWR-1-endo in biological matrices. Thus, syringe pump infusion experiments were essential to identify the most sensitive precursor and fragment ion pair for IWR-1-endo and dabrafenib (ISTD; Figure 1B).

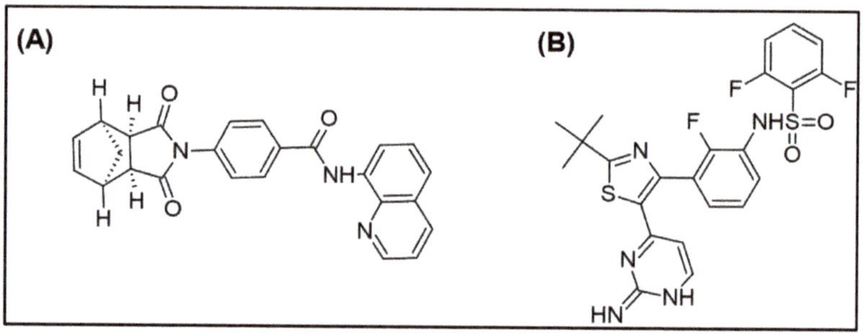

Figure 1. Structure of (**A**) IWR-1 endo and (**B**) ISTD, dabrafenib.

Initially, IWR-1-endo prepared in methanol (0.001 mg/mL) was infused in both positive and negative ionization modes on an AB Sciex 4000 Q-TRAP mass spectrometer. IWR-1-endo could be easily protonated in the positive ion mode to generate the molecular ion peak, $[M+H]^+$ at m/z 410.40. Next, the protonated precursor ion was fragmented using a collision energy of 25 eV to yield a high-intensity fragment ion at m/z 344.1. Other prominent fragment ion peaks were generated at m/z 383.4 and 274.2. Figure 2A–D demonstrates the typical precursor and product ion mass spectrum obtained for IWR-1-endo and ISTD under optimized mass spectrometric conditions in this study. Figure 3 depicts the proposed fragmentation pattern for IWR-1-endo at CE of 25 eV. The fragmentation pattern for dabrafenib has been discussed in a previous report [10].

The liquid chromatographic conditions were optimized to ensure better peak shape and separation of IWR-1-endo and ISTD, enhanced sensitivity for the analyte, and avoid matrix effect issues. Given the hydrophobic nature of IWR-1-endo (theoretical log *p*-value: 2.90), the initial method was developed using a reversed-phase C_{18} column on Shimadzu Nexera X2 HPLC equipment. By employing the gradient profiles described in Section 3.5 chromatographic conditions, acetonitrile/water and methanol/water mobile phase systems were compared, with the former showing a higher IWR-1-endo peak response and a lower background noise (Figure S1). Subsequently, the effect of acidic mobile phase additives, including formic acid (0.05 and 0.1%), acetic acid (0.05 and 0.1%), and liquid ammonia (0.05 and 0.1%), were investigated. Better peak shapes were obtained when adding 0.1%

formic acid to both mobile phases A and B, which might be due to the ease of protonation in the acidic environment. Based on the above optimization, a gradient elution comprising an acetonitrile/water system modified with 0.1% formic acid pumped at a flow rate of 0.6 mL/min was employed for further experiments (Figure S1).

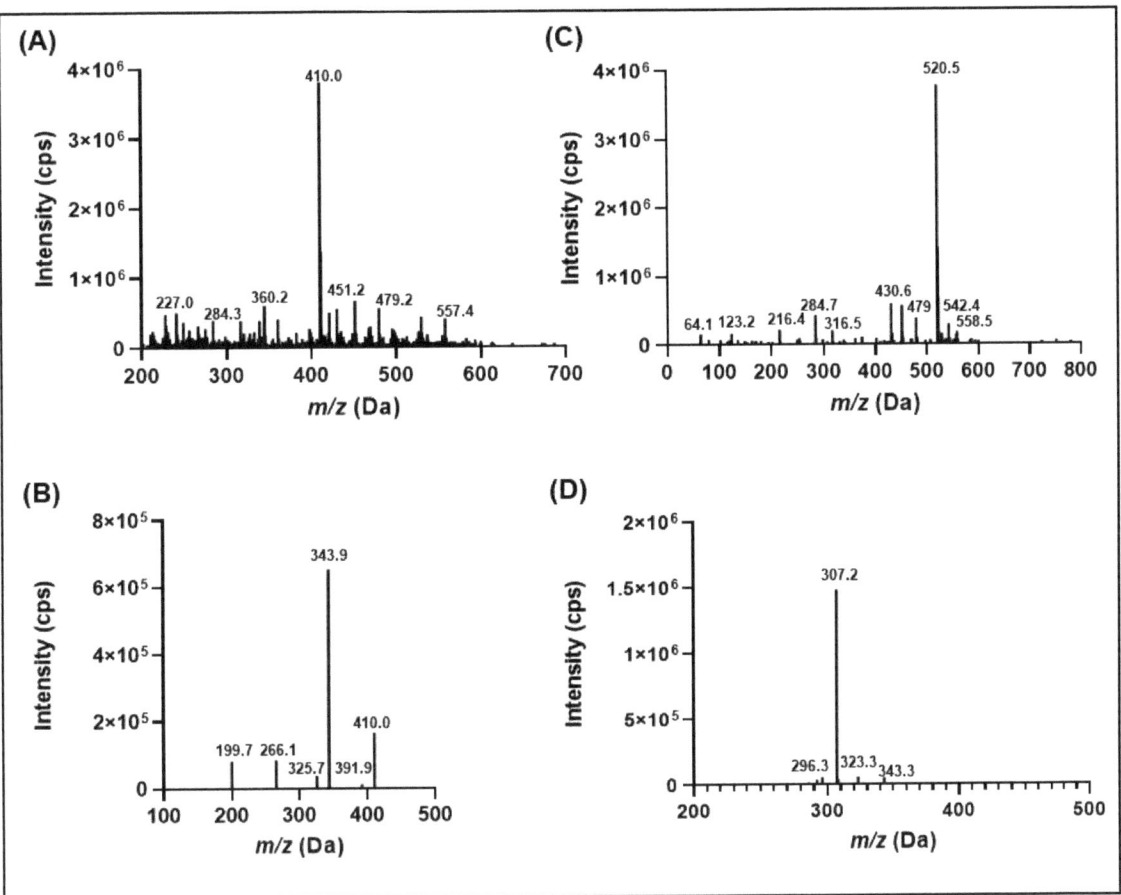

Figure 2. Representative full scan precursor (**A**) and product ion (**B**) scans for IWR-1-endo and corresponding ISTD, dabrafenib (**C**,**D**).

Next, the peak shapes on two analytical columns (i.e., Kinetex® C_{18} analytical column (100°A, 50 × 2.1 mm, 2.6 μm particle size) and Kinetex® C_8 analytical column (100°A, 50 × 2.1 mm, 2.6 μm particle size)) from Phenomenex (Torrance, CA, USA) were compared. With the C_{18} column, the ISTD eluted closer to the retention time of the analyte, which helped address the matrix effect problems when working with microdialysate samples (Figure S2). Another important contributor to peak shape and analyte response was the injection volume, which was set at 2 μL and 10 μL for the murine plasma and microdialysate methods, respectively. A higher injection volume was used for the microdialysate method, given its lower curve range compared to that of the murine plasma method. Injection volumes above 10 μL not only affected the peak shape but also introduced matrix effects from the microdialysates. Lastly, a strong needle wash of methanol/acetonitrile/water/2-propanol (1:1:1:1; *v/v/v/v*) modified with 0.1% formic acid was used between runs to reduce carry-over of the analyte. The typical experimental retention times for IWR-1-

endo and ISTD on the C_{18} column under the optimized chromatographic conditions were 1.72 ± 0.05 min and 1.65 ± 0.05 min, respectively. The total run time of this method was 5 min, which included a 1.75 min re-equilibration time for the column to return to initial chromatographic conditions between subsequent runs.

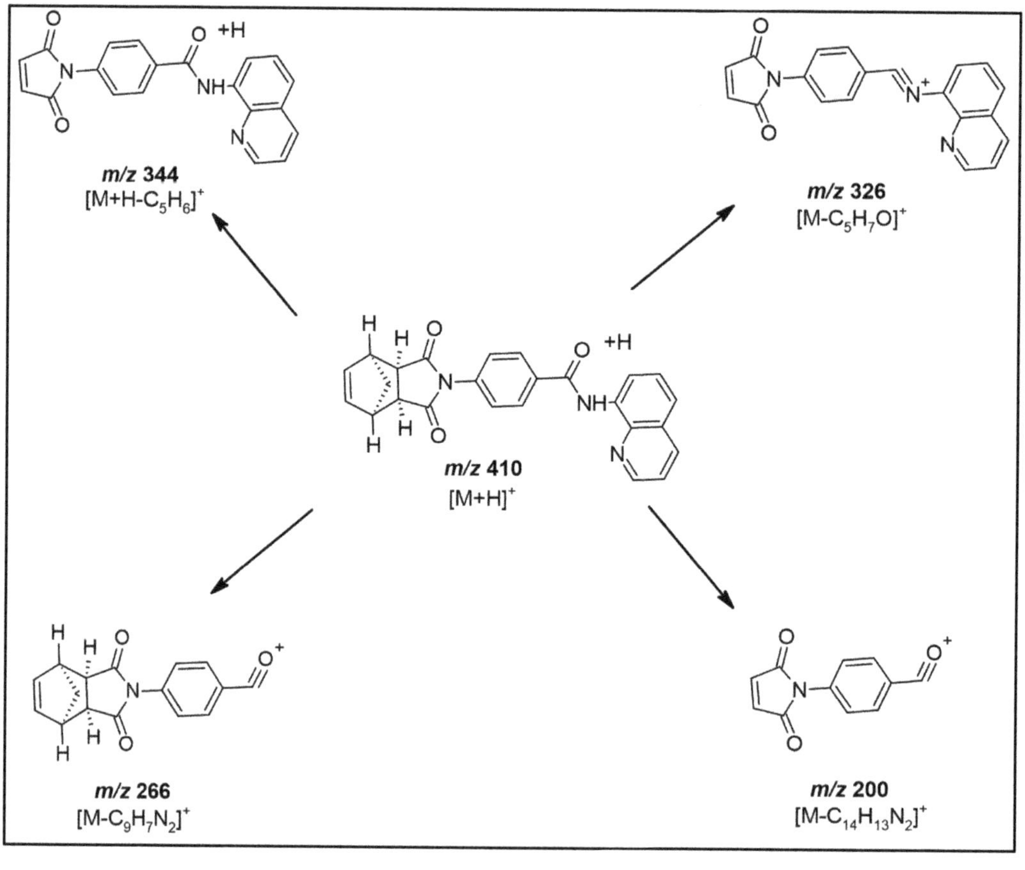

Figure 3. Proposed fragmentation pattern for the prominent product ions of IWR-1-endo.

2.2. Optimization of Sample Preparation

The optimization of the sample pretreatment protocol was a vital part of this study since high method sensitivities were required for the LC-MS/MS methods developed in murine plasma and microdialysate samples. Initially, protein precipitation was tested with methanol and acetonitrile leading to low and inconsistent extraction recoveries for IWR-1-endo in murine plasma and microdialysate samples. As a result, solid-phase extraction (SPE) was considered, which was not only instrumental in improving overall extraction recovery from plasma and microdialysate samples but also providing consistent recoveries. Waters® oasis method development 96-well µElution sorbent selection plate and HLB plate (2 mg sorbent per well, 30 µm) were evaluated for plasma and microdialysate sample cleanup. High recoveries and clean chromatograms for IWR-1-endo and ISTD were obtained by HLB µElution plates (Table S1). Additionally, the eluting capabilities of numerous solvent systems for extracting IWR-1-endo and ISTD from the SPE plates were investigated. Of these solvents, 140 µL of ACN modified with methanol (9:1, v/v) was able to disrupt all types of interactions in the case of IWR-1-endo and ISTD. Thus, it was used as an elution solvent for the final SPE extraction step. Lastly, dabrafenib was used as ISTD in the current assay as it did not interfere with the estimation of IWR-1-endo, and it resembled IWR-1-endo in terms of extractability from plasma and microdialysate samples.

2.3. Stability of IWR-1-Endo in Murine Plasma under Different StorageC

During the prevalidation runs, short-term stability studies performed at RT and 4 °C, over a time course of 24 h showed significant IWR-1-endo degradation in K_2EDTA-treated CD1 murine plasma (Figure 4). In a parallel experiment, IWR-1-endo quality control working solutions kept at RT and 4 °C had no stability-related issues, thus suggesting that IWR-1-endo was unstable in plasma. It has been previously demonstrated that the anticoagulant could impact the compound stability [11,12]. Therefore, we repeated the short-term stability studies with sodium-heparin-treated CD1 murine plasma. However, significant IWR-1-endo degradation was also observed in heparin-treated plasma for all the tested stability samples, similar to the K_2EDTA plasma results (data not shown). Since matrix-related irreproducibility is more pronounced in heparin-treated plasma, we used K_2EDTA as the anticoagulant for further optimization studies. We next tested the freeze–thaw stability as well as long-term stability of IWR-1-endo in CD1 murine plasma at −80 °C. As expected, a significant loss of drug stability was observed during both FTS and LTS studies.

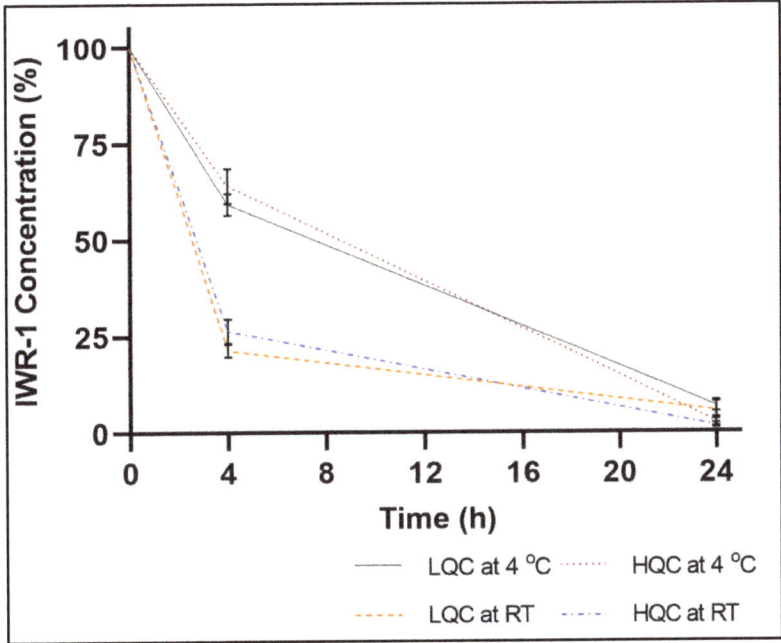

Figure 4. Time course of degradation of IWR-1 in murine plasma stored at 4 °C and room temperature over a period of 24 h (n = 3; mean ± S.D.). LQC: low-quality control; HQC: high-quality control; RT: room temperature.

To further identify whether the drug loss is due to non-specific binding (NSB) or biological sample instability, we subsequently performed STS and FTS studies in murine plasma with 0.5% Tween-80 solution to evaluate the drug's adsorption to the surface of sample storage containers. The addition of Tween-80 to the biological matrix avoids NSB effects. However, the samples fortified with Tween-80 solutions show decreased IWR-1-endo concentrations for both STS and FTS studies, similar to the untreated QCs, suggesting instability of IWR-1-endo in murine plasma.

2.4. Addressing IWR-1-Endo Instability in Murine Plasma

Previous reports have suggested that the major biotransformation of drugs in plasma is enzymatic hydrolysis, which leads to drug degradation [13,14]. The major enzymes involved in this hydrolysis include carboxylesterases, acetylcholinesterase, cholinesterase, peptidases, and nucleases. We initially hypothesized that one of the above-mentioned matrix enzymes might be the cause of IWR-1-endo instability in murine plasma and microdialysates during sample storage and processing. Since we could not identify the actual enzyme responsible for the degradation of IWR-1-endo in the biological matrix, we tested several common enzyme inhibitors such as p-chloromercuribenzoate (A-esterase inhibitor), bis-(p-nitrophenyl) phosphate (B-esterase inhibitor), and sodium fluoride (acetyl cholinesterase inhibitor). However, it was observed that IWR-1-endo was unstable in plasma despite the addition of enzyme inhibitors (Figure S3). We finally resorted to pH adjustment since most enzymes are active over a narrow pH range [15]. We used several acidic and alkaline solutions to adjust murine plasma over a range of pH values (i.e., 1–11). The data suggested that IWR-1-endo was extremely stable at pH = 1.5, obtained by the addition of 200 µL dilute hydrochloric acid (0.1 N HCl) to the 20 µL of murine plasma and microdialysate samples after the microdialysis study sample collection (Figure S4).

2.5. Validation of the LC-MS/MS Methods

Figures 5 and 6 depict the typical SRM chromatograms for the double blank, lower limit of quantitation (LLOQ), and in vivo murine plasma and microdialysate study samples collected at 1 h after intraperitoneal administration of IWR-1-endo (30 mg/kg). No interfering peaks were found at the retention of IWR-1-endo or ISTD, suggesting acceptable selectivity of the developed methods. Additionally, a negligible carry-over effect was observed in double-blank samples after the injection of the upper limit of quantitation (ULOQ) samples for both the murine plasma and microdialysate methods This was mainly due to the use of a strong flushing solvent comprising methanol/acetonitrile/water/2-propanol (1:1:1:1; v/v), modified with 0.1% formic acid between the sample runs. The %C.V. observed during method validation for system suitability samples (injected at LLOQ concentrations) for mouse plasma and microdialysate sample analysis was $\leq 8\%$ (n = 6). The typical retention time for IWR-1-endo was 1.72 ± 0.05 min, and the peak asymmetry obtained for IWR-1-endo was 1.02 ± 0.04 (n = 6).

Table 1 provides the concentration ranges, regression equations, and correlation coefficient results for the determination of IWR-1-endo in murine plasma and microdialysates. The calibration curves showed good linearity over the tested concentration ranges with $r^2 \geq 0.9989$ and 0.9990 for murine plasma and microdialysates, respectively (Figure S5). The LLOQ was 5 ng/mL and 0.5 ng/mL for IWR-1-endo in murine plasma and microdialysates, respectively, at which the S/N ratio was ≥ 10, and the accuracy (% relative error) was within $\pm 20\%$ with precision (R.S.D.) $\leq 20\%$. The LODs obtained were 0.50 ng/mL (C.V.: 29.48%, n = 6) and 0.20 ng/mL (C.V.: 20.01%, n = 6) for the murine plasma and microdialysate samples, correspondingly.

Tables 2 and 3 illustrate the summary of the within-run and between-run precision and accuracy for IWR-1-endo in murine plasma and microdialysate samples, respectively, obtained during the three days of validation by analyzing the quality control samples. The relative error (RE) ranged from -0.86% to 9.26%, with RSD below 10.10% for the murine plasma method. The RE ranged from -8.86% to 0.21%, with RSD below 7.31% for the microdialysate method, indicating that the developed method was reliable and reproducible for IWR-1-endo quantification.

The matrix effect and recovery results in murine plasma and microdialysates are illustrated in Table 4. The mean MF for IWR-1-endo ranged from 0.99 to 1.02, with R.S.D. $\leq 2.55\%$ for murine plasma and 1.05 to 1.13, with R.S.D. of $\leq 9.73\%$ for the microdialysate samples, respectively, indicating negligible matrix effect. The average extraction recoveries for IWR-1-endo in CD1 and CD1 nude murine plasma samples were 97.54% and 99.00%, respectively, with the R.S.D. $\leq 3.23\%$ for all the QC samples analyzed. Likewise, the average extraction recoveries for IWR-1-endo in microdialysate samples was 88.76%, with the R.S.D. $\leq$ 10.00% for all the analyzed QC samples. The average extraction recoveries for ISTD in murine plasma and microdialysate samples were 103.51% and 94.94%, respectively, with the R.S.D. value $\leq$ 4.78% for all the analyzed QC samples.

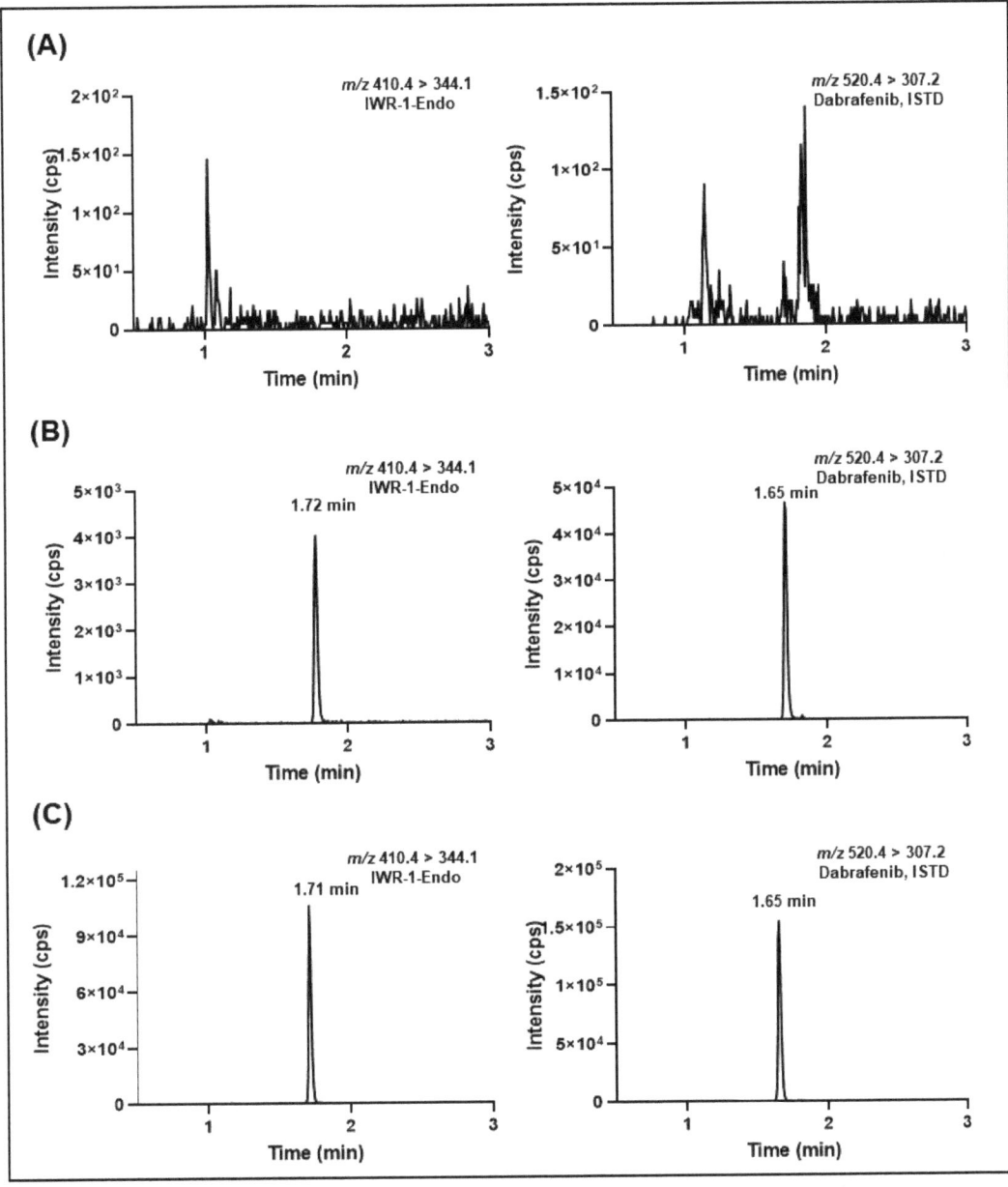

Figure 5. Representative extracted-ion chromatograms of (**A**) murine blank plasma, (**B**) IWR-1-endo-spiked murine plasma sample at LLOQ (5 ng/mL) with ISTD added, and (**C**) 1 h plasma sample from the microdialysis study (ISTD added) after administration of 30 mg/kg IWR-1-endo intraperitoneally.

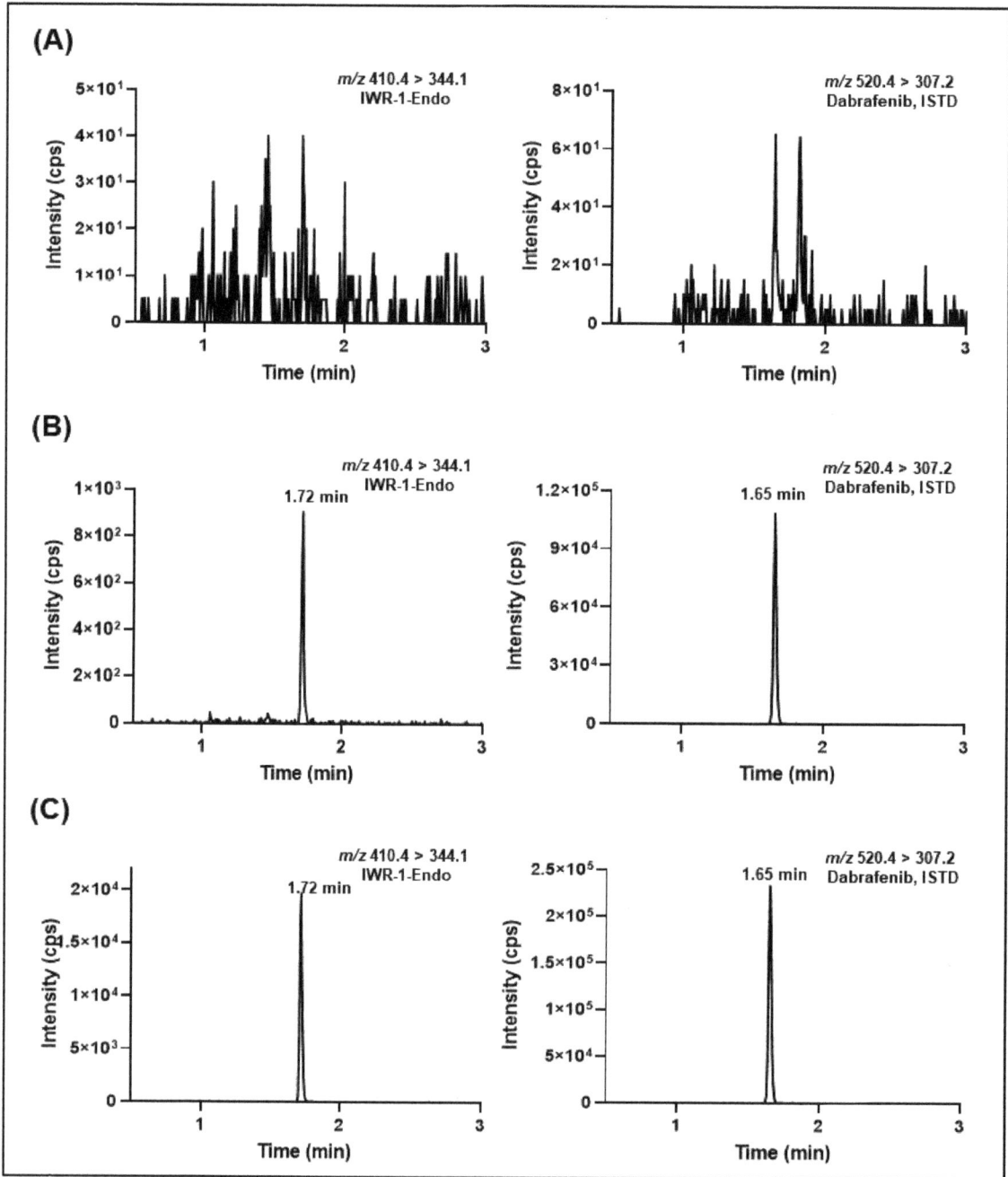

Figure 6. Representative extracted-ion chromatograms of (**A**) microdialysate blank, (**B**) IWR-1-endo-spiked microdialysate sample at LLOQ (0.5 ng/mL) with ISTD added, and (**C**) 1 h plasma sample from the microdialysis study (ISTD added) after administration of 30 mg/kg IWR-1-endo intraperitoneally.

Table 1. Parameters for calibration curves in different matrices.

Parameter	Sample Matrix	
	Murine Plasma	Microdialysates
Calibration Range (ng/mL)	5–1000	0.5–500
LLOQ (ng/mL)	5	0.5
LOD (ng/mL)	0.5	0.2
Intercept (a)	-1.50×10^{-3}	1.70×10^{-4}
Sa	1.20×10^{-3}	1.17×10^{-4}
Slope (b)	3.90×10^{-3}	1.33×10^{-2}
Sb	0.20×10^{-3}	0.07×10^{-2}
Correlation Coefficient (r^2)	0.9989	0.9990

Sa: standard deviation of intercept; Sb: standard deviation of the slope; r^2: coefficient of determination; LOD: limit of detection; LLOQ, lower limit of quantification.

Table 2. Within-run and between-run precision and accuracy for IWR-1-endo in murine plasma (n = 6 for within-run and n = 18 for between-runs).

P & A Batch Run	Replicates	Murine Plasma			
		Nominal Conc.	Calculated conc.	Accuracy	Precision
	(n)	(ng/mL)	(ng/mL)	(% R.E.)	(% R. S. D.)
Within-Run 1	6	5	5.30	6.07	5.04
	6	15	15.06	0.38	4.06
	6	300	312.81	4.27	2.89
	6	850	895.12	5.31	1.74
Within-Run 2	6	5	5.09	1.75	10.10
	6	15	14.87	−0.86	5.87
	6	300	306.98	2.33	2.29
	6	850	908.24	6.85	6.80
Within-Run 3	6	5	5.46	9.26	6.58
	6	15	15.65	4.31	4.50
	6	300	324.06	8.02	3.30
	6	850	924.98	8.82	1.03
Average of Runs 1, 2, and 3 (Between-Runs)	18	5	5.28	5.69	7.61
	18	15	15.19	1.28	5.08
	18	300	314.62	4.87	3.56
	18	850	909.45	6.99	4.08

% R.E.: percent relative error; % R.S.D.: percent relative standard deviation.

Table 3. Within-run and between-run precision and accuracy for IWR-1-endo in microdialysates (n = 6 for within-run and n = 18 for between-runs).

P & A Batch Run	Replicates (n)	Microdialysate			
		Nominal Conc. (ng/mL)	Calculated conc. (ng/mL)	Accuracy (% R.E.)	Precision (% R. S. D.)
Within-Run 1	6	0.5	0.46	−8.86	3.33
	6	1.5	1.39	−7.25	5.25
	6	150	141.88	−5.42	2.25
	6	400	372.08	−6.98	4.66
Within-Run 2	6	0.5	0.49	−1.16	6.95
	6	1.5	1.49	−0.53	7.31
	6	150	148.13	−1.24	4.11
	6	400	399.15	−0.21	3.29
Within-Run 3	6	0.5	0.49	−2.36	3.68
	6	1.5	1.44	−3.69	6.11
	6	150	144.89	−3.41	3.29
	6	400	395.57	−1.11	3.32
Average of Runs 1, 2, and 3 (Between-Runs)	18	0.5	0.48	−4.13	5.95
	18	1.5	1.44	−3.82	6.64
	18	150	144.97	−3.36	3.62
	18	400	388.94	−2.77	4.76

% R.E.: percent relative error; % R.S.D.: percent relative standard deviation.

Table 4. Matrix factors and recovery in various matrices (n = 3).

Sample Matrix	Compound	Nominal Concentration (ng/mL)	Matrix Effect		Recovery	
			Mean Calculated MF value	% R.S.D.	% Mean Recovery	% R.S.D.
Mouse Plasma CD1	IWR-1	15	1.02	2.55	97.16	1.57
		850	1.00	1.07	97.92	1.94
	ISTD	500	0.97	1.07	102.45	1.52
Mouse plasma CD1 Nude	IWR-1	15	0.99	2.51	100.67	2.40
		850	1.00	1.81	97.34	3.23
	ISTD	500	0.95	0.64	104.58	2.24
Microdialysate	IWR-1	1.5	1.05	9.73	86.56	10.00
		150	1.13	1.64	90.96	3.85
	ISTD	500	0.95	3.85	94.94	4.78

The results of the stability assessments performed for IWR-1-endo-spiked in murine plasma and microdialysates under different experimental storage conditions are summarized in Table 5. IWR-1-endo was stable for 48 h in spiked murine plasma and microdialysate samples after short-term storage at 4 °C and room temperature, indicating that the samples were stable under the laboratory handling conditions. The stability of IWR-1-endo was confirmed in spiked murine plasma and microdialysate samples after long-term storage at −80 °C for 21 and 29 days, respectively, highlighting the reliability of the developed LC-MS/MS method to handle in vivo study samples. In addition, the extracted quality control samples for IWR-1-endo in murine plasma and microdialysates were stable in the autosampler for 48 h, indicating good post-extractive stability for the analyte. Further, no stability-related concerns for IWR-1-endo were observed for murine plasma and microdialysate samples undergoing three freeze–thaw cycles. When stored at −80 °C for 83 days, the primary stock solution of IWR-1-endo was stable with an R.S.D ≤ 2.30%.

Table 5. Summary of stability evaluation for IWR-1-endo in murine plasma and microdialysates (n = 3).

Sample Matrix	Stability Study	Nominal Concentration (ng/mL)	Mean ± S.D. Calculated Concentration (ng/mL)	Precision (% R.S.D.)	Accuracy (% R.E.)	% Mean Deviation
Murine Plasma	Process [a]	15	16.33 ± 0.63	3.87	8.88	9.35
		850	916.82 ± 10.54	1.15	7.86	12.40
	Bench-Top RT [b]	15	15.87 ± 0.45	2.84	5.82	6.27
		850	915.08 ± 14.18	1.55	7.66	12.18
	Bench-Top 4 °C [c]	15	15.20 ± 0.24	1.55	1.34	1.78
		850	915.69 ± 16.12	1.76	7.73	12.26
	Freeze–Thaw [d]	15	16.01 ± 0.88	5.48	6.70	7.16
		850	917.18 ± 30.24	3.30	7.90	12.44
	Long-Term [e]	15	15.93 ± 0.21	1.35	6.17	−5.65
		850	938.89 ± 15.84	1.77	10.46	−0.33
Microdialysate	Process [a]	1.5	1.56 ± 0.042	2.68	4.16	7.79
		400	429.35 ± 7.00	1.63	7.34	6.64
	Bench-Top RT [b]	1.5	1.60 ± 0.05	3.02	6.35	9.35
		400	434.08 ± 17.01	3.92	8.52	7.61
	Bench-Top 4 °C [c]	1.5	1.60 ± 0.08	4.86	6.40	11.50
		400	415.84 ± 6.85	1.65	3.96	3.09
	Freeze–Thaw [d]	1.5	1.54 ± 0.07	4.56	2.34	5.91
		400	424.79 ± 9.18	2.16	6.20	5.31
	Long-Term [f]	1.5	1.39 ± 0.008	0.60	−7.63	−4.41
		400	390.35 ± 21.53	5.52	−2.41	−3.23

%R.S.D.: relative standard deviation; %R.E.: relative error; [a] Stability assessed after 48 h in autosampler at 4 °C; [b] Short-term stability after 48 h at room temperature; [c] Short-term stability after 48 h at 4 °C; [d] Stability evaluated after three freeze–thaw cycles; [e] 21 days long-term stability in murine plasma at −80 °C; [f] 29 days long-term stability in microdialysate at −80 °C.

2.6. Application to Cerebral Microdialysis Studies

IWR-1-endo concentrations were measured in murine plasma and Ringer's solution collected in a female non-tumor-bearing mouse dosed with 30 mg/kg IWR-1-endo intraperitoneally (Figure 7). The microdialysis probe recovery was 61%. IWR-1-endo total plasma and unbound ECF exposures were calculated as the area under the concentration–time curve (AUC), using a noncompartmental pharmacokinetic analysis. IWR-1-endo total plasma and unbound ECF AUC were 1915 and 45.4 h·ng/mL, respectively, showing a total brain ECF to plasma partition coefficient of 0.024 for IWR-1-endo.

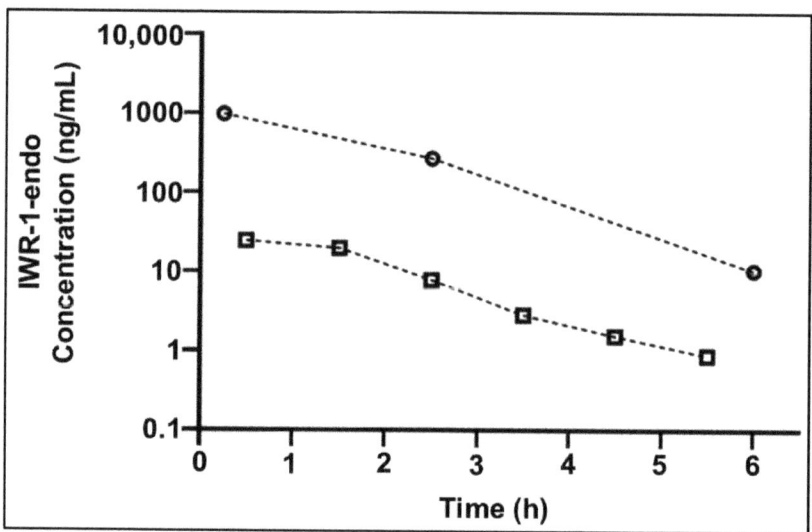

Figure 7. IWR-1-endo concentration–time profile in plasma and brain extracellular fluid (ECF) in a female non-tumor-bearing CD1 nude mouse. Open circles represent IWR-1-endo plasma concentrations, and open squares represent IWR-1-endo brain ECF concentrations.

3. Materials and Methods

3.1. Reagents and Chemicals

The reference standard for IWR-1-endo (M.W.: 409.40 g/mol; M.F.: $C_{25}H_{19}N_3O_3$; purity determined by LC-MS as 99.18%) was synthesized in-house by the Department of Chemical Biology and Therapeutics, St. Jude Children's Research Hospital (Memphis, TN, USA) [8]. The ISTD dabrafenib (M.W.: 519.56 g/mol; M.F.: $C_{23}H_{20}F_3N_5O_2S_2$; purity determined by HPLC and TLC as 100%) was obtained from Cayman Chemical (Ann Arbor, MI, USA). Blank K_2EDTA CD-1 murine plasma was purchased from BioIVT (Westbury, NY, USA). Laboratory-grade Ringer's solution (98.97% H_2O, 0.95% NaCl, 0.04 % KCl, 0.02 % $CaCl_2$, and 0.02% $NaHCO_3$) was obtained from Frey Scientific (Nashua, NH, USA) and used as the artificial microdialysate. Trappsol® 1-hydroxypropyl-β-cyclodextrin (BCD), pharmaceutical grade, was purchased from Cyclodextrin Technologies Development (Gainesville, FL, USA). Optima™ LC/MS-grade acetonitrile, Optima™ LC/MS-grade methanol, ACS-grade dimethyl sulfoxide (DMSO), J. T. Baker™ 1-methyl-2-pyrrolidinone (NMP), Medchemexpress Solutol HS-15, Fine Chemicals Biosciences d-α-tocopherol polyethylene glycol 1000 succinate (TPGS), MilliporeSigma™ polyethylene glycol 400, and Honeywell Fluka™ formic acid for mass spectrometry were purchased from Fisher Scientific (Waltham, MA, USA). All other chemicals and reagents used in this study were of analytical grade. A MilliporeSigma water purification system (Burlington, MA, USA) was used to prepare double-distilled water for LC-MS analysis.

3.2. Preparation of Stock and Working Solutions

Primary stock solutions for IWR-1-endo and the ISTD at a concentration of 1 mg/mL were prepared by independent measurements in DMSO and stored in a 4 mL amber vial at −80 °C until further use. The standard IWR-1 solution was diluted in methanol/water (1:1; v/v) to obtain calibrant working solutions at 50, 100, 500, 1000, 2500, 5000, 7500, and 10,000 ng/mL for the murine plasma standard curve. For the microdialysate standard curve, calibrant subsolutions were 5, 10, 50, 100, 500, 1000, 2500, and 5000 ng/mL. Similarly, the quality control working solutions for the murine plasma method was 150 ng/mL (low-quality control; LQC), 3000 ng/mL (middle-quality control; MQC), and 8500 ng/mL (high-quality control; HQC). For the microdialysate method, the quality control subsolutions were 15 ng/mL (LQC), 1500 ng/mL (MQC), and 4000 ng/mL (HQC). The ISTD working solution (1 µg/mL) was prepared fresh at the time of assay by serial dilution of the ISTD primary stock solution in methanol/water (1:1; v/v).

3.3. Calibration Standards and Quality Controls

For the murine plasma curve, the calibration standards were prepared by spiking 2 µL of calibrator working solutions into 20 µL aliquots of blank CD1 murine plasma to yield the following IWR-1 concentrations: 5, 10, 50, 100, 250, 500, 750, and 1000 ng/mL. Similarly, quality control (QC) samples were prepared from respective QC working solutions in blank CD1 murine plasma at IWR-1-endo concentrations of 15 ng/mL (LQC), 300 ng/mL (MQC), and 850 ng/mL (HQC).

Blank microdialysate used for validation studies was made up of Ringer's solution modified with 10% BCD, which was similar to the perfusate used for the cerebral microdialysis study. For the microdialysis curve, the calibrator standards were prepared by spiking 2 µL of calibrator working solutions into 20 µL aliquots of blank microdialysate to yield IWR-1-endo concentrations of 0.5, 1, 5, 25, 50, 125, 250, and 500 ng/mL. Likewise, the quality control samples were prepared from respective QC working solutions in blank microdialysates at IWR-1-endo concentrations of 1.5 ng/mL (LQC), 100 ng/mL (MQC), and 400 ng/mL (HQC).

3.4. Plasma and Microdialysate Sample Preparation

Plasma and microdialysate samples were pretreated by solid-phase extraction (SPE) with a Waters® Oasis HLB µElution 96-well plate, 2 mg sorbent per well, 30 µm (Milford, MA, USA). The prepared calibrator standards and QC samples in plasma or microdialysate were spiked with 10 µL of ISTD working solution (1 µg/mL). Double-blank samples (i.e., no analyte or ISTD) were prepared by adding 12 µL of methanol/water (1:1, v/v) for volume correction to 20 µL of blank CD1 plasma or microdialysate. Similarly, for blank samples (i.e., no analyte), 20 µL of blank CD1 plasma or microdialysates was spiked with 10 µL of ISTD working solution (1 µg/mL) and 2 µL of methanol/water (1:1, v/v) for volume correction. Frozen plasma and microdialysis study samples were initially thawed on ice and vortexed for 30 s. Then, 20 µL aliquots of these study plasma or microdialysate samples were mixed with 10 µL of ISTD working solution (1 µg/mL) and 2 µL of methanol/water (1:1, v/v) as volume correction to account for drug spiking in standard and QC samples. The prepared calibrators, QCs, and study sample mixtures were adjusted to a pH of 1.5 with 200 µL dilute hydrochloric acid (0.1 N). Next, acidified sample mixtures were vortex mixed for 1 min, centrifuged at 12,000× g for 5 min at 4 °C and applied to an HLB µElution plate that was earlier conditioned with 2 × 100 µL of methanol and equilibrated with 2 × 100 µL of distilled water. The loaded plasma or microdialysate samples were allowed to pass through the plate wells with minimum positive pressure. Wells were rapidly flushed with 2 × 200 µL of distilled water. Then, maximum positive pressure was applied to completely dry the plate wells, followed by elution of the analyte with 2 × 70 µL of acetonitrile: methanol solution mixture (9:1, v/v) into a clean 96-well collection plate (500 µL round-bottom well plate) from Wheaton (Millville, NJ, USA). The plate was sealed using a Wheaton® silicone cap mat (Millville, NJ, USA) after adding distilled water (60 µL) to all sample wells. Lastly, the contents of this plate were vortex mixed for 30 s, centrifuged at 4000× g for 2 min at 4 °C, and the plate was placed in the autosampler rack. Volumes of 2 µL and 10 µL of the final sample extracts were injected onto the C_{18} column for LC-MS/MS analysis of murine plasma and microdialysate samples, respectively.

3.5. Chromatographic Conditions

Liquid chromatography was performed using a Shimadzu Nexera X2 high-performance liquid chromatograph (Kyoto, Japan) equipped with a binary pump (LC-30AD), a degasser (DGU-A_5), an autosampler (SIL-30AC), a controller (CBM-20A), and a column oven (Thermasphere TS-130). A Kinetex® HPLC C_{18} column (100A, 50 × 2.1 mm, 4 µm particle size) attached to a KrudKatcher™ Ultra HPLC In-Line Filter (2 μm Depth Filter × 0.004 in ID; Phenomenex, Torrance, CA, USA) was used for chromatographic separation of IWR-1-endo and ISTD. The mobile phase for the separation of analytes included solution A (distilled water with 0.1% formic acid) and B (acetonitrile with 0.1% formic acid). The following gradient profile was used: 0–0.2 min, 5–50% B; 0.2–0.8 min, 50% B; 0.8–0.9 min, 50–75% B; 0.9–2.2 min, 75% B; 2.2–2.3 min, 75–98% B; 2.3–3.1 min, 98% B; 3.1–3.25 min, 98–5% B; 3.25–5.0 min, 5% B. This gradient phase was pumped at a constant flow rate of 0.6 mL/min throughout the sample run. The total LC-MS/MS run time per sample was 5 min. The injection volumes into the LC-MS/MS for determining IWR-1-endo from murine plasma and microdialysates were 2 μL and 10 μL per sample, respectively. The column temperature and autosampler temperature were kept at $45 \pm 1\,°C$ and $4 \pm 1\,°C$, respectively, throughout all measurements. During the sample batch run, the autosampler needle and port were rinsed using a strong flushing solution of methanol/acetonitrile/water/2-propanol (1:1:1:1; v/v) modified with 0.1% formic acid.

3.6. Mass Spectrometry Conditions

IWR-1-endo and dabrafenib (ISTD) detection and quantification were performed using the Applied Biosystems Hybrid Q-Trap 4000 mass spectrometer (Framingham, MA, USA) equipped with the Turbo Ion Spray probe. The instrument control and data processing software Sciex Analyst® (version 1.7.1) was used for spectral data acquisition, peak integration, and quantification. Mass detection was performed in the positive electrospray ionization mode and the mass transitions monitored were 410.40 > 344.10 (IWR-1-endo; Figure 2A) and 520.40 > 307.20 (ISTD; Figure 2B). Mass resolutions were set at 0.7 full width at half height (unit resolution) for both Q1 and Q3 quadrupoles. The compound-dependent parameters such as declustering potential (DP), entrance potential (EP), collision energy (CE), and collision exit potential (CXP) were set at 105.0 V, 11.0 V, 25.0 eV, and 10.0 V, respectively, for IWR-1-endo; for ISTD, it was 125.0 V, 9.0 V, 30.0 eV, and 8.0 V, respectively. The compound independent parameters including curtain gas (CUR), ion spray voltage (ISP), nebulizer gas (GS1), and heater gas (GS2) were set at 30 psi, 5500 V, 50 psi, and 50 psi, respectively, for both IWR-1-endo and ISTD. The ion source temperature was maintained at 600 °C, and the dwell time for monitoring both the transitions was set to 200 ms.

3.7. Method Validation Procedures

Using the FDA Guidance for Industry, Bioanalytical Method Validation, May 2018, this LC-MS/MS method was validated in terms of linearity, selectivity, sensitivity (lower limit of quantitation, limit of detection), precision, accuracy, matrix effects, recovery, stability, carry-over, and dilution integrity [16].

3.7.1. Linearity

The calibration curve was constructed by analyzing a double blank, a blank, and eight nonzero calibrator standards in blank murine CD1 plasma and Ringer's/BCD solution on three successive days. Standard curves were plotted using peak area ratios of analyte to ISTD (y-axis) against the nominal analyte concentrations (x-axis). A linear regression equation, with a weighing factor of $1/x^2$, was used to get the best fit for the IWR-1-endo concentration/peak area ratio relationship. The linear ranges tested for the murine plasma and microdialysate methods were 5–1000 ng/mL and 0.5–500 ng/mL, respectively. The calibration curves were considered linear/acceptable when at least 75% of the calibration standards, including the LLOQ and ULOQ, had accuracy values for the measured concentrations within a ±15% range, except ±20% for the LLOQ, and the correlation coefficient (r^2) was ≥0.99.

3.7.2. Selectivity and Sensitivity

Selectivity studies were performed to evaluate the interference in analyte estimation due to the presence of endogenous substances in murine plasma or microdialysates. Selectivity studies were conducted in two different mice plasma samples viz. CD1 mice and CD1 nude mice and Ringer's/BCD solution, with a total of six replicates in each matrix at double blank (no analyte or ISTD spiked) and LLOQ level. The acceptance criteria for demonstrating selectivity were that double-blank plasma or microdialysates should not exhibit interfering peak responses at the retention time of IWR-1-endo $\geq$20% of the LLOQ peak response and $\geq$5% ISTD peak response in the same matrix.

The sensitivity of the developed method was calculated in terms of the analyte signal-to-noise (S/N) ratio. The LOD was determined using blank CD-1 murine plasma or Ringer's/BCD solution (n = 6) spiked before extraction to concentrations that were estimated to give S/N ratios $\geq$3 based on initial method development data. The LLOQ was defined as having a signal-to-noise ratio (S/N) $\geq$10 with bias and coefficient of variation (CV) $\leq$20% by analyzing six replicates of blank CD1 murine plasma or Ringer's/BCD solution.

3.7.3. Precision and Accuracy

The within-run precision and accuracy (P & A) were determined by the QC samples at four concentration levels (LLOQ, LQC, MQC, and HQC) distributed across the calibration curve in six replicates during the same day. The between-run precision and accuracy were ascertained by assaying the QC samples on three independent days in six replicates. Precision and accuracy were expressed in terms of relative standard deviation (RSD) and relative error (RE), respectively. The acceptance criteria for within-run (n = 6) and between-run (n = 18) accuracy was that the REs obtained for each QC level were within $\pm$15% of the nominal concentrations except for the LLOQ level, where $\pm$ 20% were acceptable. The acceptance criteria for within-run (n = 6) and between-run (n = 18) precision were that the RSDs obtained for each QC level were $\leq$15% except for the LLOQ level, where $\leq$20% were acceptable. Further, for acceptance of a single P & A batch run, a minimum of 50% of QC samples and 2/3rd (67%) of the total QC samples assessed had to meet the acceptance criteria.

3.7.4. Matrix Effect and Recovery

The samples for estimating the matrix effect for IWR-1-endo and ISTD were prepared in blank microdialysate, and blank murine plasma obtained from two mice strains (CD-1 and CD-1 nude mice). The matrix effect was calculated as a matrix factor, which was the ratio of the instrument's response for post-extracted spiked samples to that of the instrument response for neat standard solutions at equivalent concentrations. For preparing post-extracted spiked samples, blank microdialysate or plasma samples were prepared as per the sample preparation protocol and later spiked at the QC levels. A value of MF close to 1 indicates no or negligible matrix effect, while MF $\leq$ 0.8 and $\geq$1.2 demonstrates ion suppression and ion enhancement, respectively [17–19]. Similarly, mean extraction recoveries for IWR-1-endo and ISTD were computed as the ratio of instrument response for the extracted samples to that of the instrument response for the post-extracted spiked samples. The matrix effect and recovery experiments were performed in triplicates at the LQC and HQC concentration levels for both the developed LC-MS/MS methods.

3.7.5. Stability

The stability studies were performed in blank plasma samples and microdialysates at the LQC and HQC concentrations (n = 3 replicates). They were studied under the following conditions: freeze–thaw cycle stability (FTS), short-term stability (STS), long-term stability (LTS), and postprocessing stability. For freeze–thaw cycle stability studies, the plasma or microdialysate samples were frozen at -80 °C for 24 h and then thawed for 1 h at room temperature and frozen again for 24 h, repeating this process until the third thawing cycle. After each freeze–thaw cycle, samples were extracted and analyzed. The STS studies were assessed for the plasma or microdialysate samples kept at ambient temperature and 4 °C for 48 h. The samples were extracted and analyzed over specified intervals to ascertain the STS for IWR-1-endo.

For establishing postprocessing stability in the autosampler, extracted samples were stored in the autoinjector at 4 °C for 48 h and then injected into the LC-MS/MS system. The LTS studies in murine plasma and microdialysate were conducted at −80 °C for 21 and 29 days, respectively. Samples were considered stable if the percentage difference in concentration was within 15% of the freshly processed quality control samples.

3.7.6. Carry-Over

Carry-over was evaluated by sequentially injecting a ULOQ sample (1000 ng/mL for the murine plasma method and 500 ng/mL for the microdialysate method) and immediately followed by the injection of two extracted double-blank samples. The concentration of the ISTD spiked in the ULOQ samples was 500 ng/mL for both the murine plasma and the microdialysate methods. A carry-over was considered negligible if the analyte peak area responses in the blank samples were ≤20% of the LLOQ peak response and ≤5% of the ISTD peak response.

3.7.7. Dilution Integrity

Dilution integrity was performed to test the ability of diluted microdialysis study samples to yield accurate results, specifically applying to study samples whose concentrations are above the upper limit of quantification (ULOQ). To study the dilution effect, murine plasma samples with initial concentrations of 4250 ng/mL and 8000 ng/mL were diluted 5- and 10-fold with blank CD1 plasma, respectively. For the microdialysate method, microdialysate samples with initial concentrations of 2000 ng/mL and 4000 ng/mL were diluted 5- and 10-fold with the blank microdialysate, respectively. Six replicates of diluted samples were processed, analyzed, and compared to a fresh calibration curve. The acceptance criteria for dilution integrity were that the concentrations obtained after applying the dilution factor were within ±15% of the nominal concentrations and % C.V. not more than 15% within replicates.

3.8. Method Application to Cerebral Microdialysis Study

A cerebral microdialysis study was performed to sample brain extracellular fluid (ECF) from a non-tumor-bearing female CD1 nude mouse (Charles River, Wilmington, MA, USA) as per a procedure previously described and approved by the St. Jude Institutional Animal Care and Use Committee [20]. A microdialysis probe with a 38 kDa MWCO semipermeable membrane (MD-2211, BASi) was inserted into a guide cannula implanted in the cerebral cortex of the animal. The microdialysis probe was perfused with Ringer's solution containing 10% BCD at a flow rate of 0.5 µL/min and equilibrated for 1 h. BCD was added to the perfusate to reduce the non-specific binding of IWR-1-endo to the microdialysis system. Then, the mouse was dosed with 30 mg/kg IWR-1-endo via intraperitoneal injection. Dosing formulation (3 mg/mL) comprised 7.5% NMP, 7.5% Solutol HS-15, 30% PEG-400, and 55% TPGS (10% *w/v* solution) and was vortexed for 2 min to obtain a clear solution. After dosing, brain dialysate fractions were collected over 1 h intervals for a total of 6 h. Blood samples were collected retro-orbitally at 0.25, 2.5, and 6 h after dosing and spun down to plasma within 2 min of collection. All plasma and dialysate samples (20 µL) were immediately transferred into siliconized tubes containing 200 µL of 0.1 N HCl and stored at −80 °C. Microdialysis probe recovery was assessed to calculate brain ECF concentrations using the zero-flow rate method, with dialysates collected at 0.5, 1, 1.5, and 2 µL/min, as previously described [21].

4. Conclusions

In this study, we developed and validated a rapid, simple, and sensitive LC-MS/MS method for the measurement of IWR-1-endo in murine plasma and microdialysate samples. Major highlights of the LC-MS/MS method include the ability to quantify IWR-1-endo over a wide dynamic range (5–1000 ng/mL for the murine plasma curve and 0.5–500 ng/mL for the microdialysate curve) in a relatively small sample volume (20 µL), coupled with a short analysis time of 5 min. To summarize, our study shows that IWR-1-endo is relatively stable in working solutions; however, the drug rapidly degrades in murine plasma and brain microdialysate samples. Furthermore, adjusting the pH of the samples with dilute hydrochloric acid to 1.5 dramatically improved the stability of IWR-1-endo in murine plasma and brain microdialysate samples. Lastly, the validated LC-MS/MS methods in murine plasma and brain microdialysates were successfully applied to cerebral microdialysis studies in CD-1 nude mice after the administration of IWR-1-endo. The findings of this study could serve as a valuable reference for future development and clinical applications of IWR-1-endo.

Supplementary Materials: The following supporting information can be downloaded at: https://www.mdpi.com/article/10.3390/molecules27175448/s1, Figure S1: Representative extracted-ion chromatograms for IWR-1-endo (1 ppm prepared in methanol/water, 1:1 v/v) in (A) acetonitrile/water system modified with 0.1 % formic acid and (B) methanol/water system modified with 0.1% formic acid. A gradient profile was employed for both the optimization, as described in Section 3.5; Figure S2: Representative extracted-ion chromatograms for IWR-1-endo and ISTD (1 ppm prepared in methanol/water, 1:1 v/v) on (A) Kinetex® C_8 analytical column (100°A, 50 × 2.1 mm, 2.6 μm particle size) and (B) Kinetex® C_{18} analytical column (100°A, 50 × 2.1 mm, 2.6 μm particle size). A gradient profile with acetonitrile/water system (0.1 % formic acid added) was employed for both the optimization, as described in Section 3.5; Figure S3: Stability evaluation of IWR-1 in murine plasma stored at 4 °C and room temperature over a period of 24 h (n = 3; mean ± S.D.) with enzyme inhibitors (A) p-chloromercuribenzoate, (B) bis-(p-nitrophenyl) phosphate, and (C) sodium fluoride. LQC: low-quality control; HQC: high-quality control; RT: room temperature; Figure S4: Stability assessment of IWR-1 in murine plasma stored at 4 °C and room temperature over a period of 24 h (n = 3; Mean ± S.D.) under different pH conditions (A) LQC level at RT, (B) HQC level at RT, (C) LQC level at 4 °C, and (D) HQC level at 4 °C. Plasma pH adjustments of 1.5, 7.0, and 11.0 were performed using 0.1 N hydrochloric acid, 1M sodium phosphate buffer, and 0.1 N sodium hydroxide, respectively. LQC: low-quality control; HQC: high-quality control; RT: room temperature.; Figure S5: Representative calibration curves for IWR-1-Endo in (A) mouse plasma and (B) microdialysates. Both curves were fitted to calibrators using weighted $1/x^2$ linear regression; Table S1: Optimization of IWR-1-endo recovery from mouse plasma and microdialysates using SPE μElution plates (n = 3).

Author Contributions: Conceptualization, S.N., A.D., O.C. and C.F.S.; methodology, S.N.; validation, S.N.; investigation, S.N., A.D., O.C. and C.F.S.; writing—original draft preparation, S.N., A.D. and O.C.; writing—review and editing, S.N., O.C., J.D.S. and C.F.S.; data curation; O.C., supervision, J.D.S. and C.F.S.; project administration; C.F.S.; Funding acquisition, J.D.S. and C.F.S. All authors have read and agreed to the published version of the manuscript.

Funding: This work was supported by a Cancer Center Support (CORE) Grant CA 021765, 2R01CA194057, and the American Lebanese Syrian Associated Charities (ALSAC).

Institutional Review Board Statement: The animal study protocol was approved by the St. Jude Institutional Animal Care and Use Committee (Protocol Number 351-100572, approved on 10/06/2021), and all procedures met the guidelines of the Association for Assessment and Accreditation of Laboratory Animal Care (AAALAC).

Informed Consent Statement: Not applicable.

Acknowledgments: This work was supported by a Cancer Center Support (CORE) Grant CA 021765, 2R01CA194057, and the American Lebanese Syrian Associated Charities (ALSAC). The content is solely the responsibility of the authors and does not necessarily represent the official views of the National Institutes of Health.

Conflicts of Interest: The authors have no conflict of interest to declare.

References

1. Northcott, P.A.; Robinson, G.W.; Kratz, C.P.; Mabbott, D.J.; Pomeroy, S.L.; Clifford, S.C.; Rutkowski, S.; Ellison, D.W.; Malkin, D.; Taylor, M.D.; et al. Medulloblastoma. *Nat. Rev. Dis. Primers* **2019**, *5*, 11. [CrossRef] [PubMed]

2. Taylor, M.D.; Northcott, P.A.; Korshunov, A.; Remke, M.; Cho, Y.J.; Clifford, S.C.; Eberhart, C.G.; Parsons, D.W.; Rutkowski, S.; Gajjar, A.; et al. Molecular subgroups of medulloblastoma: The current consensus. *Acta Neuropathol.* **2012**, *123*, 465–472. [CrossRef] [PubMed]
3. Arvanitis, C.D.; Ferraro, G.B.; Jain, R.K. The blood-brain barrier and blood-tumour barrier in brain tumours and metastases. *Nat. Rev. Cancer* **2020**, *20*, 26–41. [CrossRef] [PubMed]
4. Morfouace, M.; Cheepala, S.; Jackson, S.; Fukuda, Y.; Patel, Y.T.; Fatima, S.; Kawauchi, D.; Shelat, A.A.; Stewart, C.F.; Sorrentino, B.P.; et al. ABCG2 Transporter Expression Impacts Group 3 Medulloblastoma Response to Chemotherapy. *Cancer Res.* **2015**, *75*, 3879–3889. [CrossRef]
5. Correa, S.; Binato, R.; Du Rocher, B.; Castelo-Branco, M.T.; Pizzatti, L.; Abdelhay, E. Wnt/beta-catenin pathway regulates ABCB1 transcription in chronic myeloid leukemia. *BMC Cancer* **2012**, *12*, 303. [CrossRef]
6. Hung, T.H.; Hsu, S.C.; Cheng, C.Y.; Choo, K.B.; Tseng, C.P.; Chen, T.C.; Lan, Y.W.; Huang, T.T.; Lai, H.C.; Chen, C.M.; et al. Wnt5A regulates ABCB1 expression in multidrug-resistant cancer cells through activation of the non-canonical PKA/beta-catenin pathway. *Oncotarget* **2014**, *5*, 12273–12290. [CrossRef]
7. Gustafson, C.T.; Mamo, T.; Maran, A.; Yaszemski, M.J. Efflux inhibition by IWR-1-endo confers sensitivity to doxorubicin effects in osteosarcoma cells. *Biochem. Pharmacol.* **2018**, *150*, 141–149. [CrossRef] [PubMed]
8. Chen, B.; Dodge, M.E.; Tang, W.; Lu, J.; Ma, Z.; Fan, C.W.; Wei, S.; Hao, W.; Kilgore, J.; Williams, N.S.; et al. Small molecule-mediated disruption of Wnt-dependent signaling in tissue regeneration and cancer. *Nat. Chem. Biol.* **2009**, *5*, 100–107. [CrossRef] [PubMed]
9. Gose, T.; Shafi, T.; Fukuda, Y.; Das, S.; Wang, Y.; Allcock, A.; Gavan McHarg, A.; Lynch, J.; Chen, T.; Tamai, I.; et al. ABCG2 requires a single aromatic amino acid to "clamp" substrates and inhibitors into the binding pocket. *FASEB J.* **2020**, *34*, 4890–4903. [CrossRef] [PubMed]
10. Nair, S.; Davis, A.; Campagne, O.; Schuetz, J.D.; Stewart, C.F. Development and validation of an LC-MS/MS method to quantify the bromodomain and extra-terminal (BET) inhibitor JQ1 in mouse plasma and brain microdialysate: Application to cerebral microdialysis study. *J. Pharm. Biomed. Anal.* **2021**, *204*, 114274. [CrossRef] [PubMed]
11. Li, W.; Zhang, J.; Tse, F.L. Strategies in quantitative LC-MS/MS analysis of unstable small molecules in biological matrices. *Biomed. Chromatogr. BMC* **2011**, *25*, 258–277. [CrossRef] [PubMed]
12. Briscoe, C.J.; Hage, D.S. Factors affecting the stability of drugs and drug metabolites in biological matrices. *Bioanalysis* **2009**, *1*, 205–220. [CrossRef] [PubMed]
13. Chen, J.; Hsieh, Y. Stabilizing drug molecules in biological samples. *Ther. Drug Monit.* **2005**, *27*, 617–624. [CrossRef]
14. Hartman, D.A. Determination of the stability of drugs in plasma. *Curr. Protoc. Pharmacol.* **2002**, *19*, 7.6.1–7.6.8. [CrossRef] [PubMed]
15. Hendriks, G.; Uges, D.R.; Franke, J.P. pH adjustment of human blood plasma prior to bioanalytical sample preparation. *J. Pharm. Biomed. Anal.* **2008**, *47*, 126–133. [CrossRef] [PubMed]
16. US DHHS; FDA; CDER; CVM. *Guidance for Industry, Bioanalytical Method Validation*; US Department of Health and Human Services; Food and Drug Administration; Center for Drug Evaluation and Research; Center for Veterinary Medicine: Washington, DC, USA, 2018. Available online: https://www.fda.gov/regulatory-information/search-fda-guidance-documents/bioanalytical-method-validation-guidance-industry (accessed on 1 June 2020).
17. Guideline on Bioanalytical Method Validation, European Medicines Agency. February 2012. Available online: https://www.ema.europa.eu/en/documents/scientific-guideline/guideline-bioanalytical-method-validation_en.pdf (accessed on 1 June 2020).
18. Goswami, D.; Khuroo, A.; Gurule, S.; Modhave, Y.; Monif, T. Controlled ex-vivo plasma hydrolysis of valaciclovir to acyclovir demonstration using tandem mass spectrometry. *Biomed. Chromatogr. BMC* **2011**, *25*, 1189–1200. [CrossRef] [PubMed]
19. Nandakumar, S.; Menon, S.; Shailajan, S. A rapid HPLC-ESI-MS/MS method for determination of beta-asarone, a potential anti-epileptic agent, in plasma after oral administration of Acorus calamus extract to rats. *Biomed. Chromatogr. BMC* **2013**, *27*, 318–326. [CrossRef]
20. Jacus, M.O.; Throm, S.L.; Turner, D.C.; Patel, Y.T.; Freeman, B.B., 3rd; Morfouace, M.; Boulos, N.; Stewart, C.F. Deriving therapies for children with primary CNS tumors using pharmacokinetic modeling and simulation of cerebral microdialysis data. *Eur. J. Pharm. Sci. Off. J. Eur. Fed. Pharm. Sci.* **2014**, *57*, 41–47. [CrossRef] [PubMed]
21. Elmeliegy, M.A.; Carcaboso, A.M.; Tagen, M.; Bai, F.; Stewart, C.F. Role of ATP-binding cassette and solute carrier transporters in erlotinib CNS penetration and intracellular accumulation. *Clin. Cancer Res. Off. J. Am. Assoc. Cancer Res.* **2011**, *17*, 89–99. [CrossRef] [PubMed]

Article

Thermal and Structural Characterization of Two Crystalline Polymorphs of Tafamidis Free Acid

Norberto Masciocchi [1,*], Vincenzo Mirco Abbinante [2], Marco Zambra [1], Giuseppe Barreca [2] and Massimo Zampieri [2]

1. Dipartimento di Scienza e Alta Tecnologia e To.Sca.Lab., Università dell'Insubria, Via Valleggio 11, 22100 Como, Italy
2. Chemessentia s.r.l, Via Bovio 6, 28100 Novara, Italy
* Correspondence: norberto.masciocchi@uninsubria.it

Abstract: Tafamidis, chemical formula $C_{14}H_7Cl_2NO_3$, is a drug used to delay disease progression in adults suffering from transthyretin amyloidosis, and is marketed worldwide under different tradenames as a free acid or in the form of its meglumine salt. The free acid (CAS no. 594839-88-0) is reported to crystallize as distinct (polymorphic) crystal forms, the thermal stability and structural features of which remained thus far undisclosed. In this paper, we present—by selectively isolating highly pure batches of Tafamidis Form 1 and Tafamidis Form 4—the full characterization of these solids, in terms of crystal structures (determined using state-of-the-art structural powder diffraction methods) and spectroscopic and thermal properties. Beyond conventional thermogravimetric and calorimetric analyses, variable-temperature X-ray diffraction was employed to measure the highly anisotropic response of these (poly)crystalline materials to thermal stimuli and enabled the determination of the linear and volumetric thermal expansion coefficients and of the related indicatrix. Both crystal phases are monoclinic and contain substantially flat and π-π stacked Tafamidis molecules, arranged as centrosymmetric dimers by strong O-H···O bonds; weaker C-H···N contacts give rise, in both polymorphs, to infinite ribbons, which guarantee the substantial stiffness of the crystals in the direction of their elongation. Complete knowledge of the structural models will foster the usage of full-pattern quantitative phase analyses of Tafamidis in drug and polymorphic mixtures, an important aspect in both the forensic and the industrial sectors.

Keywords: Tafamidis; polymorphs; crystal structure; powder diffraction; thermal stability

Citation: Masciocchi, N.; Abbinante, V.M.; Zambra, M.; Barreca, G.; Zampieri, M. Thermal and Structural Characterization of Two Crystalline Polymorphs of Tafamidis Free Acid. Molecules 2022, 27, 7411. https://doi.org/10.3390/molecules27217411

Academic Editor: Franciszek Główka

Received: 3 October 2022
Accepted: 25 October 2022
Published: 1 November 2022

Publisher's Note: MDPI stays neutral with regard to jurisdictional claims in published maps and institutional affiliations.

Copyright: © 2022 by the authors. Licensee MDPI, Basel, Switzerland. This article is an open access article distributed under the terms and conditions of the Creative Commons Attribution (CC BY) license (https://creativecommons.org/licenses/by/4.0/).

1. Introduction

Crystal polymorphism of molecular and covalent compounds has recently permeated the field of solid-state organic chemistry [1] for dyes and pigments [2], organic photovoltaics [3] and, relevant here, in drug and active pharmaceutical ingredient (API) development, processing and formulation [4]. Different and important aspects of basic physico-chemical interest (structure, conformation, stability, reactivity, etc.), in applied science (crystallization, micronization and solvent incorporation/elimination effects) and industrial process control and optimization—up to patent litigation and infringing issues—are thoroughly considered during all these studies. The reason for such flourishing research activities originates from the well-known awareness that different polymorphs can possess significantly different (bio)chemical and rheological properties (hence, different therapeutic activities, tableting and formulation difficulties, shelf lives, etc.). Additionally, when a spontaneous, or poorly controlled, polymorphic change (a solid–solid phase-transition maintaining molecular integrity) occurs, e.g., during formulation and production and/or caused by prolonged storage, the phase purity of the API at the final marketing and delivery stages cannot be firmly controlled, with evident health (and economic) risks. [5]

That APIs are among the most prolific polymorphic family should not surprise, as the discovery of (so) many different solid-state forms (including hydrates, solvates, salts and cocrystals [6]) is definitely governed by the efforts and the amount of time and money spent on their isolation and use.

Once dissolved in body fluids, the individual molecules of an API do not mutually interact and reasonably behave in the same manner no matter which solid form they originate from; at variance, within each polymorph, they possess specific molecular arrangements—dictated by distinct intermolecular interactions and conformationally driven crystal packing—leading to remarkably different physico-chemical properties. In a non-exhaustive list, these include melting points, crystal polarity and chirality, solubility, thermal and chemical stability (or inertness), crystal size and size distribution, crystal shape and preferential cleavage, hardness and compressibility [7]. All these effects can significantly affect their synthesis, isolation and purification processes, as well as several formulation steps, where material rheology influences drug transport, flowability and tableting. Additionally, API's particle size and shape distributions and its chemical instability (e.g., hygroscopicity) may also affect post-formulation stability over time (e.g., shelf life of the marketed drug). More importantly, bioavailability and pharmacokinetic properties may also be dramatically changed [8], making the drug useless—as for Ritonavir [9,10]—or even toxic [11].

The thermodynamic behavior of different polymorphs can be defined as enantiotropic or monotropic, depending on the topology of their P/T and ΔH, ΔG/T phase diagrams, illustrating their stability ranges [12]. Conventional experimental methods employed to assess, at constant pressure, the relative thermal stability of polymorphs are thermoanalytical (DSC/TGA) or variable-temperature X-ray powder diffraction (VTXRD) analyses. Coupling these two techniques with structural powder diffraction analysis, which can provide atomic-scale models of polymorphs that cannot be isolated as single crystals of suitable size and quality, is now possible, and has largely benefited from the pioneering work of a few academic research groups [13], including ours [14].

In our daily work, we recently faced still-unsolved problems related to the crystal chemistry of Tafamidis, or 2-(3,5-dichlorophenyl)-6-benzoxazolecarboxylic acid (CAS no. 594839-88-0), an ingredient of several marketed drugs (mostly under the tradenames of Vyndamax or Vyndaqel). The chemical structure of the Tafamidis molecule is shown in Figure 1. Approved by the EMA and FDA in 2019, oral formulations of Tafamidis are employed in the treatment of transthyretin (TTR) amyloidosis in adult patients with early-stage symptomatic polyneuropathy to delay peripheral neurological impairment of TTR familial amyloid polyneuropathy.

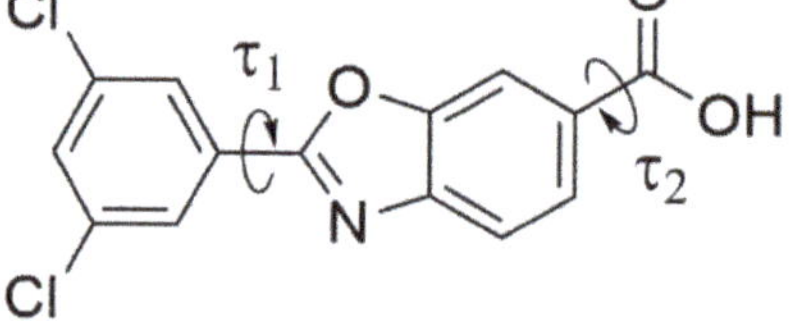

Figure 1. The molecular scheme and labeling of the freed torsion angles of the title compound.

Tafamidis free acid, in the solid state, has been the subject of several investigations, most of which are inserted in the pertinent patent literature. Suffice to say, several polymorphic forms have been isolated, or claimed, and the only firm structural report is dedicated to Form 2, the structural model of which is not publicly available. Form 1 and Form 4, which are the subject of the present study, have been proposed as pure polymorphs [15], but neither lattice metrics nor, consequently, structural models are presently available. These two forms appear to crystallize, selectively or in combination, during material processing; since the thermodynamics of their possible interconversion are not presently known, we deemed it necessary to study these aspects, additionally motivated by providing structural models making identification and quantification easy, and to interpret their crystal-chemical behavior on a sounder basis. Finally, it must be noted that there exist incomplete reports on a few Tafamidis solvates and on another anhydrous polymorph (showing a complex and uninterpretable XRD powder pattern), which, in this contribution, are not further discussed [16].

Hereafter, we report on the solid-state properties of Tafamidis Form 1 and Form 4, which include spectroscopic FT-IR characterization (a common, fast and cheap analytical method routinely used in the laboratory for the characterization of molecular polymorphs [17]), the less conventional complete structural determination by X-ray powder diffraction methods and thermal characterization by complementary TG, DSC and VTXRD analyses. The results here obtained, and discussed below, clarify the nature of polymorphic transformation of Tafamidis free acid and provide the full structural model which can be fruitfully used in further computational, structural and analytical studies. Therefore, by this study, we filled the still-existing gap between synthetic aspects (the molecular level) and the functional properties (the pharmacological performance) of Tafamidis free acid, adding relevant information on the solid-state properties of the drug.

Worthy of note, in the absence of a single crystal of suitable size and quality for conventional X-ray diffraction structural analysis, other structural methods have recently become viable. These include structural powder diffraction methods [18] (well beyond their common fingerprinting use), synchrotron X-ray single-crystal analysis of tiny specimens (down to 10 μm samples [19,20]) and, more recently, electron diffraction methods, which have been demonstrated to provide approximate, but reliable, structural models for crystalline grains with edges as low as 50–100 nm [21]. In this respect, the much wider accessibility of powder diffractometers than that of synchrotron beamlines or ED accessories for an electron microscope makes the determination of crystal structures of moderately complex molecular compounds (e.g., drugs) a viable option, at the expenses of less accurate models [22]. Accordingly, this contribution sagaciously uses the powder diffraction method in the complementary structural and thermodiffractometric modes, and opens the way to the use of XRD-based quantitative phase analysis of polymorphic mixtures [23], a problem which permeates the academic, and more relevantly, the forensic [24] and industrial [25], sectors.

2. Materials and Methods

2.1. Materials

A large batch (>100 g) of chemically pure solid Tafamidis of unknown crystal form was supplied by Química Sintética S.A., Alcalá de Henares, Spain. Crystallographically pure Tafamidis Form 1 and Tafamidis Form 4 were prepared according to the synthetic procedures described in patent WO 2016/038500 Al, in ca. 10 g batches [16]. Their phase purity was assessed by XRD after successful indexing and structure solutions were attained (*vide infra*).

2.2. Methods

2.2.1. IR Spectroscopy

Spectral data were obtained with FTIR equipment (Nicolet iS10 FTIR Spectrometer, Thermo Scientific, Rodano (Milano, Italy) using a diamond single-reflection attenuated total reflectance (ATR) device. Spectra were collected using 64 scans at 1 cm^{-1} resolution in the spectral interval range of 4000 to 650 cm^{-1}.

2.2.2. Thermal Analyses

Thermogravimetric (TG) traces were acquired from 30 to 400 °C (with a scan rate of 10 °C min^{-1}) using a Netzsch STA 409 PC Luxx® analyzer under an N2 flow and with alumina sample-holders equipped with a pierced lid. Differential scanning calorimetry (DSC) traces were acquired from 30 to 300 °C (with a scan rate of 10 °C min^{-1}) using a Mettler DSC 1 Stare System analyzer under an N2 flow (80 mL min^{-1}) and with 40 μL crimped aluminum sample-holders equipped with a pierced lid.

2.2.3. Structural X-ray Powder Diffraction Analysis

Tafamidis Form 1 and Form 4 were gently ground in an agate mortar and then deposited in the hollow of a 0.2 mm deep silicon monocrystal (a zero background plate, supplied by Assing spa, Monterotondo, Italy). Diffraction data were collected in the 5–105° 2θ range on a D8 Bruker AXS vertical sampling diffractometer operating in θ:θ mode, equipped with a linear Lynxeye position-sensitive detector, set at 300 mm from the sample (Generator settings: 40 kV, 40 mA, Ni-filtered Cu-K$\alpha_{1,2}$ radiation, λ_{avg} = 1.5418 Å). Peak search and profile-fitting allowed for the location of the most prominent, low-angle peaks, which were later used in the indexing process by the TOPAS-R software [26]. Approximate lattice parameters of primitive monoclinic cells were determined to be a = 22.99, b = 14.99, c = 3.79 Å and β = 91.0° [GOF(29) = 23.8] for Form 1 and a = 22.33, b = 15.18, c = 3.72 Å and β = 93.9° [GOF(27) = 20.4] for Form 4. Systematic absence conditions suggested $P2_1/a$ (Form 1) and $P2_1/n$ (Form 4) as the probable space groups, later confirmed by successful structure solution and refinement. The choice of maintaining non-standard, but tolerated, space-group settings is here motivated by having comparable crystal cell axis lengths and orientations. Note that the occurrence of a rather short c axis in both forms required relaxation during the solution process, as c—as well as the β angle—are determined with low accuracy (as per the well-known dominant zone problem). Density considerations indicated Z = 4, thus limiting the structural solution process of both forms to the individuation of the center-of-mass location, orientation and some conformational freedom of an otherwise-rigid Tafamidis molecule, defined by z-matrix formalism. Real-space structure modeling by the simulated annealing algorithm, coupled with a Monte Carlo search, allowed the definition of suitable models, which were later refined by the Rietveld method. Structure solution and final refinements were carried out with TOPAS-R software. The background contribution was modelled by a Chebyshev polynomial fit; atomic scattering factors for neutral atoms were taken from the internal library of TOPAS-R. Preferred orientation corrections, in the March–Dollase formulation [27], were applied on the [001] and [301] poles (both bearing an evident structural meaning, vide infra), with final magnitudes g_{001} = 1.261 (2) and g_{301} = 1.087 (2), respectively. Anisotropic peak broadening, modeled by the spherical harmonics approach, led to a significantly lower agreement factor and was, therefore, adopted. Crystal data and relevant data analysis parameters are collected in Table 1. Figure 2 shows the final Rietveld refinement plots. Fractional atomic coordinates have been deposited as CIF files within the Cambridge Crystallographic Database as publications No. CCDC 2209986-2209987.

Table 1. Crystal data and data analysis parameters for Tafamidis Form **1** and Form **4**.

Parameter	Form 1	Form 4
Formula	$C_{14}H_7Cl_2NO_3$	$C_{14}H_7Cl_2NO_3$
fw, g mol^{-1}	308.12	308.12
Crystal system	monoclinic	monoclinic
Space group	$P2_1/a$ (No. 14)	$P2_1/n$ (No. 14)
a, Å	22.976 (1)	22.364 (3)
b, Å	14.993 (1)	15.174 (2)
c, Å	3.794 (1)	3.819 (1)
β, °	90.938 (3)	95.265 (5)
V, Å^3	1306.9 (1)	1290.7 (3)
Z	4	4
V/Z, Å^3	326.7	322.27
ρ_{calc}, g cm^{-3}	1.566	1.586
μ (CuKα), cm^{-1}	45.5	46.1
F (000)	624	624
λ_{avg}, Å	1.5418	1.5418
T, K	295	295
2θ range, °	6–105	6–105
R_p, R_{wp}	0.079, 0.103	0.062, 0.083
χ^2	4.13	5.16
R_{Bragg}	0.064	0.039

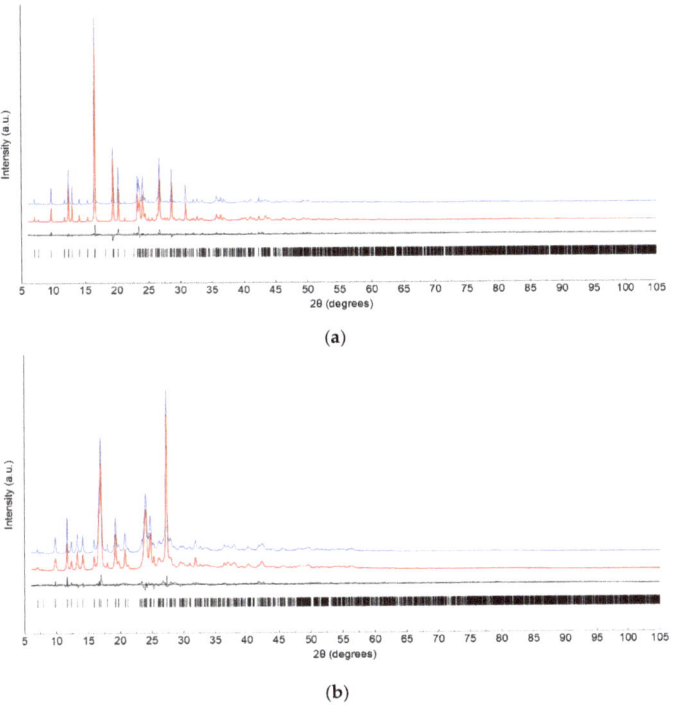

Figure 2. Rietveld refinement plots for Tafamidis Form 1 (**a**) and Tafamidis Form 4 (**b**), with peak markers and difference plot at the bottom. Observed data in blue and calculated trace in red, offset on the y axis for clarity.

2.2.4. Variable-Temperature X-ray Powder Diffractometry

Thermodiffractometric experiments were performed from 30 to 300 °C, or slightly beyond, to assess thermal and polymorphic stability. Powdered batches were deposited in the hollow of an aluminum sample-holder of a custom-made heating stage (Officina Elettrotecnica di Tenno, Ponte Arche, Italy). Diffractograms were acquired in air, in the most significant (low-angle) 5–32° 2θ range, under isothermal conditions in 20 °C steps. Since the samples in powder diffraction experiments are in direct contact with the air, and some thermal drifts/gradients are present, accurate transition temperatures are calibrated on TG/DSC measurements. Computation of the thermal strain was performed using Ohashi's method [28] as implemented by the Bilbao Crystallographic Server [29], assisted by tensor visualization [30], after extracting T-dependent cell parameters by the structureless Le Bail technique [31].

3. Results

3.1. Comparative Crystal Chemistry

Crystals of Tafamidis Form 1 and Form 4 are both monoclinic, and share space-group symmetry (standard $P2_1/c$, N. 14). However, in order to simplify the crystal-chemical comparison, their structures are here described in non-standard settings ($P2_1/a$ and $P2_1/n$, respectively), which keep the length of the **c** axes nearly unchanged, as they are strictly related to the stacking sequence of nearly planar molecules.

Laboratory X-ray powder diffraction studies of organic molecular systems of moderate complexity do not provide robust information on individual bond-distances and angles. At variance, since the final structural model is known to be more sensitive to molecular packing and conformational aspects, the only structural parameters that are relevant for the stereochemical discussion are the molecular τ_1 and τ_2 torsional angles (depicted in Figure 1) and ancillary geometrical entities, synoptically collected in Table 2. Figure 3 shows the different molecular and packing features for Tafamidis Form 1 and Tafamidis Form 4. Jointly with the values reported in Table 2, it is evident that both phases contain Tafamidis molecules in similar (though statistically distinct) conformations (not easily detectable at the drawing level, see Figure 3a,b). Through inversion centers, individual molecules give rise to hydrogen-bonded dimers, typical of monocarboxylic acids, with very similar geometries (O–H···O = 2.62–2.64 Å). A further supramolecular aspect is the presence, in both polymorphic phases, of infinite ribbons running along **b** (later discussed), generated by the alternate sequence of carboxylic dimers and centrosymmetrically related (definitely weaker) C–H···N interactions linking two oxazole rings across an inversion center (3.46–3.57 Å). This feature also explains the similarity of the **b** axis length in Tafamidis Form 1 and Form 4.

Table 2. Comparison of relevant stereochemical and geometrical parameters highlighting similarities and differences between the two polymorphic forms of Tafamidis.

	Form 1	Form 4	A Sketch of the SV, χ and ψ Parameters
τ_1 torsional angle, °	9.4 (0.2)	1.5 (0.3)	
τ_2 torsional angle, °	0.0 (0.5)	6.0 (0.5)	
O–H···O, Å	2.62	2.64	
Stacking Vector (SV), Å	3.794	3.819	
χ angle, °	78.8	70.9	
ψ angle, °	80.9	71.8	
Interplanar Distance, Å	3.51	3.41	

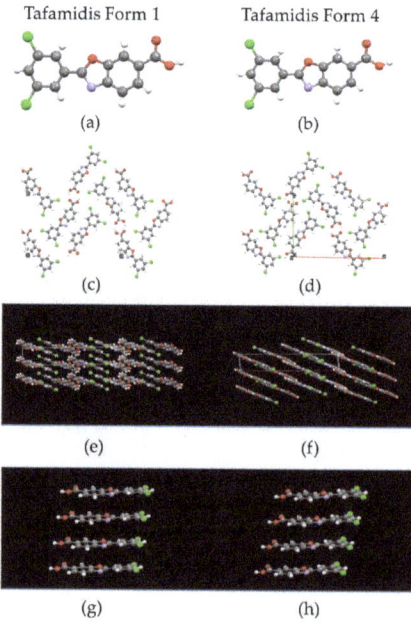

Figure 3. Comparative drawing of the most relevant molecular and packing entities of Tafamidis Forms 1 (left column) and 4 (right column) (**a,b**) the individual molecules; (**c,d**) the crystal packing viewed down **c**; (**e,f**) the crystal packing viewed down b, where the largest differences are visible; (**g,h**) the π-π stacking occurring along c, characterized by significantly different interplanar vectors: 3.51 and 3.41 Å, respectively. Color codes: C (grey); H (white), N (blue), O (red) and Cl (green).

3.2. High-Temperature Diffraction Studies

Variable-temperature X-ray diffraction was used to assess the thermal evolution of Tafamidis Form 1 and Form 4, in the form of polycrystalline powders, when heated in air from room temperature to material sublimation. The 2D thermodiffractograms, shown in Figures 4a and 5a, clearly manifest the substantial constancy of the crystal phase of Form 4. Differently, when Form 1 is progressively heated, a change of peak positions and intensities can be observed near 240 °C and can be easily explained by the quantitative formation of Form 4 through a solid–solid polymorphic transformation.

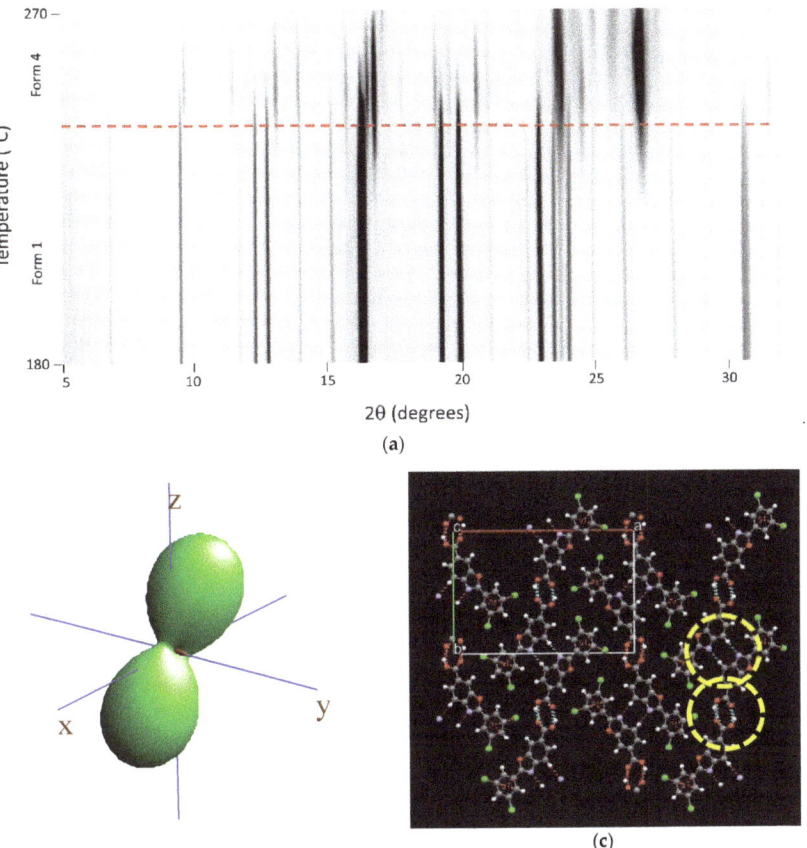

Figure 4. Variable-temperature X-ray diffraction (VTXRD) data (in the most relevant 180–270 °C range) showing the thermal evolution of the XRD pattern upon heating Tafamidis Form 1 in situ in the diffractometer cradle (**a**), showing its quantitative conversion into Form 4 at ca. 240 °C (red dashed line). (**b**) Visualization of the thermal strain tensor, showing nearly null (actually, slightly negative) thermal expansion in the y direction (see text), and (**c**), the hydrogen-bonded ribbons running along **b**. Indeed, couples of short O-H···O and C-H···N contacts, highlighted by the yellow circles, are responsible for the stiffness of the structure along **b**. Note that, for monoclinic symmetry, the y direction in the plot of the tensor isosurface is aligned with **b**, and that the xz plane corresponds to the crystallographic **ac** plane. Color codes in panel (**c**): C (grey); H (white), N (blue), O (red) and Cl (green).

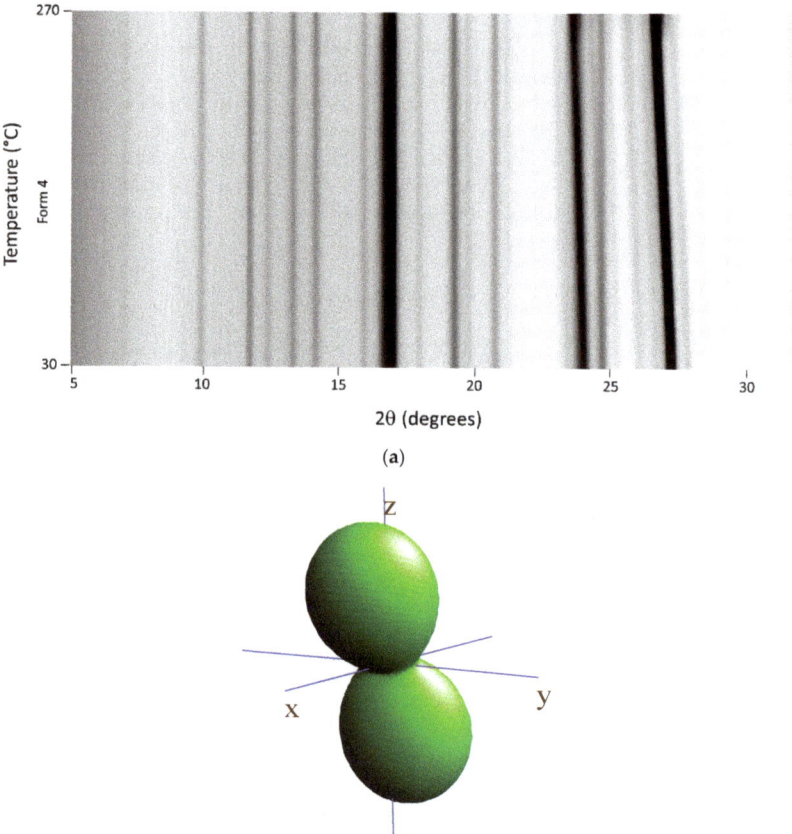

Figure 5. Variable-temperature X-ray diffraction (VTXRD) data (in the full 30–270° range) showing the thermal evolution of the XRD pattern upon heating Tafamidis Form 4 in situ in the diffractometer cradle (**a**); note that this form does not suffer from any phase transformation before melting. (**b**) Visualization of the thermal strain tensor, showing a relatively large thermal expansion in the z direction and much smaller (but still positive) components in the xy plane (see text). Note that, for monoclinic symmetry, the y direction in the plot of the tensor isosurface is aligned with **b**, and that the xz plane corresponds to the crystallographic **ac** plane.

Using the angular variation of the peak position upon heating, a significant anisotropy of the thermal-expansion tensor was derived, the isosurface plot of which is drawn in panels b of Figures 4 and 5. As expected, both isosurfaces show a prolate form pointing nearly toward z (i.e., the **c** axis). Such behavior is not unexpected, given that we deliberately chose the cell axis orientation (with the adoption of two non-standard settings of the $P2_1/c$ space group) in such a way that axes were "nearly" orthogonal, and the sequence and lengths of *all* axes were comparable. The thermal expansion values, computed in the 25–125 °C range ($\Delta T = 100$ K) using the formula $x_{125} = x_{25} (1 + 100\kappa_x)$, with x = a, b, c, β, V, are reported in Table 3. They highlight a significant stiffness in the xy plane and, particularly, along the **b** axis. The presence of supramolecular contacts generating infinite molecular ribbons, with the hinges highlighted in the yellow dashed circles in Figure 4 (for Tafamidis Form 1), is considered responsible for such thermal inertness. A nearly identical situation is present in Tafamidis Form 4 (not shown here).

Table 3. Comparison of linear and volumetric thermal expansion coefficients, calculated in the 25–125 °C range, using the formula $x_{125} = x_{25}(1 + 100\kappa_x)$, with x = a, b, c, β, V.

	Form 1	Form 4
κ_a, 10^6 K^{-1}	50	20
κ_b, 10^6 K^{-1}	−14	22
κ_c, 10^6 K^{-1}	74	93
κ_β, 10^6 K^{-1}	−48	−29
κ_V, 10^6 K^{-1}	108	141

3.3. Thermal Analyses

The thermal behavior of Tafamidis Form 1 and Form 4 were also studied by thermogravimetric and differential calorimetric analyses. Their TG and DSC traces are shown in Figure 6, and manifest a substantial chemical inertness, with material sublimation occurring only above 280 °C. The DSC curves, however, are much more informative, and provide enthalpic values in line with the occurrence of a solid-state polymorphic transformation for Form 1 (onset at 226 °C), with minimal heat absorption (ΔH = 0.8 kJ mol^{-1}) and the formation of Tafamidis Form 4. This latter form melts above 280 °C (onset at 283 °C) with a latent heat of fusion of 40.6 kJ mol^{-1} (see Figure 6a). As expected, the DSC curve of Form 4 is much flatter, and melting occurs with a ΔH$_{fus}$ of 39.5 kJ mol^{-1}, in reasonable agreement with the previously quoted value (see Figure 6b).

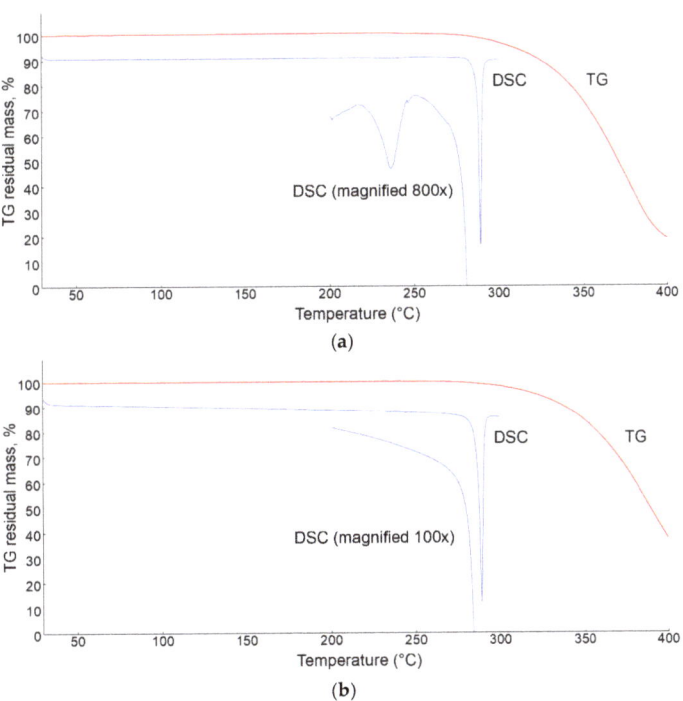

Figure 6. Thermodiffractometric (TG, in red) and calorimetric (DSC, in blue) curves for Tafamidis Form 1 (**a**) and Tafamidis Form 4 (**b**). The magnification of the DSC trace of panel (**a**) shows a broad endothermic event peaking near 240 °C, related to the Form 1 to Form 4 transformation. This peak is absent in Form 4, while melting of Form 4 (onset at ca. 283 °C) is observed in both samples.

Significantly, sublimation occurs without material decomposition. Indeed, a crystalline material could be recovered upon collecting powders sublimed in air at 325 °C on a metal blade kept at room temperature ca. 20 mm above the heating stage. Through conventional qualitative XRD fingerprinting and quantitative XRD phase analysis (enabled by the full knowledge of the structural models), such white material, of cotton-like appearance, proved to be a biphasic mixture with an approximate relative content of 13 and 87 w/w % Form 1 and Form 4, respectively.

The endothermic transformation of Form 1 into Form 4 and the subsequent melting of the latter indicate that Form 1 (at low enough temperatures) is more stable than Form 4, and that the two forms are enantiotropically related. However, Tafamidis Form 4—which should restore to Form 1 upon cooling—was found to be indefinitely stable (for months), as if a too-high energy barrier existed for the inverse transformation, which cannot be active once the temperature is (more or less) abruptly lowered.

A further experiment was also performed, aiming at proving the relative stabilities of Form 1 and Form 4. Upon heating powders of severely ground Tafamidis Form 4 on a hot metal plate set at 280 °C, progressive crystal growth of tiny needles on the cool side in contact with environmental air occurred, with the formation of a relatively homogeneous mat of intertwined crystals. This material was then characterized by quantitative XRD analysis, and proved to be nearly pure Form 1 (residual Form 4 accounted for ca. 2% w/w). Based on this result, we can safely state that if sublimation/recrystallization occurs *slowly* at high enough temperatures, the Form 1 → Form 4 phase transformation can be reverted; such observation is fully in line with the enantiotropic polymorphic relationship claimed above.

In addition, in contrast with the commonly accepted Kitaigorodskii [32] and Burger and Ramberger [33] rules, where the denser polymorph is normally considered the more stable form, here Form 1 is less dense than Form 4 (see Table 1). This aspect, however, is not worrisome, as this rule is often broken in hydrogen-bonded molecular crystals [34], as in the present case and in the widely marketed drug paracetamol [35].

3.4. IR Fingerprinting

One of the fastest, easiest and cheapest analytical methods used in API characterization and quality control is undoubtedly FTIR spectroscopy, particularly in the Attenuated Total Reflection (ATR-FTIR) mode [36]. To prove, or dismiss, the power of this technique when applied to Tafamidis polymorphs, we collected ATR-FTIR data which, in the most significant region, are graphically compared in Figure 7. In more detail, the lists of transmittance signals are here provided, with Tafamidis Form 4 values set in parentheses: 1688s (1692), 1613w (1619), 1589w (1589), 1572w (1572), 1546m (1546), 1493w (1492), 1438m (1438), 1423m (1424), 1410m (1410), 1388w (1386), 1349w (1349), 1309m (1309), 1291s (1290), 1272s (1274), 1236w (1237), 1196w (1194), 1134w (1134), 1129w (1126), 1102w (1103), 1087w (1086), 945w (944), 935w (935), 886m (887), 860s (859), 848w (850), 807w (807), 790w (792), 773s (773), 745m (745), 726s (726) and 665s (665).

This extensive list and the graphs in Figure 7 clearly demonstrate that, for this specific case, state-of-the-art ATR-FTIR cannot distinguish between the two polymorphs, as if their vibrational patterns were completely dominated by molecular modes without any contribution from intermolecular contacts. Indeed, the differences between the corresponding peaks are limited to a few cm^{-1} or even less, i.e., close to the resolution limit of any standard IR spectrometer. As our XRD study clearly proved that the basic building units of the entire crystals in both polymorphs (the Tafamidis centrosymmetric dimers) are practically identical, and that no evident "strong" interactions connect the different dimers, this result should not surprise. In such cases, Raman spectroscopy in the low-frequency range is claimed to be much more informative [37].

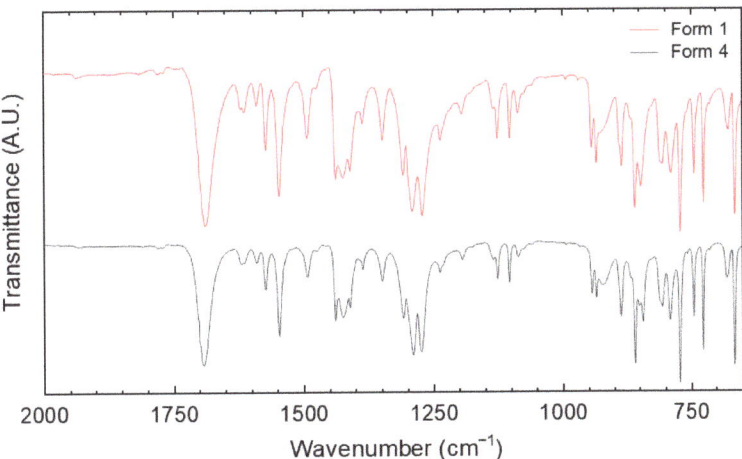

Figure 7. Plot of the ATR-FTIR transmittance spectra, in the most relevant portion (2000–650 cm^{-1} range), clearly showing the substantial identity of the FTIR traces of Tafamidis Form 1 and Form 4, preventing polymorph recognition by conventional vibrational spectroscopy (see text).

4. Conclusions

In this paper, we determined fundamental properties of a widely marketed drug (Tafamidis free acid), in the two known (polycrystalline) polymorphic Form 1 and Form 4 phases. Our study closes the existing gap between the molecular and the pharmacological levels, and provides (otherwise inaccessible) structural information through the sagacious usage of structural X-ray powder diffraction methods. Such detailed knowledge, which encompasses crystal symmetry, lattice parameter and fractional coordinates of all atoms in the structure, is indeed crucial for the identification of contaminants and enables the reliable relative quantification of the two forms in polyphasic mixtures by state-of-the-art whole-profile (Rietveld-like) quantitative analysis [38].

Additionally, we coupled common TGA/DSC methods and the less conventional thermodiffractometry, and studied the thermodynamic landscape of the polymorphic solid–solid transformation and its irreversible nature when cooling is (more or less) rapidly executed. A complete thermal strain tensor analysis was also performed, and the size and shape of the thermal strain indicatrix was determined. Finally, we proved that, in this case, the extreme similarity of the FTIR transmittance spectra—typically used for cheap and fast polymorphic recognition and fingerprinting—makes this analytical method not viable, and that resorting to XRPD becomes necessary.

Author Contributions: Conceptualization, N.M. and M.Z. (Massimo Zampier); investigation, M.Z. (Massimo Zampier), M.Z. (Marco Zambra) and V.M.A.; data analysis, N.M., M.Z. (Massimo Zampier) and M.Z. (Marco Zambra), writing—original draft preparation, N.M., writing—review and editing, G.B., N.M. and M.Z. (Massimo Zampier); project administration, M.Z. (Massimo Zampier) and G.B.; funding acquisition, G.B. All authors have read and agreed to the published version of the manuscript.

Funding: This research received no external funding.

Institutional Review Board Statement: Not applicable.

Informed Consent Statement: Not applicable.

Data Availability Statement: Data are available from the authors upon request.

Conflicts of Interest: The authors declare no conflict of interest.

Abbreviations

API	Active Pharmaceutical Ingredient
ATR	Attenuated Total Reflection
DSC	Differential Scanning Calorimetry
EMA	European Medicines Agency
FDA	Food and Drug Administration
FTIR	Fourier-transformed Infrared spectroscopy
TGA	Thermogravimetric Analysis
VTXRD	Variable-Temperature X-ray Diffraction
XRPD	X-ray Powder Diffraction

References

1. Cruz-Cabeza, A.J.; Feeder, N.; Davey, R.J. Open questions in organic crystal polymorphism. *Commun. Chem.* **2020**, *3*, 142. [CrossRef]
2. Bernstein, J. Polymorphism of dyes and pigments, Ch. 8. In *Polymorphism of Molecular Crystals*; Oxford University Press: Oxford, UK, 2020. [CrossRef]
3. van der Poll, T.S.; Zhugayevych, A.; Chertkov, E.; Bakus, R.C.; Coughlin, J.E.; Teat, S.J.; Bazan, G.C.; Tretiak, S. Polymorphism of Crystalline Molecular Donors for Solution-Processed Organic Photovoltaics. *J. Phys. Chem. Lett.* **2014**, *5*, 2700–2704. [CrossRef] [PubMed]
4. Karpinski, P.H. Polymorphism of Active Pharmaceutical Ingredients. *Chem. Eng. Technol.* **2006**, *29*, 233–237. [CrossRef]
5. Censi, R.; Di Martino, P. Polymorph Impact on the Bioavailability and Stability of Poorly Soluble Drugs. *Molecules* **2015**, *20*, 18760–18776. [CrossRef]
6. Brittain, H.G. *Polymorphism in Pharmaceutical Solids*; Informa Healthcare USA, Inc.: New York, NY, USA, 2009.
7. Lee, A.Y.; Erdemir, D.; Myerson, A.S. Crystal Polymorphism in Chemical Process Development. *Annu. Rev. Chem. Biomol. Eng.* **2011**, *2*, 259–280. [CrossRef]
8. Blandizzi, C.; Viscomi, G.C.; Scarpignato, C. Impact of crystal polymorphism on the systemic bioavailability of rifaximin, an antibiotic acting locally in the gastrointestinal tract, in health volunteers. *Drug Des. Dev. Ther.* **2015**, *9*, 1–11. [CrossRef]
9. Bauer, J.; Spanton, S.; Henry, R.; Quick, J.; Dziki, W.; Porter, W.; Morris, J. Ritonavir: An Extraordinary Example of Conformational Polymorphism. *Pharm. Res.* **2001**, *18*, 859–866. [CrossRef]
10. Chemburkar, S.R.; Bauer, J.; Deming, K.; Spiwek, H.; Patel, K.; Morris, J.; Henry, R.; Spanton, S.; Dziki, W.; Porter, W. Dealing with the Impact of Ritonavir Polymorphs on the Late Stages of Bulk Drug Process Development. *Org. Process Res. Dev.* **2000**, *4*, 413–417. [CrossRef]
11. Raza, K.; Kumar, P.; Ratan, S.; Malik, R.; Arora, S. Polymorphism: The Phenomenon Affecting the Performance of Drugs. *SOJ Pharm. Pharm. Sci.* **2014**, *1*, 10–19. [CrossRef]
12. Otto, D.P.; de Villier, M.M. Solid State Concerns During Drug Discovery and Development: Thermodynamic and Kinetic Aspects of Crystal Polymorphism and the Special Cases of Concomitant Polymorphs, Co-Crystals and Glasses. *Curr. Drug Disc. Technol.* **2017**, *14*, 72–105. [CrossRef]
13. Shankland, K.; Spillman, M.J.; Kabova, E.A.; Edgeley, D.S.; Shankland, N. The principles underlying the use of powder diffraction data in solving pharmaceutical crystal structures. *Acta Cryst.* **2013**, *C69*, 1251–1259. [CrossRef] [PubMed]
14. Abbinante, V.M.; Zampieri, M.; Barreca, G.; Masciocchi, N. Preparation and Solid-State Characterization of Eltrombopag Crystal Phases. *Molecules* **2021**, *26*, 65. [CrossRef] [PubMed]
15. Girard, K.P.; Jensen, A.J.; Jones, K.N. Crystalline Solid Forms of 6-Carboxy-2-(3,5-dichlorophenyl)-benzoxazole. Patent WO 2016/038500 Al, 17 March 2016.
16. Musanic, S.M.; Travancic, V.; Pavlicic, D. Solid State Forms of Tafamidis and Salts Thereof. Patent WO 2020/232325 Al, 19 November 2020.
17. Le Pevelen, D.D. FT-IR and Raman Spectroscopies, Polymorphism Applications. In *Encyclopedia of Spectroscopy and Spectrometry*; Lindon, J., Tranter, G.E., Koppenaal, D., Eds.; Academic Press: Cambridge, MA, USA, 2017; pp. 750–761. [CrossRef]
18. Harris, K.D. Powder diffraction crystallography of molecular solids. *Top. Curr. Chem.* **2012**, *315*, 133–177. [CrossRef]
19. Harding, M.M. Recording diffraction data for structure determination for very small crystals. *J. Synchr. Radiat.* **1996**, *3*, 250–259. [CrossRef]
20. Clegg, W. The development and exploitation of synchrotron single-crystal diffraction for chemistry and materials. *Phil. Trans.* **2019**, *A377*, 20180239. [CrossRef] [PubMed]
21. Saha, A.; Nia, S.S.; Rodriguez, J.A. Electron Diffraction of 3D Molecular Crystals. *Chem. Rev.* **2022**, *122*, 13883–13914. [CrossRef]
22. Fawcett, T.; Gates-Rector, S.; Gindhart, A.; Rost, M.; Kabekkodu, S.; Blanton, J.; Blanton, T. A practical guide to pharmaceutical analyses using X-ray powder diffraction. *Powder Diffr.* **2019**, *34*, 164–183. [CrossRef]
23. Iyengar, S.; Phadnis, N.; Suryanarayanan, R. Quantitative analyses of complex pharmaceutical mixtures by the Rietveld method. *Powder Diffr.* **2001**, *16*, 20–24. [CrossRef]
24. Kotrlý, M. Using X-ray diffraction in forensic science. *Zeit. Krist.–Cryst. Mater.* **2007**, *222*, 193–198. [CrossRef]

25. Ivanesevic, I.; McClurg, M.B.; Schields, P.J. Uses of X-ray Powder Diffraction In the Pharmaceutical Industry. In *Pharmaceutical Sciences Encyclopedia (Drug Discovery, Development, and Manufacturing)*; Gad, S.C., Ed.; Wiley: New York, NY, USA, 2010; pp. 1–42. [CrossRef]
26. *TOPAS-R, V3.0*; Bruker AXS: Karlsruhe, Germany, 2005.
27. Dollase, W.A. Correction of intensities for preferred orientation in powder diffractometry: Application of the March model. *J. Appl. Cryst.* **1986**, *19*, 267–272. [CrossRef]
28. Ohashi, Y. A program to calculate the strain tensor from two sets of unit-cell parameters. In *Comparative Crystal Chemistry*; Hazen, R.M., Finger, L.W., Eds.; Wiley: Chichester, UK, 1982; pp. 92–102.
29. Available online: https://www.cryst.ehu.es/cryst/strain.html (accessed on 30 September 2022).
30. Kaminsky, W. Wintensor, Ein WIN95/98/NT Programm zum Darstellen tensorieller Eigenschaften. *Z. Kristallogr. Suppl.* **2000**, *17*, 51.
31. Le Bail, A. Whole powder pattern decomposition methods and applications: A retrospection. *Powder Diffr.* **2012**, *20*, 316–326. [CrossRef]
32. Kitaigorodskii, A.I. *Organic Chemical Crystallography*; Consultants Bureau: New York, NY, USA, 1961.
33. Burger, A.; Ramberger, A.I. On the polymorphism of pharmaceuticals and other molecular crystals. I. *Mikroch. Acta* **1979**, *II*, 259–271. [CrossRef]
34. Perlovich, G.; Surov, A. Polymorphism of monotropic forms: Relationships between thermochemical and structural characteristics. *Acta Cryst.* **2020**, *B76*, 65–75. [CrossRef] [PubMed]
35. Nelyubina, Y.V.; Glukhov, I.V.; Antipin, M.Y.; Lyssenko, K.A. "Higher density does not mean higher stability" mystery of paracetamol finally unraveled. *Chem. Commun.* **2010**, *46*, 3469–3471. [CrossRef]
36. Salari, A.; Young, R.E. Application of attenuated total reflectance FTIR spectroscopy to the analysis of mixtures of pharmaceutical polymorphs. *Int. J. Pharm.* **1998**, *163*, 157–166. [CrossRef]
37. Larkin, P.J.; Dabros, M.; Sarsfield, B.; Chan, E.; Carriere, J.T.; Smith, B.C. Polymorph Characterization of Active Pharmaceutical Ingredients (APIs) Using Low-Frequency Raman Spectroscopy. *Appl. Spectrosc.* **2014**, *68*, 758–776. [CrossRef]
38. Gualtieri, A.F.; Gatta, G.D.; Arletti, R.; Artioli, G.; Ballirano, P.; Cruciani, G.; Guagliardi, A.; Malferrari, D.; Masciocchi, N.; Scardi, P. Quantitative phase analysis using the Rietveld method: Towards a procedure for checking the reliability and quality of the results. *Period. Miner.* **2019**, *8*, 147–151. [CrossRef]

Article

Accuracy of Citrate Anticoagulant Amount, Volume, and Concentration in Evacuated Blood Collection Tubes Evaluated with UV Molecular Absorption Spectrometry on a Purified Water Model

Nataša Gros *, Tadej Klobučar and Klara Gaber

Faculty of Chemistry and Chemical Technology, University of Ljubljana, Večna Pot 113, SI1000 Ljubljana, Slovenia
* Correspondence: natasa.gros@fkkt.uni-lj.si; Tel.: +386-1-479-8555

Abstract: Citrate anticoagulant concentration affects the results of coagulation tests. Until now, the end user had no direct insight into the quality of evacuated blood collection tubes. By introducing an easy-to-perform UV spectrometric method for citrate determination on a purified water model, we enabled the evaluation of (1) the accuracy of the anticoagulant amount added into the tubes by a producer, (2) the accuracy of the volume of anticoagulant solution in the tube at the instant of examination, (3) the anticoagulant concentrations at a draw volume. We examined the Vacuette®, Greiner BIO-ONE, Vacutube, LT Burnik d.o.o., and BD Vacutainer® tubes. The anticoagulant amount added into the tubes during production had a relative bias between 3.2 and 23.0%. The anticoagulant volume deficiency at the instant of examination expressed as a relative bias ranged between −11.6 and −91.1%. The anticoagulant concentration relative bias after the addition of purified water in a volume that equalled a nominal draw volume extended from 9.3 to 25.7%. Draw-volume was mostly compliant during shelf life. Only Vacutube lost water over time. Contamination with potassium, magnesium, or both was observed in all the tubes but did not exceed a 0.21 mmol/L level. This study enables medical laboratories to gain insight into the characteristics of the citrate blood collection tubes as one of the preanalytical variables. In situations that require anticoagulant adjustment for accurate results, this can help make the right decisions. The methodology gives producers additional means of controlling the quality of their production process.

Keywords: blood collection tubes; citrate anticoagulant; direct spectrometric determination; quality control method; anticoagulant concentration; draw volume; anticoagulant volume; magnesium contamination; potassium contamination

1. Introduction

Specimen collection, transport, storage, and processing are widely recognised preanalytical variables in coagulation testing [1]. Verification and validation protocols of blood collection tubes of different brands or lots assume a comparison to be performed on blood samples [2,3]. No method providing direct insight into the quality of the citrate anticoagulant blood collection tubes has been reported yet.

In the past, when anticoagulant solution was introduced into blood collection tubes in medical laboratories, they were in full control of this preanalytical phase which only required an accurately measured volume of a stock anticoagulant standard solution of accurate concentration and an accurate specimen volume measurement. The COVID-19 crisis caused a shortage of commercial tubes and temporarily revived this approach and its straightforwardness [4].

The introduction of evacuated blood collection tubes in the 1940s simplified phlebotomy and improved personnel safety [5]. Even though the tubes are easy to use, they are complex blood collection devices. Several previously unknown problems emerged, e.g., the material's water- and air-tightness, hydrophobicity, draw-volume accuracy, and additives as potential sources of specimen contamination [6,7].

Glass and plastic evacuated tubes were compared for coagulation tests, and although some statistically significant differences were confirmed, they were not recognised as being of clinical importance [8–12]. The 109 mmol/L citrate concentration was assumed to better support platelet aggregation than the 129 mmol/L citrate solution, and the advantages of buffered citrate were not recognised [13].

With the evacuated blood collection tubes, specimen volume is defined by the tube's draw capacity. It was suggested that routine coagulation tests had to be validated for minimal citrate tube fill volume to ensure accurate results [14]. The effect of the concentration of trisodium citrate anticoagulant on a calculation of the international normalised ratio and the international sensitivity index of thromboplastin was confirmed [15]. Prolonged prothrombin time and activated partial thromboplastin time due to underfilled specimen tubes were observed [16], and for patients with high haematocrit values, citrate concentration must be adjusted for accurate results [17]. Studies assessing thrombin generation are sensitive to a change in tube brand, and the choice of draw volume is important since low draw volumes can bias results [18].

On one side, the quality of blood-collection devices is considered a producer responsibility [19], but on the other, medical laboratories are fully responsible for the results of coagulation tests. With no direct insight into the characteristics of the citrate anticoagulant blood collection tubes, they remain unknown, which is difficult to control. Several studies confirmed that citrate concentration in a specimen affects the results of coagulation tests, and differences between brands, and even lots, were observed.

Evacuated citrate anticoagulant blood collection tubes are supposed to ensure the dilution of a specimen with the anticoagulant solution in a proportion of nine to one. To reach a correct anticoagulant level in a specimen, not only the draw volume but also the anticoagulant amount introduced into a tube during production and the volume of anticoagulant solution at the instant of venepuncture should be accurate. Anticoagulant volume accurately measured in the production setting changes during the tube evacuation process because of water evaporation. Consequently, the anticoagulant solution concentration increases.

A methodology providing direct insight into the quality of citrate anticoagulant tubes is required to properly regulate the anticoagulant volume and concentration during production and to enable medical laboratories to control the tubes as an input variable of their analytical process. No such methodology has been reported yet.

The titrimetric method suggested by the standard [20] is only suitable for the quality control of a bulk anticoagulant solution before it enters the production process. Conductometry, by which we enabled quality evaluation of K2EDTA and K3EDTA tubes [21,22], is not applicable since buffered and unbuffered citrate differ in chemical equilibrium forms [23,24]. In addition to the appropriate analytical method, the upgraded methodology is necessary since citrate tubes have several additional sources of uncertainty if compared to EDTA tubes.

The first objective of our work is to develop a low-cost, easy-to-apply methodology that would, by using purified water as a model, enable the evaluation of the following:
- the accuracy of the anticoagulant amount added into the tubes by a producer;
- the accuracy of the volume of anticoagulant solution in the tube at the instant of examination;
- the anticoagulant concentrations corresponding to the nominal draw or draw volume.

The second objective is to provide insight into the quality of the tubes of three different brands, namely Vacuette®—Greiner BIO-ONE, Vacutube—LT Burnik d.o.o., and BD—Vacutainer® that would involve the above-listed parameters together with water loss over time, draw volume during shelf life, and contamination examination.

2. Results

Citrate species absorb UV light, and spectra are influenced by chemical equilibrium. Krukowski et al., by lowering the pH of the solution below 1, shifted chemical equilibria towards the entirely protonated form and enabled citrate determination in oral electrolyte formulations [25]. To enable the quality evaluation of blood collection tubes, we adapted the method of Krukowski et al. In Section 2.1, we confirm the method's performance and fitness for the purpose. In Section 2.2, we present the results of the quality evaluation of citrate anticoagulant blood collection tubes before their intended use for specimen collection performed on a model of purified water. Water loss during the time, draw volume changes during shelf life, the accuracy of the amount of the anticoagulant added into a tube by a producer, and the accuracy of the volume of the anticoagulant solution in a tube at the instant of the examination are studied.

2.1. Spectrometric Method for Citrate or Buffered Citrate Determination

2.1.1. Method Performance

The spectra obtained for trisodium citrate or citric acid solutions prepared in HCl medium proved identical at all the examined concentration levels. Figure 1 confirms that the HCl addition is adequate to efficiently shift the citrate chemical equilibrium towards the entirely protonated citrate form. The method is consequently equally well suited for the evaluation of trisodium or buffered citrate solutions.

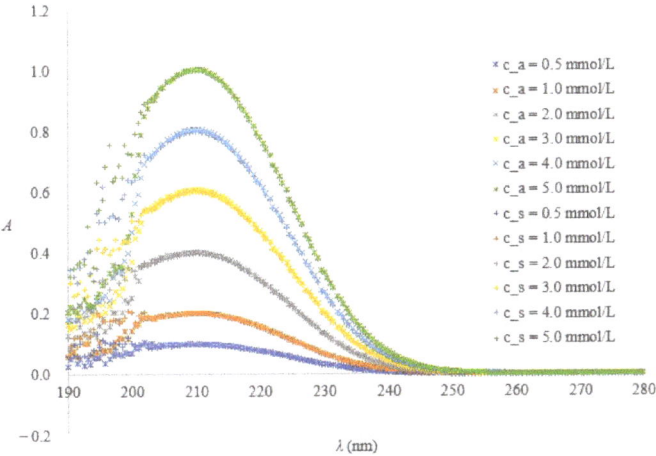

Figure 1. Citrate and citric acid model solutions-based confirmation that the suggested spectrometric method, independently of the initial equilibrium forms, ensures identical absorption spectra (c_a and c_s denote the citric acid and trisodium citrate amount concentration).

A single factor ANOVA with concentration as a controlled factor applied to the absorbance ratios obtained at 210 nm vs. 230 nm wavelengths during a month proved no significant between-groups differences for citrate or for citric acid solutions. The $F_{5/30}(p = 0.05)$ and F calculated for citrate were 2.534 and 0.7831. The $F_{5/36}(p = 0.05)$ and F calculated for citric acid were 2.477 and 1.884. Since the calculated F was lower for the former, citrate was recognised as a better choice for the preparation of calibration solutions. The 210 nm wavelength was selected for the absorbance measurements. A visual inspection of six-point calibration graphs confirmed the linearity within the targeted citrate concentration range extending from 0.25 to 4 mmol/L.

The within-day repeatability of the spectrometric procedure was tested on the 14.12 mmol/L trisodium citrate stock standard solution. The relative standard deviation of the determined citrate concentration was 0.18%, and the bias expressed relatively was 0.3%. Absorbances measured in a 1-cm cuvette at 210 nm against the HCl blank and the determined citrate concentrations for 10 replicates are given in Table 1. The symbol s denotes a standard deviation of a statistical sample.

Table 1. Within-day repeatability of the spectrometric procedure tested on the 14.12 mmol/L trisodium citrate stock standard model solution; $\bar{c} \pm s = (1.415_0 \pm 0.004_3)$ mmol/L, $n = 10$.

	1	2	3	4	5	6	7	8	9	10
$A_{210\,nm}$	0.28716	0.28839	0.28825	0.28799	0.28860	0.28827	0.28647	0.28802	0.28620	0.28868
c (mmol/L)	1.4118	1.4178	1.4171	1.4159	1.4189	1.4173	1.4085	1.4160	1.4071	1.4192

Data in Table 2 verify the within-laboratory reproducibility of calibration parameters for a six-month period. A slope a, a standard deviation of the slope s_a, an intercept b, a standard deviation of the intercept s_b, a standard error of the estimate $s_{y/x}$, and a coefficient of determination R^2 are given.

Table 2. Within-laboratory reproducibility of the six-point ($n = 6$) calibration function ($y = a \cdot x + b$) during a six-month period.

Date	a	s_a	b	s_b	$s_{y/x}$	R^2
21 February 2022	0.1992	0.0041	0.0146	0.0091	0.0135	0.99835
25 February 2022	0.2032	0.0010	0.0002	0.0021	0.0032	0.99991
28 February 2022	0.2039	0.0010	0.0006	0.0023	0.0034	0.99990
11 March 2022	0.2021	0.0006	−0.0004	0.0013	0.0019	0.99997
6 April 2022	0.2027	0.0013	0.0022	0.0029	0.0043	0.99984
5 May 2022	0.2032	0.0028	0.0055	0.0064	0.0094	0.99922
16 June 2022	0.2038	0.0008	−0.0009	0.0017	0.0026	0.99994
7 July 2022	0.2021	0.0005	0.0018	0.0011	0.0016	0.99998
17 August 2022	0.2050	0.0008	−0.0047	0.0018	0.0027	0.99994

In Table 3, minimum and maximum citrate concentrations determined on the Milli-Q water model for the blood collection tubes examined on a particular day are stated together with their combined standard uncertainties. The major source of uncertainty was the standard uncertainty of interpolation of concentration x_0 from the calibration line equation. It is denoted by s_{x_0} and evaluated by applying Equation (1). The symbols x_i, y_0, $\bar{x}$, and $\bar{y}$, not explained previously, stand for the concentrations of the citrate calibration solutions, the absorbance of the examined solution, and the coordinates of the centroid of the calibration line equation, respectively.

$$s_{x_0} = \frac{s_{y/x}}{a}\sqrt{1 + \frac{1}{n} + \frac{(y_0 - \bar{y})^2}{a^2 \sum (x_i - \bar{x})^2}} \quad (1)$$

Table 3. Minimal and maximal citrate concentrations determined on a particular day in the blood collection tubes on the Milli-Q water model and their associated combined standard uncertainties.

(mmol/L)	17 August 2022	7 July 2022	16 June 2022	5 May 2022	6 April 2022	11 March 2022	28 February 2022	25 February 2022	21 February 2022
c_{min}	11.9	11.7	11.6	11.6	11.6	11.7	11.5	13.7	14.1
u_c	±0.2	±0.1	±0.2	±0.5	±0.3	±0.1	±0.2	±0.2	±0.7
c_{max}	17.7	15.8	16.3	16.4	16.0	15.7	14.7	15.9	17.4
u_c	±0.2	±0.2	±0.2	±0.5	±0.3	±0.2	±0.2	±0.2	±0.7

The relative standard uncertainties of interpolation, which varied between 0.052 and 0.0054 depending on the day of the experiment, were combined with the 0.0087 relative standard uncertainty of the 10-fold dilution required by the spectrometric procedure. Other uncertainty contributions associated with the spectrometric procedure proved negligible. The derived combined standard uncertainties u_c vary between 0.1 and 0.7 mmol/L (Table 3).

2.1.2. Fitness for Purpose

In the wavelength range relevant to the method, the spectra of the citrate or buffered citrate solutions obtained from the tubes of different producers (Figure 2) prepared in the HCl medium and recorded in a 1-cm cuvette against the HCl blank proved very similar to the spectra of the calibration solutions (Figure 1). The only exception is a spectrum of the solution extracted from the C 1.8 tubes, which has a distinct shape within the 200 to 230 wavelength range. The letters A, B, and C denote producers, the number that follows indicates a nominal draw volume, NR means non-ridged, and numbers 105 and 129 distinguish the tubes with less common anticoagulant concentration from the rest with the 109 mmol/L concentration. More details are given in Section 4.2.

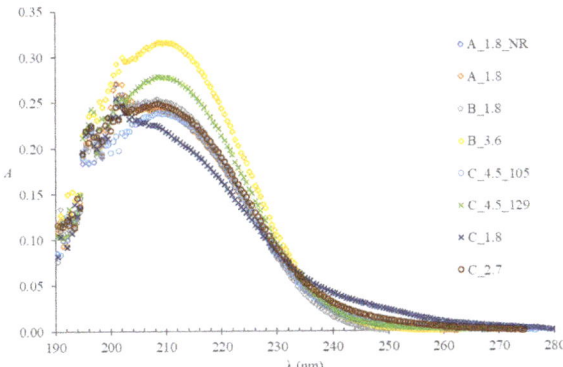

Figure 2. The spectra of citrate or buffered citrate solutions obtained from the tubes of different producers prepared in HCl medium and recorded in a 1-cm cuvette against the HCl blank.

Figure 3 confirms no apparent interference effect for the examined tube types since the anticoagulant amount expressed per a single tube and determined in composite samples proved comparable for the external standard and standard addition calibration method.

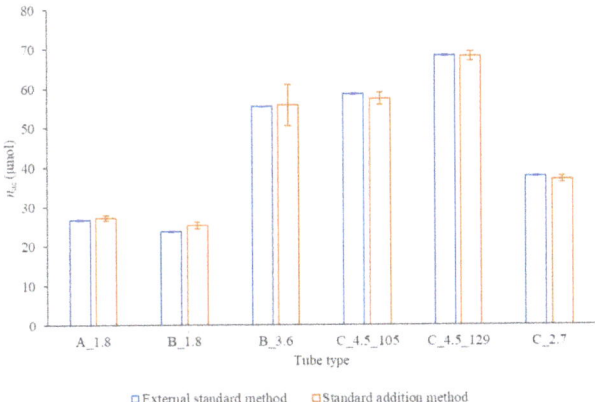

Figure 3. Anticoagulant amount per tube with 95% confidence interval determined in composite samples with the external standards (blue columns) and standard additions (orange columns) calibration method.

The spectrometric method proved reliable for the trisodium or buffered citrate concentration determination in blood collection tubes. Even though the results for C_1.8 tubes are not beyond any doubt, the inspection of the shape of several spectra implies that the absorbance changes more profoundly below 210 nm.

2.2. Evacuated Blood Collection Tubes for Coagulation Tests—Quality Evaluation

Quality evaluation of blood collection tubes comprises water loss during the time (Section 2.2.1), draw volume changes during shelf life (Section 2.2.2), the accuracy of the amount of the anticoagulant added into a tube by a producer (Section 2.2.3), and accuracy of the volume of the anticoagulant solution in a tube at the instant of the examination (Section 2.2.4).

2.2.1. Water Loss during the Time

We used one-factor ANOVA to test the hypothesis that water can evaporate from the blood collection tubes over time. The control factor was time and the examined variable was the mass of the tubes. The test was applied to the tubes of all three producers marked as already explained but extended with the expiration date. For the tubes A_1.8_9.7., C_4.5_31.7., and C_4.5_31.12. the hypothesis was not confirmed. For tubes B_1.8_31.7. and B_3.6_31.8. time proved a significant factor. The values 50.93 and 44.06 of the calculated F of the two types of tubes highly exceeded the critical value of 2.39 at the 0.05 significance level. The lowest significant differences, LSD, calculated by Equation (2) were 0.01916 and 0.02038, respectively. Symbols t, s, h, and n stand for the Student's t factor, within-sample standard deviation, number of groups, and number of within-group repetitions. Significant differences are indicated by asterisks in Table 4.

$$LSD = t_{h\,(n-1)} \cdot s \cdot \sqrt{2/n} \qquad (2)$$

Table 4. Results of the ANOVA test for the tubes of producer B, the asterisks indicate which between-groups differences proved significant.

B_1.8_31.7.	Count	Sum	Average	Variance	Differences			
25.02.	10	64.5799	6.45799	0.000517959	*			
11.03.	10	64.3702	6.43702	0.000437162		*		
06.04.	10	64.1965	6.41965	0.000752418		*		
5.05.	10	63.8514	6.38514	0.000254645			*	
16.06.	10	63.4715	6.34715	0.000271512				*
7.07.	10	63.4062	6.34062	0.000507422				*
B_3.6_31.8.	**Count**	**Sum**	**Average**	**Variance**	**Differences**			
25.02.	10	66.8741	6.68741	0.000588828	*			
11.03.	10	66.6654	6.66654	0.000599349		*		
06.04.	10	66.3714	6.63714	0.000796518			*	
5.05.	10	66.1818	6.61818	0.000428091			*	
16.06.	10	65.8877	6.58877	0.00035316				*
7.07.	10	65.5968	6.55968	0.000334517				*

* In contrast to tubes A and C, tubes B lost water during the four-month period.

2.2.2. Draw Volume during the Time

Figure 4 presents the draw volumes determined with a 5-mL Bang burette in Ljubljana at 300 m altitude at ambient temperature for the tubes of the three producers during their life cycle. Tubes A, B, and C were assigned markers of violet, blue, and green colour, respectively. The markers of the tubes are graded in size regarding their nominal draw volume, difference in shape indicate different lots or brands. The horizontal lines of a particular type define the 10% range around the targeted draw volume.

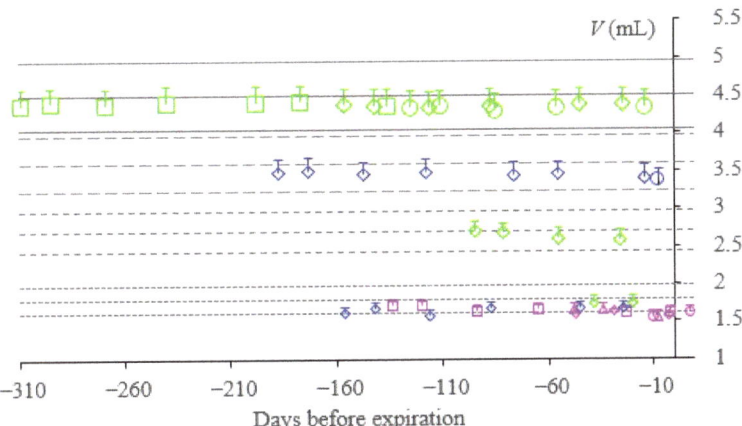

Figure 4. Draw volumes determined in Ljubljana at 300 m altitude, at ambient temperature during a life-cycle of the tubes; the error bars indicate the draw volume as assumed to be at 1013 hPa and 20 °C; the violet, blue, and green markers pertain to the tubes A, B, and C; the horizontal lines of a particular type define the 10% range around the target draw volume.

2.2.3. Accuracy of the Amount of the Anticoagulant Added into a Tube by a Producer

Table 5 presents the expected anticoagulant amount per tube (n_{ac_expt}), which is derived from the nominal anticoagulant concentration (c_{ac_nom}), together with the determined amount (n_{ac_dtmn}). The discrepancy is expressed as relative bias. The anticoagulant concentration that would have been expected in the tube ($c_{\_V\_total\_nom}$), if both the nominal anticoagulant volume (V_{ac_nom}) and the nominal draw volume ($V_{\_draw\_nom}$) hold, is calculated.

Table 5. Results of citrate determination in composite samples prepared from the blood collection tubes with Milli-Q water and expressed as an anticoagulant amount per tube (n_{ac_dtmn}) and the expected anticoagulant concentration ($c_{\_V\_total\_nom}$) if the nominal draw-volume and nominal anticoagulant volume are both compliant with the values declared by the producer.

Tubes	c_{ac_nom} (mmol/L)	n_{ac_expt} (μmol)	n_{ac_dtmn} (μmol)	n_{ac} Relative Bias (%)	$c_{\_V\_total\_nom}$ (mmol/L)
A_1.8_9.4.	109	21.8	26.3	20.6	13.2
A_1.8_9.7.	109	21.8	26.8	23.0	13.4
B_1.8_31.7.	109	21.8	23.6	8.2	11.8
C_1.8_31.3.	109	21.8	26.8	22.8 *	13.4
C_2.7_31.5.	109	32.7	38.4	17.4	12.8
B_3.6_31.8.	129	51.6	56.3	9.0	14.1
C_4.5_31.7.	105	52.5	55.0	4.8	11.0
C_4.5_31.12.	105	52.5	56.1	6.9	11.2
C_4.5_30.6.	129	64.5	66.6	3.2	13.3

* Irregularities in the absorption spectrum.

2.2.4. Accuracy of the Volume of the Anticoagulant Solution in a Tube at the Instant of the Examination

Anticoagulant concentration (c_{ac_dtmn}) determined in a tube after the addition of Milli-Q water into a blood-collection tube in a volume that equals $V_{\_draw\_nom}$ is given in Table 6. An anticoagulant volume that was present in a tube at the instant of the experiment (V_{ac}) was calculated. The relative bias of both parameters is given.

Table 6. Results of citrate determination ($c_\text{ac\_dtmn}$) after Milli-Q water addition into blood collection tubes ($n = 10$) in a volume that equals the nominal draw volume ($V_\text{\_draw\_nom}$) and other associated or derived parameters.

Tubes [1]	$V_\text{\_draw\_nom}$ (mL)	$c_\text{ac\_expt}$ (mmol/L)	$c_\text{ac\_dtmn}$ (µmol)	c_ac Relative Bias (%)	$V_\text{ac\_nom}$ (mL)	V_ac (mL)	V_ac Relative Bias (%)
A_1.8_9.4.	1.8	10.9	13.6	24.8	0.2	0.134	−33.1
A_1.8_9.7.	1.8	10.9	13.7	25.7	0.2	0.157	−21.6
B_1.8_31.7.	1.8	10.9	13.0	19.0	0.2	0.018	−91.1
C_1.8_31.3.	1.8	10.9	14.8	35.8 *	0.2	0.008	−95.9
C_2.7_31.5.	2.7	10.9	13.4	22.9	0.3	0.164	−45.3
B_3.6_31.8.	3.6	12.9	14.3	10.9	0.4	0.334	−16.6
C_4.5_31.7.	4.5	10.5	12.1	15.2	0.5	0.048	−90.4
C_4.5_31.12.	4.5	10.5	12.1	15.2	0.5	0.137	−72.6
C_4.5_30.6.	4.5	12.9	14.1	9.3	0.5	0.223	−55.5

[1] Evaluated on the 21st of March 2022. * Irregularities in the absorption spectrum.

2.3. Contaminants or Potential Additives

Table 7 summarises the potassium and magnesium concentrations, $c\,(\text{K}^+)$ and $c\,(\text{Mg}^{2+})$, respectively, determined with atomic absorption spectrometry in composite samples prepared with Milli-Q water from several tubes of the same type. Concentrations are expressed so that they represent potassium and magnesium content after a draw if $V_\text{\_draw\_nom}$ and $V_\text{ac\_nom}$ comply with the declaration.

Table 7. Potassium and magnesium concentrations determined in a composite sample prepared from the blood collection tubes of a particular type with Milli-Q water and expressed as expected if the $V_\text{\_draw\_nom}$ and $V_\text{ac\_nom}$ are compliant with the declaration.

Tubes	$c_\text{ac\_nom}$ (mmol/L)	Expiration Date	$V_\text{\_draw\_nom}$ (mL)	$V_\text{ac\_nom}$ (mL)	$c\,(\text{K}^+)$ (µmol/L)	s ($n = 3$)	$c\,(\text{Mg}^{2+})$ (µmol/L)	s ($n = 3$)
A_1.8_NR	109	3 March 2022	1.8	0.2	/	/	209.1	±1.8
A_1.8	109	9 April 2022	1.8	0.2	/	/	187.0	±1.9
B_1.8	109	31 December 2021	1.8	0.2	/	/	208.3	±1.4
B_3.6_129	129	28 February 2022	3.6	0.4	/	/	46.4	±0.1
C_1.8	109	30 September 2021	1.8	0.2	139.3	±2.6	/	/
C_2.7	109	30 November 2021	2.7	0.3	146.8	±1.9	/	/
C_4.5_105	105	31 July 2022	4.5	0.5	170.2	±0.7	167.5	±0.4
C_4.5_129	129	30 June 2022	4.5	0.5	168.4	±3.8	164.1	±0.2

3. Discussion

3.1. What Makes the Quality Evaluation of the Citrate Blood Collection Tubes More Challenging If Compared to the EDTA Tubes

In 2013, we conducted quality control of EDTA anticoagulant evacuated blood collection tubes before specimen collection by developing a methodology that can be summarised in three points [21].

(a) Conductometry was confirmed as an adequate low cost easy-to-perform analytical method for the determination of EDTA anticoagulant concentration in tubes on a deionised or distilled water model.

(b) A 5-mL-Bang burette-based method for the determination of draw volume on a model of deionised or distilled water in the laboratory setting at local experimental conditions was established.

(c) A physical model was developed, which by correcting the measurements for the ambient temperature and the unreduced air pressure influence at the altitude of the experimental setting above sea level, was able to predict a draw volume and blood anticoagulant concentration for specimen collection under the assumption of no other phlebotomy-related adverse effect.

The necessary input variables are only the tube internal volume (V_{int}), the temperature, and unreduced air pressure at the time of the laboratory experiment.

The methodology, which we applied to the Becton Dickinson, Greiner Bio-One, and Laboratorijska tehnika Burnik d.o.o. evacuated blood-collection tubes, revealed that draw-volume deficiencies are relatively rare. Noncompliance with the H1-A5 standard [20] regarding exceeding EDTA anticoagulant concentrations was much more frequent. We wished for a similar methodology to be developed for citrate tubes but recognised them as a distinct and much more challenging case [21,22].

As Figure 5 illustrates, a draw capacity of an EDTA tube depends only on internal under-pressure (p_{int}), ambient temperature, and V_{int}, which can easily be determined. The physical model for correcting the draw-volume measurements obtained under ambient conditions fully applies. Consequently, the concentration of anticoagulant can be predicted for specimen collection. Anticoagulant, if not present in a dry form, has a negligibly small volume in comparison with V_{int}. A study of the change in the draw capacity of the tubes of different producers during the time was possible [22].

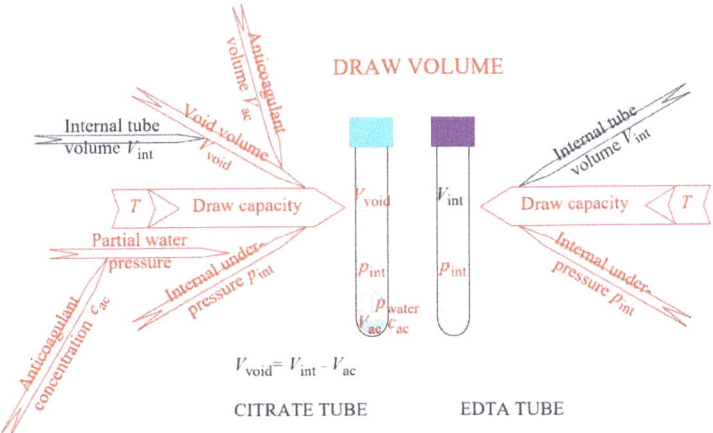

Figure 5. Parameters influencing a draw volume in the citrate (**left**) and EDTA (**right**) blood collection tubes, parameters that vary are indicated in red.

In contrast, V_{int} does not define the draw capacity of citrate tubes; since V_{ac} occupies part of V_{int}, the latter does not equal the void volume, V_{void}. The V_{ac} is an unknown, and the value declared by the producer does not necessarily hold, it can decrease over time due to water evaporation, and it is not directly determinable by simple means. Consequently, V_{void} is unknown, too. The parameter p_{water}, which is temperature and c_{ac} dependent, affects p_{int}. Determination of c_{ac} is yet another problem. For the majority of analytical methods, especially low-cost ones, V_{ac} is too small, and if it changes during time, c_{ac} changes consequently.

We had to find a way around these problems. By adding Milli-Q water in a volume that equals the $V_{\_draw\_nom}$, one does not directly get insight into the correctness of the anticoagulant amount, n_{ac} introduced into the tube at the instant of production since the total solution volume is unknown due to dilution of $V_{\_draw\_nom}$ by the unknown V_{ac}. By not knowing n_{ac} and V_{ac}, one cannot evaluate the correctness of the anticoagulant concentration for specimen collection. In the continuation, we explain the methodology for obtaining insight into n_{ac} and V_{ac}.

3.2. Quality of Evacuated Blood Collection Tubes for Coagulation Tests

The results of the citrate determination in the composite samples (c_{dtmn_comp}) provide insight into the correctness of the anticoagulant amount (n_{ac_dtmn}) added into the tubes during production (Table 5). The amount was calculated by Equation (3). V_f and N stand for the volumetric flask volume and the number of tubes used for the preparation of the composite sample, respectively.

$$n_{ac\_dtmn} = \frac{c_{dtmn\_comp} \cdot V_f}{N} \quad (3)$$

All the anticoagulant amounts exceed the producers' declarations. All three sets of the C_4.5 tubes, distinguished by their expiration dates, have the anticoagulant amount the closest to the expected value. The bias was between 3.2 and 6.9%. The tubes B_1.8 and B_3.6, by not exceeding the 10% tolerance limit, can still be considered in accordance with the standard [20]. Except for the already mentioned tube B_1.8, the rest of the tubes with 1.8 mL paediatric draw, namely A_1.8 with two distinct expiration dates and tubes C_1.8, exceed the expected citrate amount by more than 20%. The results of the latter might be overestimated due to the irregularities observed in the absorption spectrum, indicating the possible contribution of an unidentified additive or contaminant. The tubes C_2.7 are also non-compliant, with the citrate amount exceeding the declared value by 17.4%.

The tubes exposed here as potentially compliant would have been such only if the anticoagulant volume and the draw volume had been correct, too. The amount of citrate in the tubes (n_{ac_dtmn}, Table 5) together with the citrate determination (c_{dtmn}, Table 6) after the addition of Milli-Q water into the tubes in the volume corresponding to V_{draw_nom} enabled us to judge the correctness of the anticoagulant volume (V_{ac}) in the tubes at the instant of the examination (Table 6). Equation (4) was applied. For the sake of clarity, c_{dtmn} is in the equation indicated as $c_{nom\_draw\_V}$.

$$V_{ac} = \frac{n_{ac\_dtmn} - V_{draw\_nom} \cdot c_{nom\_draw\_V}}{c_{nom\_draw\_V}} \quad (4)$$

Water loss is material-dependent [7]. Even though the experimental data in Table 4 confirmed that tubes B lost water over time, tubes B_3.6 had the most accurate anticoagulant volume of all the examined tubes, with a bias of −17% (Table 6). The explanation is that the water loss experiment started five months before the expiration date and the first two dates of weighing the tubes were the 25th of February and the 11th of March. The evaluation of the correctness of the anticoagulant volume followed soon after, on the 21st of March. Water loss, even though recognised as statistically significant, could not have been yet very high in comparison with the target 0.4 mL anticoagulant volume. The 129 mmol/L target anticoagulant concentration also contributed by decreasing the partial water pressure. The impact on the B_1.8 tube with an expiration date only a month sooner but with the declared anticoagulant concentration of 109 mmol/L was entirely different. The anticoagulant volume bias was −91%, very similar to the C_1.8, 109 mmol/L tubes (−96%), and C_4.5_31.7., 105 mmol/L tubes (−90%). The anticoagulant volume bias was lower for the remaining tubes of producer C and ranged from −73% to −45%. The tubes A_1.8_9.4. and A_1.8_9.7. are by the anticoagulant volume bias −33% and −22% of a distinctively better quality regarding the already discussed tubes of the producers B and C with the paediatric draw.

To make a synthesis of the results of the two experiments, we calculated the concentrations $c_{\_V\_total\_nom}$ (Equation (5)) that could have been expected for the determined anticoagulant amount if the assumption that both $V_{\_draw\_nom}$ and V_{ac_nom} hold (Table 5). The $V_{\_total\_nom}$ stands for the total nominal volume.

$$c_{\_V\_total\_nom} = \frac{n_{ac\_dtmn}}{V_{total\_nom}} = \frac{n_{ac\_dtmn}}{V_{draw\_nom} + V_{ac\_nom}} \quad (5)$$

In Figure 6, we relate $c_{\_V\_total\_nom}$ with the $c_{nom\_draw\_V}$. The grey, blue, and red dots pertain to the tubes with 105, 109, and 129 mmol/L nominal anticoagulant concentrations, respectively. The tube labels are indicated. The diamonds in the same colours and the vertical dashed lines represent the corresponding accurate concentrations. The grey, blue, and red dotted vertical lines are the 10% upper limits of the anticoagulant concentrations. The dashed diagonal line is the line with a slope of 1.

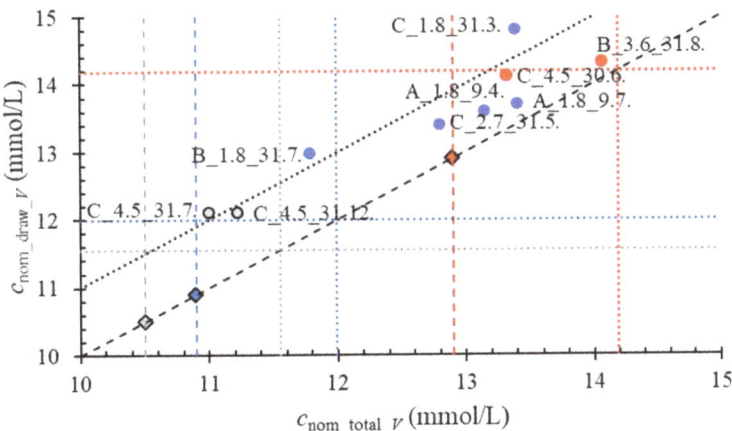

Figure 6. Relation between $c_{\text{V\_total\_nom}}$ and the $c_{\text{nom\_draw\_V}}$. The grey, blue, and red dots pertain to the tubes with 105, 109, and 129 mmol/L nominal anticoagulant concentrations, respectively. The diamonds in the same colours and the vertical dashed lines represent the corresponding accurate concentrations. The grey, blue, and red dotted vertical lines are the 10% upper limits of the anticoagulant concentrations. The dashed diagonal line is the line with a slope of 1.

Since the grey dots pertaining to the C_4.5_31.7. and C_4.5_31.12. 105-mmol/L anticoagulant tubes both lay within the grey vertical lines, the anticoagulant amount is within the 10% acceptance limit. The former set of tubes, by crossing the dotted diagonal line, exhibit more than 90% anticoagulant volume deficiency, the latter a 73% deficiency. Even with the accurate draw volume, the anticoagulant concentration would, with 12.1 mmol/L, exceed the 10% upper acceptance anticoagulant concentration limit, presented by the grey horizontal line.

The red dots correspond to the 129 mmol/L tubes C_4.5_30.6. and B_3.6_31.8. Both have anticoagulant amounts within the acceptance limits. The former of all the tubes is the only one entirely compliant with the standard [20]. The later dot laying the closest to the dashed diagonal line has the lowest anticoagulant volume deficiency of all the tubes. Since the water loss during the time was confirmed for this type of tube, the dot will be moving during the time towards the dotted diagonal line. The anticoagulant concentration, which at the instant of the experiment for the correct draw volume only slightly exceeded the upper anticoagulant limit, would rise.

Among the blue dots representing the 109 mmol/L tubes, only the B_1.8_31.7. tubes have an anticoagulant amount within the acceptance range, but their anticoagulant volume deficiency exceeded 90% since the tubes are not watertight. Consequently, the anticoagulant concentration for the nominal draw volume highly exceeds the dotted horizontal blue line. On the contrary, the dots A_1.8_9.4. and A_1.8_9.7, by being quite close to the dashed diagonal line, exhibit the lower anticoagulant volume deficiency of all the tubes with a paediatric draw, but the anticoagulant concentration at the nominal draw volume highly exceeds the 10% concentration limit, since the amount of anticoagulant in the tube is too high.

As we previously observed in EDTA tubes, citrate tubes also mostly have a draw volume within the 10% tolerance limit (Figure 4). The upper tolerance bar end is relevant for specimen collection. We did not correct the draw-volume measurements by applying the physical model due to already-explained limitations. Regarding our previous experiences, approximately a 5% increase in draw volume compensates for the altitude of our experimental setting. Lippi et al. also confirmed the adequate quality of the citrate tubes regarding their draw volume [26], and new materials are promising to limit the time-dependent decrease in draw volume [27], which we observed.

3.3. Contaminants or Potential Additives

UV absorbance spectra of the citrate anticoagulant extracted with Milli-Q water from the C_1.8, C_2.7, and C_4.5 tubes, prepared in HCl medium, and recorded against the HCl blank in 1-cm cuvette indicate possible additives absorbing in the wavelength range between 235 and 265 nm (Figure 2).

These were also the only tubes in which we were able to determine potassium with atomic absorption spectrometry. The $c_{V\_total\ nom}$ ranged from 139 to 170 µmol/L (Table 7). Even though not successfully confirmed with IR spectra due to high citrate concentration, one of the possible assumptions would be cross-contamination with K_3EDTA or K_2EDTA in a production setting. Lima-Oliveira et al., by simulating cross-contamination of citrate tubes with EDTA during phlebotomy, confirmed that 29% of K2EDTA blood causes a significant bias in the results of routine clotting assays [28]. Concentrations originating from the tube contamination, which we observed, are far below this limit.

The tubes C_1.8 raise additional concerns about other unidentified additives or contaminants absorbing in the UV range (Figure 2). No tubes were entirely free of contamination with metallic ions, as Table 7 demonstrates. Tubes A and B were contaminated with magnesium ions, tubes C_1.8 and C_2.7 with potassium ions, and tubes C_4.5 with potassium and magnesium ions.

Van der Besselaar et al. recognized a tube-stopper as a possible source of magnesium contamination influencing the prothrombin time [29]. They suggested a 1 mmol/L level as the maximal admissible concentration [30] and confirmed that low-magnesium tubes fulfil this requirement [31]. We confirmed magnesium contamination in the tubes of all three producers, but no concentration would exceed 210 µmol/L if the total solution volume corresponds to the sum of the nominal anticoagulant solution volume and nominal draw volume (Table 7). It should be mentioned that the results only reflect contamination of the inherent citrate anticoagulant solution, the volume of which is small. Determination of metallic anions was performed in composite samples prepared, as the last paragraph of Section 4.1 describes. The procedure did not involve a within-tube solution mixing. With a stopper as a source of contamination, higher concentrations are likely in such cases.

3.4. Implications and Limitations

The suggested methodology is easy to apply. It does not require any special costly instrumentation. Spectrometers for molecular absorption spectrometry are widely available. If the medical laboratory only has dedicated equipment, every university with natural science study programs certainly has a general-purpose spectrometer. Compact, low-cost spectrometers are available at affordable prices. The quality of the tubes entering the analytical process can be directly evaluated on a distilled or deionised water model. With the methodology, we suggest it is possible to predict the anticoagulant concentration for specimen collection, assuming no phlebotomy-related adverse effects. One only needs to replace $V_{\_draw\_nom}$ in Equation (5) with a measured draw volume. Further methodology development is possible. Irregularities in absorption spectra observed for 1.8-mL tubes from producer C need clarification.

The suggested methodology gives a final user a direct insight into the tubes' quality and provides the tube producers with additional means of controlling the quality of their products.

4. Materials and Methods

4.1. Spectrometric Method for Citrate or Buffered Citrate Determination

Spectrometric measurements were performed with a Varian Cary 50 UV-Vis spectrometer, Agilent Technologies, Santa Clara, CA, USA in a 1-cm cuvette against 126 mmol/L HCl solution. Absorbance was measured at 210 nm.

Deionised water was additionally purified through the Milli-Q system (Millipore, Billerica, MA, USA)—Milli-Q water in the continuation and A-class glass volumetric equipment were used for the preparation of solutions if not stated differently.

Six stock calibration solutions were prepared in 100-mL or 200-mL volumetric flasks by weighing trisodium citrate dihydrate $C_6H_5Na_3O_7 \cdot 2H_2O$ (M = 294.10 g/mol, w $\geq$ 0.98), CAS: 6132-04-3, Sigma-Aldrich, St. Louis, MO, USA. The diluent was Milli-Q water. Concentrations extended from 2.5 to 40 mmol/L in the final method.

Each working calibration solution was prepared in a 50-mL volumetric flask by measuring 5 mL of its corresponding stock solution with the glass volumetric pipette. The solution was made up to the mark with the 140 mmol/L HCl, prepared from the 37% HCl (CAS: 7647-01-0, ρ = 1.19 g/mL, Honeywell, Seelze, Germany).

For testing the repeatability of interpolation of citrate concentration from the calibration line equation, 14.12 mmol/L citrate stock solution was prepared in a 50-mL volumetric flask. Ten testing solutions were prepared from it by following the procedure as described in the previous paragraph.

For citrate determination in the evacuated blood collection tubes, the tubes were filled with Milli-Q water. The methods of filling were goal dependent. The draw volume method is described in Section 4.2.2. For the addition of a nominal volume, a Milli-Q water volume was measured with a 5-mL Bang burette directly into uncapped tubes. In both cases, solution was mixed as the producer suggested, and a 1-mL aliquot was transferred into a 10-mL volumetric flask with the A class Normax glass volumetric pipette, Duran, Portugal, and diluted with the 140 mmol/L HCl to the mark.

For citrate determination in the composite sample, the citrate or buffered citrate was quantitatively transferred from the required number of tubes of a particular type into a 50-mL volumetric flask with Milli-Q water. The solution was made up with Milli-Q water to the flask's nominal volume and further treated as the UV spectrometric method or atomic absorption spectrometry required.

4.2. Evacuated Blood Collection Tubes

The evacuated blood collection tubes were purchased from the local Slovene dealers. Two brands were global and one local. They were assigned letters A—Vacuette®, Greiner BIO-ONE, B—Vacutube, LT Burnik d.o.o., Skaručna, Slovenia, and C—BD Vacutainer®. The abbreviations, main features, and expiration dates are summarized in Table 8.

Table 8. Characteristics of the examined evacuated citrate or buffered citrate blood-collection tubes.

Abbreviation	Anticoagulant c (mmol/L)	Expiration Date	Draw Volume (mL)
B_1.8	109	31 December 2021	1.8
C_1.8	109 *	30 September 2021	1.8
C_2.7	109 *	30 November 2021	2.7
A_1.8_9.4.	109	9 April 2022	1.8
A_1.8_3.3._NR	109	3 March 2022	1.8
A_1.8_14.4._NR	109	14 April 2022	1.8
A_1.8_9.7.	109	9 July 2022	1.8
B_1.8_31.7.	109	31 July 2022	1.8
B_3.6_31.8.	129	31 August 2022	3.6
C_1.8_31.3.	109 *	31 March 2022	1.8
C-2.7_31.5.	109 *	31 May 2022	2.7
C_4.5_30.6.	129 *	30 June 2022	4.5
C_4.5_31.7.	105 *	31 July 2022	4.5
C_4.5_31.12	105 *	31 December 2022	4.5

* Buffered trisodium citrate.

4.2.1. Water Loss during a Time

The experiment was repeated for four months at regular intervals. On each occasion, 10 tubes of each type were randomly selected and weighed on an analytical balance.

4.2.2. Draw Volume during the Time

Draw volume was measured in Ljubljana at an altitude of 300 m with a 5-mL Bang burette filled to the mark with Milli-Q water. The experimental technique was more precisely described elsewhere [21,22]. The nonreduced air pressure varied between 972 and 994 hPa, and the ambient temperature was (24 ± 1) °C.

4.3. Contaminants or Potential Additives

4.3.1. Infrared Spectra

Infrared spectra were recorded on a Bruker FTIR Alpha Platinum ATR spectrometer (Billerica, MA, USA) in reflectance mode. Anticoagulant was extracted with Milli-Q water from the randomly selected tubes of the same brand. A composite sample was prepared and dried to constant mass at 50 °C under the nitrogen flux. Model solutions of sodium citrate, citric acid, citrate buffer, K_3EDTA, and K_3EDTA in the citrate matrix were treated equally. Anticoagulants extracted from the tubes and citrate model solutions formed a solid gel-type precipitate. Dried citric acid and K_3EDTA were crystalline.

4.3.2. Atomic Absorption Spectrometry

Potassium and magnesium concentrations in blood collection tubes were determined in solutions obtained by the addition of Milli-Q water in a nominal volume. The VARIAN AA240 Atomic Absorption Spectrometer was used. Experimental conditions are summarised in Table 9.

Table 9. Experimental conditions for the determination of potassium and magnesium with atomic absorption spectrometry.

Metal Ion	K^+	Mg^{2+}
Wavelength (nm)	766.5	285.2
Slot width (nm)	1	0.5
Concentration range (mg/L)	0.25 to 1	0.05 to 05
Medium	1% HCl	1% HCl

5. Conclusions

By introducing an easy-to-perform UV spectrometric method for citrate determination, we enabled for the first time a direct insight into the quality of citrate anticoagulant tubes before their intended use for specimen collection by using a purified water model. The approach reaches beyond the water loss evaluation during the time, draw-volume changes during shelf life, and contaminants determination, even though we include these aspects, too.

The major achievement is that by the complementarity use of the two methods of anticoagulant concentration determination and by using purified water as a model, we made possible the evaluation of the following:

- the accuracy of the anticoagulant amount added into the tubes by a producer;
- the accuracy of the volume of anticoagulant solution in the tube at the instant of examination;
- the anticoagulant concentrations for a nominal draw or draw volume.

The approach involves anticoagulant determination in a composite sample obtained by the anticoagulant water extraction from several tubes of the same type and dilution to a known volume with purified water on one side and anticoagulant determination in individual tubes after the addition of purified water in a volume corresponding to the nominal draw volume on the other.

We applied the developed methodology to the 105, 109, and 129 mmol/L citrate anticoagulant evacuated blood collection tubes. Nominal draw volumes ranged from 1.8 mL to 4.5 mL. Buffered and unbuffered citrate tubes were involved. The examined brands were Vacuette®, Greiner BIO-ONE, Vacutube, LT Burnik d.o.o., and BD Vacutainer®.

The results confirmed that all the producers control the draw volume well and that it mostly remains within the 10% tolerance limit during shelf life. For Vacutube, water loss during the time was observed. Contamination with potassium or magnesium or both was noticed for all the examined tubes, but the values for composite samples did not exceed 0.21 mmol/L.

Other aspects of quality that were until now not possible to control by the end user reveal the complexity of the citrate anticoagulant evacuated blood-collection tubes' production. The anticoagulant amount added into the tubes during production had a relative bias between 3.2 and 23.0%. The anticoagulant volume deficiency at the instant of examination expressed as a relative bias ranged between −11.6 and −91.1%. The anticoagulant concentration relative bias after the addition of purified water in a volume that equalled a nominal draw volume extended from 9.3 to 25.7%. For all three examined parameters, the 10% tolerance limits were exceeded.

The suggested methodology enables medical laboratories to gain better insight into the characteristics of the citrate blood collection tubes as one of the preanalytical variables. In situations that require anticoagulant adjustment for accurate results of coagulation tests, this can help make the right decisions. The methodology gives producers additional means of controlling the quality of their production process.

Author Contributions: Conceptualization, N.G.; Data curation, N.G.; Formal analysis, N.G.; Investigation, T.K. and K.G.; Methodology, N.G.; Resources, N.G.; Supervision, N.G.; Validation, T.K. and K.G.; Visualization, N.G.; Writing—review and editing, N.G. All authors have read and agreed to the published version of the manuscript.

Funding: This research was funded by the Slovenian Research Agency, grant number P1-0153.

Institutional Review Board Statement: Not applicable.

Informed Consent Statement: Not applicable.

Data Availability Statement: Data are contained within the article.

Conflicts of Interest: The authors declare no conflict of interest.

Sample Availability: Samples of the compounds are not available from the authors.

References

1. Gosselin, R.C.; Marlar, R.A. Preanalytical Variables in Coagulation Testing: Setting the Stage for Accurate Results. *Semin. Thromb. Hemost.* **2019**, *45*, 433–448. [CrossRef] [PubMed]
2. Bowen, R.A.R.; Adcock, D.M. Blood collection tubes as medical devices: The potential to affect assays and proposed verification and validation processes for the clinical laboratory. *Clin. Biochem.* **2016**, *49*, 1321–1330. [CrossRef] [PubMed]
3. CLSI. *Validation and Verification of Tubes for Venous and Capillary Blood Specimen Collection*; Approved Guideline; CLSI document GP34-A; CLSI: Wayne, PA, USA, 2010.
4. Gosselin, R.C.; Bowyer, A.; Favaloro, E.J.; Johnsen, J.M.; Lippi, G.; Marlar, R.A.; Neeves, K.; Rollins-Raval, M.A. Guidance on the critical shortage of sodium citrate evacuated collection tubes for hemostasis testing. *J. Thromb. Haemost.* **2021**, *19*, 2857–2861. [CrossRef]
5. Lippi, G.; Salvagno, G.L.; Brocco, G.; Guidi, G.C. Preanalytical variability in laboratory testing: Influence of the blood drawing technique. *Clin. Chem. Lab. Med.* **2005**, *43*, 319–325. [CrossRef]
6. Bowen, R.A.R.; Remaley, A.T. Interferences from blood collection tube components on clinical chemistry assays. *Biochem. Med.* **2014**, *24*, 31–44. [CrossRef]
7. Bush, V.; Cohen, R. The evolution of evacuated blood collection tubes. *Lab. Med.* **2003**, *34*, 304–310. [CrossRef]
8. Gosselin, R.C.; Janatpour, K.; Larkin, E.C.; Lee, Y.P.; Owings, J.T. Comparison of samples obtained from 3.2% sodium citrate glass and two 3.2% sodium citrate plastic blood collection tubes used in coagulation testing. *Am. J. Clin. Pathol.* **2004**, *122*, 843–848. [CrossRef]
9. Flanders, M.M.; Crist, R.; Rodgers, G.M. A comparison of blood collection in glass versus plastic vacutainers on results of esoteric coagulation assays. *Lab. Med.* **2003**, *34*, 732–735. [CrossRef]

10. van den Besselaar, A.; Chantarangkul, V.; Tripodi, A. A comparison of two sodium citrate concentrations in two evacuated blood collection systems for prothrombin time and ISI determination. *Thromb. Haemost.* **2000**, *84*, 664–667.
11. Yavas, S.; Ayaz, S.; Kose, S.K.; Ulus, F.; Ulus, A.T. Influence of Blood Collection Systems on Coagulation Tests. *Turk. J. Hematol.* **2012**, *29*, 367–375. [CrossRef]
12. Tripodi, A.; Chantarangkul, V.; Bressi, C.; Mannucci, P.M. How to evaluate the influence of blood collection systems on the international sensitivity index. Protocol applied to two new evacuated tubes and eight coagulometer/thromboplastin combinations. *Thromb. Res.* **2002**, *108*, 85–89. [CrossRef] [PubMed]
13. Germanovich, K.; Femia, E.A.; Cheng, C.Y.; Dovlatova, N.; Cattaneo, M. Effects of pH and concentration of sodium citrate anticoagulant on platelet aggregation measured by light transmission aggregometry induced by adenosine diphosphate. *Platelets* **2018**, *29*, 21–26. [CrossRef] [PubMed]
14. Elst, K.V.; Vermeiren, S.; Schouwers, S.; Callebaut, V.; Thomson, W.; Weekx, S. Validation of the minimal citrate tube fill volume for routine coagulation tests on ACL TOP 500 CTS (R). *Int. J. Lab. Hematol.* **2013**, *35*, 614–619. [CrossRef] [PubMed]
15. Duncan, E.M.; Casey, C.R.; Duncan, B.M.; Lloyd, J.V. Effect of concentration of trisodium citrate anticoagulant on calculation of the international normalized ratio and the international sensitivity index of thromboplastin. *Thromb. Haemost.* **1994**, *72*, 84–88.
16. Reneke, J.; Etzell, J.; Leslie, S.; Ng, V.L.; Gottfried, E.L. Prolonged prothrombin time and activated partial thromboplastin time due to underfilled specimen tubes with 109 mmol/L (3.2%) citrate anticoagulant. *Am. J. Clin. Pathol.* **1998**, *109*, 754–757. [CrossRef]
17. Marlar, R.A.; Potts, R.M.; Marlar, A.A. Effect on routine and special coagulation testing values of citrate anticoagulant adjustment in patients with high hematocrit values. *Am. J. Clin. Pathol.* **2006**, *126*, 400–405. [CrossRef]
18. Salvagno, G.L.; Demonte, D.; Gelati, M.; Poli, G.; Favaloro, E.J.; Lippi, G. Thrombin generation in different commercial sodium citrate blood tubes. *J. Med. Biochem.* **2020**, *39*, 19–24. [CrossRef]
19. CLSI. *Tubes and Additives for Venous and Capillary Blood Specimen Collection*, 6th ed.; CLSI document GP39-A6; CLSI: Wayne, PA, USA, 2010.
20. NCCLS. *Tubes and Additives for Venous Blood Specimen Collection*, 5th ed.; NCCLS document H1-A5; NCCLS: Albany, NY, USA, 2003.
21. Gros, N. Evacuated blood-collection tubes for haematological tests-a quality evaluation prior to their intended use for specimen collection. *Clin. Chem. Lab. Med.* **2013**, *51*, 1043–1051. [CrossRef]
22. Gros, N. Pre-Analytical Within-Laboratory Evacuated Blood-Collection Tube Quality Evaluation. In *Biochemical Testing: Clinical Correlation and Diagnosis*; Varaprasad, B.E., Ed.; IntechOpen: London, UK, 2018.
23. Lito, M.; Camoes, M.; Covington, A.K. Effect of citrate impurities on the reference pH value of potassium dihydrogen buffer solution. *Anal. Chim. Acta* **2003**, *482*, 137–146. [CrossRef]
24. Bates, R.G.; Pinching, G.D. Resolution of the dissociation constants of citric acid at 0-degrees to 50-degrees, and determination of certain related thermodynamic functions. *J. Am. Chem. Soc.* **1949**, *71*, 1274–1283. [CrossRef]
25. Krukowski, S.; Karasiewicz, M.; Kolodziejski, W. Convenient UV-spectrophotometric determination of citrates in aqueous solutions with applications in the pharmaceutical analysis of oral electrolyte formulations. *J. Food Drug Anal.* **2017**, *25*, 717–722. [CrossRef]
26. Lippi, G.; Salvagno, G.L.; Biljak, V.R.; Kralj, A.K.; Kuktic, I.; Gelati, M.; Simundic, A.M. Filling accuracy and imprecision of commercial evacuated sodium citrate coagulation tubes. *Scand. J. Clin. Lab. Investig.* **2019**, *79*, 276–279. [CrossRef] [PubMed]
27. Weikart, C.M.; Breeland, A.P.; Wills, M.S.; Baltazar-Lopez, M.E. Hybrid Blood Collection Tubes: Combining the Best Attributes of Glass and Plastic for Safety and Shelf life. *SLAS Technol. Transl. Life Sci. Innov.* **2020**, *25*, 484–493. [CrossRef] [PubMed]
28. Lima-Oliveira, G.; Salvagno, G.L.; Danese, E.; Favaloro, E.J.; Guidi, G.C.; Lippi, G. Sodium citrate blood contamination by K-2-ethylenediaminetetraacetic acid (EDTA): Impact on routine coagulation testing. *Int. J. Lab. Hematol.* **2015**, *37*, 403–409. [CrossRef] [PubMed]
29. van den Besselaar, A.; van Dam, W.; Sturk, A.; Bertina, R.M. Prothrombin time ratio is reduced by magnesium contamination in evacuated blood collection tubes. *Thromb. Haemost.* **2001**, *85*, 647–650.
30. van den Besselaar, A.; van Zanten, A.P.; Brantjes, H.M.; Elisen, M.; van der Meer, F.J.M.; Poland, D.C.W.; Sturk, A.; Leyte, A.; Castel, A. Comparative Study of Blood Collection Tubes and Thromboplastin Reagents for Correction of INR Discrepancies A Proposal for Maximum Allowable Magnesium Contamination in Sodium Citrate Anticoagulant Solutions. *Am. J. Clin. Pathol.* **2012**, *138*, 248–254. [CrossRef]
31. van den Besselaar, A.; van Vlodrop, I.J.H.; Berendes, P.B.; Cobbaert, C.M. A comparative study of conventional versus new, magnesium-poor Vacutainer (R) Sodium Citrate blood collection tubes for determination of prothrombin time and INR. *Thromb. Res.* **2014**, *134*, 187–191. [CrossRef]

Disclaimer/Publisher's Note: The statements, opinions and data contained in all publications are solely those of the individual author(s) and contributor(s) and not of MDPI and/or the editor(s). MDPI and/or the editor(s) disclaim responsibility for any injury to people or property resulting from any ideas, methods, instructions or products referred to in the content.

Article

Quantification and Validation of an HPLC Method for Low Concentrations of Apigenin-7-*O*-Glucuronide in *Agrimonia pilosa* Aqueous Ethanol Extract Topical Cream by Liquid–Liquid Extraction

Jin Seok Lee [1], Yu Ran Nam [2,3], Hyun Jong Kim [2,3,*] and Woo Kyung Kim [1,3,*]

1. Department of Internal Medicine, Graduate School of Medicine, Dongguk University, 27 Dongguk-ro, Goyang 10326, Republic of Korea
2. Department of Physiology, College of Medicine, Dongguk University, 123 Dongdae-ro, Gyeongju 38066, Republic of Korea
3. Channelopathy Research Center (CRC), College of Medicine, Dongguk University, 32 Dongguk-ro, Goyang 10326, Republic of Korea
* Correspondence: designed_hj@naver.com (H.J.K.); wk2kim@naver.com (W.K.K.)

Abstract: In this study, we aimed to develop and validate a pretreatment method for separating and analyzing the small amounts of biomarkers contained in topical cream formulations. Analyzing semisolid formulations that contain low concentrations of active ingredients is difficult. Cream formulations containing an aqueous ethanol extract of 0.1% Agrimonia pilosa is an example. Approximately 0.0013% of apigenin-7-O-glucuronide(A7OG) was contained as a biomarker in the cream. To determine the A7OG content present in the cream formulation, liquid–liquid extraction using dichlormethane was applied. In addition, the volume of the distribution liquid was measured using the peak ratios of the indicator component, A7OG, and an internal standard, baicalin. Subsequently, the A7OG content in the cream formulation was calculated. Using this time-saving method, A7OG can be simply analyzed without additional pretreatment steps, such as evaporation and reconstitution. Moreover, the validation results confirmed that this analytical method met all of the criteria. Consequently, A7OG was successfully isolated from the cream, analyzed, and quantified using the developed method.

Keywords: *Agrimonia pilosa*; apigenin-7-*O*-glucuronide; cream; HPLC-DAD; validation

1. Introduction

Agrimonia Pilosa, which belongs to the Rosaceae family, is a herb indigenous to the mountains and fields of Asia and Europe [1,2]. Additionally, this herb contains various tannins, such as agrimoniin and ellagic acid, and flavonoids, such as apigenin-7-*O*-glucuronide (A7OG) and luteolin-7-*O*-glucuronide. This herb, which has a slightly bitter and astringent taste, has been traditionally used to treat hemostasis, intestinal disorders, and inflammation [3,4]. Recent studies have reported the herb as having anticancer, antioxidant, and anti-inflammatory properties [5–8]. Therefore, it has emerged as a potential candidate for a new class of drugs with a wide range of benefits. As an efficacious natural resource, it is currently being researched and developed in both the cosmetic and pharmaceutical fields [9,10]. However, the quality control of herbal products is more challenging than that of synthetic chemical products. For example, identifying biomarkers that are stable and safe over long periods can be difficult. In addition, some biomarkers are present in trace amounts and may be subject to interference from various components, thereby hindering instrumental quantification [11–14]. Therefore, a growing demand exists for a quality control protocol for herbal medicines and cosmetics. Quality control can be achieved for the management of biomarkers and the evaluation of stability. There are ongoing studies to analyze and quantify these products, and many methods use high-performance liquid chromatography diode-array detector (HPLC-DAD) after increasing the sensitivity through sample pretreatment methods to remove the matrix [15–17]. Among the various techniques reported, liquid–liquid extraction (LLE) is a simple and economical method in which components are moved from one phase to another such that the two phases are not miscible with each other. This method enables the separation of unwanted matrices and other components. However, this process has certain limitations; for instance, a large volume of solvent is required to separate the matrix and the component. Moreover, this method may require several distribution processes. Although LLE can be easily used for simple component analyses, its application is difficult for quantitative purposes. Because the volume of the solvent in which the desired component is distributed cannot be accurately determined, additional preprocessing steps, including evaporation and reconstruction, are required. The more complex the process, the more time consuming, and a difference in the results between people can occur [18–22]. Therefore, to quantify the components present in the matrix, it is necessary to apply other pretreatments or to optimize the liquid–liquid extraction process as much as possible. In this study, a method of separating the unnecessary ingredients and components to be analyzed using liquid–liquid extraction with a specific organic solvent was investigated. LLE was used for the separation of A7OG from the cream matrix, because it is an easy and quick method. In addition, the volume of the separated solution containing A7OG was determined using baicalin, an internal standard substance showing the same distribution pattern as A7OG. This standard was also used to calculate the content of A7OG present in the cream formulation. The content of A7OG (chemical structure shown in Figure 1) in the previously prepared 0.1% *Agrimonia pilosa* aqueous ethanol extract (AE) was approximately 1.3%. Therefore, the cream formulation containing 0.1% AE contained approximately 0.0013% of A7OG. Therefore, the 0.1% AE cream contained 0.0013% A7OG, which is a considerably low concentration to be detected via DAD. In order for the 0.1% AE cream to be used as an investigational drug, it is necessary to be able to analyze the biomarker since strict quality control is required. Consequently, we aimed to establish and validate a liquid–liquid extraction-based HPLC-DAD method for analyzing and quantify the cream.

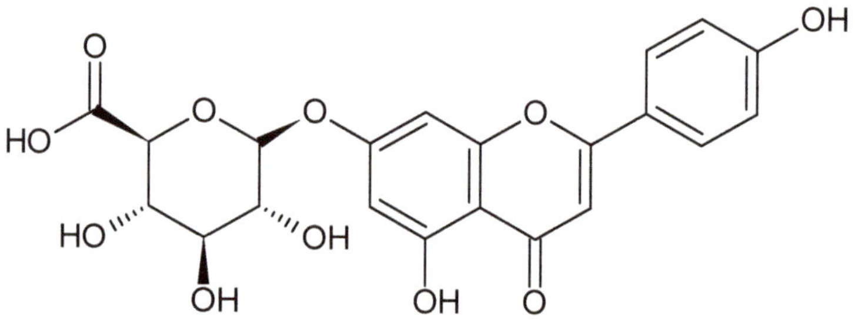

Figure 1. Structure of apigenin-7-*O*-glucuronide.

2. Results

2.1. Optimization of the Sample Preparation

When the 0.1% AE cream was dissolved in acetonitrile or methanol, the A7OG peak was barely observed. The concentration of A7OG in the diluted cream was approximately 2.4 µg/mL (assuming that 1 g of cream is 1 mL), which is higher than the LOQ of 0.873 µg/mL. Therefore, we assumed that if the concentration of A7OG was lowered, the detection of A7OG would be inhibited due to the matrix interference effect of the cream base. A peak that could not be quantified was confirmed even when the theoretical concentration was higher than the LOQ. Moreover, reducing the amount of diluting solvent makes it difficult to dissolve the cream. Even if the dilution factor is lowered, it is difficult to observe a peak with a sensitivity close to the LOQ. Therefore, the dilution of the 0.1% AE cream was not suitable for this experiment. In addition, the liquid–liquid extraction using organic solvents, such as acetone, hexane, and dichloromethane, was employed as an alternative method. As a result, the A7OG peak was observed only in dichloromethane; for the other solvents, the A7OG peak was barely detectable. Therefore, subsequent experiments were carried out by employing liquid–liquid extraction with dichloromethane. The analyzed chromatograms are shown in Figure 2.

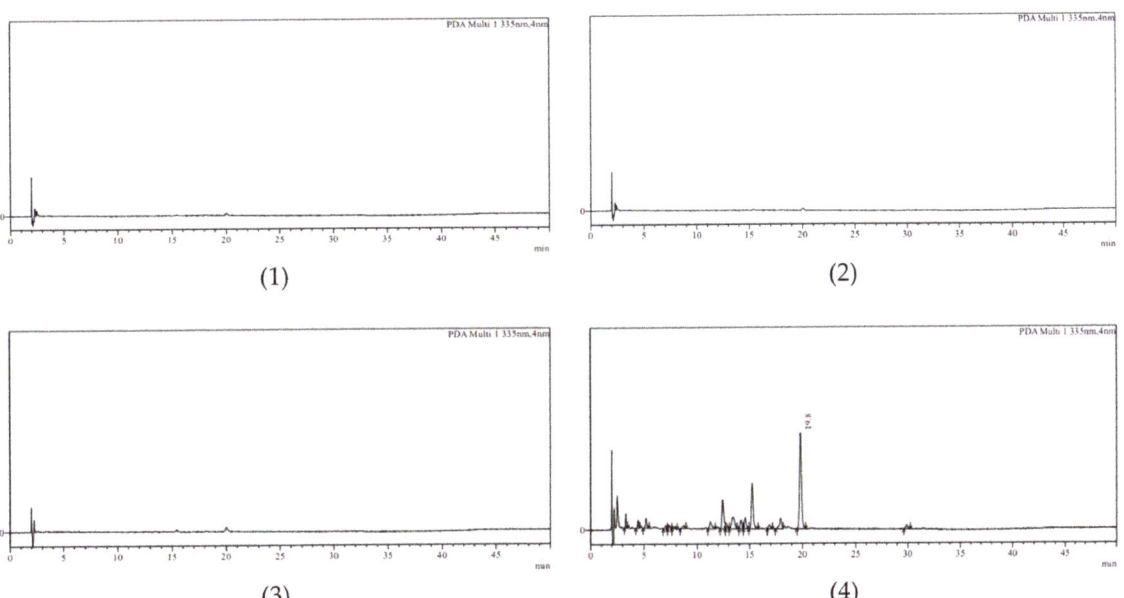

Figure 2. HPLC-DAD chromatogram according to the pretreatment method: (**1**) 0.1% AE cream dissolved in acetonitrile; (**2**) analysis after liquid–liquid extraction using acetone; (**3**) analysis after liquid–liquid extraction using hexane; (**4**) analysis after liquid–liquid extraction using dichloromethane.

2.2. Summary of the Validation

The validation results are summarized in Table 1.

Table 1. System suitability.

Parameter		Result	
System Suitability		%RSD of A7OG AREA %RSD of Baicalin AREA	0.4% 0.2%
Linearity (µg/mL)		10.2–23.8	
Slope		25,289	
Intercept		−9184	
Regression coefficient (R^2)		0.9999	
%Recovery	60%	97 ± 0.3	
	100%	99.8 ± 0.5	
	140%	102 ± 0.4	
Precision (%RSD)		0.20%	
Intermediate Precision (%RSD)		0.15%	
LOD (µg/mL)		0.28	
LOQ (µg/mL)		0.87	

2.3. System Suitability

A system suitability test was performed to evaluate whether the analysis system consisting of equipment worked as intended. The concentration of the standard solution corresponded to 100% of the A7OG target concentration in the 0.1% AE cream pretreatment solution. The solution was injected six times (the concentration of A7OG was 17.0 µg/mL and that of baicalin was 30 µg/mL). Consequently, the peak average area for A7OG was

413,138.3 and the percentage of the relative standard deviation (%RSD) was 0.4%. The peak average area and %RSD for baicalin were 323,756.5 and 0.2%, respectively.

2.4. Specificity

A blank matrix cream pretreatment solution and the 0.1% AE cream pretreated with the internal standard were analyzed to compare the obtained peak patterns at 335 nm (Figure 3). In the chromatogram, the retention time was 19.8 min for A7OG and 24.7 min for baicalin. The ultraviolet-visible (UV-vis) spectra of A7OG and baicalin were also compared in the standard and sample (Figure 4).

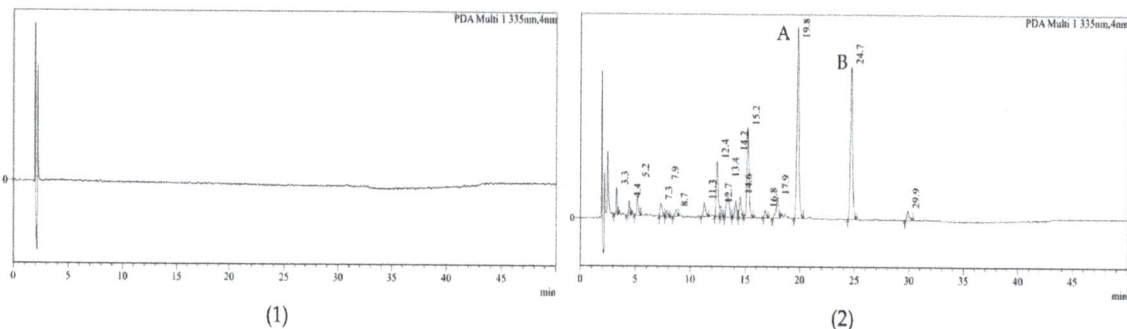

Figure 3. HPLC-DAD chromatogram of the (**1**) blank matrix cream; (**2**) 0.1% AE cream pretreated solution with internal standard. Peak A represents A7OG, and peak B represents baicalin, which is the internal standard.

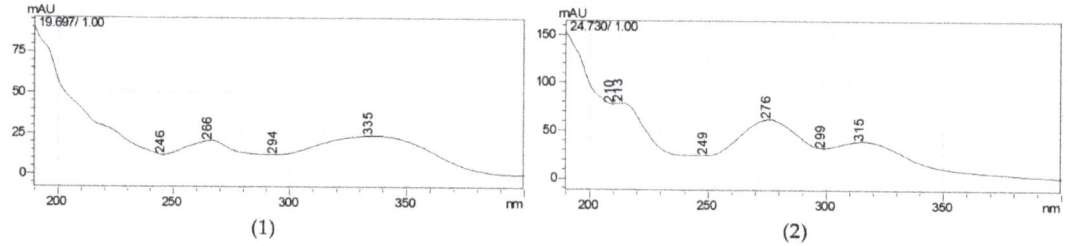

Figure 4. UV-vis spectrum of (**1**) A7OG and (**2**) baicalin from the 0.1% AE cream pretreated solution with internal standard and standard solution with internal standard.

2.5. Linearity, Limit of Detection (LOD), and Limit of Quantitation (LOQ)

Three sets of solutions of 60, 80, 100, 120, and 140% target concentrations were prepared and analyzed. The linearity was examined with the concentration of the solution plotted on the x-axis and the peak area of the component plotted on the y-axis. The mean slope (S) was 25,289 and the mean intercept was 2206. As a result, the coefficient of determination (R^2) of each calibration curve was higher than 0.995. Therefore, the linearity was confirmed within the concentration range. The average peak area ratio of the A7OG/internal standard was 1.27.

The LOD was 0.288 µg/mL, and the LOQ was 0.873 µg/mL.

2.6. Accuracy

As described in the sample preparation, three concentrations (60, 100 and 140%) of samples prepared in triplicate via a pretreatment process containing A7OG and internal standards in the blank cream, as well as the separation with dichloromethane, were analyzed and quantified. We compared the content of the sample with the theoretical content contained in the cream using the quantitative equation provided in Section 4.8. At a 60%

concentration, the average recovery rate was 97% and the %RSD was 0.3%, whereas at a 100% concentration, the average recovery rate was 99.8% and the %RD was 0.5%. Further, the average recovery rate was 102% and the %RD was 0.4% at a 140% concentration.

2.7. Precision (Repeatability)

First, when six standard solutions with a 100% concentration were analyzed, the average ratio of the peak area of A7OG to the peak area of the internal standard was 1.28 and the %RSD was 0.56%. The theoretical value of the concentration of the A7OG obtained by analyzing the standard solution was 13.9 µg/g. This value represents the A7OG content in the cream converted inversely based on a standard solution of a 100% target concentration (17 µg/mL). The average recovery rate of the sample solution was 101.4%, and the %RSD was 0.20%.

2.8. Intermediate Precision

The intermediate precision in the laboratory was analyzed by different testers applying different test equipment over different test days. The experiment was conducted in the same way as described in Section 2.7. When six standard solutions with a 100% concentration were analyzed, the average ratio of the peak area of the A7OG peak to the peak area of the internal standard was 1.3 and the %RSD was 1.09%. The average recovery rate of the sample solution was 100.5%, and the %RSD was 0.15%.

2.9. Solution Stability

The 17 µg/mL concentration of six samples were stored at a room temperature of approximately 15 °C for approximately 43 h and then analyzed and compared with the initial analysis value. The average change rate of the peak area of the A7OG was −0.6%.

2.10. Quantification of the 0.1% AE Cream

The content of A7OG in the 0.1% AE cream in each of the three batches was determined, and an average content value of 12.2 µg/g and %RSD of 1.1% were obtained. Since the A7OG content in the *Agrimonia pilosa* extract was known in advance, the theoretical value was obtained as 12.7 µg/g. The actual value was approximately 4% lower than the theoretical value. The dichlormethane layer obtained in the pretreatment process was also analyzed, and the peak of the A7OG did not appear. Therefore, we assumed that the manufacturing process affected the content of A7OG. However, this aspect needs to be researched further. The analyzed concentrations were within the range of linearity verified values through validation. Therefore, it was confirmed that the A7OG in the 0.1% AE cream could be quantified by this method.

3. Discussion

Generally, cream formulations are analyzed after dissolving the cream in an organic solvent. In such cases, the dissolved cream must contain a sufficient amount of the analyte for HPLC-DAD analysis. Other methods should be considered when interference from the base matrix may occur or when the concentration of the analyte itself is low. In order to analyze the components that are present in the finished product, these components have to be separated from the matrix. This separation can be carried out using liquid–liquid extraction. The reason for adopting this method is that most of the cream bases are nonpolar and can distribute the water and index components present in the cream into two immiscible layers. Liquid–liquid extraction is a very simple and economical method; however, this method can be difficult to quantify, because it is difficult to determine the volume of solvent. In general, a method for volatilizing the solvent in which the components are dissolved is required, and another solvent can be subsequently added to dissolve them again. However, this method takes time and labor, and the reproducibility of the results depends on the skill level of the person performing the experiment. Therefore, in this study, the optimal solvent for separating A7OG from the cream base was selected, and an internal standard was used.

Moreover, we found a suitable solvent for liquid–liquid extraction. We simultaneously determined the volume of the solution in which the A7OG was dissolved using an internal standard, bacicalin. Furthermore, we successfully measured the A7OG content in the 0.1% AE cream formulation. The target compound, A7OG, and other materials, such as cetanol and stearyl alcohol, were separated using dichloromethane. When the cream was mixed with dichloromethane, and the solution was separated, A7OG was distributed in the supernatant. The volume of the distributed supernatant and the concentration of A7OG can be calculated using an internal standard material with the same distribution pattern as that of A7OG. Subsequently, this analytical method was validated, and all of the validation criteria were met. In addition, this method was verified once more by quantifying the commercially produced 0.1% AE cream.

It is becoming increasingly important to perform quality control and consistently manage various products, such as pharmaceuticals and cosmetics. In addition, there are many commercial topical products, such as cosmetics and medicines, which contain natural product extracts. However, it is difficult to develop innovative, simple, time saving, and reproducible test methods for natural product components. In this respect, the methods developed this time can be helpful for related practitioners.

Moreover, we further plan to conduct a long-term stability test and quality control of this cream by applying the technique developed in this study. In addition, investigations for developing an optimal separation technology suitable for the characterization of formulations and compound will be undertaken.

4. Materials and Methods

4.1. Chemicals and Reagents

The A7OG standard and baicalin were purchased from Biopurify (Chengdu, China) and Alladin-e (Shanghai, China), respectively. The acetonitrile, methanol, and water were HPLC-grade. Extra-pure grade dichloromethane (J.T.Baker) and phosphoric acid (Daejung, Korea) were used in this experiment.

4.2. Preparation of the Agrimonia pilosa Extract

The *Agrimonia pilosa* herb (10 kg) was added to 70% ethanol at a weight ratio of 20 times (200 L). The resultant mixture was refluxed at 90 °C for 4 h. Subsequently, the mixture was filtered and concentrated under reduced pressure at approximately 40 °C. Finally, the concentrate was lyophilized at −80 °C for approximately one day to obtain a powder. The yield was approximately 10%. Moreover, the content of A7OG in this extract was approximately 12.6 mg/g.

4.3. Preparation of 0.1% AE Cream

The extract cream consisted of AE (1% by weight), water, propylene glycol, liquid paraffin, cetanol, stearyl alcohol, polysorbate 60, monostearate sorbitan, benzyl alcohol (preservative), and citric acid hydrate. First, the *Agrimonia pilosa* herbs were extracted using water with ethanol, freeze-dried, and powdered. The water phase was prepared by mixing purified water and citric acid. The oil phase was then prepared by stirring and dissolving the remaining ingredients, including the extract, at approximately 90 °C. Subsequently, the water and oil phases were mixed and stirred uniformly under heat. Thereafter, the heat was rapidly lowered to prepare a cream formulation.

4.4. Instrument and Chromatographic Condition

The HPLC system with a Shimadzu SPD-M20A photodiode array detector (Shimadzu, Kyoto, Japan) was used for the analysis and validation. The Waters e2695 alliance (Waters Corporation, Milford, MA, USA) HPLC system was used for the inter-laboratory validation. The temperature of the column (YoungJin Biochrom INNO-P C18 5 μm, 4.6 × 150 mm) was set at 30 °C, and the flow rate was maintained at 1 mL/min. The injection volume was 10 μL, and the detection wavelength was 335 nm. The mobile phase consisted of

(A) distilled water with 0.1% phosphoric acid and (B) acetonitrile. For the separation of A7OG and baicalin, the following gradient was used: solution B was changed from 90% to 75% between 0 and 30 min; then, B was 75% to 90%, from 30 to 40 min, and isocratic from 40 to 50 min.

4.5. Optimization of the Sample Preparation Method

To develop an appropriate pretreatment method, the following experiments were conducted. The first method was to take five grams of 0.1% AE cream and dissolve it in 20 mL of acetonitrile or methanol. After being dissolved, it was filtered and analyzed by HPLC-DAD. For the second method, liquid–liquid extraction was performed. Five grams of the 0.1% AE cream was weighed, and 20 mL of organic solvent was added to it. Examples of organic solvents that were included: acetone, hexane, dichloromethane, and ethyl acetate. When the cream was divided into two layers, the layer that was immiscible with the organic solvent was taken and analyzed after being filtered.

4.6. Sample Preparation for the Validation

Seventeen milligrams of the A7OG standard were dissolved in methanol in a 100 mL volumetric flask to create the standard stock solution (Solution A). Secondly, 25 mg of baicalin standard was dissolved in methanol and adjusted to 100 mL in a volumetric flask as an internal standard solution. Furthermore, 6, 8, 10, 12 and 14 mL of the standard stock solution were mixed with 12 mL of the internal standard solution, and water was added to result in exactly 100 mL. These solutions were used as the standard solutions (concentrations of 10.2, 13.6, 17.0, 20.4 and 23.8 µg/mL, respectively). The concentration of the internal standard in the standard solutions was 30 µg/mL.

For validating the accuracy and precision, blank matrix sample preparation was also needed. Five grams of blank matrix cream were weighed; placed in a 100 mL volumetric flask; and mixed with 20 mL of dichloromethane, 500 µL of the internal standard solution, and 0.3, 0.5 and 0.7 mL of solution A as the standard stock solution (60, 100 and 140% of the target concentration, respectively). After being sufficiently dissolved, the supernatant was collected and centrifuged at 13,500 rpm for 10 min. Once clear, the supernatant was filtered and used for quantification using the HPLC system.

4.7. Validation

In this study, system suitability, specificity, linearity, accuracy, precision, and robustness (stability) were validated for the assay. A blank matrix of the cream formulation was used to validate the specificity, accuracy, precision, and solution stability. The criteria for the validation test were followed by considering the A7OG concentration contained in the formulation: "AOAC Guideline for Single Laboratory Validation of Chemical Methods for Dietary Supplements and Botanicals" [23]. As the A7OG content unit in the formulation was µg/g, the recovery rate and relative standard deviation range of the precision and accuracy were set within 90~110%, and the relative deviation of the peak area ratio of the system suitability was set within 10%. In addition, the limit of detection and limit of quantitation were calculated according to the following formula.

Each calculation formula is as follows:

$$\text{Limit of Detection} = 3.3 \times \sigma/S,$$

$$\text{Limit of Quantitation} = 10 \times \sigma/S,$$

where S is the mean of the slopes, and σ is the standard deviation of the intercepts.

4.8. Quantification of the 0.1% AE Cream

The following procedure was followed to quantify the A7OG content in the previously prepared 0.1% cream. Please refer to Section 4.6 to prepare the following solutions. Seventeen milligrams of the A7OG standard was dissolved in methanol in a 100 mL volumetric flask (standard stock solution). Subsequently, twenty-five milligrams of baicalin standard were dissolved in methanol and adjusted to 100 mL in a volumetric flask (internal standard solution). Next, 10 mL of the standard stock solution was mixed with 12 mL of the internal standard solution, and the resultant solution was diluted with water to result in a volume that was exactly 100 mL. These solutions were used as the standard solutions. The concentrations of the A7OG and baicalin were 17 and 30 µg/mL, respectively.

Five grams of the 0.1% AE cream were mixed with 20 mL of dichloromethane and 500 µL of the internal standard solution. After sufficiently dissolving the cream, the supernatant was taken aside and filtered before the analysis. The equation for obtaining the content of the A7OG in the 0.1% AE cream is given below.

The quantitative equation is given by:

$$= \frac{(5 \times WI.S \times S.s \times Rc)}{(Rs \times S.I \times W)}$$

WI.S: Weight of the internal standard (i.e., baicalin);
S.s: The concentration of A7OG in the standard solution (concentration correction required according to the purity of the standard product);
Rc: The ratio (A7OG peak area/internal standard peak area) in the sample solution;
Rs: The ratio (A7OG peak area/internal standard peak area) of the standard solution;
S.I: The concentration of the internal standard in the standard solution (concentration correction required according to the purity of the standard product);
W: Weight of the cream.

5. Conclusions

In this study, a method for quantifying the A7OG content in a natural *Agrimonia pilosa* extract was successfully optimized and verified. Liquid–liquid extraction was performed using a dichloromethane solution to separate A7OG, a marker component, for analysis. In addition, the A7OG content in the cream formulation could be measured by using the peak area ratio of the A7OG and the internal standard. Since this method did not require additional pretreatment, it was possible to measure the A7OG content simply and accurately. The results obtained from this experiment indicate that the developed method is suitable for the quality control of topical cream formulations.

Author Contributions: Conceptualization, W.K.K. and H.J.K.; methodology, J.S.L.; software, J.S.L.; validation, J.S.L.; formal analysis, J.S.L.; investigation, J.S.L. and Y.R.N.; resources, J.S.L.; data curation, J.S.L.; writing—original draft preparation, J.S.L.; writing—review and editing, W.K.K. and H.J.K.; visualization, J.S.L.; supervision, W.K.K. and H.J.K.; project administration, W.K.K.; funding acquisition, W.K.K. All authors have read and agreed to the published version of the manuscript.

Funding: This research was supported by the Bio & Medical Technology Development Program of the National Research Foundation (NRF), funded by the Ministry of Science & ICT (2018M3A9F3081729). This work was supported by the National Priority Research Center Program Grant (NRF-2021R1A6A1A03038865), funded by the Korean government.

Institutional Review Board Statement: Not applicable.

Informed Consent Statement: Not applicable.

Data Availability Statement: The data that support this study are available from the corresponding author upon reasonable request.

Acknowledgments: The authors are thankful to the Department of Internal Medicine Graduate School of Medicine at Dongguk University for this study.

Conflicts of Interest: The authors declare no conflict of interest.

Sample Availability: Samples of the compounds are not available from the authors.

References

1. Le, Q.U.; Joshi, R.K.; Lay, H.L.; Wu, C.W. Agrimonia pilosa Ledeb: Phytochemistry, ethnopharmacology, pharmacology of an important traditional herbal medicine. *Int. J. Pharmacogn. Phytochem. Res.* **2018**, *7*, 3202–3211.
2. Krasnopolska, D.; Evarts-Bunders, P.; Svilāne, I. Contribution to knowledge of genus agrimonia L.(rosaceae) in Lavita. *Acta Biol. Univ. Daugavp.* **2020**, *20*, 47–53.
3. Fei, X.; Yuan, W.; Jiang, L.; Wang, H. Opposite effects of Agrimonia pilosa Ledeb aqueous extracts on blood coagulation function. *Ann. Transl. Med.* **2017**, *5*, 157. [CrossRef] [PubMed]
4. Wen, S.; Zhang, X.; Wu, Y.; Yu, S.; Zhang, W.; Liu, D.; Yang, K.; Sun, J. Agrimonia pilosa Ledeb.: A review of its traditional uses, botany, phytochemistry, pharmacology, and toxicology. *Heliyon* **2022**, *8*, e09972. [CrossRef] [PubMed]
5. Jin, T.; Chi, L.; Ma, C. Agrimonia pilosa: A phytochemical and pharmacological review. *Evid. Based Complement Alternat. Med.* **2022**, *2022*, 3742208. [CrossRef]
6. Kim, T.Y.; Koh, K.S.; Ju, J.M.; Kwak, Y.J.; Bae, S.K.; Jang, H.O.; Kim, D.S. Proteomics analysis of antitumor activity of Agrimonia pilosa Ledeb. In human oral squamous cell carcinoma cells. *Curr. Issues Mol. Biol.* **2022**, *44*, 3324–3334. [CrossRef]
7. Kim, H.S. Physiological activities of Agrimonia pilosa extract. *Korean J. Food Preserv.* **2015**, *22*, 261–266. [CrossRef]
8. Jin, X.; Song, S.; Wang, J.; Zhang, Q.; Qiu, F.; Zhao, F. Tiliroside, the major component of Agrimonia pilosa Ledeb ethanol extract, inhibits MAPK/JNK/p38-mediated inflammation in lipopolysaccharide-activated RAW 264.7 macrophages. *Exp. Ther. Med.* **2016**, *12*, 499–505. [CrossRef]
9. Kim, N.C. Need for pharmacopeial quality standards for botanical dietary supplements and herbal medicines. *Food Suppl. Biomater. Health* **2021**, *1*, e10. [CrossRef]
10. Indrayanto, G. Recent development of quality control methods for herbal derived drug preparations. *Nat. Prod. Commun.* **2018**, *13*, 1599–1606. [CrossRef]
11. Balekundri, A.; Mannur, V. Quality control of the traditional herbs and herbal products: A review. *Future J. Pharm. Sci.* **2020**, *6*, 67. [CrossRef]
12. Rossi Forim, M.; Perlatti, B.; Soares Costa, E.; Facchini Magnani, R.; Donizetti de Souza, G. Concerns and considerations about the quality control of natural products using chromatographic methods. *Curr. Chromatogr.* **2015**, *2*, 20–31. [CrossRef]
13. Länger, R.; Stöger, E.; Kubelka, W.; Helliwell, K. Quality standards for herbal drugs and herbal drug preparations–appropriate or improvements necessary? *Planta Med.* **2018**, *84*, 350–360. [CrossRef] [PubMed]
14. Wolfender, J.-L. HPLC in natural product analysis: The detection issue. *Planta Med.* **2009**, *75*, 719–734. [CrossRef]
15. Govindarajan, R.; Tejas, V.; Pushpangadan, P. High-performance liquid chromatography (HPLC) as a tool for standardization of complex herbal drug. *J AOAC Int.* **2019**, *102*, 986–992. [CrossRef]
16. Chan, W.F.; Lin, C.W. High-pressure liquid chromatography: Quantitative analysis of Chinese herbal medicine. *J. Chem. Educ.* **2007**, *84*, 1982–1984. [CrossRef]
17. Alqarni, M.H.; Alam, P. Highly Sensitive and Ecologically Sustainable Reversed-Phase HPTLC Method for the Determination of Hydroquinone in Commercial Whitening Creams. *Processes* **2021**, *9*, 1631. [CrossRef]
18. Lee, J.; Kumar, J. Liquid-Liquid extraction general principles A review. *J. Korean Inst. Resour. Recycl.* **2009**, *18*, 3–9.
19. Raikos, N.; Spagou, K.; Vlachou, M.; Pouliopoulos, A.; Thessalonikeos, E.; Tsoukali, H. Development of a Liquid-Liquid Extraction Procedure for the Analysis of Amphetamine in Biological Specimens by GC-FID. *Open Forensic. Sci. J.* **2009**, *2*, 12–15.
20. Yusiasih, R.; Marvalosha, R.; Suci, S.D.S.; Yuliani, E.; Pitoi, M.M. Low volume liquid-liquid extraction for the determination of benzene, toluene, and xylene in water by GC-FID and HPLC-UV. *Environ. Earth Sci.* **2019**, *277*, 012019. [CrossRef]
21. Roque, L.R.; Morgado, G.P.; Nascimento, V.M.; Ienczak, J.L.; Rabelo, S.C. Liquid-liquid extraction: A promising alternative for inhibitors removing of pentoses fermentation. *Fuel* **2019**, *242*, 775–787. [CrossRef]
22. Stone, J. Sample preparation techniques for mass spectrometry in the clinical laboratory. In *Mass Spectrometry for the Clinical Laboratory*; Academic Press: Cambridge, MA, USA, 2017; pp. 37–62.
23. AOAC International. *AOAC Guidelines for Single Laboratory Validation of Chemical Methods for Dietary Supplements and Botanicals*; AOAC International: Gaithersburg, MD, USA, 2002; Available online: http://www.aoac.org/dietsupp6/Dietary-Supplement-web-site/slv_guidelines.pdf (accessed on 5 October 2021).

Disclaimer/Publisher's Note: The statements, opinions and data contained in all publications are solely those of the individual author(s) and contributor(s) and not of MDPI and/or the editor(s). MDPI and/or the editor(s) disclaim responsibility for any injury to people or property resulting from any ideas, methods, instructions or products referred to in the content.

MDPI
St. Alban-Anlage 66
4052 Basel
Switzerland
Tel. +41 61 683 77 34
Fax +41 61 302 89 18
www.mdpi.com

Molecules Editorial Office
E-mail: molecules@mdpi.com
www.mdpi.com/journal/molecules

www.ingramcontent.com/pod-product-compliance
Lightning Source LLC
LaVergne TN
LVHW070226100526
838202LV00015B/2097